ROOFING
CONSTRUCTION & ESTIMATING
REVISED

by Daniel Atcheson

- Turn your estimate into a bid.
- Turn your bid into a contract.
- ConstructionContractWriter.com

Craftsman

Craftsman Book Company
6058 Corte del Cedro / Carlsbad, CA 92011

Acknowledgements

The author wishes to thank the following individuals and companies for providing general assistance, information, and various illustrations used in the preparation of this book.

American Plywood Association (APA) 7011 South 19th Street, Tacoma, WA 98466-0700
ASC Profiles. 2110 Enterprise Boulevard, W. Sacramento, CA 95691-3493
Asphalt Roofing Manufacturers Association (ARMA) 2331 Rock Spring Road, Forest Hill, MD 21050
Tim Atcheson, **Atcheson Building Systems** 3306 61st Street, Lubbock, TX 79413
James Edward Atcheson, 1906-1986
Tami Danielle Atcheson, Lubbock, TX
Mike Atcheson, **AIA,** Lubbock, TX
Cedar Shake & Shingle Bureau 355 Lexington Ave, 15th Floor, New York, NY 10017
The Celotex Corporation 4010 Boy Scout Boulevard, Tampa, FL 33607-5750
Allan B. Jones, Crowder Brothers Appliance Repair 11523 Palmbrush Trail, Lakewood Ranch, FL 34202
Gaco Western, LLC Nashville, TN
GAF Building Materials Corporation 1 Campus Drive, Parsippany, NJ 07054
Gladding McBean LLC 601 7th Street, Lincoln, CA 95648
Haag Engineering Co. 1401 Lakeside Parkway, #100, Flower Mound, TX 75028
Inland Buildings 2141 2nd Ave SW, Cullman, AL 35055
Koppers Industries, Inc. 436 Seventh Avenue, Pittsburgh, PA 15219-1800
Steve Lewis, **Engineering Consultant** Allentown, PA
Randy Hooks, **Lydick-Hooks Roofing** 1501 Central Freeway East, Wichita Falls, TX 76302
Manville Sales Corporation, Roofing Systems Division P.O. Box 5108, Denver, CO 80217
Mayes Brothers Division of Great Neck Saw Manufacturers, Inc. 165 East Second Street, Mineola, NY 11501
MM Systems Corporation 50 M M Way, Pendergrass, GA 30567
Monier Roof Tile P.O. Box 19792, Irvine, CA 92713
Owens-Corning 1 Owens Corning Pkwy, Toledo, OH 43659
Petersen Aluminum Corporation 1005 Tonne Road, Elk Grove Village, IL 60007
Rubatex Corporation Adams Street, Bedford, Virginia, 24523
Shakertown Cedar Siding, Shakertown Corporation 1200 Kerron Street, Winlock, WA 98596
David Carlson, **Southwest Florida Roofing** 4566 Frankfort Clarksburg Pike, Clarksburg, OH 43115
Stanley Tools Division of The Stanley Works New Britain, CT 06050
U.S. Department of Housing and Urban Development (HUD) 451 Seventh Street SW, Washington, DC 20410
Vermont Structural Slate Co. 3 Prospect Street, Fair Haven, VT 05743
Zappone Manufacturing N. 2928 Pittsburg, Spokane, WA 99207

To my grandchildren Sophie & Cole Painter
And to my Lord and Savior, Jesus Christ, who gave me eyes to see, hands to type and a mind to think.

Looking for other construction reference manuals?
Craftsman has the books to fill your needs. **Call toll-free 1-800-829-8123**
Visit our Web site: http://www.craftsman-book.com

Library of Congress Cataloging-in-Publication Data
Names: Atcheson, Daniel Benn, author.
Title: Roofing construction & estimating / by Daniel Atcheson.
Other titles: Roofing construction and estimating
Description: Revised edition. I Carlsbad, CA : Craftsman Book Company, 2022.
 I Includes bibliographical references and index.
Identifiers: LCCN 2022045408 (print) I LCCN 2022045409 (ebook) I ISBN
 9781572183803 (pdf) I ISBN 9781572183803 (paperback) I ISBN
 9781572183803¬q(paperback) I ISBN 9781572183803¬q(pdf)
Subjects: LCSH: Roofs--Design and construction. I Roofs--Estimates.
Classification: LCC TH2401 (ebook) I LCC TH2401 .A83 2022 (print) I DDC
 690/.15 23/eng/20221--dc13
LC record available at https://lccn.loc.gov/2022045408

First edition ©1978 Craftsman Book Company
Second edition ©2022 Craftsman Book Company

Third printing 2023

Contents

1 Measuring and Calculating Roofs 5
Level Roofs 7
Sloped Roofs 10
How to Measure Roof Slope 12
Perimeter of a Sloped Roof 17
Net Versus Gross Roof Area..................................... 17
Calculating Total Net Roof Area 18
Roof Overhangs, Hips and Valleys.......................... 21
Length of Ridge (Hip Roofs)..................................... 22

2 Roof Sheathing, Decking and Loading 23
Check the Framing 23
Solid Roof Sheathing... 24
Spaced Board Sheathing.. 29
Roof Decking 32
Loading the Roof... 32
Estimating Roof Sheathing.. 34

3 Underlayment on Sloping Roofs 35
Saturated Felt Underlayment.................................... 36
Saturated Fiberglass Underlayment......................... 36
Synthetic Underlayment.. 37
Underlayment Requirements.................................... 38
Drip Edge 41
Installing Underlayment 43
Estimating Underlayment Quantities 49
Interlayment (Lacing).. 56
Eaves Flashing (Ice Shield or Water Shield) 60
Valley Flashing.. 63

4 Asphalt Shingles 71
UL Ratings for Shingles 72
Deck Requirements 74
Asphalt Strip Shingles 75
Flashing at Chimneys and Other
    Vertical Structures 93
Fasteners... 102
Number of Shingles Required per Square 104
Number of Shingle Courses 105
Estimating Asphalt Strip Shingle Quantities........ 109
Ridge and Hip Units..110
Estimating Asphalt Shingle Roofing Costs 119

5 Mineral-Surfaced Roll Roofing 121
Installing Mineral-Surfaced Roll Roofing........... 123
Valley Flashing... 124
Estimating Mineral-Surfaced Roll Roofing........ 134
Waste from Non-conforming Roof Layout 136
Estimating Mineral-Surfaced Roll Roofing Costs .. 147

6 Wood Shingles and Shakes 149
Installing Wood Shingles and Shakes.................. 154
Covering Capacity of Shakes 164
Covering Capacity of Wood Shingles 164
Estimating Wood Shingle
    and Shake Quantities... 166
Staggered Patterns 175
Sidewall Shakes and Wood Shingles................... 175
Roof Junctures.. 182
Estimating Wood Shingle Roofing Costs 186

7 Tile Roofing 187
Underlayment Under Tile Roof Coverings 189
Installing Roof Tiles.. 190
The Starter Course... 192
Fastening Roofing Tiles.. 194
Flashing at Vertical Walls 207
Replacing Broken Tiles 214
Estimating Tile Quantities.................................... 215
Estimating Total Tile Roofing Costs................... 219

8 Slate Roofing 221
Slate Size, Color and Texture 221
Underlayment... 224
Installation on a Sloping Roof.............................. 226
Fasteners... 234
Flashing 235
Estimating Slate Quantities 239
Estimating Slate Roofing Costs............................ 244

9 Metal Roofing and Siding — 245
- Modern Metal Panel Systems 246
- Installing Metal Roofing Panels 248
- Job-Fabricated Seams 253
- Estimating Metal Roofing and Siding 259
- Steel Roofing and Siding Quantities 260
- Ribbed Metal Panel Quantities 261
- Labor for Installing Metal Panels 275

10 Built-Up Roofing — 277
- Roof Slopes 278
- Substrate Design 278
- Metal Decks 279
- Back Nailing 283
- Base Sheets (Vapor Retarders) 284
- Roofing Membranes 286
- Hot Bitumens 289
- Cold-applied Bitumens 291
- Surface Aggregate 293
- Smooth-surface Roofing 294
- Cap Sheets 295
- Aluminum Roof Coatings 296
- Cant Strips 299
- Temporary Roofs 300
- Water-retaining Roofs 301
- Flashing on Flat Roofs 302
- Roof Expansion Joints 307
- Estimating BUR Systems 310
- Built-up Roofing Warranties 314
- Built-up Roofing Repairs and Re-roofing 315

11 Elastomeric Roofing — 319
- The Advantages of Elastomeric Systems 320
- Liquid-applied Elastomers 321
- Single-Ply Roofing Systems 322
- EPDM Elastomeric Roofing 324
- CPE and CSPE Elastomeric Roofing 328
- PVC Elastomeric Roofing 329
- Composite Roofing Systems 329
- TPO Roofing 330
- Flashings for Elastomeric Roofs 333
- Estimating Elastomeric Roofing 334

12 Insulation, Vapor Retarders and Waterproofing — 337
- Insulation Materials 337
- Reducing Heat Loss 344
- Insulation Values 350
- Vapor Retarders 352
- Weatherproofing Existing Homes 353
- Caulking and Sealants 355
- Wall Flashing 359
- Waterproofing 360
- Dampproofing 367

13 Roof Coatings — 369
- Advantages of Roof Coatings 371
- How to Apply Elastomeric Roof Coating 372
- Sprayed Polyurethane Foam (SPF) Roofing 375
- Liquid EPDM Rubber Roofing 376
- Silicone Roofing 377
- Estimating Elastomeric Roof Coating 381
- Tools of the Trade 384

14 Roofing Repair and Maintenance — 387
- Finding the Source of Leaks 387
- Repairing Leaks 390
- Roof Maintenance 392
- Assessing Hail Damage 394
- Roofing Demolition 396
- Re-Roofing 400
- Estimating Re-Roofing 406
- Attic Ventilation 407
- Gutters and Downspouts 409

15 Estimating (and Maximizing) Production Rates — 413
- Labor Unit Prices 413
- Estimating with Published Prices 417
- Roofing Labor Tips 422

Appendix A Roof-Slope Factors 430
Appendix B Valley Length Factors 431
Appendix C Equations Used in This Book 432
Index 437

1 Measuring and Calculating Roofs

▶ If you're like some roofing contractors, you estimate roofing quantities by calculating the area of a roof, then adding 10 percent for waste. That might be OK in a fat building market, but in a tight market you'll need a sharper pencil to compete successfully for the good jobs, and then make money on them. In this book, I'm going to show you how to make a quick and accurate takeoff for any kind of roof.

You'll also learn the latest and most acceptable roofing methods in an industry where installation practices are closely related to warranties. That's because material warranties may be invalid if you don't follow the manufacturer's recommendations for installation. Look here for general guidelines, but always follow the manufacturer's instructions to the letter.

New products come on the market every day to solve the complex roof covering requirements presented by modern building technology. Your job is to know as much as you can about those products. You also have to know how to install them so the job passes inspection and presents no future repair and maintenance problems. Callbacks are hard on your profit margin — and they don't do your reputation any good either. Know as much as you can about your roofing business, and you'll avoid them.

This book is more than an estimating book for roofing contractors. It develops a *system*, beginning with Chapter 1, for all types of roofing materials and installation methods. We'll cover the entire roofing trade, including how to manage your crews and keep them safe. So let's get started.

Before you can bid any job, you have to figure your costs. And before you can figure the costs, you have to know the size of the job. So you have to do two things: First, measure the roof and calculate the total area. Then find the lengths of the eaves, gables (or rakes), ridges, hips and valleys.

When you construct a roof on a new building, you can get these measurements from the plans. On repair or replacement jobs, you'll probably have to take your pencil, clipboard and tape measure, haul out your ladder, climb onto the roof, and start measuring.

To avoid mistakes, or a second trip to the job site, develop a system for taking measurements. Use a 100-foot flexible tape which has a $1/2$-inch grout hook at the "stupid" end of the tape. Flexible tapes are made of metal, or fiberglass-reinforced nylon fabric. Find a tape that's marked with highlights at 5-inch intervals to match the exposure of most composition shingles.

There is no cardinal rule for the sequence you use to measure a roof, as long as you don't miss anything. Here's a system that works for me:

Start by measuring the length of the eaves. On a gable roof, you only have to measure in one direction. On a hip roof, you'll have to measure the eaves in two directions.

Next, measure the width of the roof. On a gable roof, hook the tape over one of the eaves, and run it over the ridge to the opposite eave. On a hip roof, measure the width the same way. To measure the length, hook the tape to the eaves at the ridge rafter (look ahead to Figure 1-16 on page 13 for an illustration of the parts of a roof), run the tape the length of the ridge and down the opposite ridge rafter. Measure the ridge at the same time.

Now, measure the hips and valleys by hooking the tape to a building corner and running the tape to the ridge. You use these measurements to calculate material requirements such as valley flashing and hip-covering material.

When you measure, some dimensions need to be more accurate than others. For instance, you could miss the length of ridge, hip or valley by a foot or more, and the error wouldn't affect your total bid price too much. But don't make a mistake in the length and width, because that error could be substantial. For example, assume you measure a roof at 100 feet by 200 feet, while the actual measurements are 100'6" by 200'6". The difference between the two measurements is 150 square feet, or $1\frac{1}{2}$ squares of material.

Always make a sketch of the roof layout, including dimensions, roof slopes, location of penetrations and any unusual circumstances such as rotten deck areas, ventilation problems, or overhanging tree branches or other obstructions.

Once you have the measurements, you'll use them to calculate areas, slopes, angles, and allowance factors. Let's begin with an easy example.

Measuring and Calculating Roofs

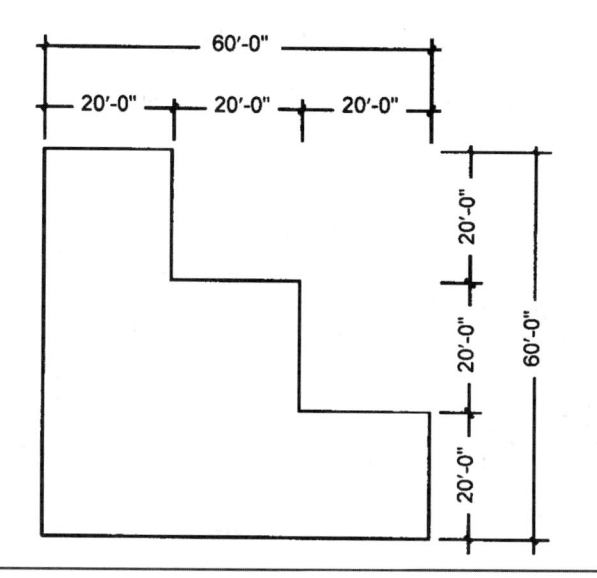

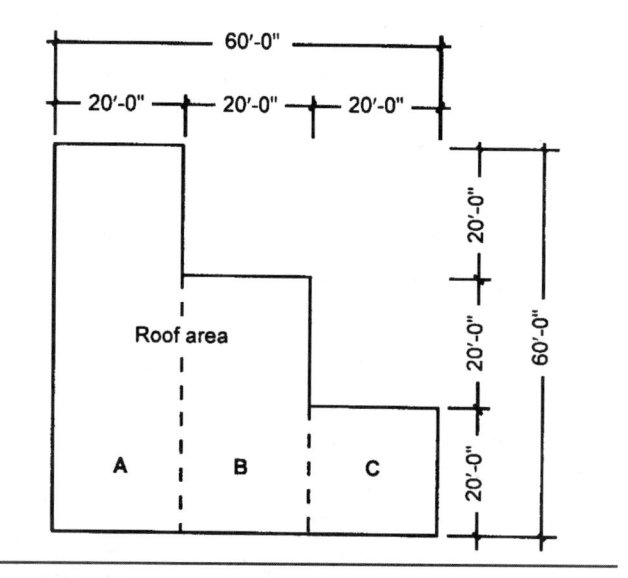

Figure 1-1 Roof plan of level roof

Figure 1-2 The positive method

Level Roofs

The dimensions on the plans give you the actual measurements for a level roof. To get the area of a rectangular roof, multiply its length by its width.

Area of a level rectangular roof = L x W

Equation 1-1

where L is the length and W is the width.

Of course, not every roof you work on will be a single rectangle. You may need to figure the area of a roof like the one in Figure 1-1. There are two ways to calculate this area:

1) The positive method

2) The negative method

In the *positive* method, you divide the roof into rectangular areas, then add the parts to get the total area. See Figure 1-2.

With the *negative* method, you extend the roof lines to form a single rectangle. Calculate the area of this rectangle, and subtract the areas of the rectangular spaces which lie outside the actual roof. Figure 1-3 illustrates this.

▼ **Example 1-1:** The Positive Method

Divide the roof into rectangles as shown in Figure 1-2. Calculate the area of each rectangle, then add them together:

Area A = 20 feet by 60 feet, or 1,200 square feet
Area B = 20 feet by 40 feet, or 800 square feet
Area C = 20 feet by 20 feet, or 400 square feet

Then, the total area = 1,200 SF + 800 SF + 400 SF, or 2,400 SF

▼ **Example 1-2:** The Negative Method

Extend the roof lines to form one rectangle, as in Figure 1-3. Calculate the total area of that rectangle, then subtract the areas of any rectangles which aren't in the actual roof:

Extended rectangle = 60' x 60' = 3,600 SF

Area A = 40 feet by 20 feet, or 800 square feet
Area B = 20 feet by 20 feet, or 400 square feet

Total area outside the roof layout is 800 SF + 400 SF, or 1,200 SF. Subtract that from the extended area to get the total area:

3,600 SF - 1,200 SF = 2,400 SF

You get the same answer both ways. So you might as well use the easiest method — the one that requires the fewer calculations. For example, in Figure 1-4 you'd have to calculate three areas, then add them together. But in Figure 1-5 you only have to calculate two areas, and then subtract one from the other.

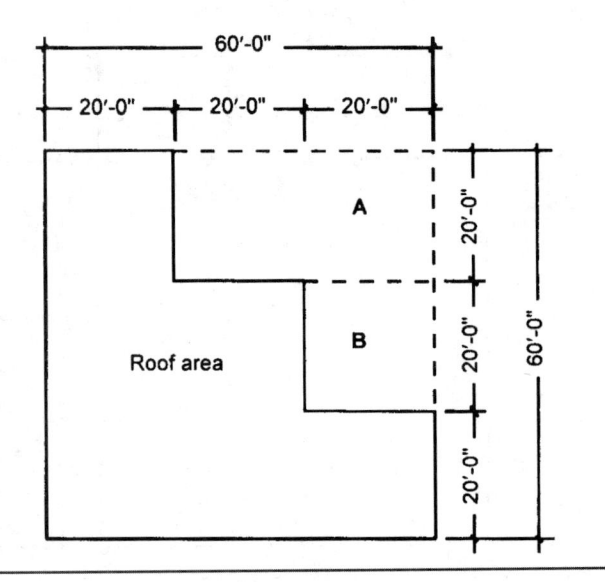

Figure 1-3   The negative method

Perimeter of a Level Roof

The *perimeter* (also called the perimetry or periphery) of a level roof is the total distance around the roof, measured from outside of roof to outside of roof. For example, in Figure 1-6, the perimeter is:

L + W + L + W + R + R, or 2L + 2W + 2R, or 2(L + W + R) | Equation 1-2 |

where L is the roof length, W is the width, and R is the depth of the recess.

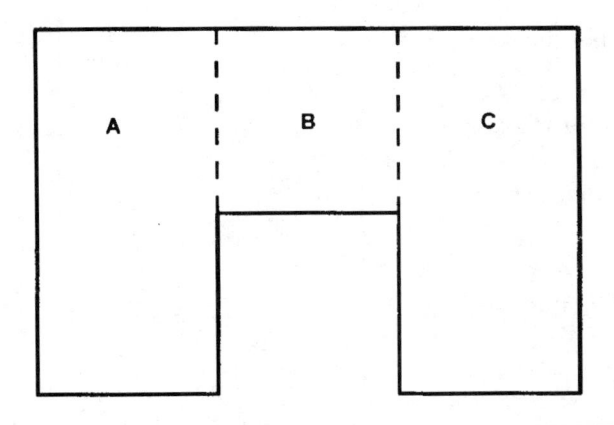

Figure 1-4   The positive method requiring three area calculations

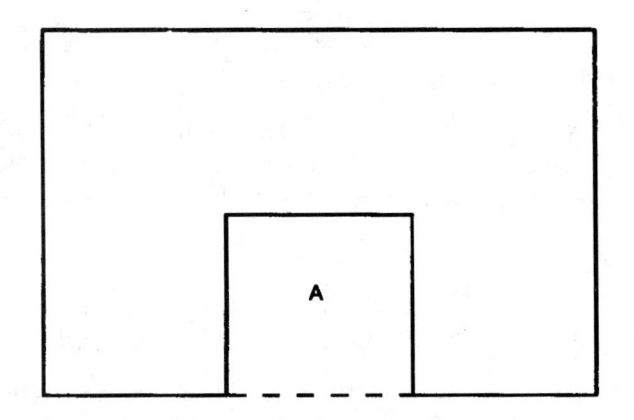

Figure 1-5   The negative method requiring two area calculations

Measuring and Calculating Roofs

If a building doesn't have any recesses, the equation is simply:

Perimeter = 2(L + W) | Equation 1-3 |

or 2 times the total of length plus width.

▼ **Example 1-3:** Find the perimeter of the level roof shown in Figure 1-7

The perimeter of the roof = 2 x (40' + 45' + 12')
= 194 linear feet

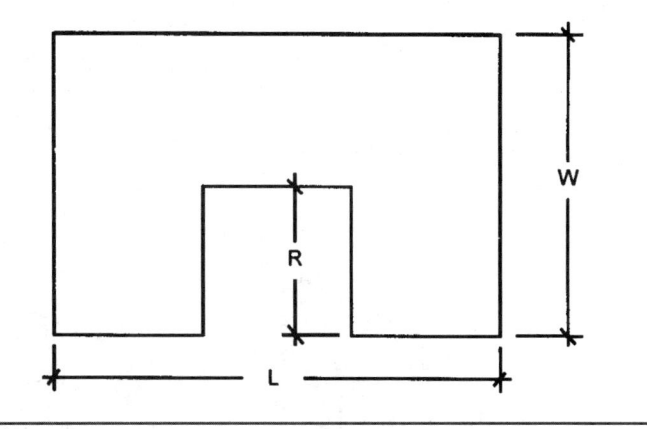

Figure 1-6   Roof perimeter

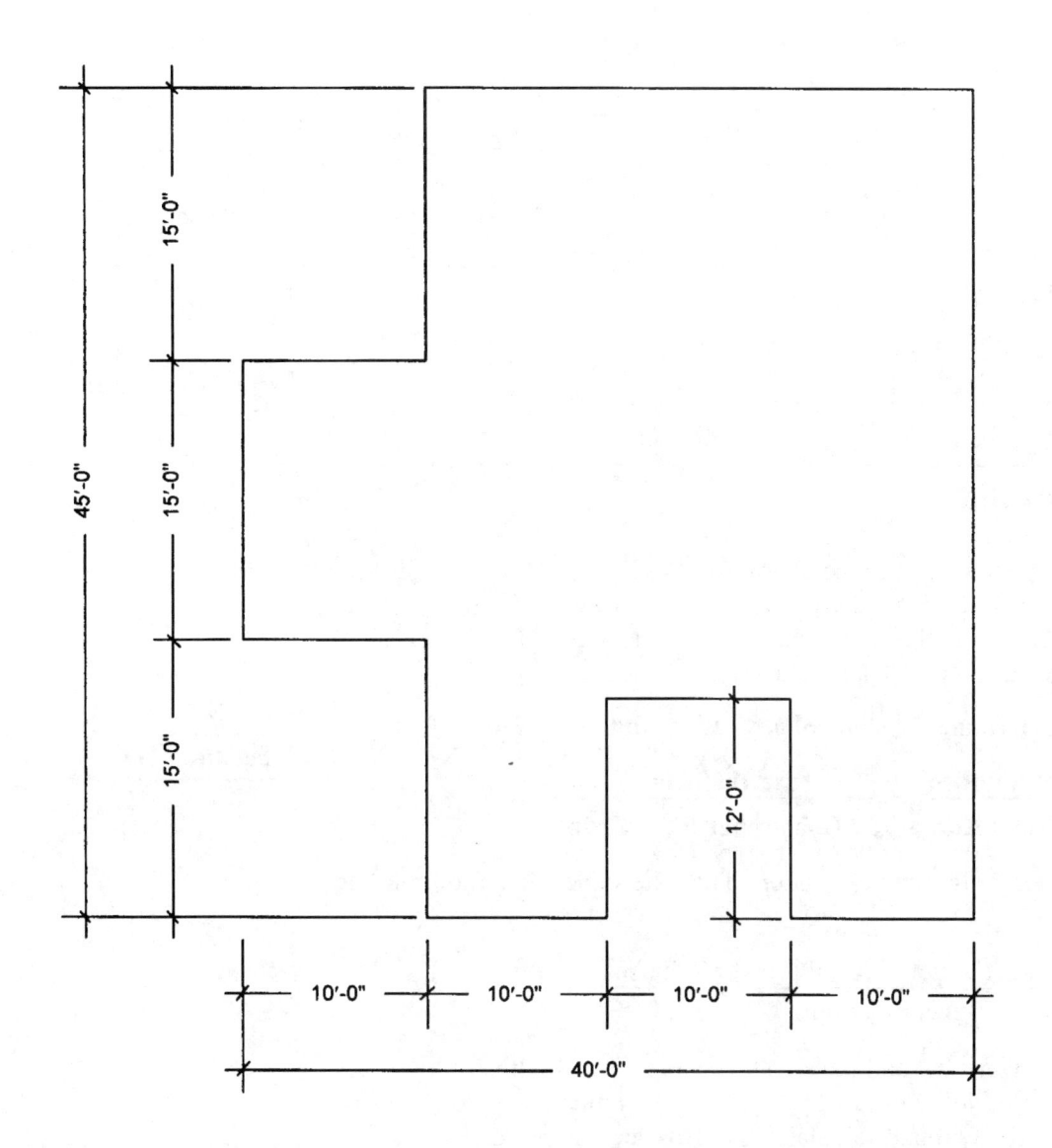

Figure 1-7   Roof perimeter example

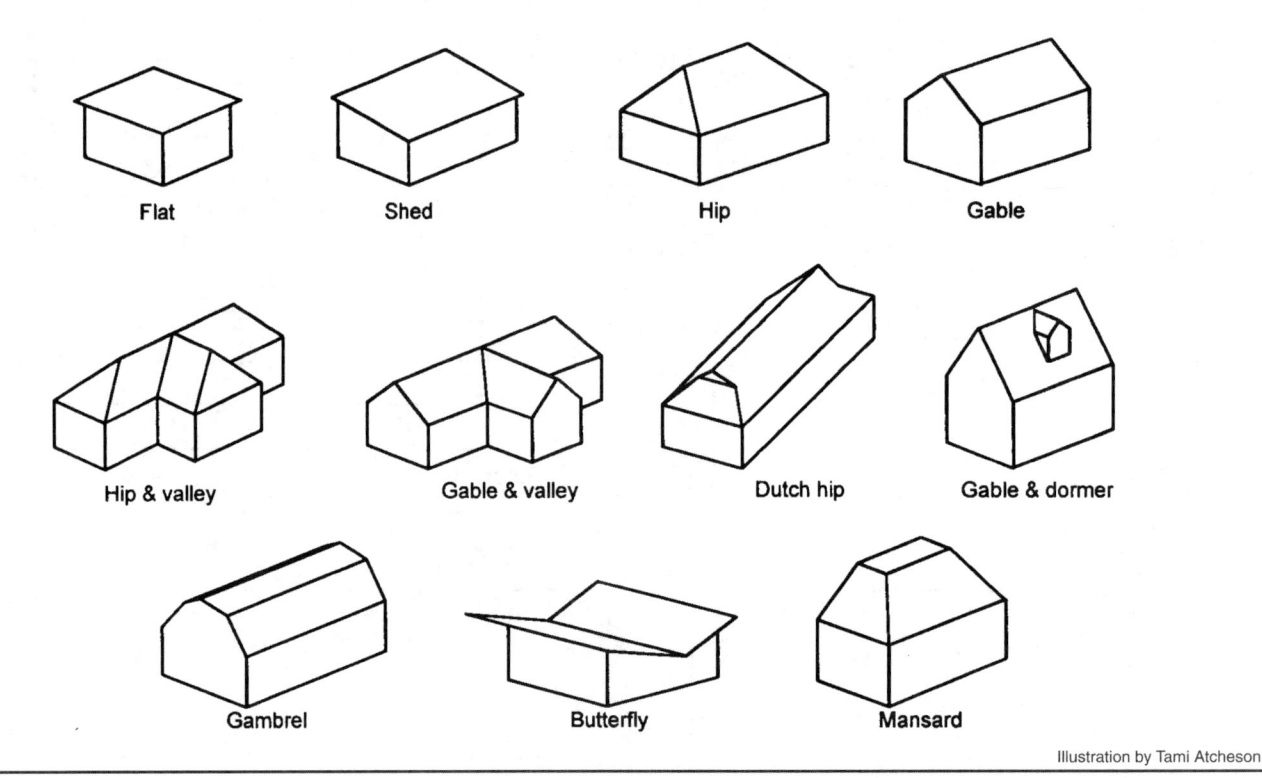

Figure 1-8 Roof types

Sloped Roofs

Figure 1-8 shows a few of the almost limitless types of sloped roofs. You define the *slope* of a roof in terms of rise (in inches) per 12 inches of run. For example, a "6 in 12 roof" is a roof that rises 6 inches for every 12 inches of horizontal run. That's illustrated in Figure 1-9.

You can determine the slope of any roof with the equation:

Equation 1-4

$$\text{Slope} = \frac{\text{Total Rise}}{\text{Total Run}} = \frac{\text{Rise (in inches)}}{12 \text{ (inches per foot of run)}}$$

▼ **Example 1-4:** Use the above equation to find the slope of the roof in Figure 1-10.

$$\text{Slope} = \frac{5 \text{ feet } 5 \text{ inches (total rise)}}{13 \text{ feet (total run)}} = \frac{? \text{ inches}}{12}$$

Step 1: Convert feet and inches into feet and hundredths of a foot. To do that you divide 5 inches by 12 (inches). Notice that throughout the book, we usually round calculations to two decimal places.

$5 \div 12 = 0.4166 = 0.42'$ (rounded off)

Now the equation reads:

$$\text{Slope} = \frac{5.42' \text{ (total rise)}}{13' \text{ (total run)}} = \frac{? \text{ inches}}{12}$$

Step 2: To solve for "? inches," multiply both sides of the equation by 12:

$$\text{Slope} = \frac{5.42}{13} \times 12 = 0.417 \times 12 = 5.004$$

You'll round that answer down to 5, so the slope of the roof is 5 in 12.

You can change the original formula to find total rise if you already know the slope and the total run:

$$\text{Slope} = \frac{\text{Total Rise}}{\text{Total Run}}$$

$$\begin{aligned}\text{Total Rise} &= \text{Slope} \times \text{Total Run} \\ &= \frac{5}{12} \times 13 \\ &= 5.42' \text{ or } 5'5''\end{aligned}$$

You can also find the total run if you know the slope and the total rise:

Total Run = Total Rise ÷ Slope = 5.42 ÷ 0.417 = 13

(Remember, the 0.417 is 5 divided by 12.)

▼ **Example 1-5:** Calculate the total rise for a roof with a slope of 5 in 12 and a run of 20 feet.

$$\begin{aligned}\text{Total Rise} &= 0.417 \times 20 \\ &= 8.34' \text{ or } 8'4''\end{aligned}$$

Roof Pitch

The total *span* of a roof is the horizontal distance, from one eave to the other, as shown in Figure 1-11. You can use that information in a formula to find roof slope, if you know the roof *pitch*, by:

$$\text{Pitch} = \frac{\text{Total Rise}}{\text{Total Span}} \quad \boxed{\text{Equation 1-5}}$$

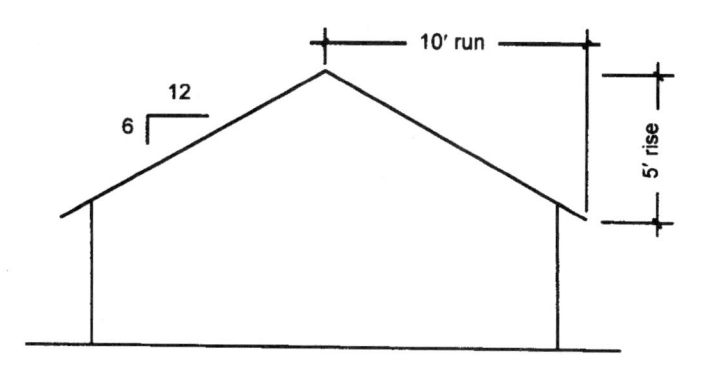

$$\text{Slope} = 2 \times \text{Pitch} \quad \boxed{\text{Equation 1-6}}$$

Figure 1-9 Roof rise and run

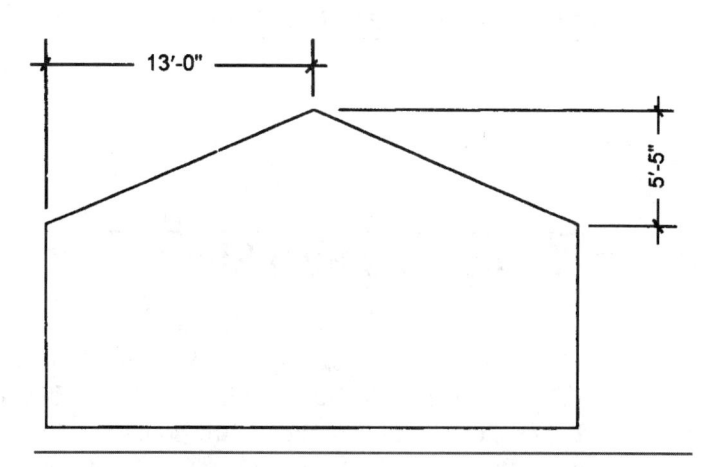

Figure 1-10 Roof slope example

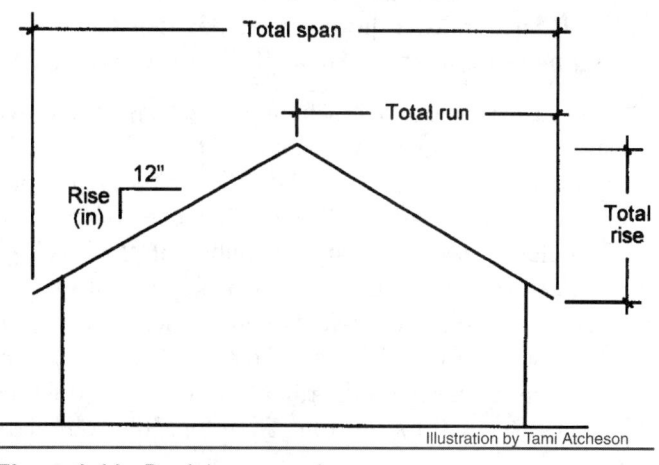

Figure 1-11 Roof rise, run and span

Occasionally a roof is described in terms of pitch, although the term means more to the framer than to a roofing estimator. But sometimes the pitch is the only information you have. Here's how to convert roof pitch to roof slope:

▼ **Example 1-6:** Convert a $1/3$ pitch into terms of roof slope.

Slope $= 2 \times 1/3$
$= 2/3$

From Example 1-4, Step 2, you have:

Slope $= \dfrac{2}{3} \times 12 = 8$

Therefore, the roof slope is 8 in 12.

Roof Slope in Degrees of an Angle

Sometimes roof slope is described in terms of degrees of an angle. When it is, you can use Figure 1-12 to convert roof slope to degrees, and vice versa.

How to Measure Roof Slope

You can determine the slope of a roof with an adjustable device called a *Squangle*®. You simply place the Squangle® against an exposed rafter tail or a block placed over the fascia board, adjust the square so that it lines up with the slope of the roof and read the scale. Figure 1-13 shows a Squangle®.

You can also use a *sliding T-bevel* to size the angle between the fascia and roof deck. That's shown in Figure 1-14. Then transfer the angle to a board or sheet of paper and measure it with a Squangle® or protractor.

You can also place a bubble level and ruler over a straight board on the roof slope as shown in Figure 1-15. Since roof slopes are expressed in terms of rise per 12 inches of run, mark the level at 12 inches from one end. To determine slope, center the bubble in the level, place a ruler vertically so that its scale is lined up at your mark 12 inches from the up-slope end of the level, then read the distance to the bottom of the level. If, for instance, you read 4 inches on the ruler, the roof rises (or falls) 4 inches for each foot of run. Therefore, the roof slope is 4 in 12.

Roof slope (rise & run)	Roof Slope (degrees)
1 in 12	4.76
2 in 12	9.46
3 in 12	14.04
4 in 12	18.43
5 in 12	22.62
6 in 12	26.57
7 in 12	30.26
8 in 12	33.69
9 in 12	36.87
10 in 12	39.81
11 in 12	42.51
12 in 12	45.00
13 in 12	47.29
14 in 12	49.40
15 in 12	51.34
16 in 12	53.13
17 in 12	54.78
18 in 12	56.31
19 in 12	57.72
20 in 12	59.04
21 in 12	60.26
22 in 12	61.39
23 in 12	62.45
24 in 12	63.43

Figure 1-12 Converting roof slope to degrees

Courtesy of Mayes Brothers Division of Great Neck Saw Manufacturers Inc.

Figure 1-13 Squangle ®

Measuring and Calculating Roofs

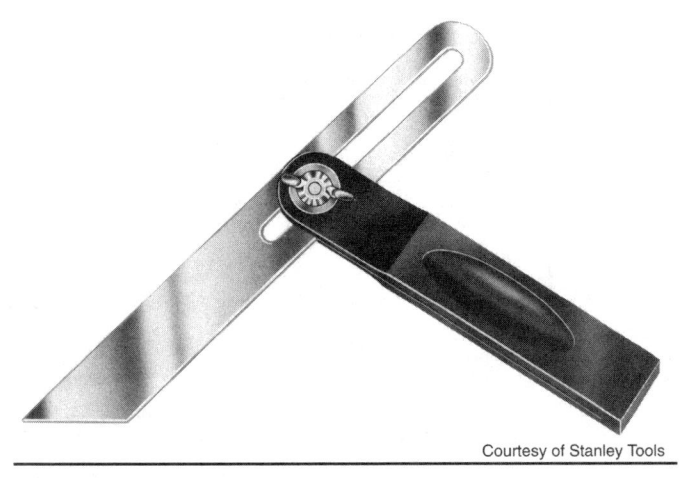

Figure 1-14 T-bevel

Figure 1-15 Determining roof slope

Rafters

Rafters are the inclined members of the roof frame. Figure 1-16 illustrates these rafter types.

- A rafter that extends perpendicularly from the top of an outside wall to the ridge board is called a *common rafter*.

- A common rafter that runs parallel to the ridge board is called a *ridge rafter*.

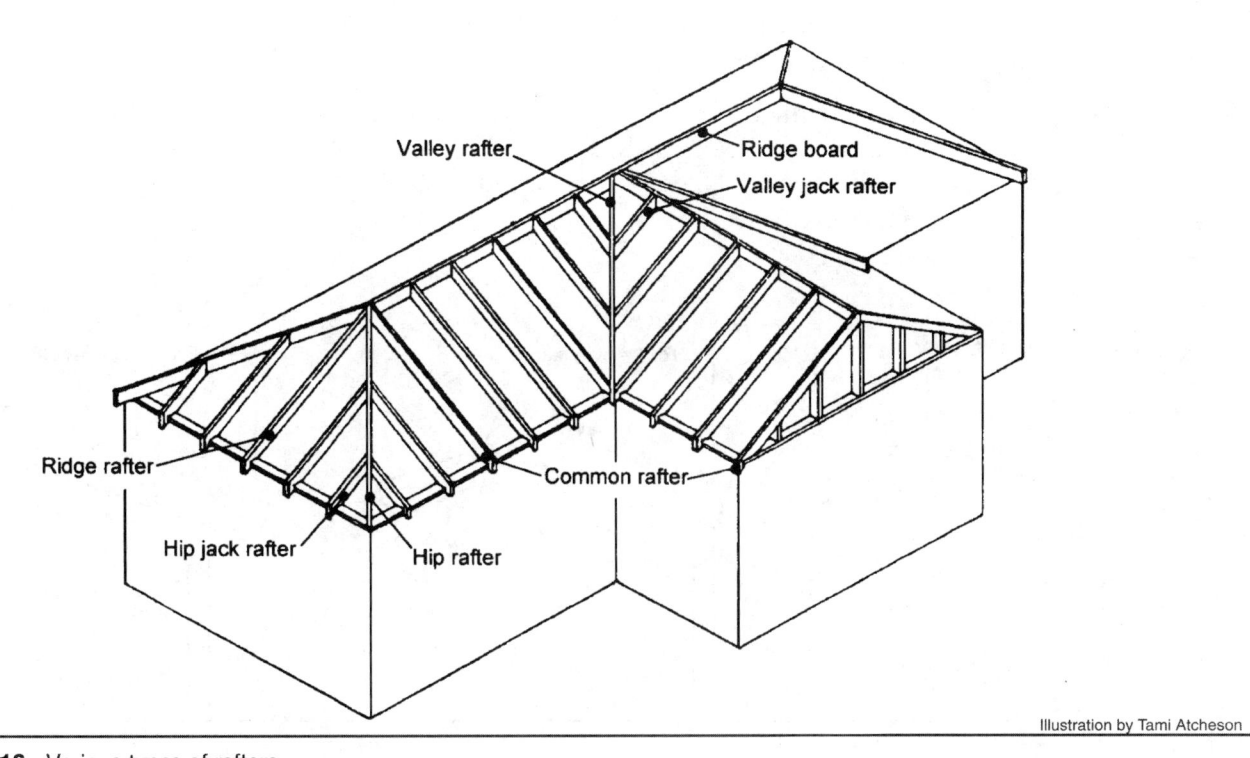

Figure 1-16 Various types of rafters

- A rafter that extends diagonally from an outside corner of a building to the ridge board is called a *hip rafter*.

- A rafter that extends diagonally from an inside corner of a building to the ridge board is called a *valley rafter*.

- A rafter that extends from an outside wall to a hip rafter is called a *hip jack rafter*.

- A rafter that extends from the ridge board to a valley rafter is called a *valley jack rafter*.

Rafter Length

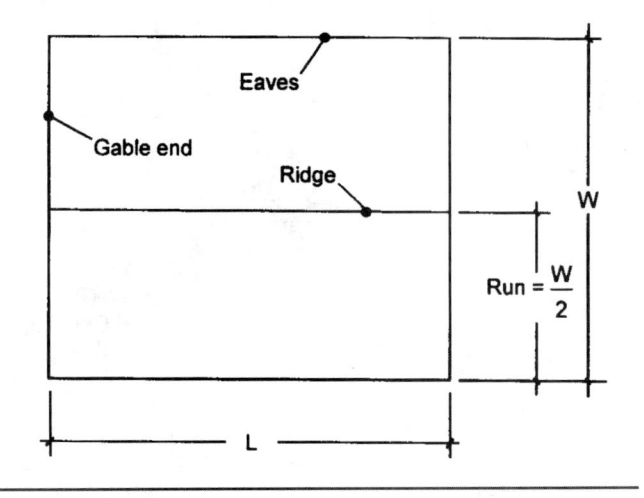

Figure 1-17 Plan view of gable roof

Figure 1-17 is a plan view of a gable roof. The length (L on Figure 1-17) of the eaves edge (the roof dimension perpendicular to the run of the rafters) is horizontal. Therefore, you can read that dimension directly from the plans. But you can't see the exact size of the width (W in the figure) because the roof slopes. The *plan length* of a common rafter is called the *run* of the rafter. Figure 1-18 illustrates this for three kinds of rafters.

Sometimes you can scale the lengths of common rafters from an elevation or cross section drawing. But it's safer and more convenient to convert the plan dimensions. Figure 1-19 is a table which gives you the appropriate conversion factors. Column 2 of the table gives factors for common rafters, and columns 3 and 4 are for hip or valley rafters.

The values in column 2 are called *roof-slope factors*. The values in columns 3 and 4 are called *hip/valley-slope factors*. The conversion factors in Figure 1-19 assume that all hips and valleys are framed at an angle of 45 degrees with respect to the eaves line.

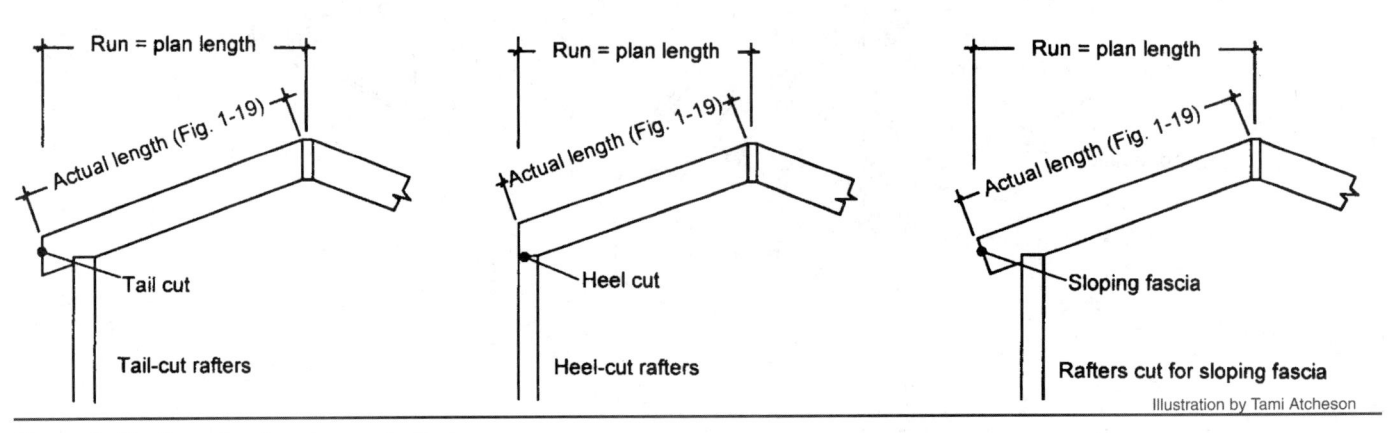

Figure 1-18 Run of various types of rafters

(1) Roof Slope	(2) Common or jack rafters (factor x run = actual length)	(3) Hips or valleys (factor x run = actual length)	(4) Hips or valleys (factor x plan length = actual length)
1 in 12	1.004	1.417	1.002
2 in 12	1.014	1.424	1.007
3 in 12	1.031	1.436	1.015
4 in 12	1.054	1.453	1.027
5 in 12	1.083	1.474	1.042
6 in 12	1.118	1.500	1.061
7 in 12	1.158	1.530	1.082
8 in 12	1.202	1.564	1.106
9 in 12	1.250	1.601	1.132
10 in 12	1.302	1.642	1.161
11 in 12	1.357	1.685	1.191
12 in 12	1.414	1.732	1.225
13 in 12	1.474	1.782	1.260
14 in 12	1.537	1.833	1.296
15 in 12	1.601	1.888	1.335
16 in 12	1.667	1.944	1.375
17 in 12	1.734	2.002	1.416
18 in 12	1.803	2.062	1.458
19 in 12	1.873	2.123	1.501
20 in 12	1.944	2.186	1.546
21 in 12	2.016	2.250	1.591
22 in 12	2.088	2.315	1.637
23 in 12	2.162	2.382	1.684
24 in 12	2.236	2.450	1.732

Figure 1-19 Roof-slop factors for determining rafter lengths

Use roof-slope factors from column 2 of Figure 1-19 to determine the actual length of a common rafter or jack rafter.

You'll refer to this table again in later chapters. There's another copy, Appendix A, in the back of the book.

▼ **Example 1-7**: Look at the diagram in Figure 1-20. Assume a roof slope of 10 in 12, then find the actual length for the typical common rafters.

Actual Length (common rafter) = 10' x 1.302 (from column2, Figure 1-19)
= 13.02 linear feet

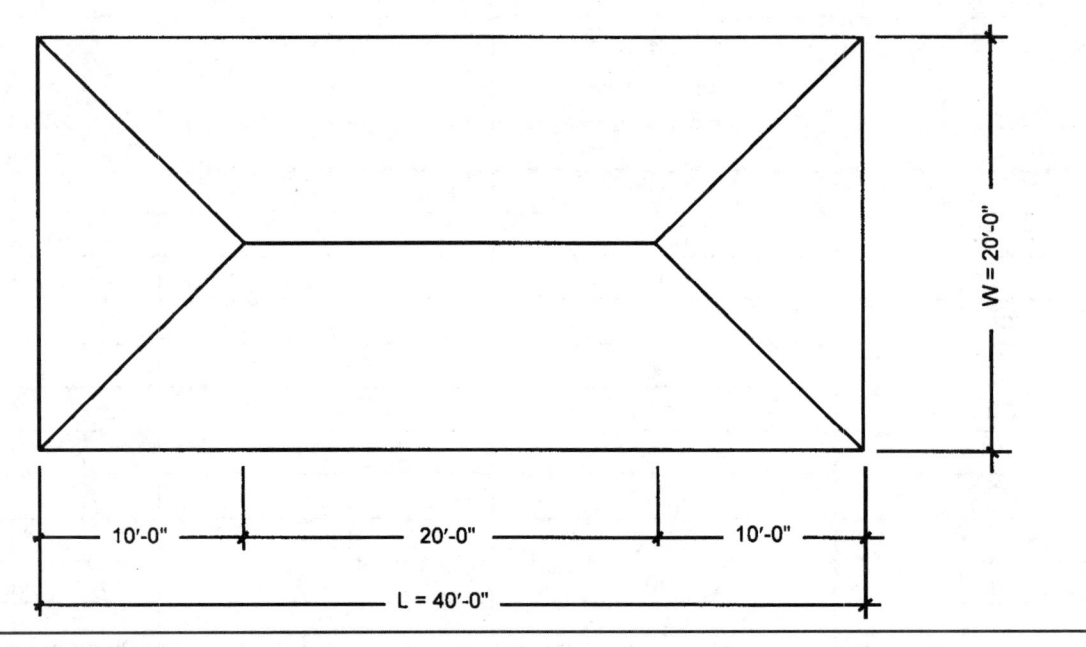

Figure 1-20 Hip roof example

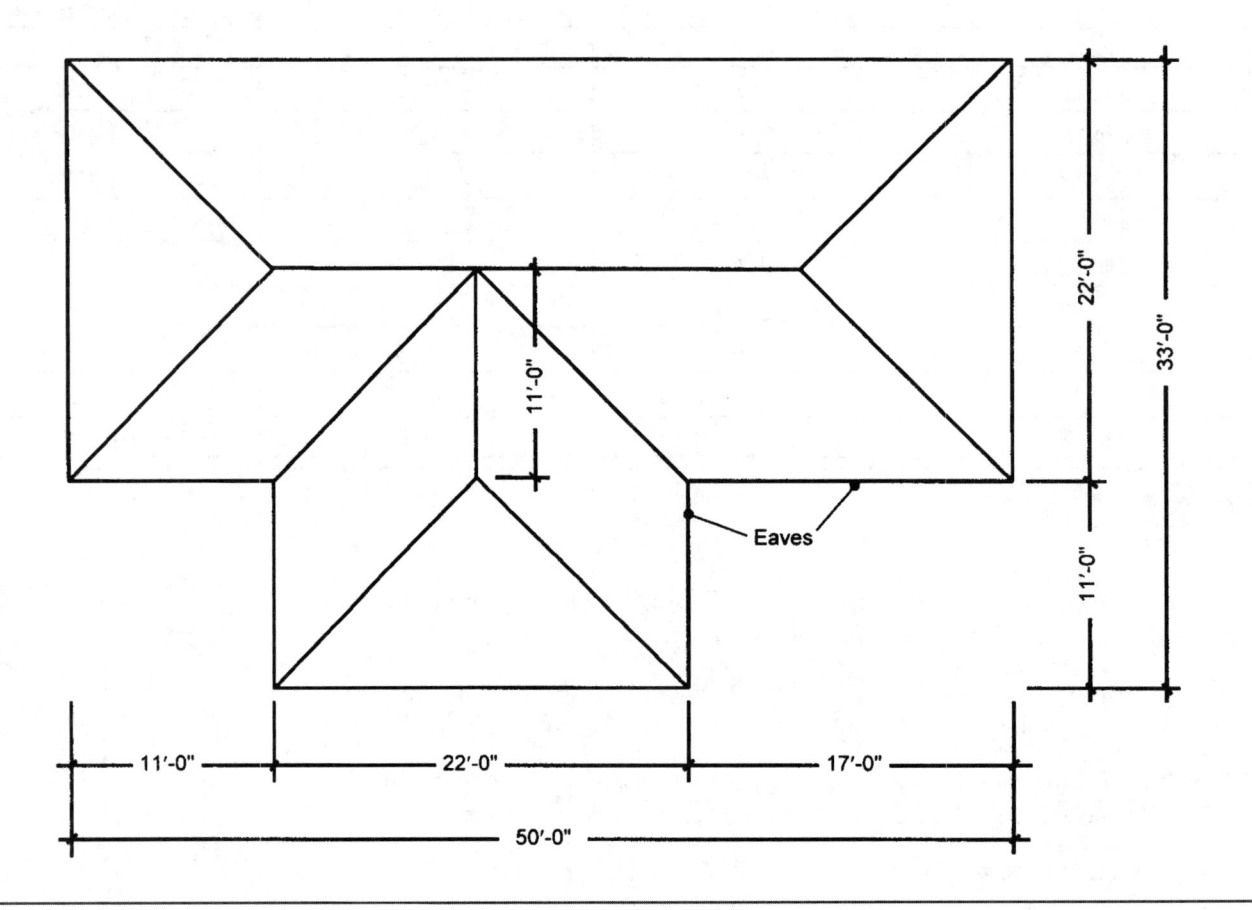

Figure 1-21 Hip-and-valley roof example

Measuring and Calculating Roofs

Perimeter of a Sloped Roof

The eaves of a hip roof (Figure 1-20) or a hip-and-valley roof (Figure 1-21) run horizontally all the way around the building, so you can determine the perimeter from the dimensions on the roof plan. The formula is the same as that for a level roof.

P = 2(L + W + R)   *Equation 1-7*

where L is the roof length, W is the roof width, and R is the depth of the recess.

If the building has no recess, the formula for the perimeter is simply:

P = 2(L + W)   *Equation 1-8*

▼ **Example 1-8:** Find the perimeter of the hip roof in Figure 1-20.

P = 2(40' + 20') = 120 linear feet

To find the perimeter of a gable roof like the one in Figure 1-17, the formula is:

Perimeter = 2(Length + Actual Width)   *Equation 1-9*

Actual Width = 2(Run x Roof-Slope Factor)   *Equation 1-10*

$$= 2\left(\frac{W}{2} \times \text{Roof-Slope Factor}\right)$$

= W x Roof-Slope Factor (from column 2 of Figure 1-19)

Thus, the perimeter of a *gable roof* is:

Perimeter = 2[L + (W x Roof-Slope Factor)]   *Equation 1-11*

▼ **Example 1-9:** Find the perimeter of the gable roof in Figure 1-22 if the roof slope is 8 in 12.

   P = 2(31' + [26' x 1.202])
      = 2(31' + 31.25')
      = 2 x 62.25'
      = 124.5 linear feet (For estimating purposes, round this to 125 feet.)

Net Versus Gross Roof Area

The *net* area of a roof is the area of roof sheathing that will be covered with roofing material. But you have to provide materials for an area much larger than the net roof area. You have to allow for such things as:

- Additional felt underlayment at the ridge, hips, and valleys

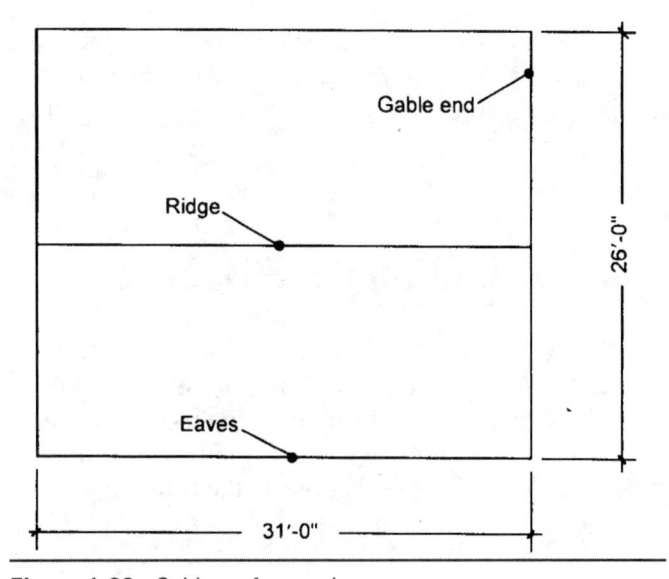

Figure 1-22 Gable roof example

- A starter course
- Hip and ridge units
- Cutting allowances at rakes, hips, and valleys (for shingles)

This larger roof area is called the *gross* area of the roof. For example, the net area of a roof might be 10 squares. However, you might have to provide additional material equal to a roof area requiring 12 squares. A roofing *square* is 100 square feet.

Allowance Factors

The simplest way to account for material required for overcutting and lapping is to use an *allowance factor*. An allowance factor is the ratio of the actual amount of material required to cover the roof (gross roof area) to the net area of roof deck covered:

$$\text{Allowance factor} = \frac{\text{Area Covered (including allowances)}}{\text{Net Roof Area}}$$

Equation 1-12

You can figure the percentage of material overrun by using the allowance factor. Here's an example:

▼ **Example 1-10:** Assume that it will require 12 squares of material (including allowance for waste and lapping) to cover 10 squares of roof deck. Calculate the allowance factor and the percentage of material allowance.

$$\text{Allowance factor} = \frac{12 \text{ squares}}{10 \text{ squares}}$$
$$= 1.20, \text{ or a 20 percent material allowance factor}$$

In later chapters, you'll see that allowance factors can be predicted, based on the roof type, roof size, roof slope, roofing material exposure, and type of roof construction.

Calculating Total Net Roof Area

Since the eaves and ridge of a roof run horizontally, their plan lengths are their actual lengths. And, as you've seen, you can find the actual length of any common rafter by multiplying its plan length by the roof-slope factor in Figure 1-19. So you can use a universal formula to calculate the actual (net) area of any roof that meets the following conditions:

- All roof planes have the same slope
- All hips and valleys are framed at 45 degrees with respect to the eaves

Here's the formula:

Actual (Net) Roof Area = Roof Plan Area x Roof-Slope Factor | Equation 1-13 |

where Roof Plan Area equals roof area as seen in plan view. The Roof-Slope Factor is from column 2 of Figure 1-19.

▼ **Example 1-11:** Assuming a roof slope of 5 in 12, find the net area of the roof shown in Figure 1-22.

Net Roof Area   = 31' x 26' x 1.083
                    = 873 SF ÷ 100 SF/square
                    = 8.73 squares

▼ **Example 1-12:** Assume a roof slope of 4 in 12, then find the net area of the roof shown in Figure 1-21.

Total Roof Plan Area   = (50' x 22') + (22' x 11')
                              = 1,342 SF

Net Roof Area   = 1,342 SF x 1.054
                    = 1,415 SF ÷ 100
                  = 14.15 squares

When the slope of a roof changes from one section to another, you have to do a separate take-off for each area with a different slope. Here's an example:

▼ **Example 1-13:** Compute the net area of the roof in Figure 1-23.

Step 1: Section off the drawing, as shown in Figure 1-24, to isolate the two different slopes. Begin with the large 6 in 12 section. Notice you must deduct the triangle formed by the section of 4 in 12 roof (labeled ABC on the drawing). Multiply the length by the width and subtract the area of the triangle:

Roof Plan Area (6 in 12)   = (100' x 48') - $\left(\dfrac{48' \times 8'}{2}\right)$
                                   = 4,800 SF - 192 SF
                                   = 4,608 SF

Step 2: Use the Roof-Slope factors from column 2 of Figure 1-19 to find the net roof area for the 6 in 12 roof:

Net Roof Area (6 in 12) = 4,608 SF x 1.118 = 5,152 SF

Step 3: Find the area of the 4 in 12 section. Notice that you *add* the area of the triangle to this section:

Roof Plan Area (4 in 12)   = (48' x 16') + $\left(\dfrac{48' \times 8'}{2}\right)$
                                   = 768 SF + 192 SF
                                   = 960 SF

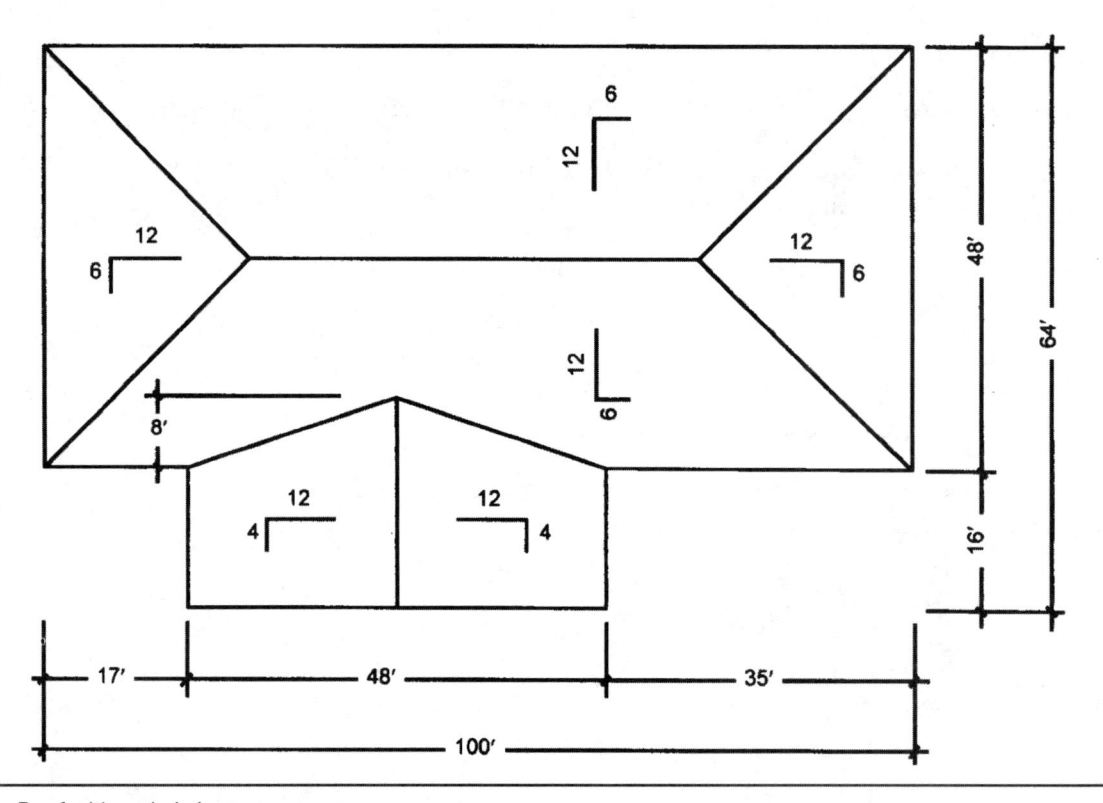

Figure 1-23 Roof with varied slopes

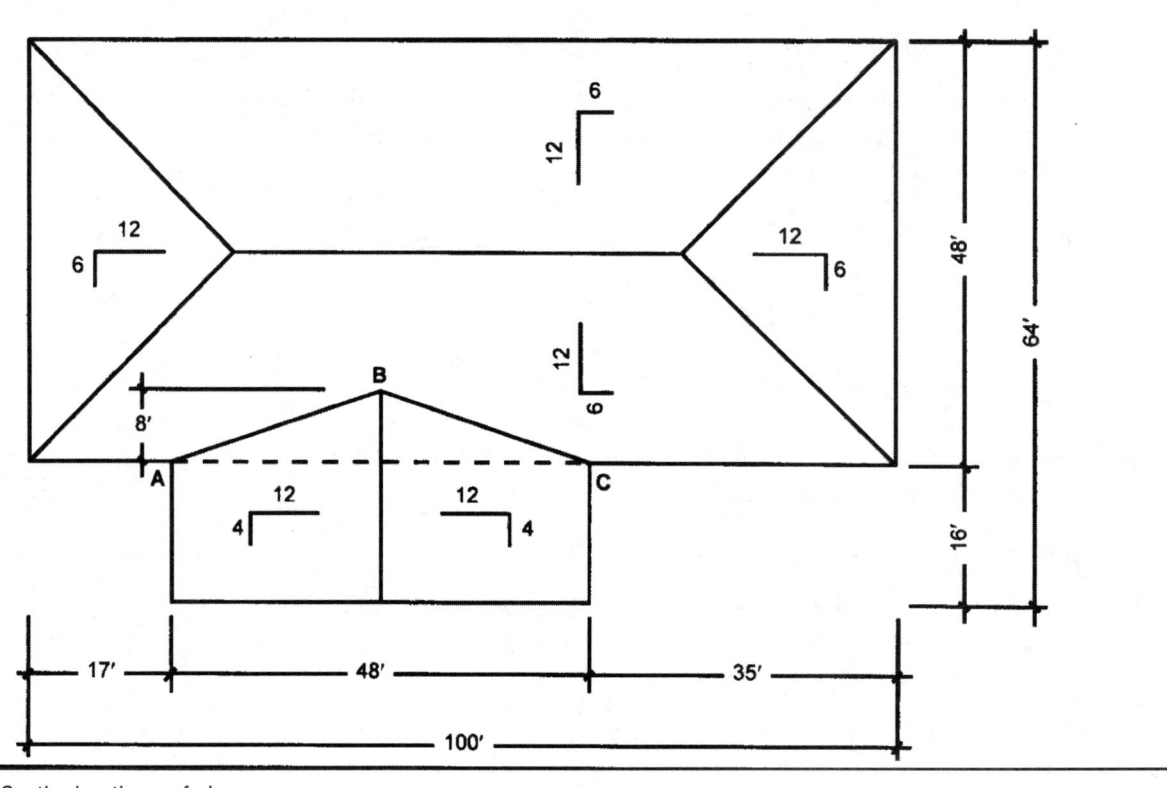

Figure 1-24 Sectioning the roof plan

Step 4: Repeat Step 2 for the 4 in 12 section:

Net Roof Area (4 in 12) = 960 SF x 1.054
= 1,012 SF

Step 5: Add together the two net areas:

Total Net Roof Area = 5,152 SF + 1,012 SF
= 6,164 SF

Roof Overhangs, Hips and Valleys

When you calculate the net area of a roof, be careful you don't omit the roof overhang that extends beyond the walls of the building. (See Figure 1-25.) Also watch for overhangs at interior gable end walls on multi-level roofs like the one in Figure 1-26, and on dormers.

You also need accurate measurements for hips and valleys, which require a variety of roofing materials. Again, refer to Figure 1-19 for conversion factors you can use to calculate the actual lengths for hip and valley rafters. Use the slope factors in column 3 if the hip or valley rafter dimensions are based on the run. Use column 4 if measurements are taken from the plan length. See Figure 1-27.

If the hips and valleys are framed conventionally at a 45-degree angle to the outside walls, you can calculate the plan length with this formula:

Plan Length (Hip or Valley) | **Equation 1-14**
= 1.414 x Run

▼ **Example 1-14:** Assume a roof slope of 10 in 12, then find the actual length of any hip rafter for the roof in Figure 1-20.

In this illustration, the run of the hip rafter is 10 feet. Refer to column 3 in Figure 1-19 and you see the conversion factor for a 10 in 12 slope is 1.642.

Rafter length = 10' x 1.642 = 16.42 linear feet

You can also use the formula above to calculate the plan length based on the run, then use the factor from column 4 in Figure 1-19 to get the actual length:

Plan length = 1.414 x 10' = 14.14 linear feet

Rafter length = 14.14 x 1.161 = 16.42 linear feet

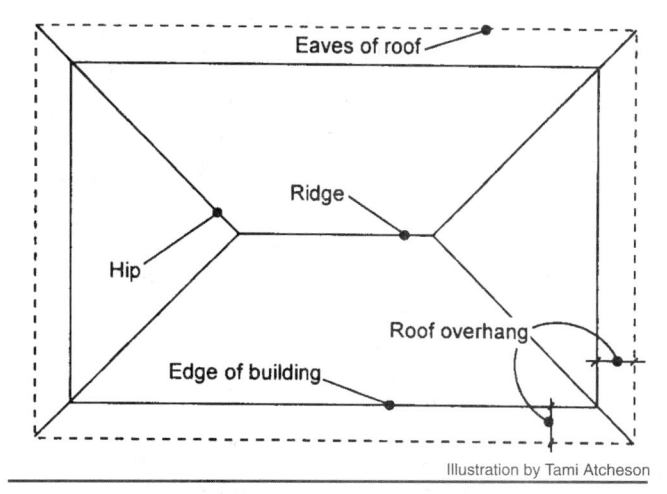

Figure 1-25 Roof overhangs

Figure 1-26 Multi-level roof overhang

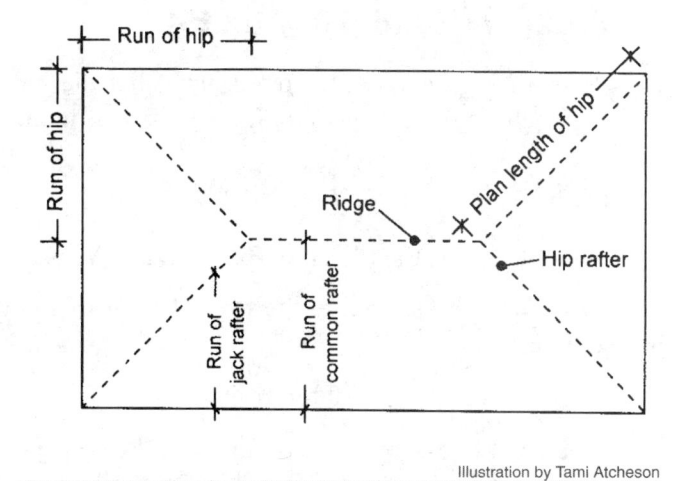

Figure 1-27 Lengths of hip rafters

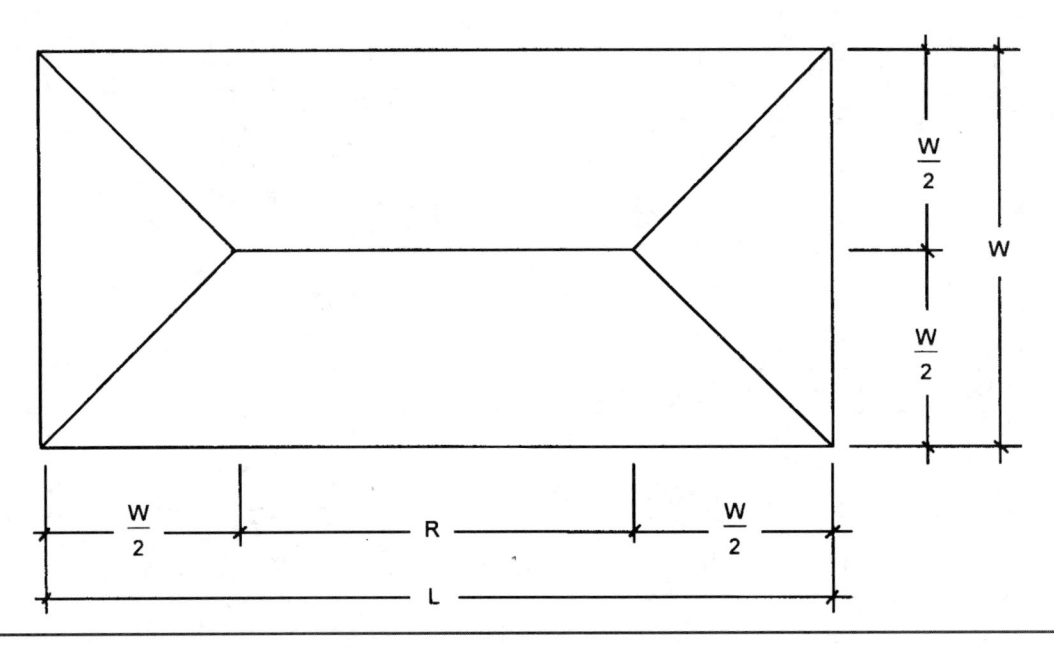

Figure 1-28 Ridge length

Now, what if a building is built with roofs of unequal slopes, such as the one shown in Figure 1-23? You can find the actual length of a valley where the roofs intersect by multiplying the run of the low-sloping roof by the appropriate factor from the table in Appendix B.

▼ **Example 1-15:** Determine the actual length of each valley of the roof diagrammed in Figure 1-23. The run of the low-sloping roof is 24 feet.

So the actual valley length (using the factor from Appendix B) is:

Valley length = 24' x 1.247 = 29.93 linear feet

Length of Ridge (Hip Roofs)

Refer to Figure 1-28. If you assume that the hips are conventionally framed, you find the ridge length on a hip roof with:

$$\text{Ridge} = L - \left[2 \times \left(\frac{W}{2}\right)\right] = L - W$$

Equation 1-15

where L equals the length of the roof, and W is the width.

▼ **Example 1-16:** Determine the ridge length of the roof diagrammed in Figure 1-20.

Ridge = 40' - 20' = 20 linear feet

You can't do the example problems in later chapters if you don't know the formulas in this chapter. Don't go on until you're sure you know how these equations work and how to apply them.

2 Roof Sheathing, Decking and Loading

▶ In the last chapter, you learned how to measure a roof and figure out how much roofing material you need to cover it. The next step is to decide what type of roof deck or sheathing you'll need to support the roof covering. For a sloping roof, that depends on several things:

- The type or roofing material you'll install
- The roof slope
- The local climate
- The building code requirements in your area

Check the Framing

One problem you may face is faulty construction of the roof frame, especially at the overhang along the rakes. The correct method of framing this overhang is to construct a "ladder" of *lookout rafters*. That's shown in Figure 2-1. If the lookout rafters aren't there, the overhang will sag. When that happens, mineral granules can pop off the roof covering, leaving it discolored and deteriorated. If you run into this situation, tell the owner why the roof looks the way it does. Quote a price to repair the overhangs, assuming the rest of the roof frame is sound.

If you're replacing a relatively lightweight roof covering with a heavier one, or covering over an existing roof, make sure the roof frame will support the extra load. An overloaded roof can make the deck, rafters, and ridge sag. If the roof frame sags, so will the ceiling, since the mid-span of the rafters in a conventional roof frame is supported by struts and uprights fastened to the ceiling joists. See Figure 2-2.

A sagging deck will also stretch asphalt shingles, bending them over the rafters so the mineral granules pop off. This will leave light-colored lines over the rafters where the shingles are stressed and darker areas where the deck has sagged. And who does the customer blame for that? You guessed it — the roofing contractor. Most homeowners' insurance policies don't cover damage due to ignorance — whether it's the owner's or the contractor's. And if the roof fails, there may be injuries or even death. That's bad for your peace of mind and not real good for your insurance premiums either.

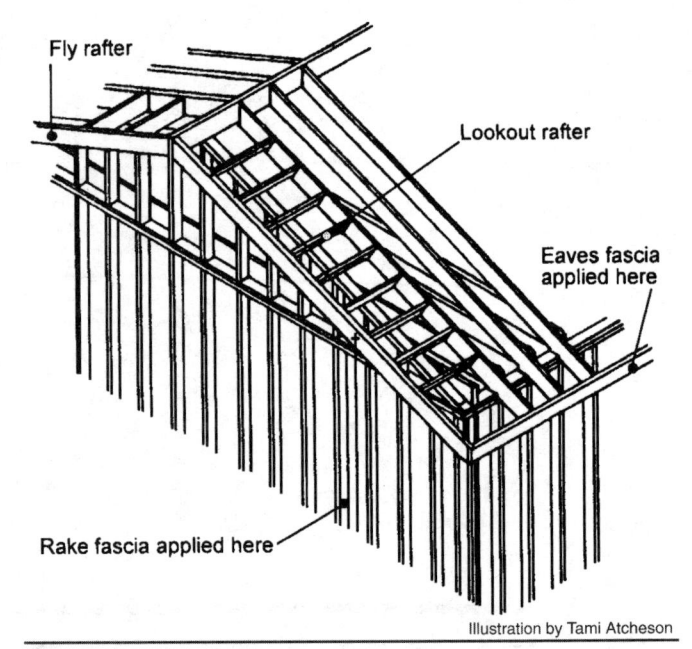

Figure 2-1 Lookout rafters

Solid Roof Sheathing

You always need solid roof sheathing, whether it's boards, plywood or waferboard, under asphalt, fiberglass and metal shingles, mineral-surfaced roll roofing, or built-up roofing. You must also use solid sheathing under tile, slate and metal roofing (except corrugated steel), or any other roof covering that requires continuous support. Building codes in some seismic zones require that you install solid roof sheathing under *all* roofs. Solid sheathing is also recommended (and sometimes required) in climates where wind-driven snows or rains occur.

Should you ever install shakes or wood shingles, I recommend that you do it over solid sheathing where:

- The outside design temperature (the average expected low temperature) is 0° F or colder. The map in Figure 2-3 shows those areas.

- The January mean temperature is 25° F or less.

- There's a possibility that ice will form along the eaves and cause leaks from a backup of water.

You can use 1 x 4, 1 x 6, 1 x 8, 1 x 10 or 1 x 12 boards for solid roof sheathing. But I don't recommend you use anything wider than a 1 x 6 because larger boards tend to swell and shrink excessively after they're installed.

Roof Sheathing, Decking and Loading

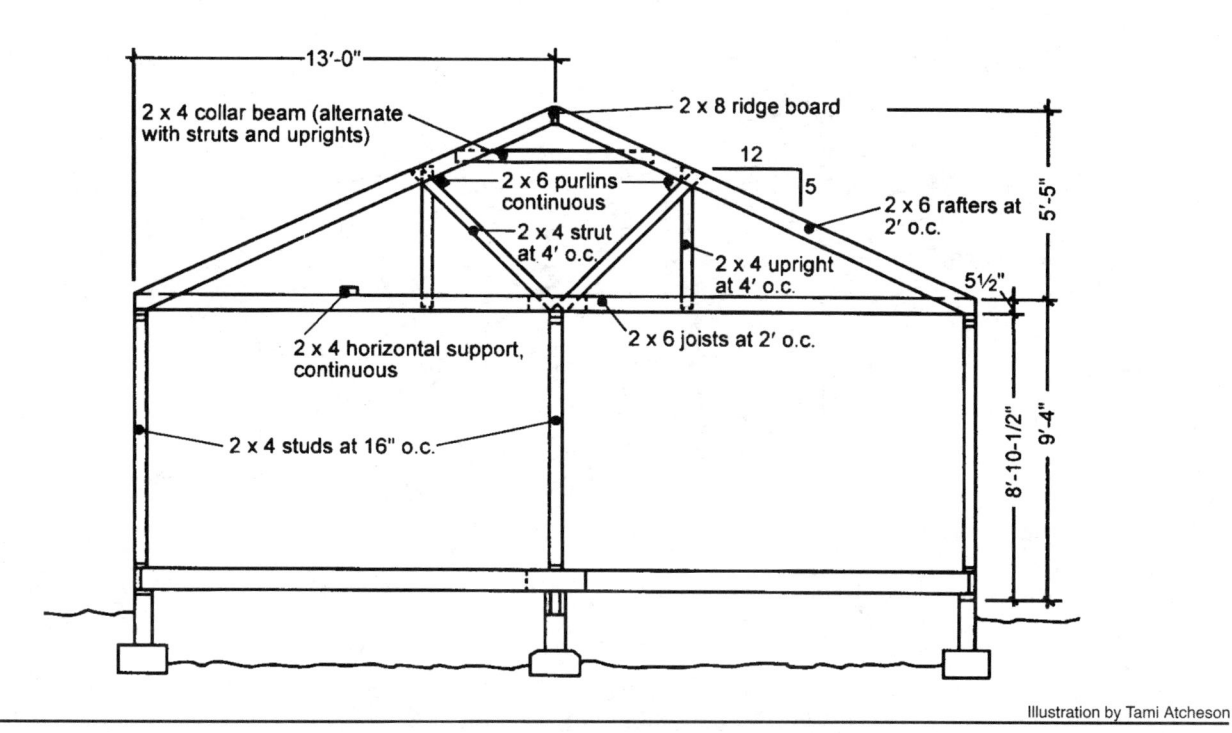

Figure 2-2 Roof supports

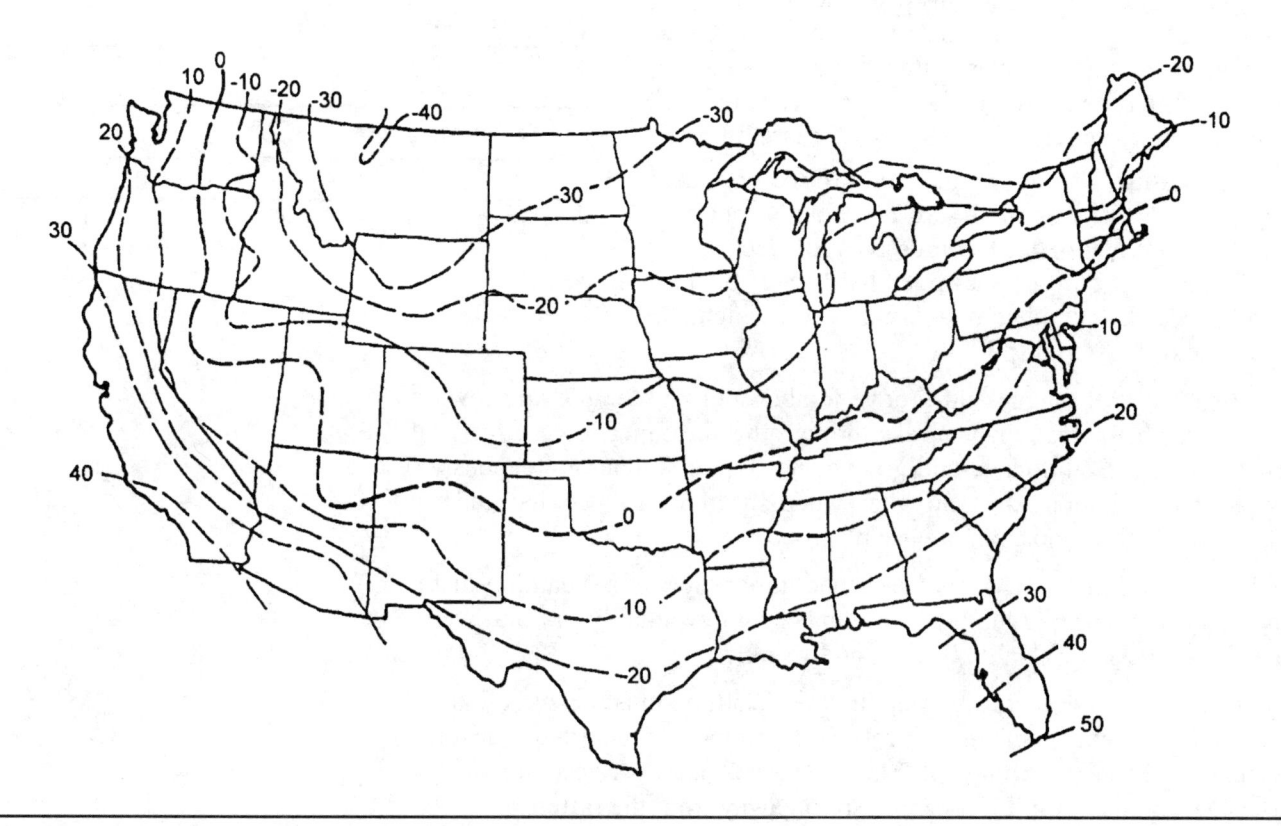

Figure 2-3 Outside design temperatures, U.S. Weather Bureau

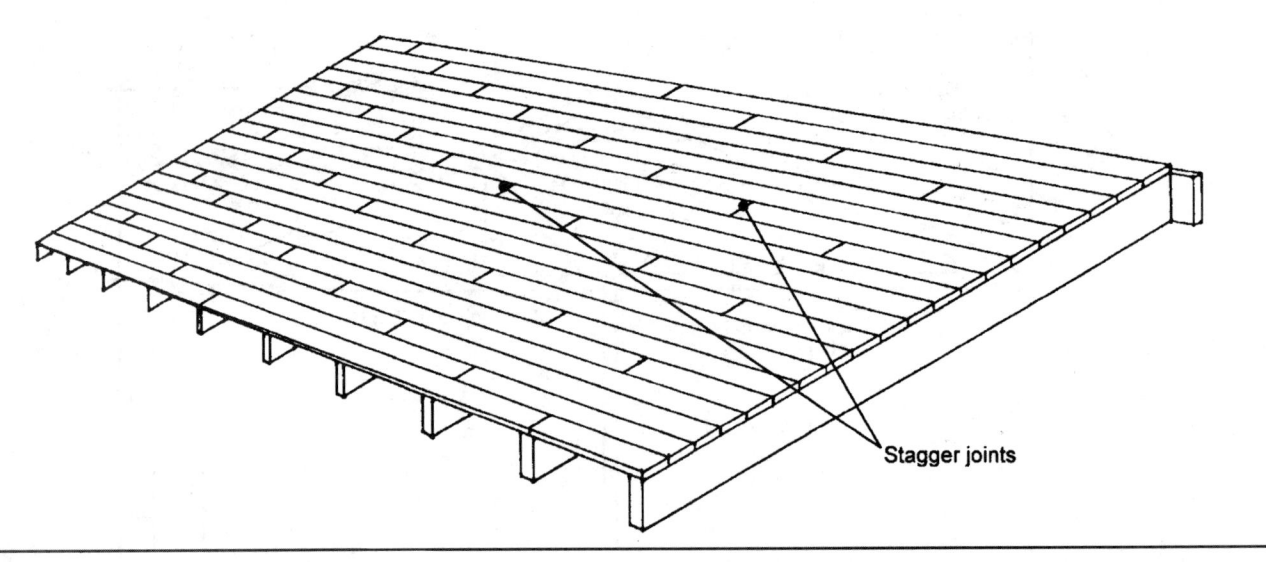

Figure 2-4 Solid board sheathing

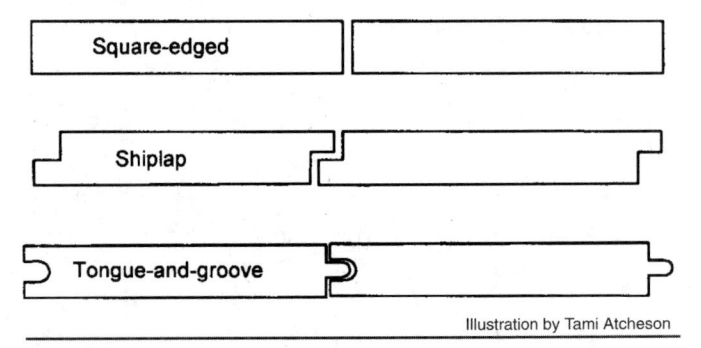

Figure 2-5 Sheathing boards

Install sheathing boards with a fairly tight edge joint, but allow at least ⅛ inch of space at the end joints for expansion. If it's possible the sheathing will be rained on before you get the felts installed (which can cause warping), you might consider not drawing the edge joints up too close. Stagger the end joints of the boards and make sure they're centered over rafters. Figure 2-4 shows how. Face-nail boards up to 1x8s over each rafter with two 8d common nails. Use three nails for larger boards.

You can use shiplap, tongue-and-groove or square-edged boards like those in Figure 2-5. But it's not a good idea to use square-edged boards on roof decks when you'll install a built-up or mineral-surfaced roll roof. That's because individual boards can warp, which may make the roll material split.

Always install tongue-and-groove lumber with the tongue laid toward the ridge. This helps prevent leaks through the sheathing in case the roof covering fails. Both tongue-and-groove and shiplap lumber are expensive, and so is installation labor. So when wood sheathing is specified under a flat, built-up roof, most people use plywood.

Never use "green" or poorly-seasoned lumber for roof sheathing material. Choose kiln-dried lumber. Wet lumber will eventually dry and warp, which can make the shingles buckle and wrinkle.

For the roof covering to be smooth, the sheathing must be smooth also. Never install scrap lumber, concrete form lumber, or boards of unequal thickness for sheathing material. Also, be sure all boards are securely nailed so the sheathing doesn't loosen and spring away from the rafters.

Don't use lumber with loose knots or areas where pitch runs over the board surface, because pitch (tree sap) reacts with asphalt, dissolving the asphalt. If you're covering an existing roof which has these defects, cover the affected area with sheet metal patches before you install the roof covering.

Plywood Sheathing

Most solid roof sheathing is made of plywood. The American Plywood Association recommends a minimum plywood grade of C-D Interior-and-Exterior Glue, or another equally durable exterior grade of plywood. And while the APA recommends it, most building codes *require* it.

Consider the rafter spacing (span) and anticipated roof loads when you decide on what thickness plywood. Deck deflection due to concentrated loads should be limited to 1/240th of the total span (Figure 2-6) so a 4-foot span should support plywood strong enough to limit the deflection to:

$$\text{Allowable deflection} = \frac{4}{240} = 0.0167 \text{ feet}$$
$$= 0.20 \text{ inches}$$

Figure 2-6 gives the American Plywood Association's specifications for plywood roof sheathing. Roofing material manufacturers recommend a minimum thickness of ½ inch for roof sheathing under a built-up roof. Be sure the index number of the plywood isn't less than 32/16 (unsanded) or Group 1 (sanded). The index number is a rating that says the plywood can span a minimum of 32 inches between rafters, or 16 inches between floor joists. This rating is based on plywood that's installed with the face grain of the plywood perpendicular to the supports and the sheets spanning at least two supports. That's shown in Figure 2-7.

Notice footnote "d" of Figure 2-6. You see that this data includes a 10-pound *dead load*. The dead load on a roof is the weight of the roof deck, underlayment, and shingles. Some roof coverings (clay tile for instance) weigh as much as 16 pounds per square foot. When you install that over a wood deck (3 pounds per square foot), 15-pound felt (0.15 pounds per square foot) and 90-pound roll roofing (0.9 pounds per square foot), the total dead load is more than 20 pounds.

If the dead load of the roof is more than 10 pounds per square foot (PSF), you must deduct the excess dead load from the allowable live loads. *Live loads* are loads caused by snow or workers — any temporary loads that may be added to the dead load. Keep in mind that building codes allow a minimum anticipated live load of 30 PSF for designing the required strength of a roof deck. For example, if you have a 20-pound dead load and a rafter spacing of 24 inches, the minimum permissible panel span rating is 24/16. That's because the allowable live load (40 PSF in Figure 2-6, Tables 21 and 22 of the American Plywood Association standards), less the excess dead load (10 PSF), is 30 PSF, which is the minimum anticipated roof live load allowed by the building code.

Use 6d common nails to fasten ½-inch panels, and 8d nails for thicker panels. Drive nails 6 inches o.c. along panel edges and 12 inches o.c. over intermediate supports. If you're using staples, use 16-gauge wire staples with

Table 21

Recommended uniform roof live loads for APA rated sheathing[c] and APA rated Sturd-I-Floor with long dimension perpendicular to supports[e]

Panel span rating	Minimal panel thickness (in.)	Maximum span (in.) With edge support[a]	Maximum span (in.) Without edge support	Allowable live loads (psf)[d] — Spacing of supports center-to-center (in.)							
				12	16	20	24	32	40	48	60
APA rated sheathing [c]												
12/0	5/16	12	12	30	----	----	----	----	----	----	----
16/0	5/16	16	16	70	30	----	----	----	----	----	----
20/0	5/16	20	20	120	50	30	----	----	----	----	----
24/0	3/8	24	20[b]	190	100	60	30	----	----	----	----
24/16	7/16	24	24	190	100	65	40	----	----	----	----
32/16	15/32	32	28	325	180	120	70	30	----	----	----
40/20	19/32	40	32	----	305	205	130	60	30	----	----
48/24	23/32	48	36	----	----	280	175	95	45	35	----
60/32	7/8	60	48	----	----	----	305	165	100	70	35
APA rated Sturd-I-Floor [f]												
16 oc	19/32	24	24	185	100	65	40	----	----	----	----
20 oc	19/32	32	32	270	150	100	60	30	----	----	----
24 oc	23/32	40	36	----	240	160	100	50	30	25	----
32 oc	7/8	48	48	----	----	295	185	100	60	40	----
48 oc	1-3/32	60	48	----	----	----	290	160	100	65	40

[a] Tongue-and-groove edges, panel edge clips (one midway between each support, except two equally spaced between suppors 48 inches on center), lumber blocking, or other. For low slope roofs, see Table 22.
[b] 24 inches for 15/32-inch and 1/2-inch panels.
[c] Includes APA RATED SHEATHING/CEILING DECK.s
[d] 10 psf dead load assumed.
[e] Applies to panels 24 inches or wider.
[f] Also applies to C-C Plugged grade plywood.

Table 22

Recommended maximum spans for APA panel roof decks for low slope roofs [a]
(long panel dimension perpendicular to supports and continuous over two or more spans)

Grade	Minimal nominal panel thickness (in.)	Minimum span rating	Maximum span (in.)	Panel clips per span[b] (number)
APA rated sheathing	15/32	32/16	24	1
	19/32	40/20	32	1
	23/32	48/24	48	2
	7/8	60/32	60	2

[a] Low slope roofs are applicable to built-up, single-ply and modified bitumen roofing systems. For guaranteed or warranted roofs contact membrane manufacturer for acceptable deck.
[b] Edge support may also be provided by tongue-and-groove edges or solid blocking.

Courtesy of American Plywood Association

Figure 2-6   Recommended roof live loads and maximum spans

a minimum crown of ³/₈-inch and 1 inch longer than the thickness of the panel. Drive staples 4 inches o.c. along the panel edges and 8 inches o.c. at intermediate supports. On thicker panels, drive staples at 2 inches o.c. along the edges and 5 inches o.c. at intermediate supports.

Support for Panel Sheathing

For a built-up roof with ½-inch plywood sheathing, roofing material manufacturers recommend a maximum rafter spacing of 24 inches. Unless the plywood is tongue-and-groove or you use edge clips, install blocking at the unsupported edges of the panels. Use 2 x 4s (or larger) installed edgewise between rafters for blocking. Space the edges of the plywood panels at least ¹/₁₆ inch between end joints and ⅛ inch between side joints to allow for expansion. That's shown in Figure 2-7. Double this spacing in climates where the relative humidity is high. (See Figure 6-1 on page 149.) Nail plywood panels to a wood structure with 8d annular-threaded or ring-shank nails spaced at 6-inch centers along the edges and at 12-inch centers for the rest of the roof.

You can also use waferboard for roof sheathing. Figure 2-8 shows the specifications. If you're installing OSB, use a minimum ¹⁵/₃₂-inch nominal thickness.

Install any wood sheet roofing material with staggered end joints and be sure that all end joints lie on the tops of rafters. (Again, see Figure 2-7.) Whenever you install solid sheathing, make sure the attic is well-ventilated to reduce the possibility of shingle and deck deterioration, Here's why:

Today's better-built homes are fairly airtight, so water vapor is trapped in the house. Eventually, that moisture passes through the ceiling and accumulates in the attic. In cold weather, that warm, moist air will condense on the underside of the sheathing in a poorly-ventilated attic. When the weather is hot, the moist air will try to escape through gaps in the roof covering. The end result is a buckled, rotten deck, deteriorated underlayment and buckling shingles. With proper ventilation, air circulates freely in the attic so the water vapor dissipates before it can condense and cause trouble.

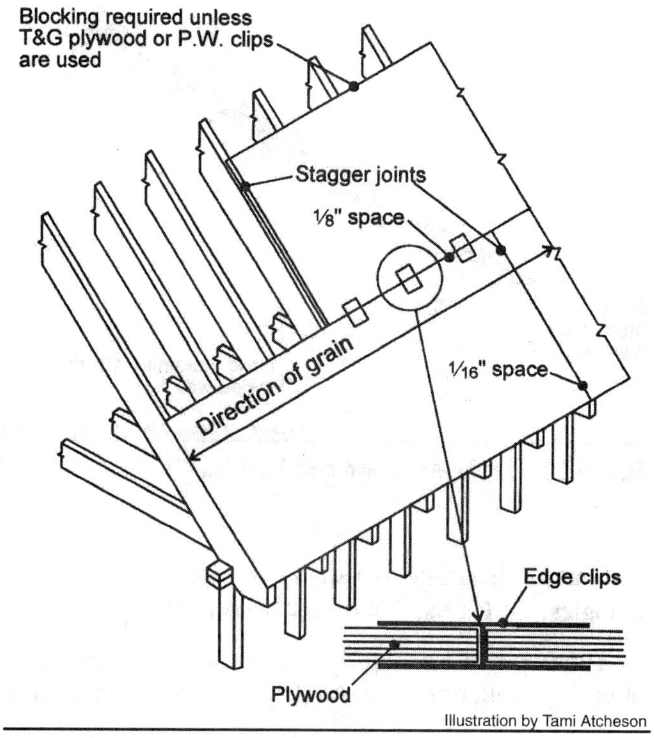

Figure 2-7 Plywood installation

Roof sheathing (1)	
Thickness (in.)	Maximum spam (in.)
3/8	16
7/16	24
1/2	24

(1) Refer to appropriate research reports for specific installation information. There are some difference among the code agencies and the applicable requirements are to be followed.

Figure 2-8 Waferboard roof sheathing

Spaced Board Sheathing

In milder climates, you can install spaced sheathing (skip sheathing) under wood shingles or shakes. In fact, some roofing contractors recommend that you always use spaced sheathing under wood shingles or shakes

Figure 2-9 Shingles applied over 1 x 4 boards

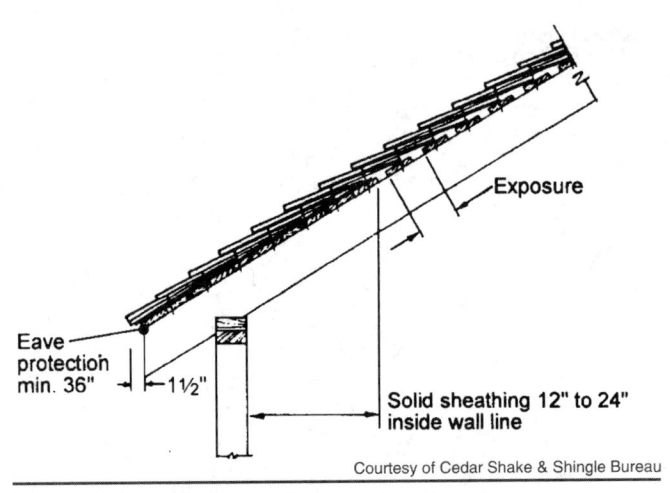

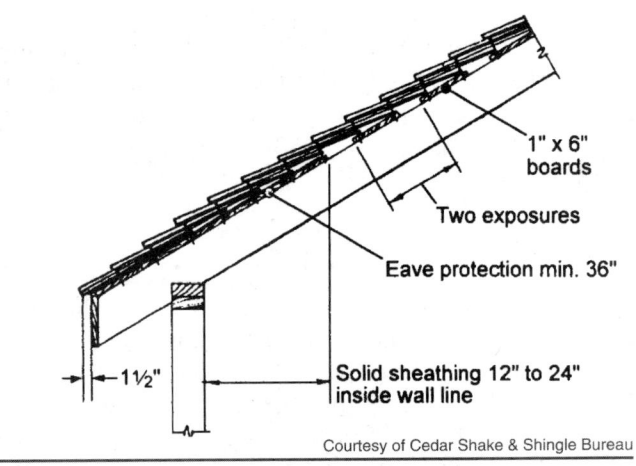

Figure 2-10 Shingles applied over 1 x 6 boards

to permit air circulation. You can also use spaced sheathing under corrugated metal panel roofs, if the local building code allows it.

Spaced sheathing is usually square-edged 1 x 4s and wider for wood shingles, or 1 x 4s, 1 x 6s and wider for shakes.

There are two acceptable ways to install spaced sheathing under wood shingles. With one, you install 1 x 4s spaced on centers equal to the shingle exposure. If you lay the shingles with $5^1/_2$ inches to the weather, space the sheathing boards at $5^1/_2$-inch centers. Nail each shingle at the center of each board, as shown in Figure 2-9. Here's something to watch for. Some building codes specify that the board spacing can't exceed the nominal width of the board.

In the second application method, you install 1 x 6s, and nail two courses of shingles to each 1 x 6, up to and including $5^1/_2$ inches weather exposure. That's shown in Figure 2-10. With $7^1/_2$-inch exposures, the center-to-center spacing of the sheathing boards is equal to the weather exposure.

If you're installing spaced sheathing under shakes, you'll normally use 1 x 6s. Make the spacing between the shakes equal to their exposure to the weather, as shown in Figure 2-11. Don't space the sheathing more than $2^1/_2$ inches in order to prevent interlayment from sagging into the attic.

When you use spaced sheathing, install solid sheathing from the eaves of the roof up to 12 to 24 inches beyond the interior face of the exterior wall line. That's also shown in Figures 2-9 through 2-11. Check the local building code for the required distance. In some locations, the code may specify 36 inches on slopes that are less than 4 in 12.

If your roof has a gable overhang with an open cornice (an overhang with the rafters exposed from beneath), you also need to install solid sheathing over the exposed overhang. I recommend that you solidly cover the roof at least 18 inches on each side of the ridges as well. That gives you backing to adjust the exposure of the last few shingles, as shown in Figure 2-12.

Some building codes allow spaced sheathing under corrugated metal roofs. However, most sheet metal roofs require a solid deck for support.

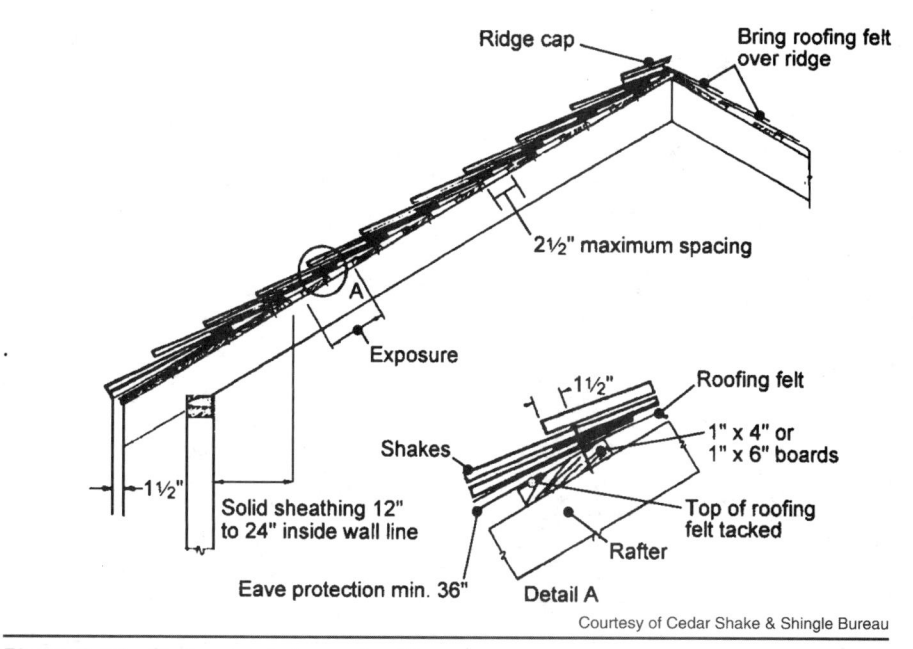

Figure 2-11 Shakes applied over 1 x 6 boards

Spaced Sheathing over Solid Sheathing

The minimum roof slope recommended for wood shingles or shakes is 4 in 12. However, you can install wood shingles and shakes on lower slopes if you install 2 x 4 spacers over a built-up roof covering applied to the sheathing. Then install 1 x 4 or 1 x 6 nailing strips as shown in Figure 2-13.

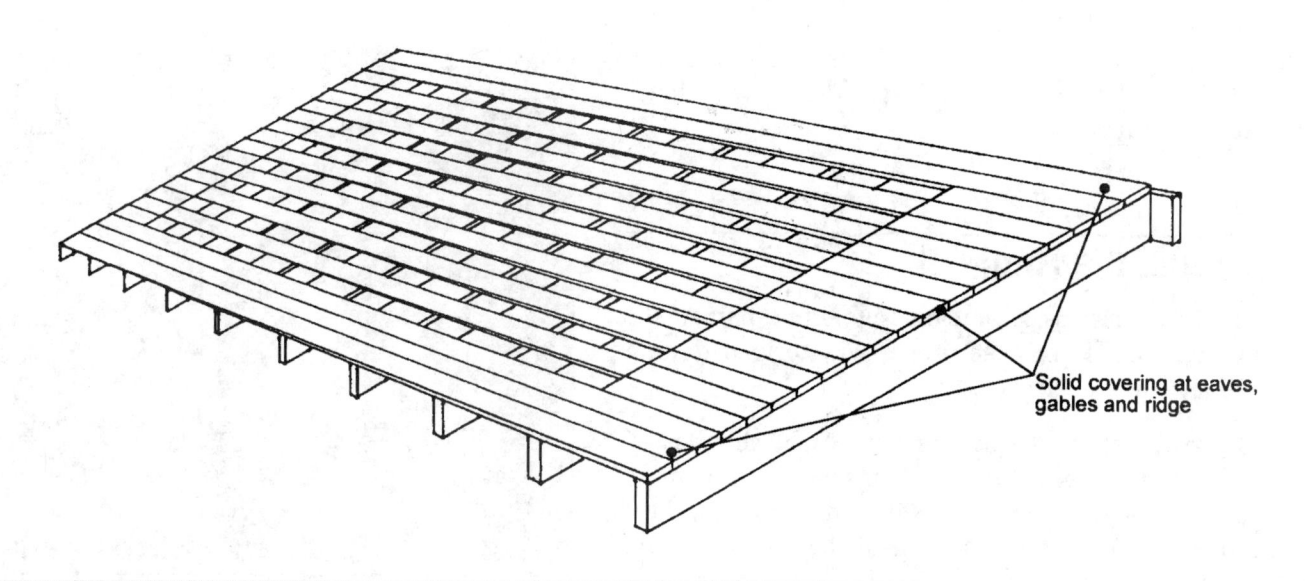

Figure 2-12 Spaced sheathing

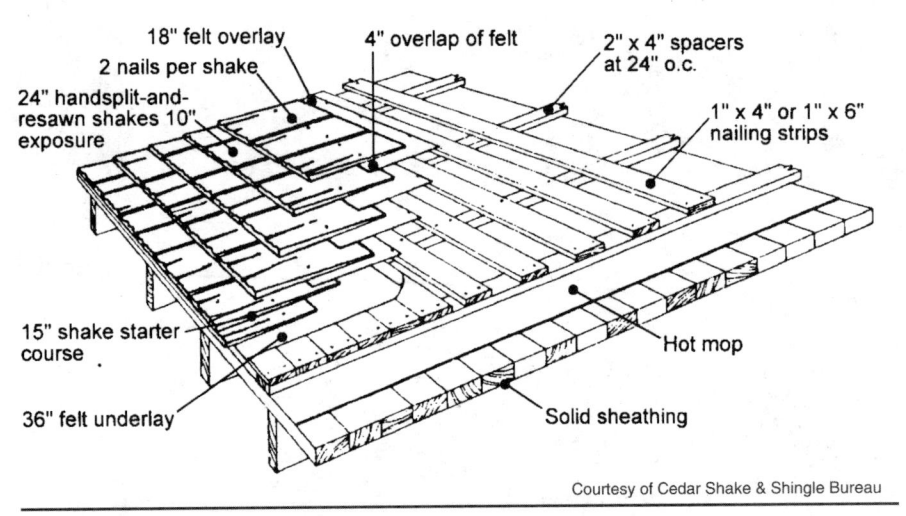

Figure 2-13 Spaced sheathing over a solid sheathing

Roof Decking

Roof decking serves both as a solid roof deck and wood ceiling. Use it on projects which call for an attractive exposed wood ceiling. You can get roof decking in solid tongue-and-groove planks from 1 to 5 inches (nominal) thick and 4 to 12 inches (nominal) wide. You can also get laminated tongue-and-groove roof decking that's 2 to 5 inches thick and 6 and 8 inches wide. See Figure 2-14. Roof decking is sold in random lengths of 6 to 20 feet. Decking is manufactured in commercial grades (where appearance and strength aren't important) and select grades. Many patterns are available for the ceiling side of the decking. Laminated decking is available with the same, or different, lumber species for the top, intermediate, and bottom layers.

Figure 2-14 Roof decking

Loading the Roof

The roof deck is an integral supporting part of the roof structure. Here's the proper way to load roofing materials onto the deck:

Load the roof by stacking bundles of shingles or roofing felt along the ridge. Distribute the weight as evenly as possible on both sides of the ridge. This keeps the roofing materials out of the way until you need them, and eliminates uneven and concentrated loads. Figure 2-15 shows an even load.

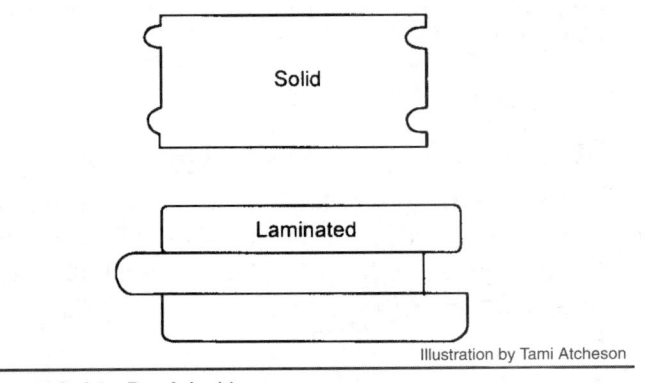

Figure 2-15 Loading the roof along the ridge

Roof Sheathing, Decking and Loading

Figure 2-16 Distributing the load throughout the roof

Distribute roofing tiles throughout the roof and take special care not to overload the roof at the mid-span of rafters. This is the weakest area of the roof framing system. See Figure 2-16.

Load roofing tiles on a gable roof in stacks of eight along every fourth course, except along the ridge where you load in stacks of four. Figure 2-17 is a diagram for tile loading on a gable roof.

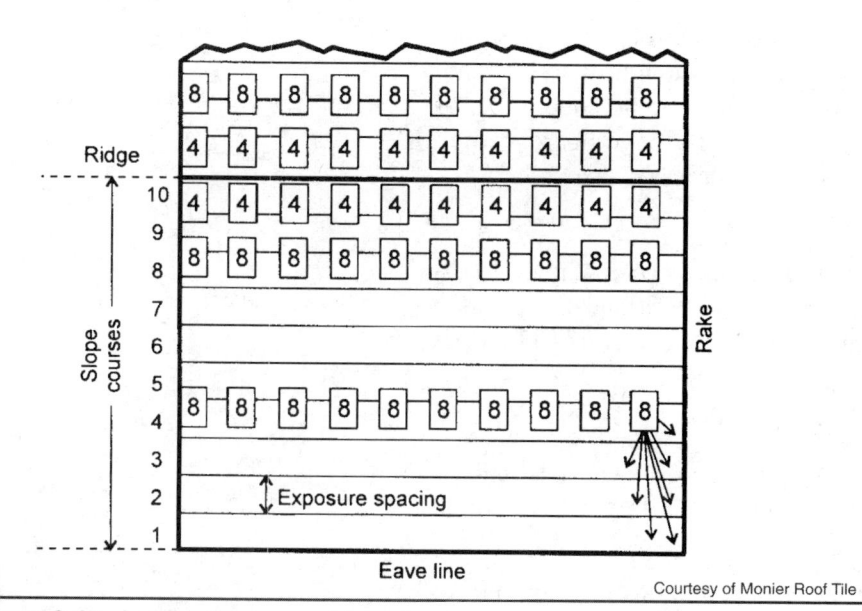

Figure 2-17 Loading tiles on a gable roof

Figure 2-18 Loading tiles on a hip roof

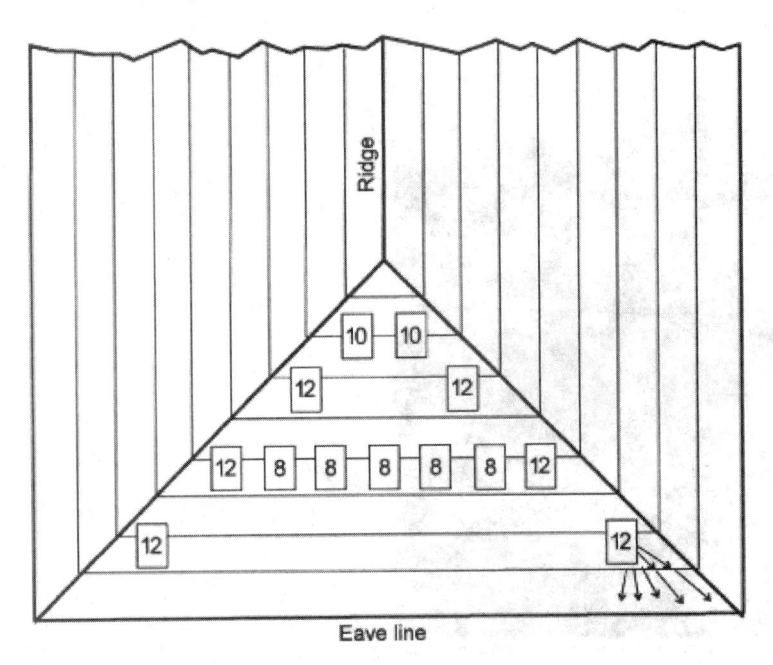

Load roofing tiles on a hip roof as shown in Figure 2-18. On a gable or hip roof, allow a horizontal space of about 1 foot between stacks. Also, pick tiles from different pallets as you load them to avoid color patterning. On a re-roof, load shingles before starting the job to allow the building to adjust to the weight.

Estimating Roof Sheathing

In most cases, the carpenters will have installed the roof sheathing before you arrive on the site. But it won't hurt for you to know how to estimate this part of the job. Assume you're required to install 3,000 square feet of roof deck (including waste) using 1 x 8 tongue-and-groove boards which cost $3.18 per square foot, including fasteners and sales tax.

3000 x $3.18 = $9,540

According to the *National Construction Estimator* (see Chapter 15), a crew of one carpenter and one laborer can install board sheathing at the rate of 1 square foot per 0.026 manhour. At an average cost of $33.85 per manhour for this operation, the labor cost is:

3000 SF x 0.026 x $33.85 = $2,640.30

The total cost for the job is $12,180.30 (the total of materials and labor). The manhour rate to install plywood sheathing (for the same two-man crew) is 0.013 manhours per square foot.

In the next chapter, we'll discuss underlayment and flashings before we go on to specific finish roofing materials.

3 Underlayment on Sloping Roofs

▶ In this chapter, we'll cover underlayment and metal items such as drip edge and valley flashing. Other flashings, such as those for chimneys, vents and parapet walls, appear in chapters where they apply to particular kinds of roofing. By *underlayment*, I mean any roll-type waterproofing material you install under the finished roof covering on a sloping roof. (We'll cover underlayment and drainage for level roofs in Chapter 10 on built-up roofing.) When the code says you need additional weather protection, underlayment also includes fortified eaves flashing. Installing underlayment is often called *capping in*, *drying in* or *pre-felting*.

Underlayment serves several purposes:

1) It protects roof sheathing before the roof covering is applied.

2) It acts as added weather protection for the building. In fact, in the case of shake and tile roofs, underlayment *is* the weather protection.

3) It separates asphalt shingles from areas where surface resins in wood sheathing may damage the shingle.

4) It provides additional insulation, especially over roofs with nailing strips that create air space between the sheathing and roof covering.

5) Felt underlayment cushions heavy roofing units such as tile or slate. While it doesn't absorb a lot of shock, underlayment does help to prevent breakage of these brittle materials under foot traffic.

Type	Length (ft.)	Width (ft.)	Factory squares per roll [1]	Weight (lb/square) (single coverage)
#15	144	3	4	15
#20	108	3	3	20
#30	72	3	2	30

[1] A factory square is 108 square feet

Figure 3-1 Saturated felt

6) When low-slope roofs are mopped with hot asphalt, underlayment prevents the asphalt from leaking into the building through joints in the roof deck.

Underlayment must be water-resistant, but not vapor-resistant. Vapor-resistant materials allow moisture or frost to accumulate between the roof sheathing and the underlayment. Then, unless the attic is well-ventilated, the deck will deteriorate.

Saturated Felt Underlayment

One type of underlayment is made of asphalt- or tar-saturated felt. The felt is primarily rag, shredded wood, cellulose and animal fibers. A saturated, organic felt is saturated with bitumen. An *asphalt felt* is saturated with an asphalt flux. A *tarred felt* is saturated with coal tar. These saturated materials allow vapor to pass through them. *Coated felts* differ from saturated felts in that they are actually coated with a glossy, vapor-resistant finish.

Your choice of underlayment depends on the roof slope, local temperatures, and type of material being applied over it. There's more information about asphalt and tar saturants in Chapter 10.

Asphalt-saturated felts are sold in quantities shown in Figure 3-1.

Saturated Fiberglass Underlayment

This type of underlayment is made of asphalt-saturated fiberglass. The glass fiber mats are made of continuous or random glass fibers bonded with plastic binders reinforced with chopped, continuous-random, or parallel glass fiber strands. Fiberglass underlayment comes in three types:

1) Base sheets (the first membrane in a built-up roofing system)

2) Ply-sheets (intermediate layers in a built-up system, or underlayment under shingle, slate or tile roofs)

3) Cap sheets (a heavy membrane, sometimes used instead of aggregate in a built-up roof system)

Type	Length (ft.)	Width (ft.)	Factory squares per roll	Weight (lb/square) (single coverage)
Base sheet	108	3	3	25
Ply-sheet	180	3	5	12
(Type IV) cap sheet	36	3	1	75

Figure 3-2 Saturated fiberglass

Asphalt-saturated fiberglass felts are sold in quantities as shown in Figure 3-2.

Organic felts can deteriorate due to oxidation and wicking (absorption of water by capillary action). Fiberglass felts don't rot or absorb water, so they last longer than organic felts, and they're fire resistant. But the down side is they cost about 65 percent more (for the same weight material) than organic felts.

Fiberglass felts don't buckle as much as organic felts, but they're less resistant to tearing, so they're more prone to wind damage. Use fiberglass felts on roofs with a long life expectancy, such as slate. Lower-cost organic felts are adequate under composition shingles.

Synthetic Underlayment

Fully synthetic material, made of polypropylene fabric, usually with closely laid fibers in both directions, is now the more-commonly used material in most parts of the country. It boasts a tear-resistance many times that of felt, provides better water-resistance than felt, allows air to pass through, and because there's nothing organic in it, it isn't subject to mold growth. Felt underlayment is subject to wrinkling caused by expansion (as water permeates the felt) and deteriorating or weakening caused by the volatile compounds in its makeup.

Purely from the roofer's perspective, it's much lighter than felt; so carrying a roll up on the roof is much easier, and it's easier to roll it out and handle overall. One roll may contain as much as 10 squares, compared with four for 15-pound felt and two for 30-pound. It also provides a better grip for safe walking, and in the event of a delay, where the underlayment is down but the shingles didn't arrive, or there was a problem with them, it can be left exposed to the sun for a month or even months. With felt, five days is about the limit, plus, if it rains or gets windy before the shingles are installed, it may have to be replaced.

Cost-wise, while synthetic may cost a little more per square foot, because it's so much lighter and a roll covers a much-larger area, the labor cost is less, in most cases outweighing the higher material cost.

On the negative side, when synthetic is used under brittle roofing such as tile or slate, it doesn't provide any cushioning. Felt does provide a little. Synthetic underlayment is also slightly more susceptible to wicking at laps than asphalt felt, which can lead to moisture damage to the roof.

And synthetic is somewhat new; possibly viewed with some suspicion by old-timers or in more-conservative areas of the country. Saturated felt works just as well as it ever did, and has effectively protected millions of roofs for decades. Discussion of saturated felts is included here because roofers will be installing it for many years to come, and in some situations it's a better choice.

Underlayment Requirements

You must use underlayment with all shingle roofs — with two exceptions. First, you don't use it under a wood shingle roof because it'll keep air from circulating and let moisture collect. That in turn reduces the life of the roof covering and sheathing. Second, you don't have to install underlayment between the two layers of roof coverings when you re-roof over existing shingles.

The type and amount of underlayment required depends on the roof slope and the type of roof covering you install. See Figure 3-3 for those requirements. In that table, solid sheathing is assumed, unless otherwise noted. These underlayment specs are the *minimums* required. Local building codes may have stricter requirements, so be sure to check with your inspector.

Keep in mind that some roofing materials (like heavy laminated fiberglass shingles) have a life expectancy of 20 to 30 years, while other roof coverings (like tile or slate) should last 50 years or more. When a long-life roof fails, it's usually because the flashing or underlayment failed, and not the roof covering. For that reason, I recommend a heavier underlayment than that recommended in Figure 3-3 whenever you install a long-life roof covering.

Many roofing contractors don't recommend the use of underlayment under asphalt or fiberglass shingles on roof slopes steeper than 4 in 12. They install it only because the building code requires it. They argue that it does more harm than good because the underlayment can buckle, causing the shingles to buckle. If underlayment is required, these contractors install heavy fiberglass felts. I've personally never seen shingles buckle due to warped underlayment. Since roofers sometimes disagree on the best installation methods, I've included options. Follow the guidelines in this book, and use your own judgment and experience where application methods aren't specified by codes or manufacturer's warranty requirements.

Underlayment on Sloping Roofs

Roof type	Description	Min. slope	On slopes	Underlayment recommended [1][2][13]
Asphalt shingles	2- or 3- tab strip (self sealing)	2 in 12	2/12	Double coverage with 15 lb felt (nailed)[3]
			3/12 and steeper	Single coverage with 15 lb felt (nailed)
	3-tan strip (standard)	2 in 12	2/12-3/12	Double coverage with 15 lb felt (nailed)[3]
			4/12 and steeper	Single coverage with 15 lb felt (nailed)
	Individual Dutch lap or American	4 in 12	4/12 and steeper	Single coverage with 15 lb felt (nailed)
Fiberglass shingles	Light weight (to 260 lbs)	2 in 12	2/12-3/12	Double coverage with 15 lb felt (nailed)[3]
			4/12 and steeper	Double coverage with 15 lb felt (nailed)
	Medium weight (260-280 lbs)	2 in 12	2/12-3/12	Double coverage with 30 lb felt (nailed)[3][4]
			4/12 and steeper	Double coverage with 30 lb felt (nailed)
	Heavy weight (280 lbs and more)	2 in 12	2/12-3/12	Double coverage, 1st ply is 15 lb felt (nailed), 2nd ply is Type IV fiberglass (nailed)
			4/12 and steeper	Single coverage with 30 lb felt (nailed)[4]
Mineral-surfaced roll roofing	Single coverage	1 in 12[18]	1/12-3/12	Double coverage with 15 lb felt (nailed)[3][10]
			4/12 and steeper	Single coverage with 15 lb felt (nailed)[10]
	Double coverage	1 in 12	1/12 and steeper	None required [6][10]
	Pattern edge	4 in 12	4/12 and steeper	None required [6][10]
Clay tile	All types	3 in 12	3/12 and steeper	Triple coverage. 30 lb felt (nailed) followed by two plies of Type IV fiberglass (mopped)[4]
Cement tile	All types	2.5 in 12	2.5/12 and steeper	Double coverage. 30 lb felt (nailed) followed by one ply of 90 lb mineral-surfaced roll roofing (nailed)
Slate	Standard	4 in 12	4/12 and steeper	Single coverage with 15 lb felt (nailed)[9]
	Textural	4 in 12	4/12 and steeper	Single coverage with 15 lb felt (nailed)[9]
	Graduated	4 in 12	4/12 and steeper	Single coverage with 45 lb mineral-surfaced roll roofing (nailed)[9][11]
Aluminum	All types	4 in 12	4/12 and steeper	Single coverage with 30 lb felt (nailed)
Porcelain	All types	3 in 12	3/12 and steeper	30 lb felt underlayment (nailed) with 30 lb felt interlayment

Figure 3-3 Underlayment recommendations

Roof type	Description	Min. slope	On slopes	Underlayment recommended[1][2][13]
Wood	Shakes, spaced sheathing	4 in 12	4/12 and steeper	Single coverage. 30 lb felt underlayment starter course with 30 lb felt interlayment
	Shakes, solid sheathing	3 in 12	3/12-4/12	Double coverage. 30 lb felt underlayment (nailed) with 30 lb felt interlayment
			4/12 and steeper	Single coverage. 30 lb felt underlayment starter course with 30 lb felt interlayment
	Shingles, solid or spaced sheathing	3 in 12	3/12 and steeper	None required[6][7]
Copper	Standing seam, pan or roll method	2.5 in 12	2.5/12 and steeper	Single coverage with 15 lb felt and rosin paper
	Batten seam	3 in 12	3/12 and steeper	Single coverage with 15 lb felt and rosin paper
	Flat seam	0.25 in 12	0.25/12 and steeper	Single coverage with 15 lb felt and rosin paper
Copper-bearing steel	Standing seam, roll method	2 in 12	2/12 and steeper	None required[6]
	Pressed standing seam	0.25 in 12	0.25/12 and steeper	None required[6]
Lead	All types	0.25 in 12	0.25/12 and steeper	Single coverage with 30 lb felt and rosin paper
Copper-zinc alloy	All types	3 in 12	3/12 and steeper	Single coverage with 15 lb felt and rosin paper
Stainless steel	All types	0.25 in 12	0.25/12 and steeper	None required[6]
Terne plate	Batten seam or standing	2.5 in 12	2.5/12 and steeper	Rosin paper[12]
	Flat seam	0.25 in 12	0.25/12 and steeper	Rosin paper[12]

Footnotes:

[1] Apply underlayment horizontally. Exception: Hot-mopped mineral-surfaced roll roofing underlayment can be installed parallel to the slope.
[2] Check your local building code. It may require heavier underlayment than those given.
[3] To help keep water from leaking through the nail holes on roof slopes less than 3 in 12, first nail on a 15-pound felt, and follow that by mopping on a 15-pound felt (or Type IV fiberglass ply-sheet).
[4] Use heavy, durable underlayment for shingles with long life spans (such as slate or tile) or the shingle will outlast the underlayment.
[5] As an alternate to interlayment over solid decks (see Figure 3-36)
    a) Roof slopes less than 6 in 12: Nail one 30-pound felt followed by mopping two 15-pound felts (or two Type IV fiberglass ply-sheets).
    b) Roof slopes equal to 6 in 12: Nail two 30-pound felts.
    c) Roof slopes greater than 6 in 12: Nail one 30-pound felt
[6] Underlayment might be desireable for protection of the sheathing.
[7] Some roofing contractors recommend:
    a) Roof slopes less than 6 in 12: Nail one 30-pound felt, followed by nailing one Type IV fiberglass ply-sheet.
    b) Roof slopes of 6 in 12 and greater: Nail one 30-pound felt.
[8] Minimum slope varies, depending on the installation method. (Seee Figure 5-3).
[9] Some roofing contractors recommend the underlayment as specified for tile roofs.
[10] Some roofing contractors recommend that you never install underlayment under mineral-surfaced roll roofing.
[11] On graduated roofs, you usually use 30-pound felt under slates that are ¾ inches or less thick, and 50-pound underlayment under slates 1 inch thick and thicker. In some areas, it's common to install a 30-pound felt followed by a 15-pound felt under any graduated roof.
[12] Terne coated stainless steel can be underlaid with a single 15-pound felt, with a slip sheet of rosin paper.
[13] Solid sheathing is assumed, unless otherwise noted.

Figure 3-3 (cont.) Underlayment recommendations

Underlayment on Sloping Roofs

Drip Edge

Install a drip edge (nosing or roof edging) at the eaves and rake to help shed water and prevent runoff water from getting to the underlayment and sheathing where it can lead to deterioration. In Figure 3-4 the drip edge is installed under the felt along the eaves and over the underlayment along the rakes.

You install drip edge *under* the felt at the eaves. Otherwise, water might get under the shingles and seep under the drip edge. You don't have that problem at the rakes because of rapid runoff. In that case, put the drip edge over the felt to help secure the felt. If the drip edge is installed under the felt along the rakes, wind-driven water can penetrate between the felt and the drip edge.

Make drip edges from 28-gauge (minimum) galvanized metal or other non-corrosive, stain-resistant material such as copper or terne metal. Extend the drip edge back over the roof deck no less than 3 inches and nail it on 8- to 10-inch centers as shown in Figure 3-5. In high-wind areas, nail the drip edge on 4-inch centers. If you need more than one piece to cover the eaves, lap the joint at least 3 inches.

Run the drip edge down the eaves 4 to 6 inches beyond the corner of the roof, then make an angle cut in the metal and extend the drip edge up the rake. Place roofing cement between the laps where you made the angle cut. That's shown in Figure 3-6.

Figure 3-4 Installing the drip edge along rake

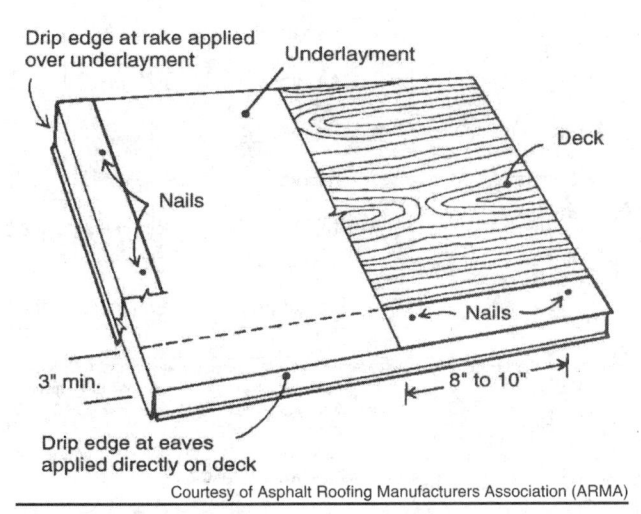

Figure 3-5 Application of drip edge at rake and eaves

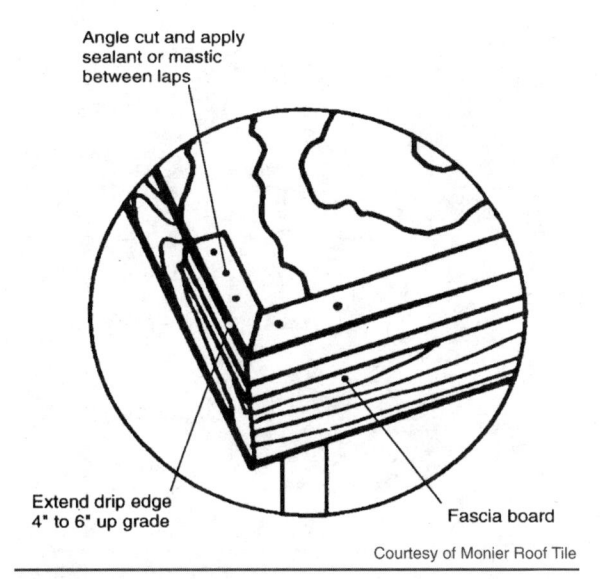

Figure 3-6 Application of drip edge at corners

At gable ends, install the drip edge beginning at the lower end of the rakes. If you need more than one piece to cover the rake, lap the upper unit over the lower by at least 3 inches.

You can install a sheet metal product called a *gravel stop* at the edge of aggregate-surfaced roofs to help prevent loose gravel from falling off the edge of the roof. Gravel stop is similar to the drip edge in Figure 3-7, except the gravel stop has a more pronounced lip at the edge of the roof to retain the gravel.

After the drip edge was installed, we used to place a bundle of shingles at the eaves, protruding a little over the edge to support the top of the ladder. This kept the ladder from slipping and prevented it from denting the drip edge. Now there are metal devices that clamp the upper part of the ladder to the fascia as well as protect the drip edge.

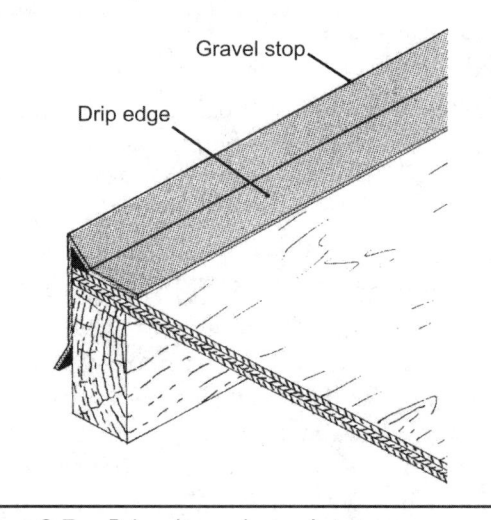

Figure 3-7 Drip edge and gravel stop

Calculating Drip Edge

Here's how to calculate how much drip edge you need:

▼ **Example 3-1:** Assume a 10 percent material cutting allowance and find the total linear feet of drip edge required for the roof of the building in Figure 3-8.

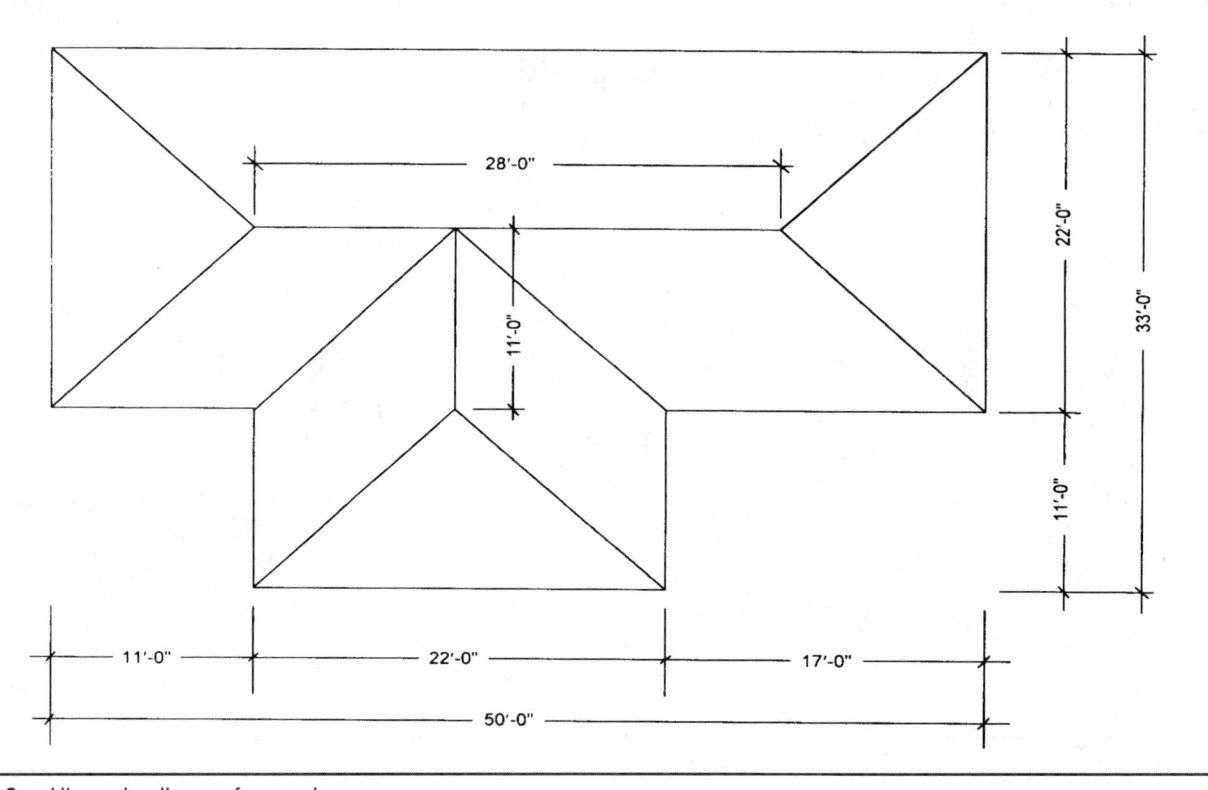

Figure 3-8 Hip-and-valley roof example

First, you remember from Chapter 1 that the formula for the perimeter of a building is 2(L + W). The perimeter of this building is 2 x (50' + 33'), or 166 linear feet. Multiply that by 1.10 to add the 10 percent allowance factor, and round your answer up to 183 linear feet.

Use 166 linear feet to figure manhours at the rate of 1.4 manhours per 100 linear feet.

Installing Underlayment

Underlayment is applied single-coverage (Figure 3-9) or double-coverage (Figure 3-10). In either case, install felts *over* the drip edge at the eaves and *under* the drip edge at the rakes. Roll out the felt, nail down one end, and stretch the felt. Roll out only as much felt as you can install in about an hour. Otherwise, the loose felt may buckle because heat causes the felt to relax. The hotter the weather, the more the felt will buckle. Wind is also a factor.

Figure 3-9 Application of single-coverage underlayment

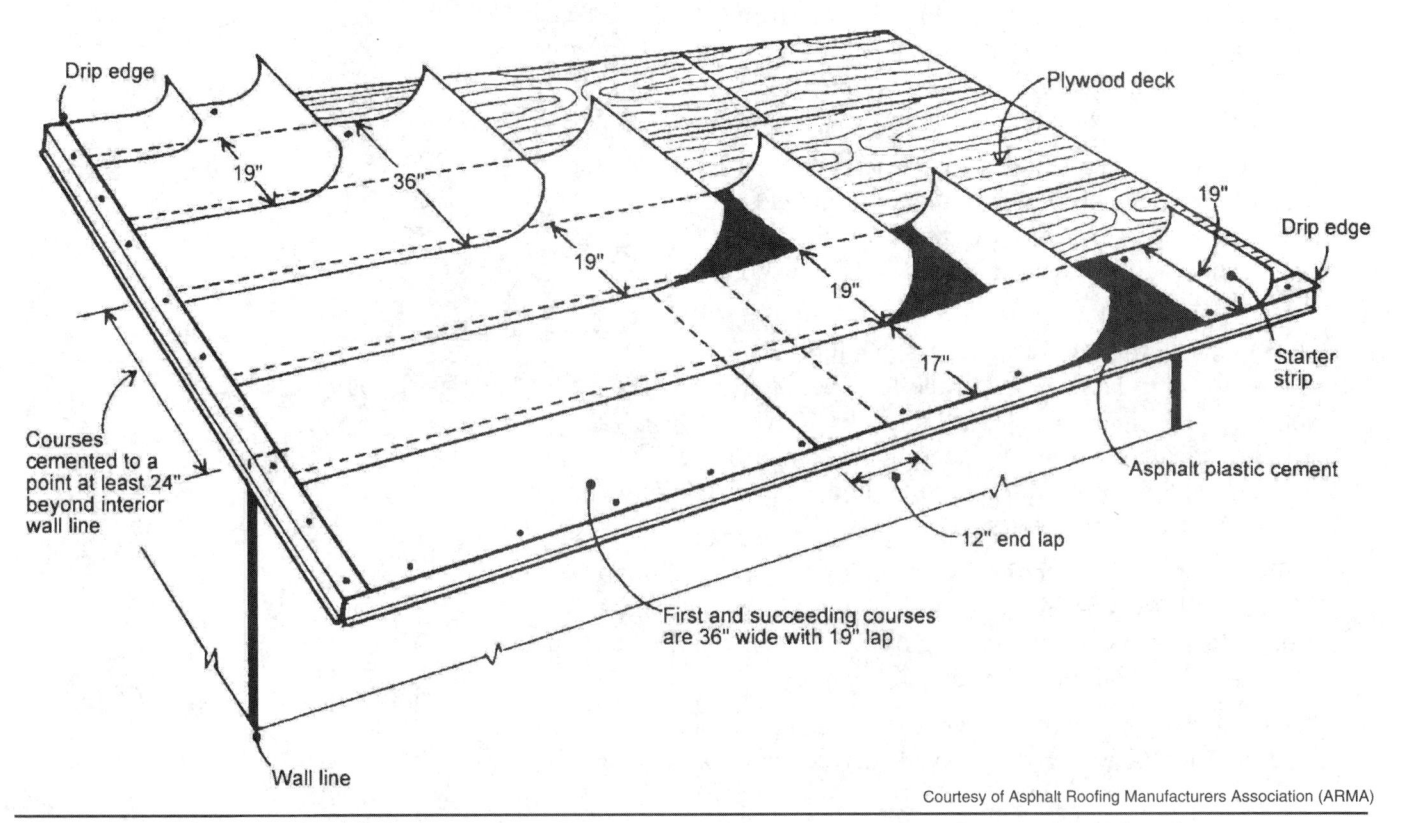

Figure 3-10 Application of double-coverage underlayment on low slopes where icing along the eaves is anticipated

Figure 3-11 Installing felt in a valley

Figure 3-12 Extended valley underlayment beyond the ridge

Moisture on the deck can also cause buckling. (This problem is worst in early morning.) If you can't pull a wrinkle out, cut it, lap the edges, and nail the felt flat. Never install felt over a wet deck.

Install felt underlayment (sweat sheet) in the valleys before you install felts over the main roof deck. Nail the valley felt on 24-inch centers in rows along the edges of the felts as shown in Figure 3-11. Drive the nails through tin caps, or use Simplex nails. If you use tin caps (tin tags), they should be 32-gauge caps that are at least 1⁵⁄₈ inches in diameter. Use galvanized nails at least ¾-inch long, or long enough to go all the way through the sheathing.

On any roof with a valley, extend the valley underlayment at least 6 inches beyond the ridge and cut it so that it conforms to the surface of the sheathing with no bulges. Then nail it down. It should look like Figure 3-12.

Install all felt parallel to the eaves, beginning at the eaves. Be sure to let the edge of the felt overhang the drip edge at the eaves by at least ³⁄₈ inch. Better yet, let

Figure 3-13 Nail the felt as you roll it out

the felt overhang the drip edge the same distance as the final roof covering. That prevents water from seeping under the drip edge where it can damage the deck or rot the fascia.

Align the first felt with the lower left corner of the roof and secure it with one nail. Roll out the felt, straighten it, and nail it randomly as you go, on 3- to 4-foot centers. See Figure 3-13. Don't nail just at opposite ends or you'll get humps in the felt.

Then, go back and nail in rows along the edges on 12- to 18-inch centers, and along the center on 18- to 24-inch centers. Nail end laps on 6- to 8-inch centers as shown in Figure 3-14.

Underlayment on Sloping Roofs

Figure 3-14 Running felts into a valley

Figure 3-15 Lapping the felt over the ridge

Figure 3-16 Lapping the felt over a hip

Continue up the roof the same way. When you reach the ridge, fold the last felt over the ridge at least 6 inches and nail it on the other side. Make sure the felt covers at least 6 inches on both sides of the ridge. See Figure 3-15.

You must also roll and lap the felt at least 6 inches beyond the centerlines on both sides of hips as shown in Figure 3-18. Overlap the 36-inch-wide felt in the valley by 6 inches from both sides, as shown in Figure 3-14. Extend the felt 3 to 4 inches up any vertical surface such as a chimney or wall. That's illustrated in Chapter 4, as Figure 4-62.

Use a 2-inch top lap (allowing a 34-inch exposure) for single underlayment coverage, as shown in Figure 3-9. For double coverage, use a 19-inch top lap and a 17-inch exposure, as in Figure 3-10. You'll find guidelines for various top laps printed on the rolls. End laps vary, depending on the type of underlayment, the slope of the roof, the severity of the climate, and local code requirements. Figure 3-17 shows end-lap requirements for various underlayment materials. Be sure to stagger end laps.

To work around a vent pipe, temporarily roll the felt out so that the edge of the roll just touches the pipe. Mark the location on the edge of the roll and cut a slit to the location where the pipe will penetrate the roll. You can see that in Figure 3-13. Slide the felt into final position and cut the felt around the vent pipe for a smooth fit. Then drive nails along each side of the slit.

Material type	End lap (inches)
15, 20 or 30# saturated felt (single coverage)	4
15, 20 or 30# saturated felt (double coverage)	6[1]
Fiberglass base sheet	6
Fiberglass ply-sheet	12
Fiberglass cap sheet	6
Mineral-surfaced roll	6
Mineral-surfaced roll (double coverage)	6
Coated roll	6

[1] 12 inches is recommended by some manufacturers on low-slope roofs in colder climates

Figure 3-17 End-lap requirements

Underlayment for Tile Roof Coverings

When you choose which type of underlayment to use under a tile roof, you should consider your confidence in the roofing crew to install a watertight roof covering, as well as the:

- Local climate
- Method of fastening the tiles
- Roof slope
- Weatherproofing integrity of the tile
- Local building code

Except in highly humid areas, you can install a *non-sealed underlayment system* as in Figure 3-18 when you direct-nail tiles to the sheathing or battens on roof slopes of 4 in 12 and steeper. This system is a single 43-pound coated base sheet nailed to the deck. Install sheets with a 2-inch top lap and a 6-inch end lap. Drive nails along the eaves at 24-inch centers and along the top laps at 12-inch centers.

When you direct-nail tiles on slopes of 4 in 12 and steeper in more severe climates, install the underlayment as described above, but seal the laps with roofing cement on slopes up to 6 in 12, as shown in Figure 3-19. This method is called a *sealed underlayment system*. You can omit the roofing cement on slopes of 6 in 12 and steeper.

In high-wind and hurricane areas, nail down a 30-pound felt followed by 90-pound mineral-surfaced roll roofing embedded in a solid hot mopping of Type IV (steep) asphalt. That's shown in Figure 3-20. You can also use a flood coat of cold-process roofing cement (Figure 3-21). These methods are called the *30/90 hot-mop* and *2-ply sealed systems,* respectively.

Underlayment on Sloping Roofs

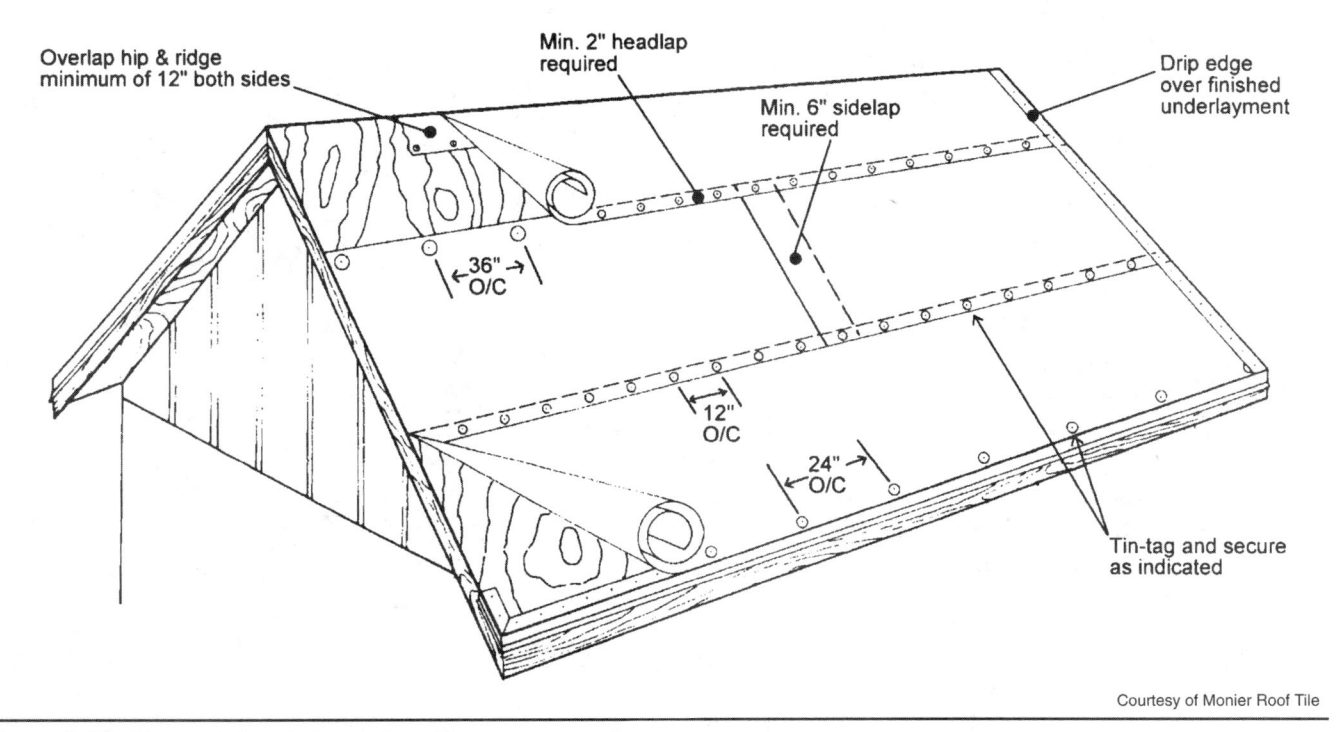

Figure 3-18 Non-sealed underlayment system

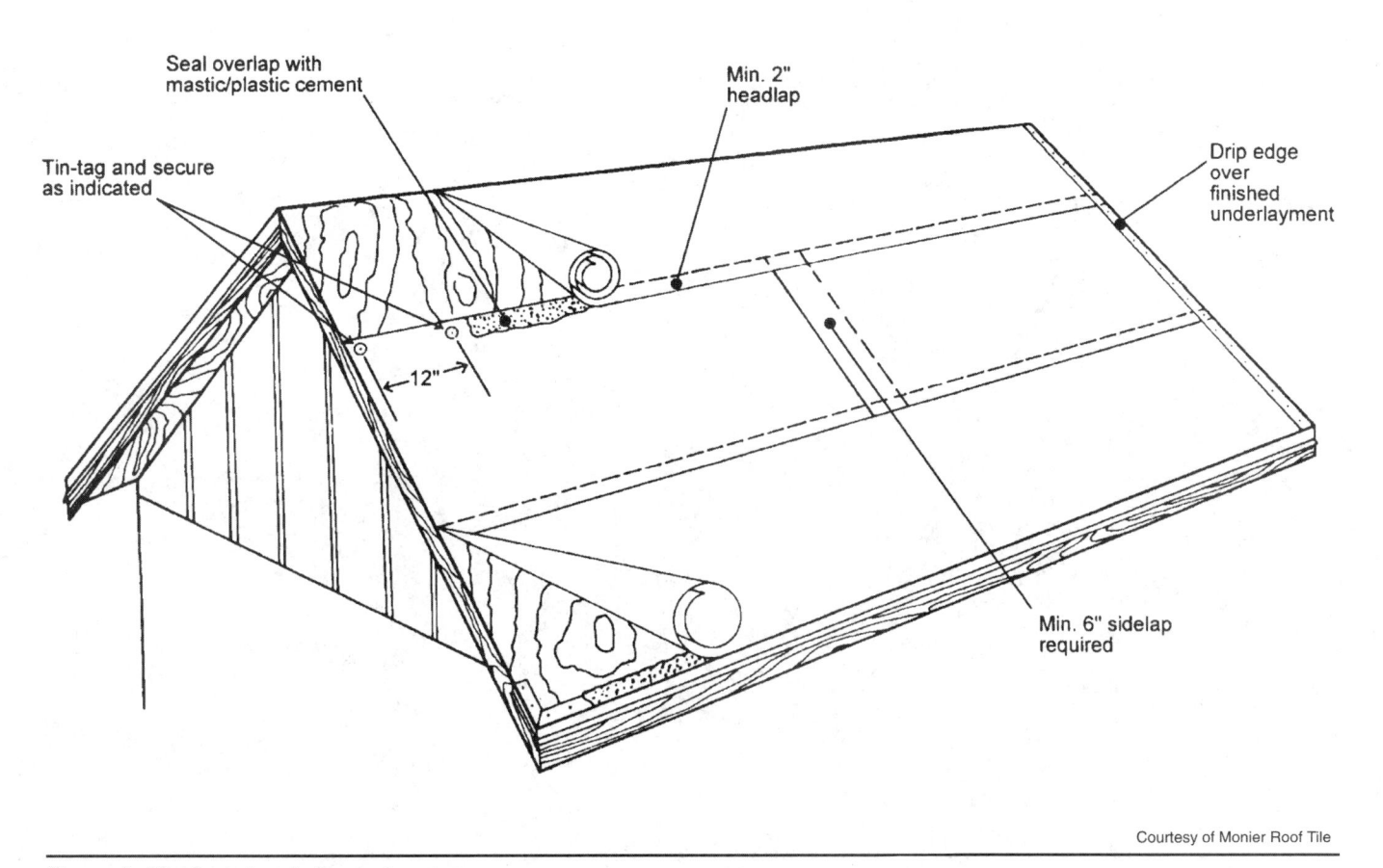

Figure 3-19 Sealed underlayment system

47

Roofing Construction & Estimating

Use the 30/90 hot-mop method whenever you'll be installing the tiles with mortar (mortar-set or mud-on system). You can also use it on roof slopes between $2^1/_2$ and 4 in 12 if you're going to use nails to install tiles. Use hot asphalt on roofs 5 in 12 to 6 in 12. You can use cold-applied roofing cement instead of hot asphalt (2-ply sealed system) on roof slopes steeper than 6 in 12.

When you hot-mop the 90-pound material along the eaves, nail the top of the sheet and pull back the loose edge. Mop the back of the sheet (Figure 3-22) and flip it over to its original position. Then walk along the sheet to help stick it to the underlying felt. That's called *walking in*. Mopping the back of the sheet as opposed to mopping the roof surface and laying the sheet on top helps prevent hot bitumen from running off the edge of the roof.

You can use a similar method to mop the 90-pound material into a valley by mopping half of the valley at a time. That's shown in Figure 3-23. With this type of

Figure 3-20 Installing mineral-surfaced roll roofing over hot asphalt

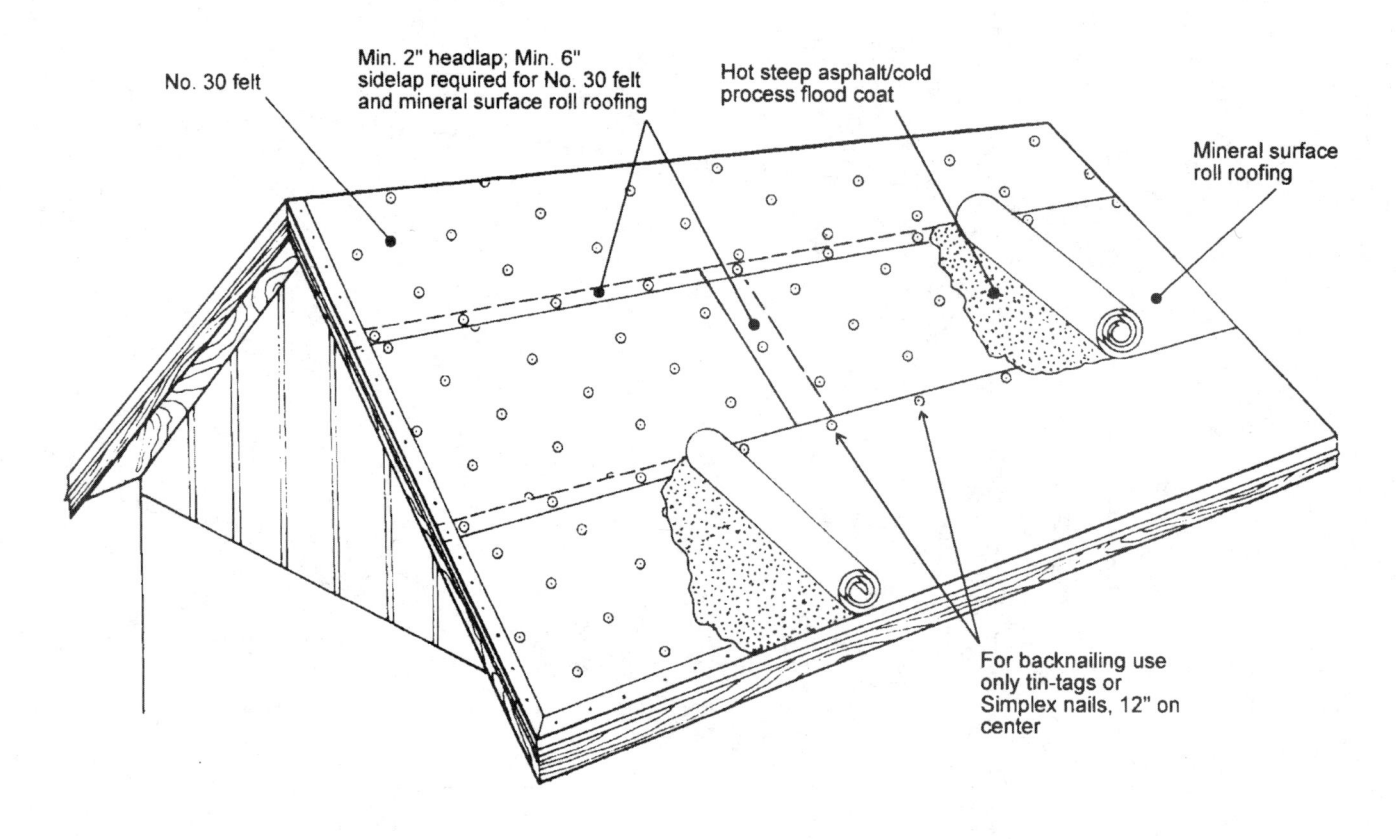

Courtesy of Monier Roof Tile

Figure 3-21 Two-ply sealed system

48

installation you can install the 90-pound material down slope, as well as perpendicular to the slope (Figure 3-24).

When you come to a roof penetration, cut the 90-pound material so it slips over the penetration, as in Figure 3-25. Then seal the edges with roofing cement.

Estimating Underlayment Quantities

Before we get into takeoffs and estimating, let's discuss *rounding off*. In the estimating calculations in this book, I generally round up if the decimal is 0.5 or greater, and down otherwise. The theory is that you'll round up about as often as you round down, and the final answer is close enough even if it's not accurate to four decimal places. But there are exceptions, depending on variations in material cost and manhours.

If valley flashing metal was pure gold, you'd want to be very precise in your takeoff. And you'll want to figure slate more closely than the felt which goes under it. But you don't want to waste time figuring materials down to the nearest inch if that won't make more than a

Figure 3-22 Mopping the back of the first sheet

Figure 3-23 Mopping mineral-surfaced roll roofing into a valley

Figure 3-24 Installing mineral-surfaced roll roofing down slope

couple of dollars' difference on a $20,000 job. You're better off spending your time negotiating with subcontractors, finding the most economical material source, or training your crews to be more efficient. So use your judgment and common sense before you spend needless time figuring everything down to the last nickel.

Roofing felt is relatively inexpensive and it's eventually concealed, so roofing contractors don't worry too much about waste or beauty when they install it. Accurately estimating it is impossible, and much depends on the experience of the crew and the effort they put into planning. But you need some type of estimate so you can price the job and don't risk running out. You certainly can't estimate it with any precision.

You begin estimating underlayment by calculating square feet of coverage. But this material is sold by the roll, so you have to convert square feet to rolls.

- After you deduct the material lost due to a 2-inch top lap and a 4-inch end lap, one factory square (108 square feet) will actually cover only 101 square feet (single coverage).

- After you deduct material lost due to a 19-inch top lap and a 6-inch end lap, one factory square will cover only 49.5 square feet (double coverage).

To make estimating easy (but still accurate enough), you can figure that one factory square of underlayment will cover 100 square feet (single coverage) or 50 square feet (double coverage) of roof area as shown in Figure 3-26.

The quantities in Figure 3-26 don't allow for waste due to crew error or the material you need for overlap at the ridge, hips, and valleys.

Figure 3-25 Rolling over a roof penetration

Lap Allowance

Remember, you have to lap underlayment at least 6 inches over each side of the centerlines of ridges and hips. You can see that in Figures 3-15 and 3-16. Also, you must install underlayment running into valleys with two 6-inch laps as shown in Figure 3-14. So, add 1 square foot to the roof area to be covered for each linear foot of ridge, hip, and valley.

Lap Area (ridge, hip, and valley) = Total Length x 1 SF/LF | Equation 3-1

For added protection, I recommend that you apply a 6-inch-wide strip of 30-pound felt over the hips and ridges before you install the hip and ridge units. That's relatively inexpensive leak insurance.

Underlayment on Sloping Roofs

No. of plies	Coverage (factory squares per square)	Coverage (square feet per factory square)
1	1	100
2	2	50
3	3	33
4	4	25

Figure 3-26 Covering capacity of underlayment

Figure 3-27 Underlayment waste along the rake

Overcut Allowance

On a gable roof, cut the underlayment about 4 inches beyond the rake edge as shown in Figure 3-27. Trim it flush with the edge later. If you're going to cover the gable roof with tiles and trim it with rake tiles, it's customary to wrap the underlayment down over the rake fascia and nail it to the barge board on 6-inch centers. The rule for this is:

Overcut Area (gables) = LF of Rake x 0.34 SF/LF

Equation 3-2

In this case, overlaps are covered using the information in Figure 3-26.

A Shortcut Allowance Calculation

As a rule of thumb, some contractors add 10 percent to the net underlayment quantities on small gable roofs and 15 percent on small hip roofs. This additional material allowance is for lapping at the ridges, hips and valleys, overcutting at rakes, and crew error. They allow a larger percentage for more complex roofs and a smaller percentage for large, simple roofs.

I don't like that method for figuring allowances. Without considering crew errors or job conditions, actual allowances depend on the roof type, the ratio of building length to building width, and the roof slope. Use Figures 3-28 and 3-29 to get a ballpark percentage for underlayment allowances for single or double coverage. Crew-error waste isn't included in the figures. Figure 3-28 shows the percentage to add to the net roof area because of the overlap at the ridge and the 4-inch overcut at the rakes. Figure 3-29 shows the percentage you add to the net roof area for overlaps at the hips and ridge.

The following examples demonstrate the reliability of Figures 3-28 and 3-29. You can use them to find underlayment quantities for any roof size and slope.

Roofing Construction & Estimating

Building dimensions (L x W)	Roof slope 3/12	6/12	12/12
30 x 20	7	7	6
40 x 30	5	5	4
45 x 30	5	5	4
50 x 30	4	4	3
60 x 30	4	4	3
70 x 30	4	4	3
80 x 30	4	4	3
50 x 40	3	3	3
60 x 40	3	3	3
70 x 40	3	3	3
80 x 40	3	3	3
90 x 40	3	3	3
60 x 50	3	3	2
70 x 50	3	3	2
80 x 50	3	3	2
90 x 50	3	3	2

Figure 3-28 Underlayment material overrun percentage (gable roof)

Building dimensions (L x W)	Roof slope 3/12	6/12	12/12
30 x 20	10	10	9
40 x 30	8	8	7
45 x 30	7	7	6
50 x 30	7	7	6
60 x 30	6	6	5
70 x 30	6	6	5
80 x 30	6	5	5
50 x 40	6	6	5
60 x 40	6	5	5
70 x 40	5	5	4
80 x 40	5	5	4
90 x 40	5	4	4
60 x 50	5	5	4
70 x 50	5	4	4
80 x 50	4	4	4
90 x 50	4	4	3

Figure 3-29 Underlayment material overrun percentage (hip roof)

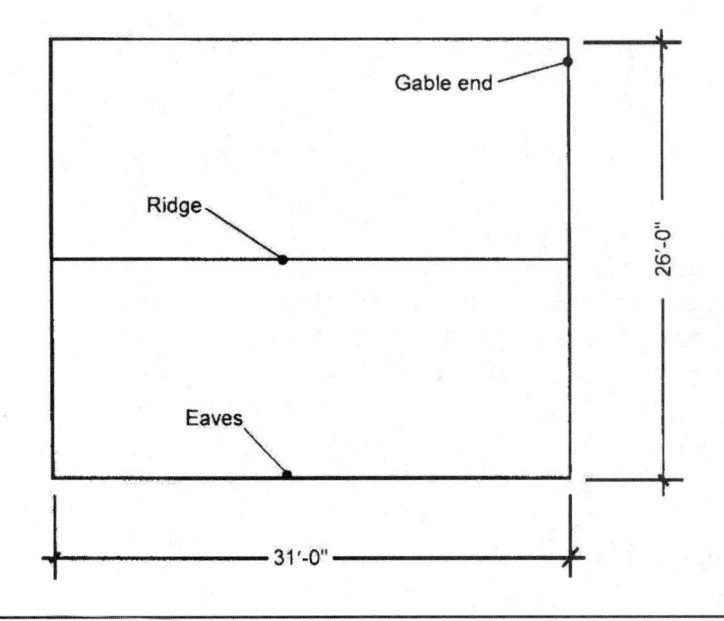

Figure 3-30 Gable roof example

Underlayment on Sloping Roofs

▼ **Example 3-2:** Find the quantity of 15-pound single-coverage underlayment required for the roof shown in Figure 3-30. Assume a roof slope of 5 in 12 and a 4-inch overcut at each rake.

Use the equation for net roof area from Chapter 1, with the roof-slope factor from the slope factor conversion chart, Appendix A.

Actual (Net) Roof Area = Roof Plan Area x Roof-Slope Factor

Net Roof Area = 31' x 26' x 1.083
= 873 square feet ÷ 100
= 8.73 squares (Use this quantity to estimate labor costs.)

The ridge length is 31 feet, so you add an additional 31 square feet (from Equation 3-1):

Lap Area (Ridge) = 31 feet x 1 SF/LF
= 31 square feet

The total length of roof on the plan view is 52 feet (4 x 13 feet). To convert that to actual length, use the equation from Chapter 1 and Column 2 of Appendix A for common rafter length:

Actual Length = Plan Length x Roof-Slope Factor
= 52 feet x 1.083
= 56 linear feet

Now, use Equation 3-2 to find how much additional underlayment to order for the rakes:

Overcut Area (Gables) = LF of Rake x 0.34 SF/LF
= 56 feet x 0.34 SF/LF
= 19 square feet

Now, add the three answers together to find the total underlayment required:

Gross Roof Area = 873 SF + 31 SF + 19 SF
= 923 square feet ÷ 100
= 9.23 squares

Next, convert that figure to rolls. Look back at Figure 3-1. There are 4 factory squares (FS) per roll of 15-pound material.

Rolls (#15) = $\frac{9.23 \text{ FS}}{4 \text{ FS/Roll}}$ = 2.3 rolls

The allowance factor is the area covered (including cutting allowance) divided by the net roof area.

Allowance Factor = $\frac{9.23 \text{ squares}}{8.73 \text{ squares}}$ = 1.06

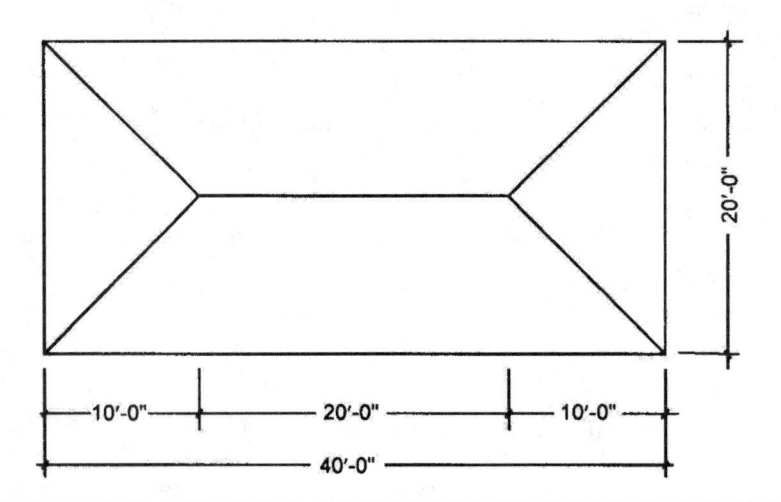

Figure 3-31 Hip roof example

That's equal to 6 percent material lap and overcut allowance. Now, look back at Figure 3-28. A 30- x 20-foot roof with a slope of 6 in 12 is about like the one in this example, and that table shows a material overrun of 7 percent. Our calculations agree with that. Now, try another one:

▼ **Example 3-3**: Assume single coverage and a roof slope of 6 in 12, then find how much 15-pound underlayment is required for the roof of the building in Figure 3-31.

First, use the formula for net roof area from the last example:

Net Roof Area = 40' x 20' x 1.118
             = 894 square feet ÷ 100
             = 8.94 squares (Use this quantity to estimate labor costs.)

You can see from Figure 3-31 the total ridge length is 20 feet. Also, using the run of any hip (10 feet), and the slope factor from Column 3, Appendix A (1.5), each hip is 15 feet long. The total for the four hips is 60 feet. That makes the total length of ridge and hips 80 feet. So, you add 80 SF (1 SF per linear foot of ridge and hip) to the net roof area for lapping at the ridge and hips:

Gross Roof Area = 894 SF + 80 SF
               = 974 square feet ÷ 100
               = 9.74 squares

Now, use Figure 3-1 to see how much felt to order:

$$\text{Rolls (\#15)} = \frac{9.74 \text{ FS}}{4 \text{ FS/Roll}} = 2.4 \text{ rolls}$$

The allowance factor is:

$$\text{Allowance Factor} = \frac{9.74 \text{ squares}}{8.94 \text{ squares}} = 1.09$$

That's in line with Figure 3-29, where a roof of similar size and slope requires from 8 to 10 percent allowance.

We'll do one more example, this time for a hip-and-valley roof.

▼ **Example 3-4:** Assuming single coverage, a crew-error waste of 3 percent, and a slope of 6 in 12, calculate the amount of 15-pound underlayment required for the roof of the building in Figure 3-8.

First, calculate the roof plan area:

Roof Plan Area = (50' x 22') + (22' x 11') = 1,342 square feet

The actual net roof area is the plan area times the roof slope factor from Appendix A:

Net Roof Area = 1,342 SF x 1.118
= 1,500 square feet ÷ 100
= 15 squares

Use this quantity to estimate labor cost at 0.36 manhours per square.

From Figure 3-9, the total ridge length is 28 + 11, or 39 linear feet.

Each hip or valley is 16.5 feet long (11 x 1.5, the slope factor from Appendix A).

There are six hips and two valleys for a total length of 132 linear feet (8 x 16.5 feet).

The total length of ridge, hips and valleys is 171 linear feet (39 + 132). At 1 square foot per linear foot of ridge, hip and valley, the amount to add to the roof plan area for lap allowance is 171 square feet. Add that to the gross roof area to see how much material to order:

Gross Roof Area = 1,500 SF + 171 SF
= 1,671 square feet ÷ 100 = 16.71 squares

Multiply that by 1.03 for the 3 percent crew waste allowance:

16.71 x 1.03 = 17.21 squares, which you convert to rolls using Figure 3-1:

Rolls (#15) = $\dfrac{17.21 \text{ FS}}{4 \text{ FS/Roll}}$ = 4.3 rolls

The allowance factor is:

Allowance Factor = $\dfrac{17.21 \text{ squares}}{15 \text{ squares}}$ = 1.15, or 15 percent allowance.

If you want to use double coverage for the previous example, refer back to Figure 3-26. The material quantity required is:

FS = 17.21 squares x 2 FS/Square = 34.42 FS

Use Figure 3-1 to convert this to rolls:

Rolls (#15) = $\dfrac{34.42 \text{ FS}}{4 \text{ FS/Roll}}$ = 8.6 rolls

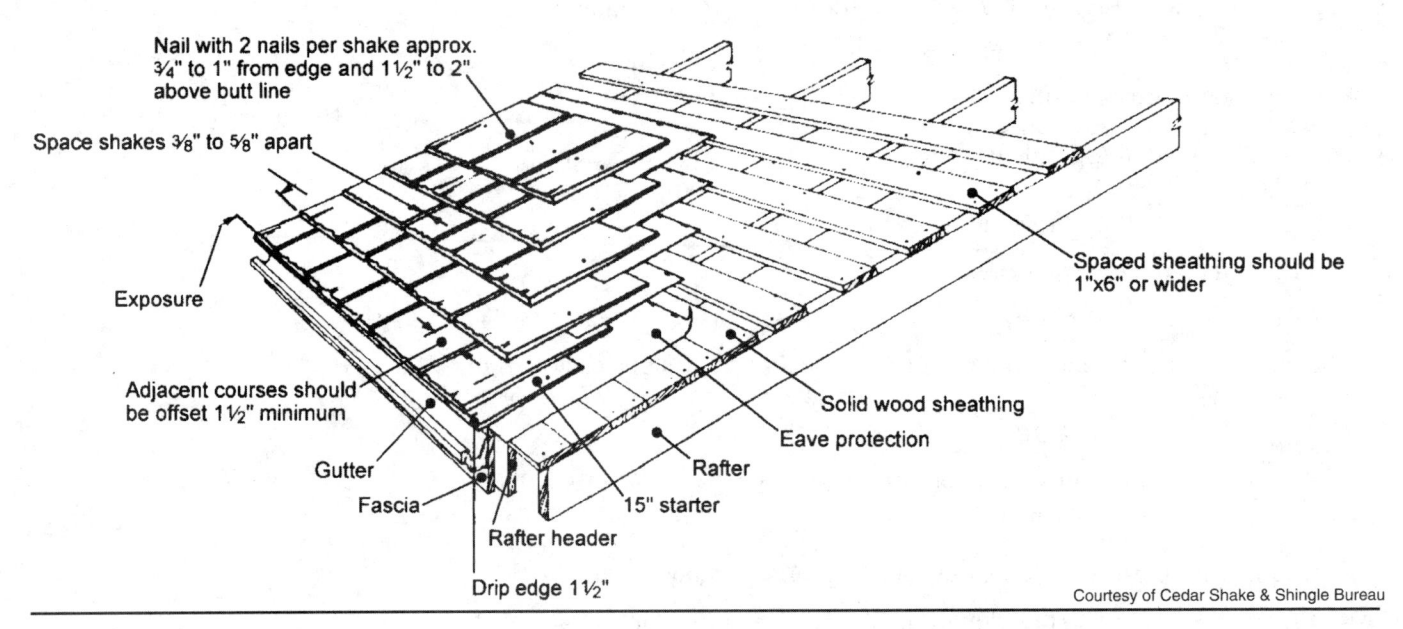

Figure 3-32 Interlayment

Interlayment (Lacing)

Whether you use solid or spaced sheathing for shake roofs, you're normally required to put felt *interlayment* between courses (see Figure 3-32). The interlayment acts as a baffle to keep wind-driven snow or other foreign material from getting into the attic cavity during extreme weather conditions. It also increases the roof's insulating value.

To be an effective baffle, the top edge of the interlayment felt must rest on the sheathing. Space open sheathing carefully so that you can properly attach the top of the felt to the sheathing board.

Code permitting, you can omit felt interlayment when you apply straight-split or taper-split shakes in snow-free areas at weather exposures less than one-third the total shake length (3-ply roof).

Usually you apply shakes over a 30-pound, 36-inch-wide underlayment starter course installed at the eaves (eaves protection). Follow that with 30-pound, 18-inch-wide interlayment courses. Some building codes allow 15-pound felt, Figure 3-32 shows interlayment between shake courses.

Install the underlayment course under the double starter-course shakes at the eaves with the lower edge aligned with the butts of the shakes.

Install the first 18-inch interlayment course so the bottom edge of the felt is twice the exposure distance above the butts of the starter course. For example, for 24-inch shakes laid with a 10-inch exposure, apply the first interlayment felt so the bottom edge is 20 inches above the

Underlayment on Sloping Roofs

Exposure (in.)	FS per square	Coverage (SF/FS)
5½	3.04	32.9
7½	2.23	44.8
8	2.09	47.8
8½	1.97	50.8
10	1.67	59.8
11½	1.45	68.8
12	1.39	71.8
14	1.20	83.7
16	1.05	95.7
18	0.93	107.6

Figure 3-33 Interlayment coverage (30-pound felt)

butts of the starter-course shingles, including the overhang. The felt will cover the top 4 inches of the shakes and extend 14 inches onto the sheathing. In heavy snow areas, I recommend that you install two strips of felt for the first course of interlayment.

Install subsequent interlayment courses so the distance between their bottom edges is equal to the weather exposure. Nail the top edges of the felts to the sheathing at about 1- to 2-foot centers.

▼ **Example 3-5:** Assume an exposure of 7½ inches, and a 1½-inch overhang at the eaves. To find the distance from the eaves to the bottom edge of the first course of interlayment, multiply the exposure by 2, then subtract the overhang from the total:

(2 x 7½") - 1½" = 13½"

Then, the distance between the bottom edges of the succeeding courses of underlayment is 7½ inches (equal to the exposure).

Estimating Interlayment Quantities

Use Figure 3-33 to find interlayment quantities. Data in Figure 3-33 doesn't include:

- The 36-inch-wide underlayment starter-course felt installed at the eaves
- The additional felt required due to lapping at ridges, hips and valleys. Take off the starter-course felt separately, then add 1 square foot per linear foot of ridge, hip and valley to be covered.
- The cutting waste, which depends on roof complexity and the crew's experience.

To take off the 36-inch-wide underlayment starter course, assume that the length of underlayment you need is equal to the total eaves length. On a hip roof, this estimate is a bit long on the underlayment starter-course quantities. So, don't add material for lap or cutting allowance involved with the underlayment starter course to your estimate.

▼ **Example 3-6:** Assume a 5 in 12 roof slope, 30-pound interlayment laid at a 10-inch exposure and a 30-pound underlayment starter course at the eaves. Find the interlayment and underlayment quantities required for the roof of the building in Figure 3-31. Assume a 4 percent cutting allowance for the interlayment and no allowance for the underlayment starter course.

Use the formula:

Net Roof Area = Length x Width x Slope Factor

The net roof area of this building is 8.66 squares (40' x 20' x 1.083). That's also the area you use to estimate labor.

The area requiring interlayment begins at a point above the eaves equal to twice the shingle exposure (2 x 10" = 20"). You deduct 20 inches from each edge of the roof, so the interlayment area is:

Interlayment Area = (40' - 40") x (20' - 40") x 1.083

Remember to convert inches to hundredths of a foot, then proceed:

Interlayment Area = (40' - 3.33') x (20' - 3.33') x 1.083
           = 36.67 x 16.67 x 1.083 = 662 SF

From Figure 3-31 you can see the ridge length is 20 feet.

You calculate the hip length using Column 2 of Appendix A (plan length times slope factor)

Length (Hip) = 10' x 1.474 = 14.74 linear feet

There are four hips, so the total length is 4 x 14.74 = 58.96 or 59 linear feet.

The total length of ridge and hips is 20 + 59 = 79 linear feet.

At 1 square foot per linear foot, the total allowance for lapping is 79 SF.

Including material required for lapping at the ridge and hips, the gross interlayment area is the same as the interlayment area plus the lap allowance:

Gross Interlayment Area = 662 SF + 79 SF
              = 741 SF ÷ 100
              = 7.41 squares

Now, from Figure 3-33, the interlayment quantity is:

FS (30-pound interlayment) = 7.41 squares x 1.67 FS/Square
= 12.37 FS

Now, multiply by 1.04 for a 4 percent cutting allowance:

12.37 x 1.04 = 12.9 factory squares

The roof perimeter (twice the total of length plus width) taken at the eaves, is:

Perimeter = 2 x (40' + 20') = 120 linear feet

The area covered by the underlayment starter course is the perimeter times 3 feet:

Area of Underlayment = 120' x 3'
= 360 square feet ÷ 100
= 3.6 squares

The total quantity of 30-pound felt required for all interlayment and underlayment is 16.5 factory squares (12.9 FS + 3.6 FS)

Look back at Figure 3-1 to find that 30-pound felt yields 2 FS per roll, so you need to order 8.3 rolls of felt:

$$\text{Rolls (30-pound felt)} = \frac{16.5 \text{ FS}}{2 \text{ FS/Roll}} = 8.3 \text{ rolls}$$

Figure 3-33 gives interlayment coverage data for common shake exposures. But remember, it doesn't allow for eaves underlayment, laps at ridges, hips and valleys, or crew waste. Use the following equations to find the interlayment coverage based on any exposure:

$$\text{Coverage (SF/FS)} = \frac{\text{(Roll Length - End Lap) x Exposure}}{\text{Number of FS per Roll}}$$

Equation 3-3

where:

$$\text{Exposure} = \frac{\text{Roll Width - Top Lap}}{\text{Number of Plies}}$$

Equation 3-4

The coverage, in factory squares installed per square (100 SF) of roof surface covered, is:

$$\text{FS/Sq.} = \frac{\text{100 SF/Square}}{\text{Coverage/FS}}$$

Equation 3-5

▼ **Example 3-7:** Assume an exposure of 6 inches and a 6-inch end lap, then find the coverage of 30-pound felt interlayment, using Equation 3-3 on the previous page.

You see in Figure 3-1 that 30-pound felt is normally sold in 72-foot-long rolls. You split each roll in half for 18-inch-wide strips, so the total roll length is 2 x 72, or 144 linear feet. Use Equation 3-3 to find the coverage in terms of square feet covered per factory square:

$$\text{Coverage (SF/FS)} = \frac{(144' - 0.5') \times 0.5'}{2} = 35.9 \text{ SF/FS}$$

Now, use Equation 3-5 to find the number of factory squares required per square of roof surface:

$$\text{FS/Square} = \frac{100 \text{ FS/Square}}{35.9 \text{ SF/FS}} = 2.79 \text{ factory squares per square}$$

Eaves Flashing (Ice Shield or Water Shield)

In snow-belt areas, there's sometimes enough heat loss from inside a building to melt snow on the roof. When that happens, water can leak into the walls and ceiling areas due to the freeze-thaw cycle of the snow. That cycle causes *ice dams* (freezebacks) to form on the roof. Water backs up under the shingles and down through the roof sheathing. This is shown in Figure 3-34. From there, the water drips down onto the ceiling or enters through the top of the walls where it finally ends up on the floor under the carpet.

This particular interior heat loss is caused by insufficient ceiling insulation. Ice dams can also occur over windows with large headers where there isn't enough wall insulation over the windows. Heat escapes up the wall and into the attic immediately above. This causes ice on the shingles to melt from the bottom up. Then, the weight of the ice remaining above presses down, forcing the water up the slope and under the shingles.

The action you take to prevent leaks caused by ice dams will depend on the roof slope and the local building code.

On slopes of 4 in 12 and steeper, in addition to underlayment, install an ice shield made of one ply of 50-pound smooth-coated roll roofing parallel to the eaves. Provide a $1/4$- to $3/8$-inch overhang beyond

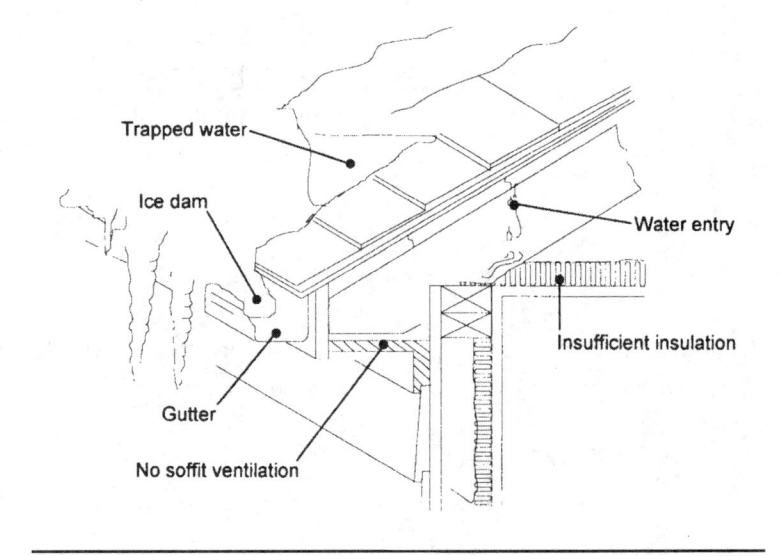

Figure 3-34 Ice dam caused by insufficient insulation and ventilation

Underlayment on Sloping Roofs

the drip edge. Install the flashing 12 inches (some building codes require 24 inches) beyond the interior face of the exterior wall line as shown in Figure 3-35. If it takes more than one width of flashing to reach that point, locate the lap outside the exterior wall face. Overlap the flashing at least 2 inches and cement the entire length of the lap. If you have to use more than one roll over the length of the eaves, lap the ends at least 12 inches and cement the end lap.

On slopes less than 4 in 12, you can use double-coverage underlayment for the eaves flashing. However, you have to install the flashing 24 inches (some building codes require 36 inches) beyond the interior face of the exterior wall line. Also, you must embed each course of the flashing material into roofing cement, which you've applied at the rate of 2 gallons per 100 square feet. That's shown back in Figure 3-11.

I believe a heavyweight roll-roofing ice shield under asphalt shingles does more harm than good. The heavy material can expand and buckle during hot weather, forcing shingles up and making them more susceptible to hail damage and foot traffic. You only need two layers of 15-pound felt under double-coverage asphalt shingles. If a freezeback does force moisture under the shingles, no leak will occur, provided the felts are smooth and watertight. But remember that you have to follow your local building code.

W. R. Grace Co. manufactures a rubberized product you can use as ice-shield material. It comes in 3-foot-wide rolls with an adhesive back covered with release paper. Remove the release paper and stick the material down to the sheathing, underlayment, or existing asphalt shingles. This material won't buckle, even in hot weather. You can also stick this material to flanges on flashing to help waterproof the flashing.

To cut down on ice dams, check the design of the building. The roof should be well ventilated, as shown in Figure 3-36. There should be a constant flow of cold air between the ceiling insulation and the roofing material. The vent space must let air flow freely from eaves to roof top. The steeper the roof, the better the venting system works.

Discourage wide overhangs at the eaves. They make cold areas where snow and ice can build up. Warn against putting doors and large windows with large headers and no wall insulation at the bottom of a roof slope. Instead, locate them beneath the gable ends of the roof. Be sure insulation doesn't block the soffit vents.

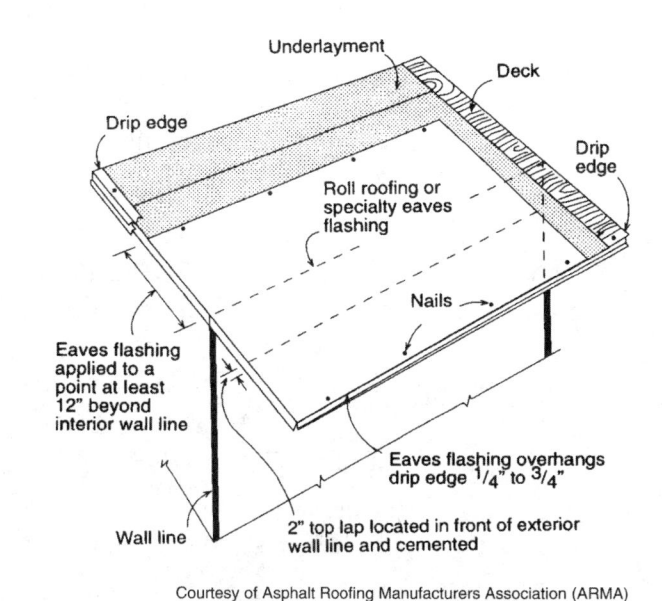

Courtesy of Asphalt Roofing Manufacturers Association (ARMA)

Figure 3-35 Application of eaves flashing

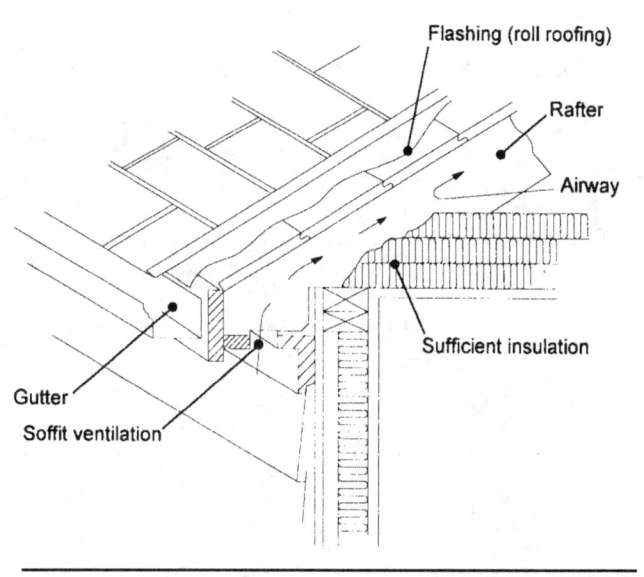

Figure 3-36 Good ventilation, insulation and roof flashing

Ice dams don't form just at the eaves. They can occur anywhere heat escapes to the underside of the roof. It's best if chimneys are located at the ridge or gable ends. Recommend that plumbers stack out plumbing vent pipes so they go through the roof near the ridge, and use cast iron or galvanized iron vent pipes because sliding snow can move plastic ones. Make sure all vent pipes are securely anchored to the structure.

Ice dams can also occur at the lower end of a valley the same way they do at the eaves. To minimize or prevent this problem, building owners can install an electric thermal wire in the valley in a zigzag pattern, and turn on the current when it snows.

▼ **Example 3-8:** Assume a potential for ice dams, a roof slope of 6 in 12, and that the interior face of the exterior wall line is 30 inches (measured horizontally) from the eaves line. Find the amount of eaves flashing required for the hip and valley roof in Figure 3-8.

Since the roof slope is 6 in 12, you need to put a 50-pound coated felt flashing along the eaves 24 inches beyond the inside face of the outside wall line. The actual flashing width is:

Flashing Width = (30" + 24") x 1.118
    (the slope factor from Column 2, Appendix A)
    = (2.5' + 2') x 1.118 = 5.59' = 6 feet (rounded up)

The formula (from Chapter 1) for the building perimeter is:

Perimeter = 2(L + W)

The total length of flashing required (using the formula for building perimeter) is:

Flashing Length = 2 x (50' + 33') = 166 linear feet

The total area of flashing required is:

Flashing Area = 166' x 6'
    = 996 square feet ÷ 100
    = 9.96 or 10 squares

(Use this figure to estimate manhours at the rate of 0.738 manhours per square.)

Since the length of flashing required is equal to the total eaves length, you don't need to add for cutting allowance.

Valley Flashing

If a roof leaks, it's most likely to leak at a valley. So you need to flash valleys for added weather protection. The type of valley flashing you use depends on the type of valley you'll construct. That in turn depends on the type of roof covering you'll install.

Closed and Open Valleys

In a closed valley, you lap the roof covering from both sides of the valley to cover the valley flashing. In an open valley, there's space between the roof covering on adjacent roofs. Shingles don't extend across the valley, so the valley flashing is exposed as in Figure 3-37. You can use open valleys with all types of asphalt or fiberglass shingle products. In fact, this type of valley is recommended for mineral-surfaced roll roofing.

Open valleys tend to leak more than other types of valleys because heavy rain or debris collected in the valley can force water up under shingles adjacent to the valley.

If you have a choice, avoid installing open valleys where the two roofs forming the valley have considerably different slopes or areas which let very different amounts of water flow into the valley. For example, at a dormer, there would be a much higher volume of water flowing into the valley off the main roof than from the dormer. The heavier flow might back up under the shingles on the opposite side of the valley.

Figure 3-37 W-type metal valley on an asphalt shingle roof

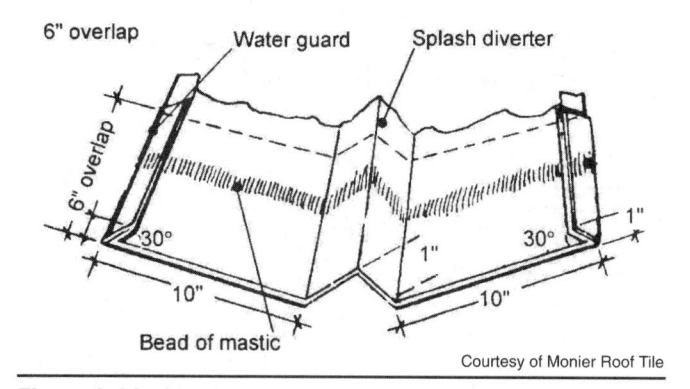

Figure 3-38 Metal valley with splash diverter and water guards

On strip-shingle roofs with varying slopes, you can reduce the problem by installing a closed-cut valley, and shingling the lower-sloped roof surface first. This is the slope that will receive most of the problem water. When you have to install an open valley on this type of roof, be sure to use metal valley flashing with a splash diverter and water guards as shown in Figure 3-38 to break the force of the water from the steeper or longer slope to keep it from being driven up under the shingles on the opposite side.

You are required to use open valleys on tile, wood shingle, or shake roofs. You can use a closed valley on a slate roof, but the open valley is more common. Chapter 8 has more information about that. You can use the closed-cut (half-lace) valley (Figure 3-42) with all types of strip shingles. The woven (full-lace) closed valley is recommended only with 3-tab shingles.

Valley Flashing Materials

Use mineral-surfaced roll roofing material to flash valleys on roofs covered with 3-tab and laminated shingles, and mineral-surfaced roll roofing. You can also use metal flashing in these valleys, but this method (shown in Figure 3-37) isn't common because the metal expands and contracts when the temperature changes.

Use metal flashing on open valleys or roofs covered with slate, wood shingles, and shakes. Use metal, or a combination of metal overlaid with mineral-surfaced roll roofing, to flash open valleys on tile or metal roofs.

Here are some of the many types of metal valley flashing materials you can get:

- Galvanized steel
- Aluminum
- Copper
- Lead
- Zinc

Galvanized steel and aluminum are the least expensive, so they're used the most. You'll usually find the other (more expensive) metals only on certain commercial or long-lasting roofs such as tile or slate.

Galvanized steel valley flashing (roll valley metal) is made of 26- or 29-gauge galvanized steel or other corrosion-resistant metal. It comes in 50-foot rolls 12, 14, 16, 18, and 20 inches wide (Figure 3-39). I recommend you use (and some building codes require) 26-gauge metal because rainwater from a roof carries erosive roof dust and grit that'll quickly undermine lesser metal gauges. For this reason, use at least 16-ounce copper (16 ounces per square foot), 12-ounce tin, 3-pound hard lead and 11-gauge (0.024-inch thick) zinc whenever those materials are specified for any type of flashing.

Some roofing manufacturers recommend that you paint galvanized steel on both sides with metal paint or bituminous paint. Paint tin valley metal on the underside. If you have to bend the flashing at a sharp angle, apply the paint after you bend it. The paint, especially on the underside, adds another layer to help protect the metal from corrosion.

Figure 3-39 Roll metal valley flashing

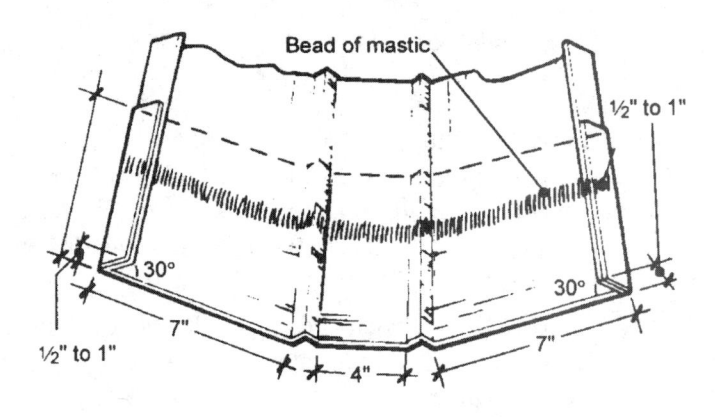

Courtesy of Monier Roof Tile

Figure 3-40 Metal valley with a double crimp

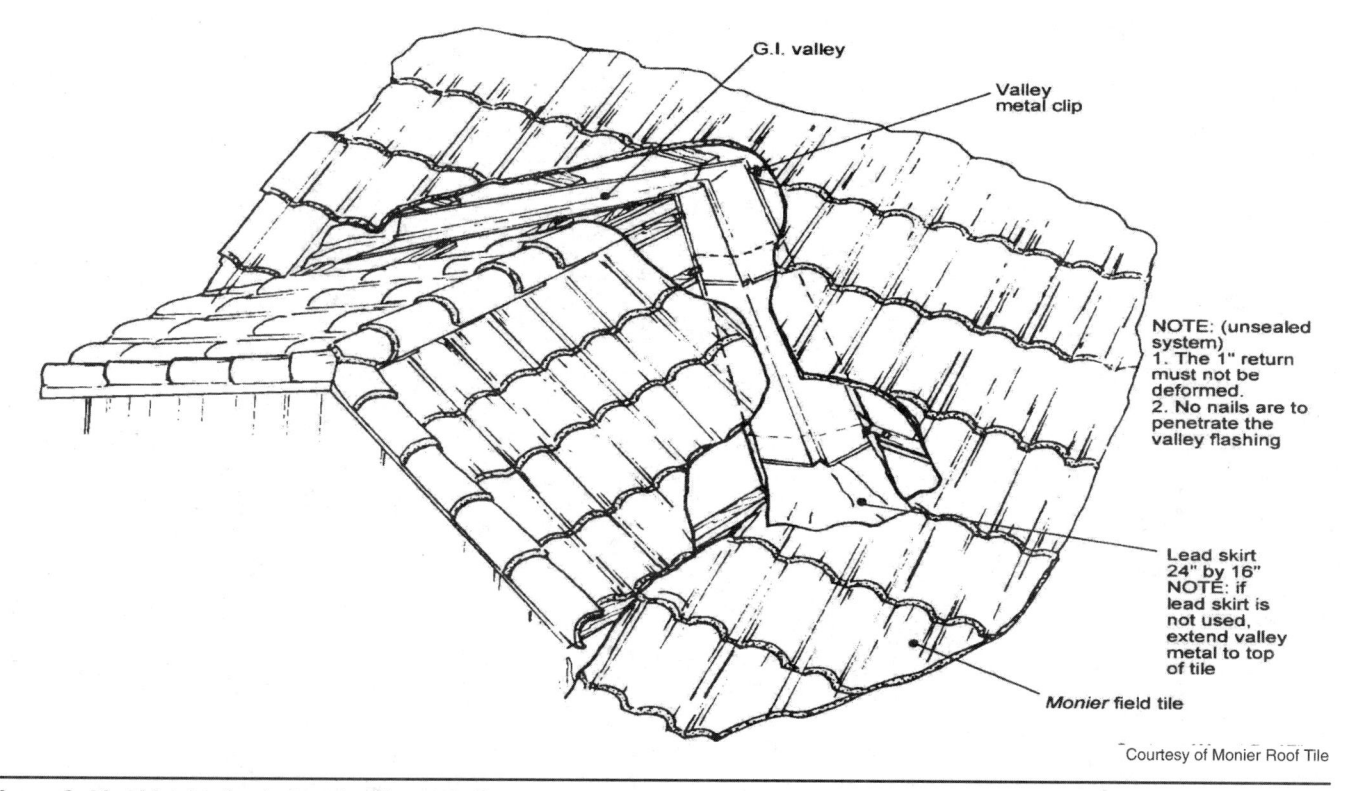

Figure 3-41 Metal valley fastened with metal clips

To help keep water from getting under the shingles adjacent to the valley, install metal with a W-shaped crimp (splash diverter rib) in the center and water guards along the edge. See Figure 3-38. In valleys likely to accumulate debris, install flashing with a double crimp like the one in Figure 3-40. If you fabricate the flashing with water guards, you'll need to anchor the flashing to the sheathing with metal clips, as shown in Figure 3-41, on 12-inch centers. Some manufacturers recommend using metal clips rather than driving nails into the valley metal because the clips can expand and contract when the temperature changes and they pivot to move with the valley metal.

You can leave out the splash diverter ribs and water guards where a valley takes a sharp turn, because those features make it hard to change direction. Instead, install metal from a smooth stock metal roll, or use an open mineral-surfaced roll-roofing valley.

Installing Non-Metal Valley Flashing

In a closed valley, in addition to the required felt underlayment just described, flash the valley with a 36-inch-wide layer of No. 15 or No. 30 asphalt-saturated felt. Many building codes allow this; however, I recommend using mineral- or smooth-surfaced roll roofing (50 pounds or

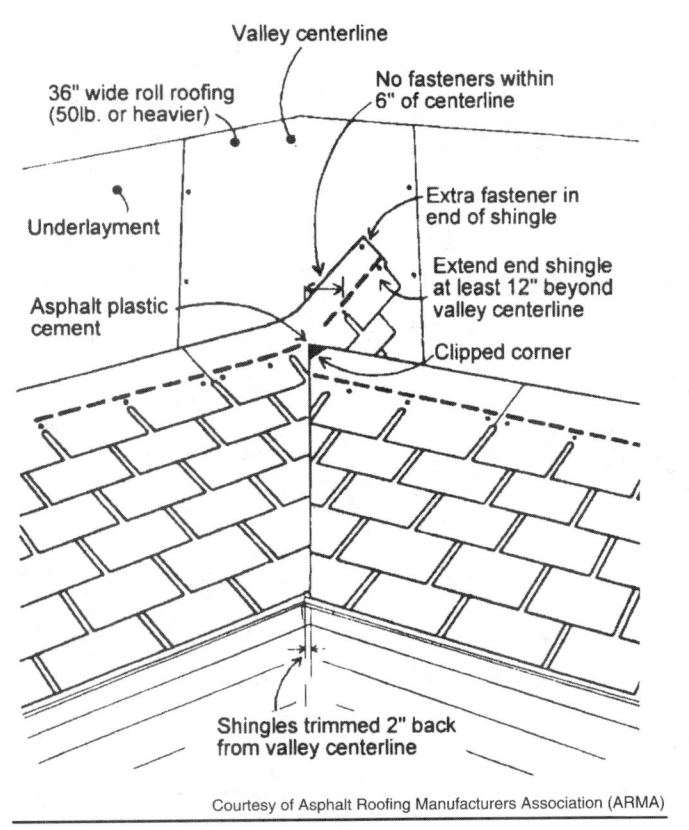

Figure 3-42 Flashing a closed-cut valley

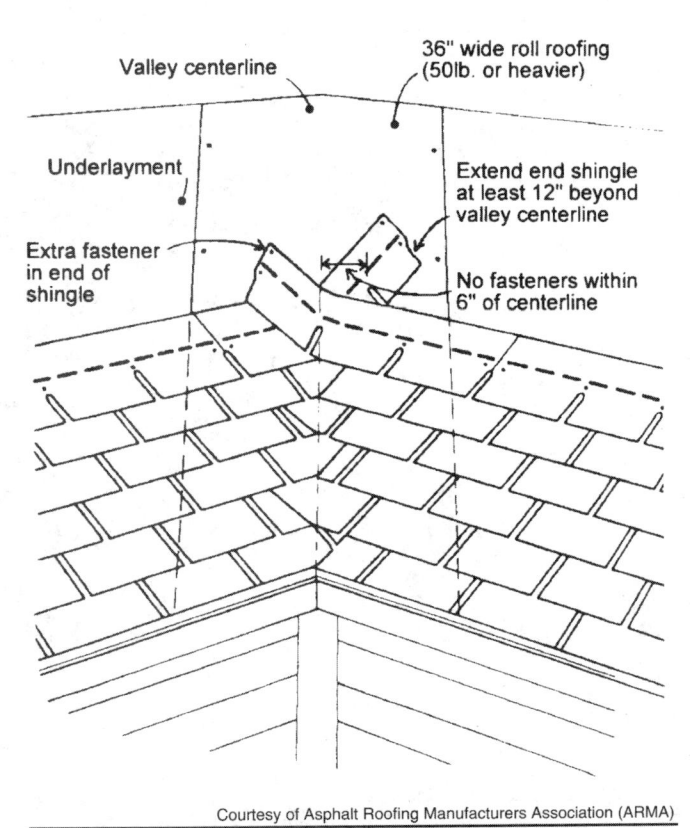

Figure 3-43 Flashing a woven valley

heavier). That's shown in Figures 3-42 and 3-43. If you need more than one strip of roll roofing over the length of the valley, lap the upper strip over the lower by at least 12 inches and bond the lap with roofing cement. Don't apply too much cement because it could cause blistering. Don't drive nails through the lap within the valley. Nail the flashing strip to the deck along a line 1 inch from the edge of the flashing, using only enough nails to hold the strip in place. Extend the flashing beyond the ridge and cut it so it lays close to the surface of the sheathing with no bulges. Then nail it down. See Figure 3-44.

In an open valley, in addition to the felt underlayment previously discussed, flash the valley with two layers of mineral-surfaced roll roofing (90 pounds). Cement the layers together with the bottom 18-inch-wide layer mineral surface down, and the top 36-inch-wide layer with the mineral surface facing up. See Figure 3-45. If you need to splice the roll roofing over the length of the valley, lap the upper strip over the lower by at least 12 inches and bond the lap with roofing cement. Don't drive nails through the lap

Figure 3-44 Extend the valley flashing beyond the ridge

within the valley. Nail both strips to the deck along a line 1 inch from the edge of the flashing, driving only enough nails to hold each strip in place. Install a top layer of roll roofing that's the same color as the shingles, or a neutral color.

Installing Metal Valley Flashing

Metal valley flashing should never be less than 16 inches wide. Figure 3-46 gives the minimum allowable gauges, thicknesses or weights of the flashing.

To install light-gauge roll metal valley flashing, roll the flashing down the valley. Step into the center of the sheet so that it conforms to the valley, nailing as you walk up the valley. Drive the nails on 12-inch centers along rows located 1 inch from the edges of the flashing. Use metal clips if the flashing has water guards.

Don't fabricate heavy-gauge metal flashing pieces longer than 10 feet. If you need more than one piece or roll of flashing over the length of a valley, lap the upper layer over the lower by at least 6 inches, as shown in Figure 3-46 and bond the lap with roofing cement. Don't drive nails through the lap within the valley.

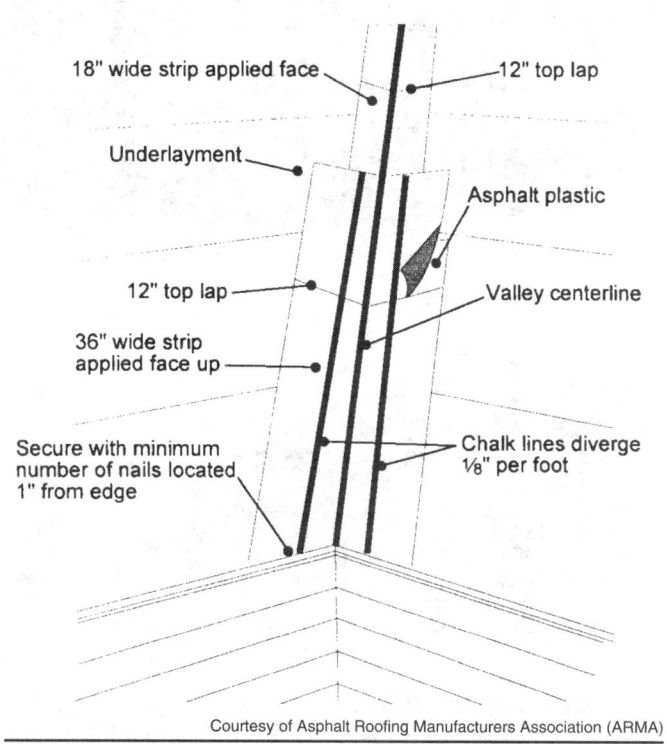

Figure 3-45 Application of 90-pound roll roofing as flashing for an open valley

Material	Minimum thickness	Gauge	Weight
Aluminum	0.024 in	--	--
Cold-rolled Copper	0.0216 in	--	ASTM B 370, 16 oz per SF
High-yield Copper	0.0162 in	--	ASTM B 370, 12 oz per SF
Lead-coated Copper	0.0216 in	--	ASTM B 101, 16 oz per SF
Lead-coated high-yield Copper	0.0162 in	--	ASTM B 101, 12 oz per SF
Lead	--	--	2.5 lbs
Stainless steel	--	28	--
Galvanized steel	0.0179 in	26 (zinc-coated G90)	--
Zinc alloy	0.027 in	--	--

Figure 3-46 Valley lining material

Roofing Construction & Estimating

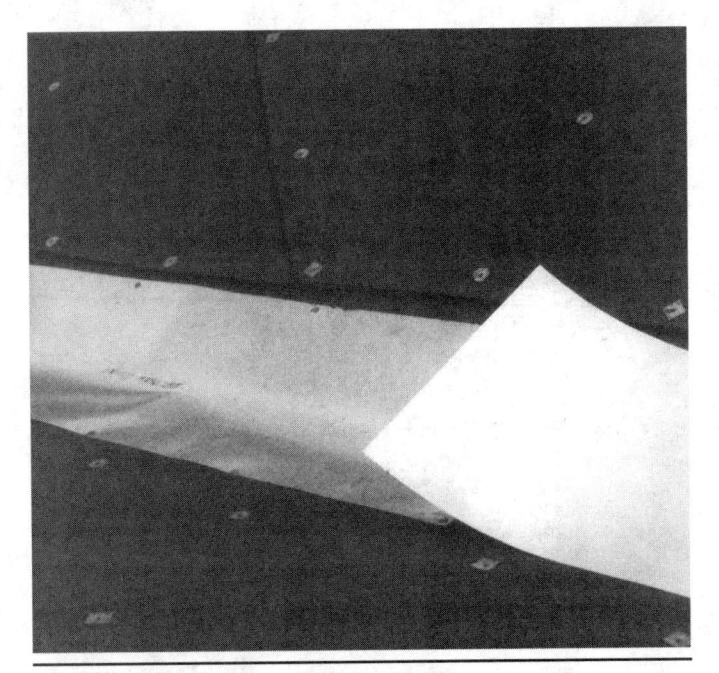

Figure 3-47 Lapping metal valley flashing

Figure 3-48 Extend the metal valley flashing beyond the ridge

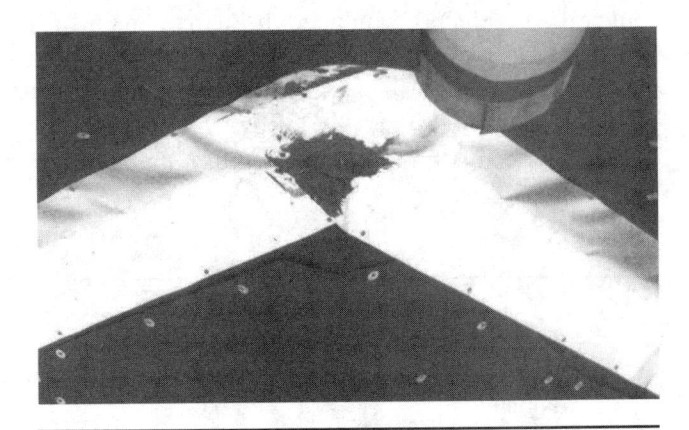

Figure 3-49 Metal valley flashing in two intersecting valleys

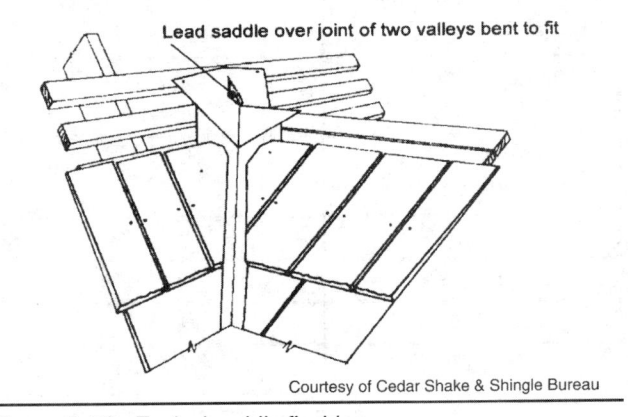

Figure 3-50 Typical saddle flashing

Extend roll metal flashing beyond the ridge and cut it so that it conforms with the surface of the sheathing. Then nail it down as shown in Figure 3-47. Lap the metal where two valleys intersect and seal the joint with roofing cement, as in Figure 3-48. For added leak protection, install a lead saddle at the juncture of two metal valleys. That's shown in Figure 3-49.

On wood-shingle roofs, extend metal valley flashing at least 10 inches beyond each side of the centerline of the valley on roofs with slopes up to 6 in 12, and 7 inches beyond the centerline of roofs with slopes 6 in 12 and steeper. Snap a chalk line as a shingling guide 2 to 4 inches on either side of the centerline, depending on the anticipated water volume. See Figure 3-50.

You don't need underlayment beneath the metal valley on a wood-shingle roof. But you should use a 36-inch-wide strip of 15-pound felt, especially in cold-weather areas where there are wind-driven snows. The underlayment helps to prevent leaks. It also helps to prevent condensation that would eventually cause corrosion on the underside of the metal. That's also shown in Figure 3-50.

On shake roofs, extend the metal valley flashing at least 10 inches beyond each side of the centerline of the valley. Install a 36-inch-wide roll of 15-pound felt (minimum) beneath the metal flashing. Code permitting, you can omit the valley felt when you install the shakes over spaced sheathing.

Underlayment on Sloping Roofs

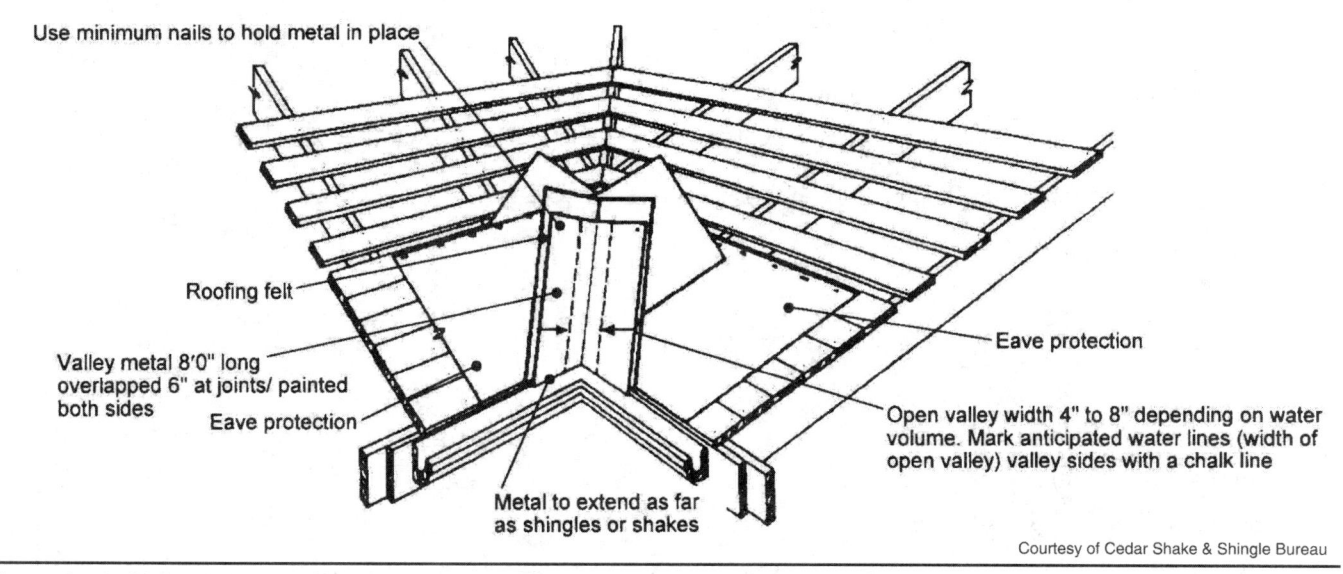

Figure 3-51 Valley flashing for wood shingle and shake roofs

Some roofing manufacturers don't recommend using felt beneath any valley metal except copper or lead since the felt keeps condensed water beneath the metal from running off or evaporating. When you use felt to head off wind-driven precipitation, it's very important to paint the underside of a metal valley.

Estimating Valley Flashing Material

For estimating, you take off roll roofing flashing material by the square foot, then convert to squares. Take off metal flashing material by the linear foot. You have to account for end laps and overcuts for valley ends at inside roof corners. Figure 3-51 shows that the length of additional material you need is one-half the width of the flashing material. In most cases, add 1 foot to the valley length to allow for this additional material. Some roofing contractors cut the material off at the lower end, forming a small triangular extension that helps shed runoff water away from the building. That's shown in Figure 3-52.

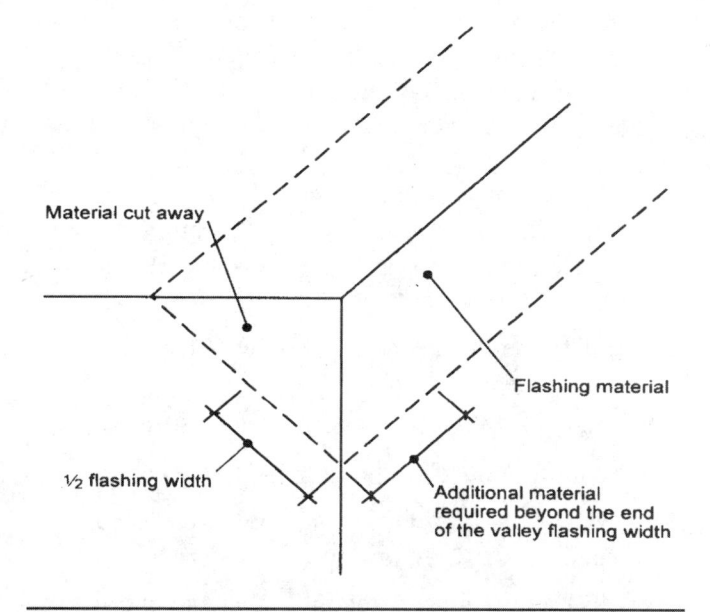

Figure 3-52 Additional flashing material required at the lower end of a valley

Here's how to estimate valley flashing quantities:

▼ **Example 3-9:** Assume you're applying 3-tab asphalt shingles on a 6 in 12 roof. Find the quantity of valley flashing required for the roof in Figure 3-8, for both a closed and open valley. Assume the flashing ends where the valleys meet.

Using the slope factor conversion table from Column 3 of Appendix A, the actual length of material required at either valley (including 1 foot of additional material required beyond the lower end) is:

Length (Valley) = 12' x 1.5 = 18 linear feet

Multiply by 2 for the total length of both valleys, 36 linear feet.

For a closed valley, the total area to be covered with 50-pound roll roofing is:

Flashing Area = 36' x 3'
= 108 square feet ÷ 100
= 1.08 squares

For an open valley, the total area to be covered with 90-pound mineral-surfaced roll roofing is:

Flashing Area = 36' x (3' + 1.5')
= 162 square feet ÷ 100
= 1.62 squares

The manhour rate for the above example is 1.03 manhours per square.

When you're estimating roll roofing for flashings, be alert to unusual colors. If the roof is a color you seldom install, you might never be able to use leftover roll roofing material. In that case, figure the entire roll into your estimate, including any excess you won't use on the current job.

▼ **Example 3-10:** Assuming a roof slope of 6 in 12, find the total square feet of 18-inch-wide metal valley flashing material required for the roof of the building in Figure 3-8. Assume 10-foot lengths of metal installed with a 6-inch end lap.

The actual length of either valley (using the slope factor from Column 3 of Appendix A) is:

Length (Valley) = 11' x 1.5 = 16.5 linear feet

Multiply by 2 for the total length of both valleys, 33 linear feet

Also, add 0.5 foot for material at each lap. Since there are two joints, add 2 x 0.5 foot, or 1 foot.

The metal must extend 9 inches beyond the end of the valley, so add 0.75 foot for each valley, rounded up to 2 feet total.

The total is:

Total Valley Metal = 33 feet + 1 foot + 2 feet = 36 feet

Now that we've covered underlayment and sheathing, let's go on to roof coverings. In the next chapter, we'll begin with asphalt shingles.

4 Asphalt Shingles

Asphalt roofing materials have been manufactured since the early 1890s. Today, asphalt shingles cover about 70 to 80 percent of all roofs in the United States. Those roofs are attractive, versatile, and fire- and wind-resistant. Asphalt shingle roofs are relatively inexpensive, easy to install, and require little maintenance. The normal life expectancy of an organic asphalt shingle roof is 15 to 20 years. Heavyweight laminated fiberglass shingles will last 20 to 30 years.

Organic Shingles

The base mat of organic-based asphalt shingles (organic shingles) was originally composed of cellulose fibers made from recycled paper or wood chips, and cotton or wool fibers made from rags. Now, it's made of a tough, asphalt-saturated roofing felt, coated on both sides with asphalt, as shown in Figure 4-1.

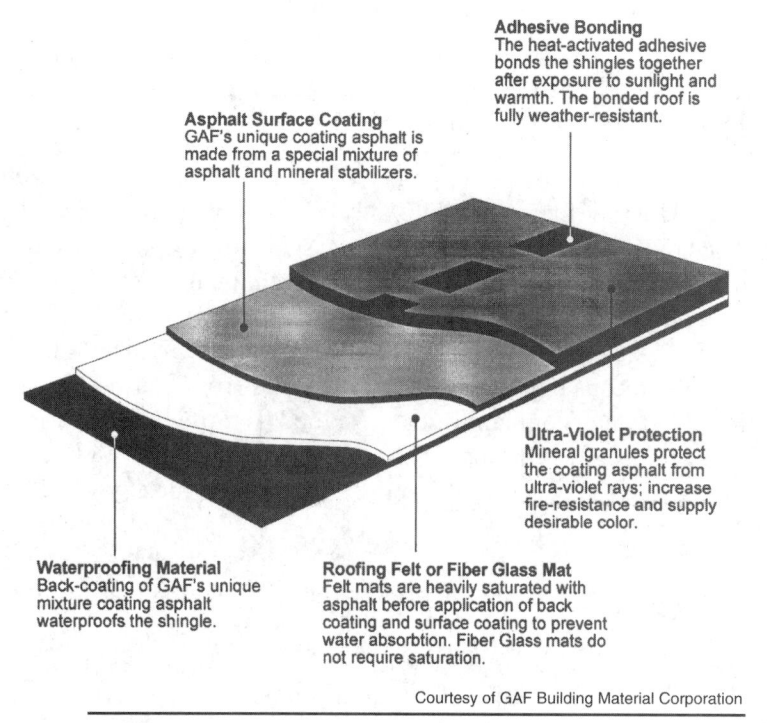

Asphalt Surface Coating
GAF's unique coating asphalt is made from a special mixture of asphalt and mineral stabilizers.

Adhesive Bonding
The heat-activated adhesive bonds the shingles together after exposure to sunlight and warmth. The bonded roof is fully weather-resistant.

Ultra-Violet Protection
Mineral granules protect the coating asphalt from ultra-violet rays; increase fire-resistance and supply desirable color.

Waterproofing Material
Back-coating of GAF's unique mixture coating asphalt waterproofs the shingle.

Roofing Felt or Fiber Glass Mat
Felt mats are heavily saturated with asphalt before application of back coating and surface coating to prevent water absorbtion. Fiber Glass mats do not require saturation.

Courtesy of GAF Building Material Corporation

Figure 4-1   Asphalt shingle components

The base mat gives shingles their strength. The base material is saturated and covered with a high-melting-point flexible asphalt called a *saturant*. The saturant is reinforced with mineral stabilizers such as ground limestone, slate, trap rock (weathered volcanic rock) or other inert materials such as ceramic-coated rock granules. Coarse mineral granules are pressed into the asphalt coating on the exposed face. This gives the shingle its color and helps it resist weather and fire.

The materials most often used for coarse mineral surfacing are natural-colored slate, natural-colored rock granules, or ceramic-coated rock granules. The back of each shingle is covered with talc, sand or mica to prevent shingles from sticking together in the bundle.

Fiberglass Shingles

Fiberglass shingles first appeared in the late 1950s. By the late 1970s, they had improved so much they were as good as traditional asphalt shingles. Fiberglass shingles have a fiberglass base mat saturated and covered with flexible asphalt and surfaced with mineral granules. The weight and thickness of a fiberglass mat is usually much less than a cellulose-fiber mat. Fiberglass shingles contain more asphalt than organic-based asphalt shingles.

Nowadays, organic shingles aren't used very often. They soak up water from underneath, which makes the corners at the bottom of the tabs curl up.

Throughout this chapter, the term "asphalt shingle" means an organic- or fiberglass-based shingle saturated with asphalt. I'll distinguish between the two products only when it's necessary for the sake of accuracy.

UL Ratings for Shingles

The Underwriters' Laboratory (UL) is a non-profit organization founded in 1894 under the sponsorship of the National Board of Fire Underwriters. The UL has the most widely-accepted standards for fire resistance of building materials. The UL classifies a fire-resistant shingle as A, B, or C. Class A shingles withstand severe fire exposure. Class B shingles withstand moderate fire exposure and Class C, light fire exposure. With all three ratings, "exposure" means exposure to fire that comes from sources outside the building. Insurance companies may use the UL rating to determine insurability of a building. If you install shingles that don't have a good UL or similar rating, they may decline coverage. To qualify for any UL classification, a shingle must not:

a) disintegrate and fall off the roof as glowing brands (airborne embers)

b) break, slide, warp, or crack, exposing the deck

c) allow the roof deck to fall away as glowing particles

d) allow continued flaming beneath the roof deck.

Asphalt Shingles

To bear a UL "wind-resistant" label, a shingle must withstand winds up to 63 miles per hour for two hours without a single tab being uplifted. Figure 4-2 shows a UL shingle label. Look for this label on each bundle of shingles, and be sure to install them according to the manufacturer's instructions.

To increase their wind resistance, many asphalt shingles come with a self-sealing thermoplastic adhesive strip (tar strip) above the cutouts on the face of the shingle. That's shown in Figure 4-3. Heat from the sun makes the strip sticky and helps to bond each shingle to the one above it. The adhesive strip takes longer to bond in cold weather or when the roof is shaded, has a low slope, or faces north or east.

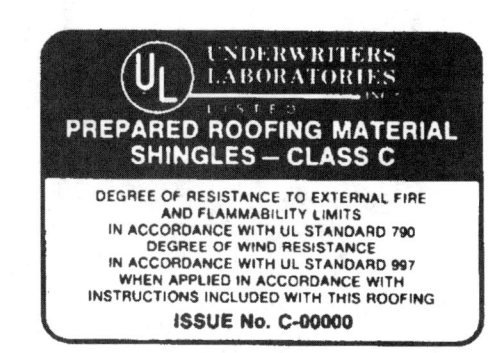

Courtesy of Asphalt Roofing Manufacturers Association (ARMA)

Figure 4-2 Underwriters' Laboratory label

The best temperature range for installing asphalt shingles is between 40° F and 85° F. Before you install asphalt shingles during cold weather, store the shingles in a warm location or lay them in the sun until they soften up.

If you will be storing shingles, keep them in a cool dry area in stacks no more than 4 feet high. Your local roofing materials supplier can advise you about how long it's safe to stockpile asphalt shingles. Rotate the bundles so the shingles stored the longest will be the first ones you'll use. That's so the shingles at the bottom of the stacks don't become discolored. Light-colored shingles may darken because oils in the asphalt move. Dark-colored shingles may show light smudges when backing materials (such as talc, which helps keep shingles from sticking together in the bundle) transfer to adjacent shingles.

Figure 4-3 Shingle with adhesive strip

If you store shingles outdoors, place them on a raised platform so they don't touch the ground. Cover the shingles to protect them from wet weather. Don't store shingles in the hot sun because heat makes them stick together.

Deck Requirements

Asphalt shingles require a solid roof deck. As a general rule, you can install asphalt shingles on roof slopes ranging from 4 in 12 through 21 in 12 using standard application methods. You can also install asphalt shingles on slopes as flat as 2 in 12, or steeper than 21 in 12, but you'll have to follow special application procedures. We'll cover this later in this chapter (see Figure 4-27 and related text). Figure 4-4 gives minimum roof slope requirements for various asphalt roofing materials.

Shingle Colors

The color of the shingles you use can dramatically affect the appearance of a building. For example, a light-colored roof directs the eye upward and gives the illusion of spaciousness. Dark colors create the opposite effect. In the case of a large, steep roof, you can use that illusion to scale down the roof structure and make the building look more proportional and attractive. Use Figure 4-5 as a guide for choosing shingle colors that go with various colors of siding, trim, shutters and doors.

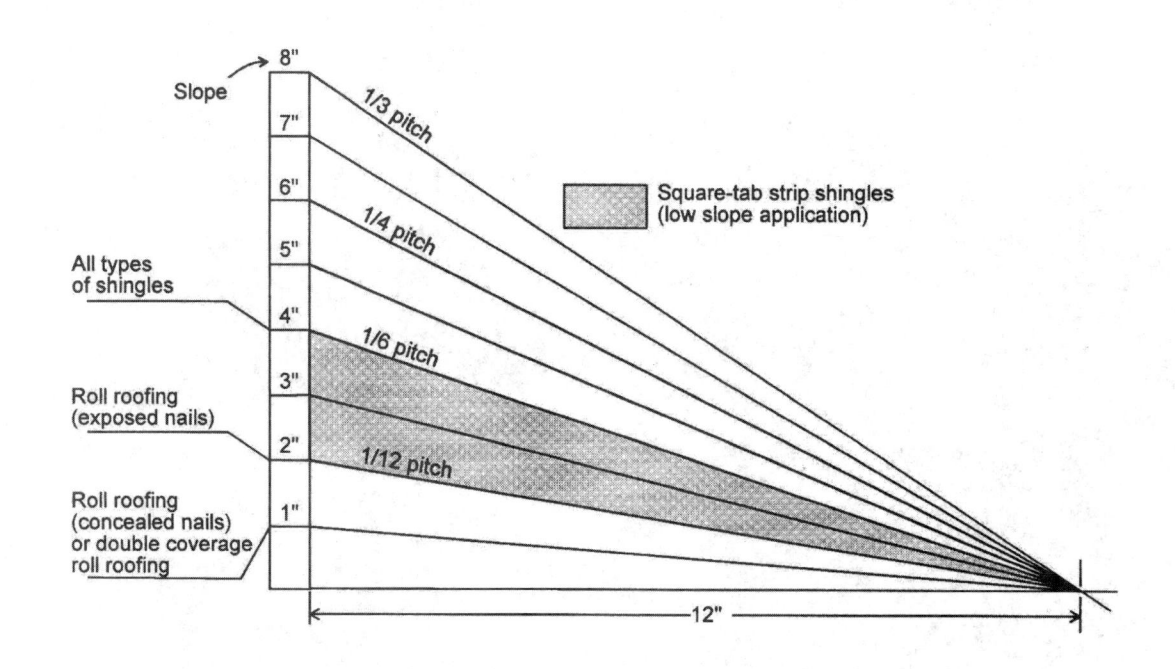

Courtesy of Asphalt Roofing Manufacturers Association (ARMA)

Figure 4-4   Minimum pitch and slope requirements for various asphalt roofing products

Asphalt Shingles

Roof Shingles	Siding	Trim	Shutters and Doors	Roof Shingles	Siding	Trim	Shutters and Doors
White	White	White	Deep Gold, Maroon	**Brown**	White	White	Dark Brown, Terra Cotta
	White	Gray	Charcoal		Green	White	Dark Brown, Dark Green
	Green	White	Dark Brown, Dark Green		Yellow	White	Dark Brown, White
Black	White	White	Black, Maroon	**Green**	White	White	Dark Green, Black
	Yellow	White	Black, Deep Olive Green		Yellow	White	Dark Green
	Gold	White	Black, Deep Olive Green		Lt. Green	White	Dark Green, Terra Cotta
Gray	Red	White	Black, White	**Blue**	White	White	Blue
	Yellow	White	Gray, Charcoal, Green		Yellow	White	White
	Coral Pink	Lt. Gray	Charcoal		Lt. Blue	White	Dark Blue, White
Red	White	Gray	Charcoal				
	White	White	Red				
	Beige	White	Dark Brown				

Courtesy of Asphalt Roofing Manufacturers Association (ARMA)

Figure 4-5   Asphalt shingle color guide

Asphalt Strip Shingles

The most widely used type of asphalt shingle is the 3-tab (triple-tab) strip shingle like the one in Figure 4-6. This is also called a square-butt or thick-butt shingle because some manufacturers make a shingle that's thicker at the butt edge. Cutouts (also called keys, water lines, bond lines, tab notches, and water jackets) make a roof look like it's finished with many smaller units.

You can also find 2-tab (twin-tab) strip shingles, strip shingles with no cutouts, and shingles with as many as five tabs (random-tab strip shingles). Some shingles have staggered butt lines. Shingles whose tabs are all the same size are called square-tab shingles. Strip shingles with more than one layer of tabs are called laminated, dimensional, or three-dimensional shingles. These shingles create extra thickness and give a three-dimensional effect. See Figure 4-7.

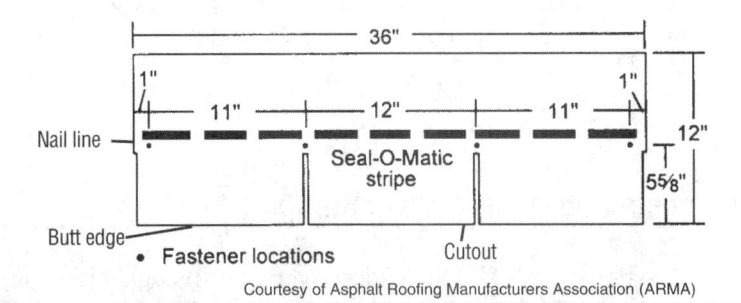

Figure 4-6   Three-tab Fire-Glass III fiberglass shingle

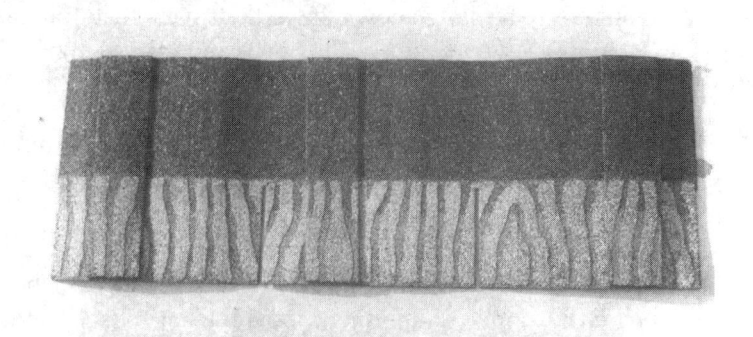

Figure 4-7   Laminated shingle

75

Asphalt strip shingles weigh from 135 to 390 pounds per square depending on:

- the shape of the shingle
- the thickness of the base mat
- the amount of asphalt absorbed by the base mat
- the thickness of the asphalt coating
- the amount of surface material pressed into the exposed face.

Shingle type	Approximate weight per square (pounds)
Two-tab strip	300
Three-tab strip	235
Individual Dutch lap	165
Individual American	330

Figure 4-8 Approximate asphalt shingle weights

You can use Figure 4-8 to estimate approximate asphalt shingle weights.

The chart in Figure 4-9 is a quick reference to the specs and coverage for several typical asphalt shingles. Asphalt shingles are from $35^{5}/_{16}$ to 42 inches long, and from 12 to $14^{1}/_{8}$ inches wide. They have recommended exposures of 4 to 6 inches. (Exposure is the part of the shingle not covered by the next course of shingles.) The most common asphalt strip shingle is 3 feet by 1 foot laid at a 5-inch exposure.

The first edition of this manual included description of and installation methods for some types of individual shingles no longer in use or even available today. Individual "hex" shingles, interlocking (T-lock) shingles and giant individual shingles have for the most part gone the way of the dinosaurs, though some may still be available in Europe. You almost certainly won't be installing any, though if you work in an area with older houses, you may be tearing them off.

Installing Asphalt Strip Shingles

After you've laid the underlayment, drip edge and valley flashing, you're ready to install a starter course at the eaves of the roof. The starter course protects the eaves of the roof by filling in the spaces under the cutouts and joints of the first course of shingles. Without a starter course, there would only be single coverage at the eaves. Install the starter course with a $^{1}/_{4}$- to $^{3}/_{8}$-inch overhang at the eaves and rakes.

■ **Roll Roofing Starter Course** Although I don't recommend it, you can install a starter course by using a 7-inch-wide (minimum) strip of mineral-surfaced roll roofing whose color matches the shingles (Figure 4-10). Place the starter roll along the eaves with a $^{1}/_{4}$- to $^{3}/_{8}$-inch overhang and nail the strip on 12-inch centers. Drive the nails along a line 3 to 4 inches above the eaves. If you're installing the starter roll over board sheathing, *stagger nail* to prevent splitting a board. That means don't hammer nails in a straight line along the grain of the board.

PRODUCT	Configuration	Per Square Approximate Shipping Weight	Per Square Shingles	Per Square Bundles	Size Width	Size Length	Exposure	ASTM* fire and wind ratings
Self-sealing random-tab strip shingle 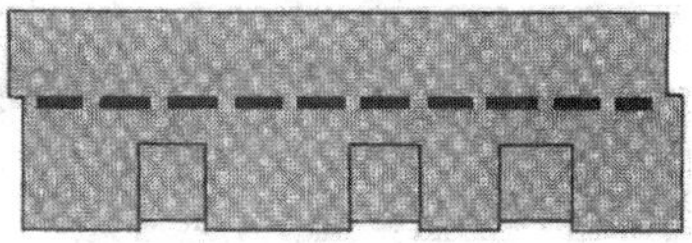 Multi-thickness	Various edge, surface texture and application treatments	240# to 360#	64 to 90	3, 4 or 5	11½" to 14"	36" to 40"	4" to 6"	A or C - Many wind resistant
Self-sealing random-tab strip shingle 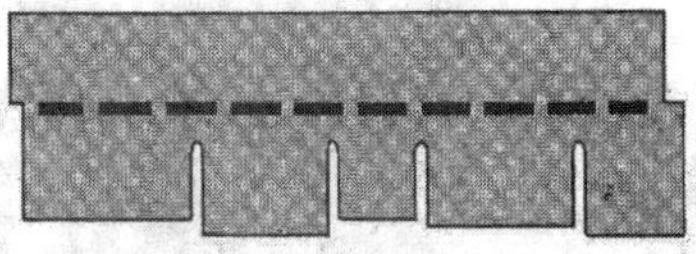 Single-thickness	Various edge, surface texture and application treatments	240# to 300#	65 to 80	3 or 4	12" to 13¼"	36" to 40"	4" to 5⅝"	A or C - Many wind resistant
Self-sealing square-tab strip shingle 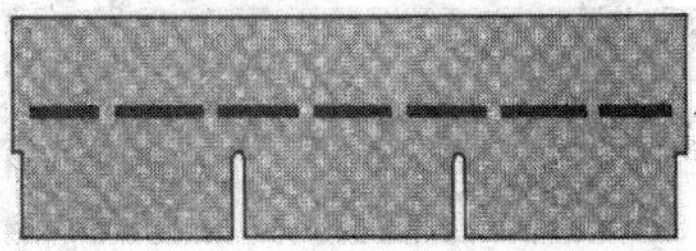 Three-tab	Three-tab or Four-tab	200# to 300#	65 to 80	3 or 4	12" to 13¼"	36" to 40"	5" to 5⅝"	A or C - All wind resistant
Self-sealing square-tab strip shingle 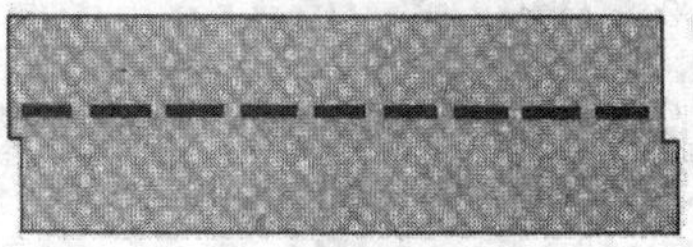 No-cutout	Various edge and surface texture treatments	200# to 300#	65 to 81	3 or 4	12" to 13¼"	36" to 40"	5" to 5⅝"	A or C - All wind resistant

*American Society for Testing and Materials

Courtesy of Asphalt Roofing Manufacturers Association (ARMA)

Figure 4-9 Typical asphalt shingles

Roll roofing comes in 36-foot lengths. If you need more than one strip to cover the length of the eaves, lap the end joint at least 2 inches. Nail the underlay, then embed the overlap in roofing cement and nail it in place with three nails.

I recommend you use shingles rather than roll roofing for the starter course. That way you don't have to worry about matching colors, and the laps won't show through the overlaying shingles.

■ **Shingle Starter Course** Most strip shingle manufacturers recommend that you make the starter course by cutting off the shingle tabs and installing the shingles with the factory-applied adhesive along the eaves, as shown in Figure 4-11.

Trim about 3 inches from the end of the first starter-course shingle to keep the joints of the first course of shingles from lining up with the joints of the starter-course shingles. That's shown in Figure 4-12. Position the starter-course shingles along the eaves with a $1/4$- to $3/8$-inch overhang. Drive nails into the shingles along a line 3 to 4 inches above the eaves. Position the nails so that they won't be exposed under the cutouts of the shingles in the first course. Stagger nail the starter-course shingles over board sheathing.

If you use roll roofing or shingles that don't have a factory-applied adhesive strip (free-tab shingles) for the starter course, bond the tabs of each shingle in the first course to the starter strip. Use a spot of roofing cement about the size of a quarter beneath each shingle tab. Figure 4-13 shows this. Install *all* free-tab shingles this way in high-wind areas. That includes the starter course, even when you use shingles that *do* have factory-applied adhesive strips.

■ **Start with a Straight Line** It's very important that you install the starter course and first course of asphalt shingles straight. To align asphalt shingles, nail down a shingle with the correct overhang on each end of the eaves. Snap a chalk line along the top edges of the shingles as shown in Figure 4-14. Then line up the top edges of intervening shingles along the chalk line. Repeat this alignment every third or fourth course. Measure from the eaves up to the butt position for the next course of shingles at the rakes. Install the end shingles, then snap another chalk line to align that course. The first course of random-tab strip shingles is installed the same way, except that you first need to trim the longer tabs to the length of the shortest tab so the bottom edge is level.

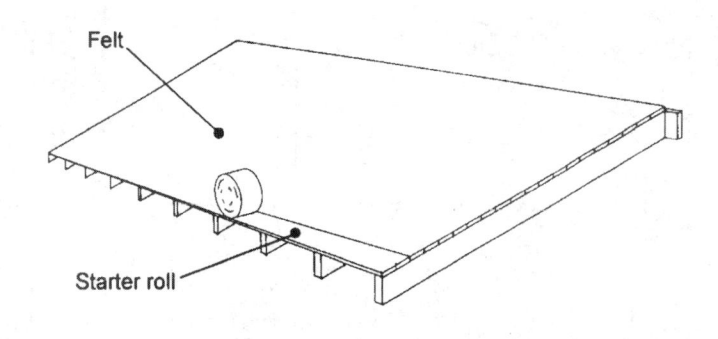

Figure 4-10 Mineral-surfaced starter roll

Figure 4-11 Field-fabricated starter course

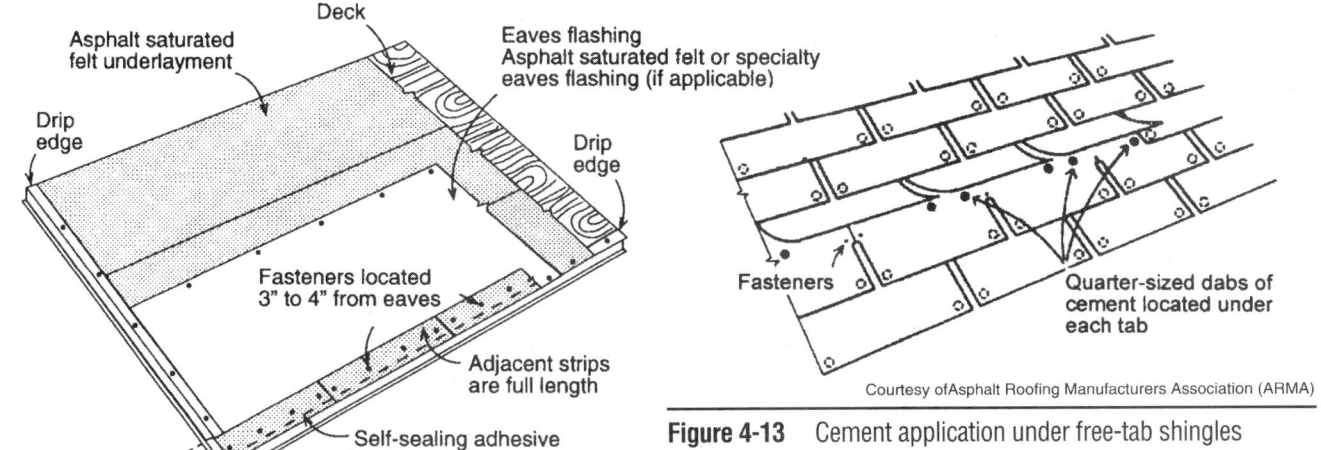

Asphalt Shingles

Figure 4-12 Application of starter strip

Figure 4-13 Cement application under free-tab shingles

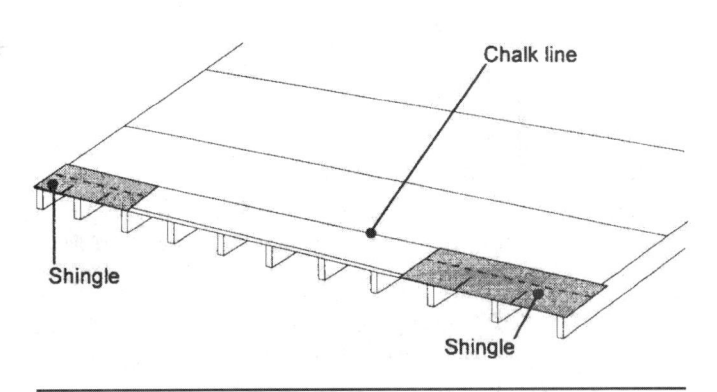

Figure 4-14 Lining up the first course

You can save time by snapping all the chalk lines before you install any shingles. Snap horizontal chalk lines on 5-inch centers (assuming a 5-inch shingle exposure), allowing for the overhang at the eaves. Then snap vertical chalk lines on 6-, 12- or 36-inch centers, depending on how good you are at eyeballing a straight line. Once you've done this, you can line up shingles at the correct positions with the proper exposure without having to use the exposure gauge on your hatchet. Be sure to allow for the required overhang when you snap your chalk lines.

For example, if your exposure is 5 inches with a 1/4-inch overhang, snap your first horizontal chalk line at 11 3/4 inches from the edge of the eaves. Snap succeeding chalk lines 5 inches apart.

To maintain the correct exposure for square-tab strip shingles, align the butts with the top of the cutouts in the course below, since the cutouts in these shingles are 5 inches deep.

■ **Shingle Patterns** There are three basic shingling patterns used to install 3-tab asphalt strip shingles.

a) joints broken into halves, or the 6-inch pattern (half pattern)

b) the 5-inch pattern (random pattern)

c) joints broken into thirds, or the 4-inch pattern

To install the 6-inch pattern, start the first course with a full-length shingle. Remove 6 inches from the first shingle of the second course. Then remove 12 inches from the first shingle of the third course. Continue,

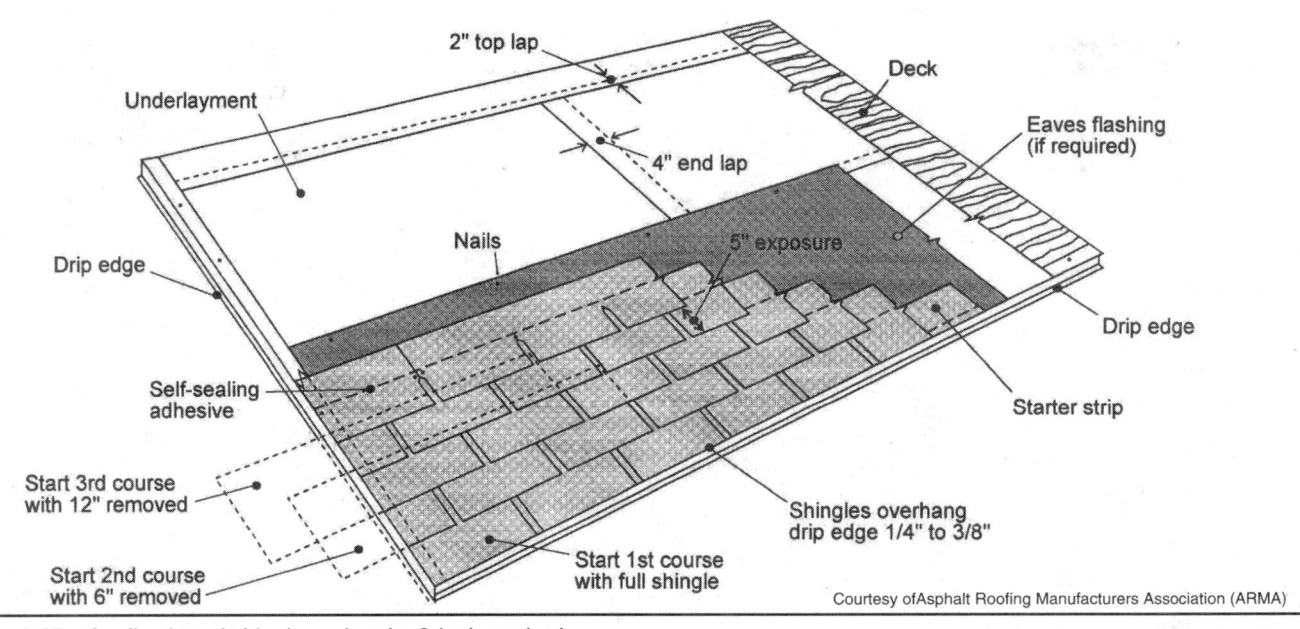

Figure 4-15 Application of shingles using the 6-inch method

removing an additional 6 inches from the first shingle of each course until you begin with a full shingle again on the seventh course. You can see how this works in Figure 4-15.

Save the full tabs you cut off and use them for hip and ridge units, filler tabs adjacent to valleys, and at the opposite ends of a gable-framed roof. The 6-inch pattern is the simplest style to install. But, because you align the cutouts every other course and the shingles vary slightly in size, you must snap chalk lines up the roof slope so you can align the edges of the shingles to keep the cutouts lined up vertically. The easiest way to install the 6-inch pattern on a gable roof is to shingle up the rake and install each course only far enough out over the deck to establish a pattern. When you get to the ridge, return to the bottom and finish out each course across the roof, working your way up the slope. This method of shingling up the rakes followed by shingling across the roof is called the *diagonal* method.

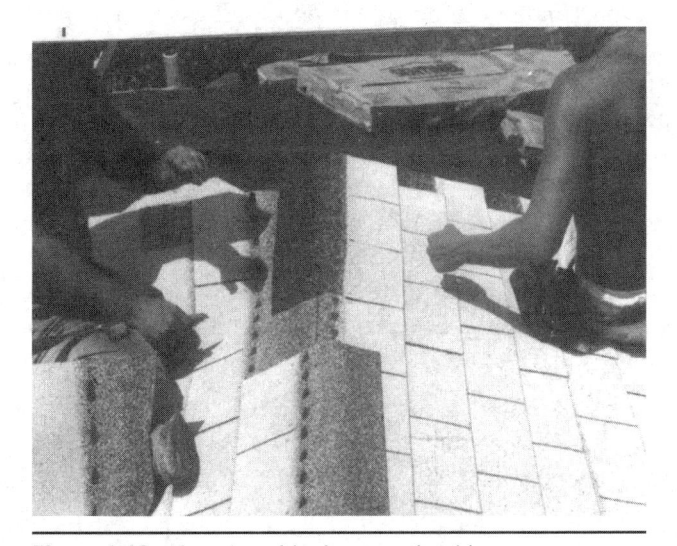

Figure 4-16 Lapping shingles over the ridge

If the top of a shingle extends beyond the centerline of the ridge, lap the shingle over the ridge and nail it on both sides of the ridge, as in Figure 4-16.

The 5-inch pattern is often called a random pattern. This pattern gives you some flexibility when you align the cutouts up the roof slope. Start the first course with a full-length shingle. Remove 5 inches from the first shingle of the second course. Then remove 10 inches from the first shingle of the third course. Continue, removing an additional 5 inches from the first shingle of each course until you begin with a full shingle again on the eighth course. (You don't start the eighth course with a 1-inch section.) Use the exposure gauge on your hatchet to measure the 5-inch increments. Figure 4-17 shows the shingle pattern this method produces. On gable roofs, shingle up the rake to the ridge, as with the 6-inch pattern, then return to the bottom and finish out each course across the roof, working your way up the slope.

You'd usually use the 4-inch pattern only on low-slope roofs ranging from 2 through 3 in 12. Trim the first shingle of each course in a multiple of 4 inches, beginning again with a full-width shingle at the 10th course. Figure 4-18 shows this pattern.

The 5-inch pattern is sometimes used on hip roofs, while the 6-inch pattern is generally used on gable roofs. Never install a shingle pattern less than 4 inches because the cutouts and joints would be so close on adjacent courses that leaks could occur.

Shingle Application

The order you follow to install shingles depends on the roof style. On gable roofs broken by dormers or valleys, start shingling at the rake and proceed toward the breaks. On simple gable roofs, start shingling at the gable end that's most visible to passers-by. On hip roofs and roofs where both gable ends are equally visible, start shingling at the center of the roof and proceed in both directions. In this case, set all your chalk lines (for the offset pattern you'll use) before you begin shingling.

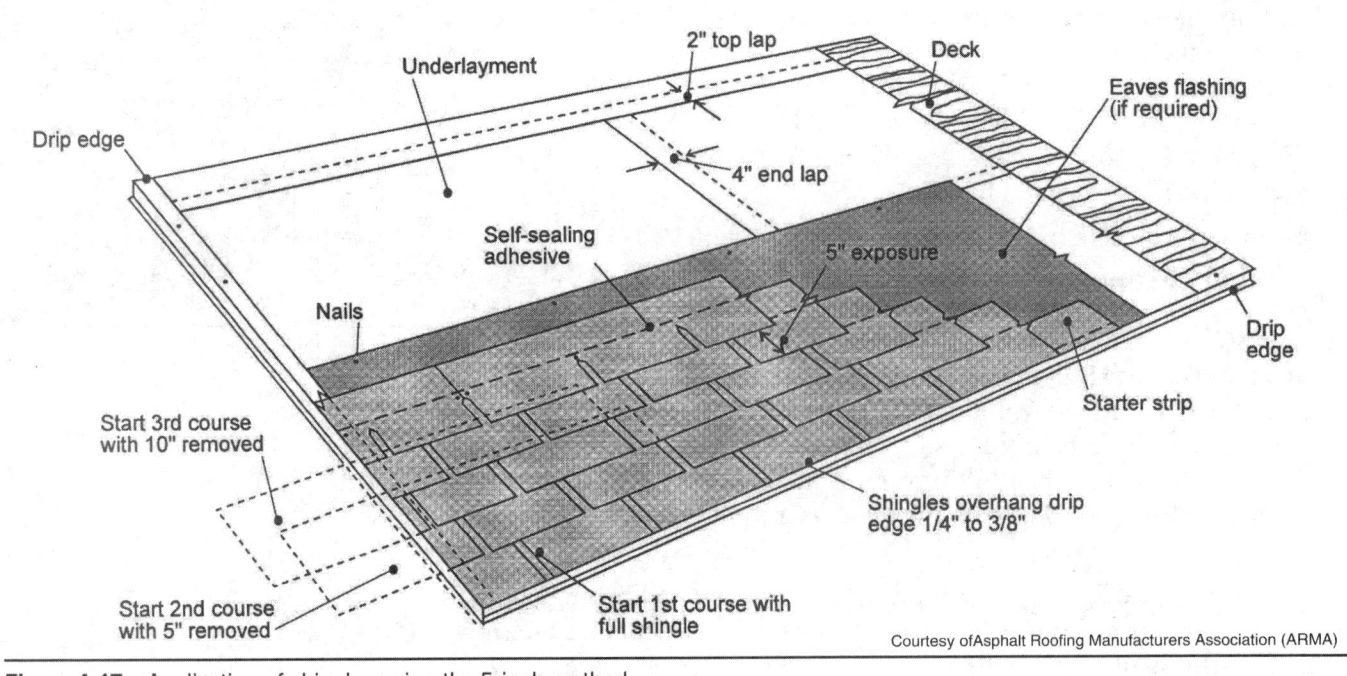

Figure 4-17 Application of shingles using the 5-inch method

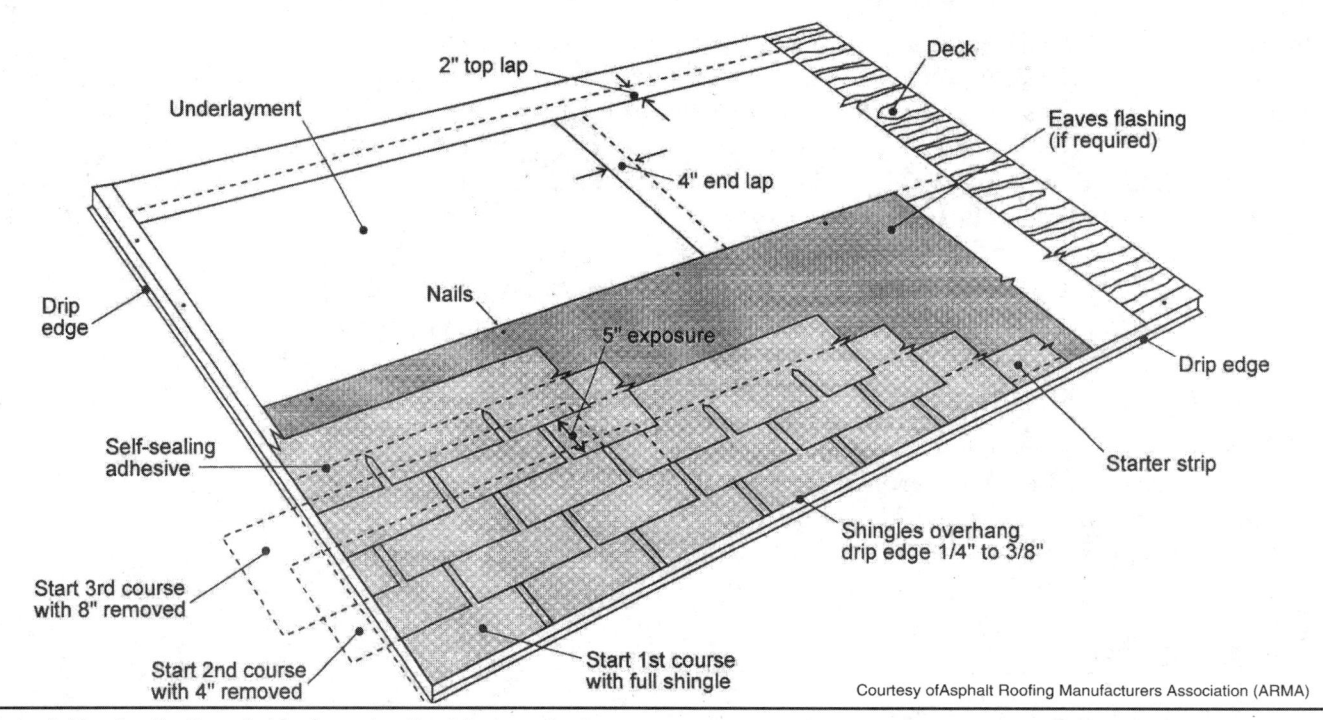

Figure 4-18 Application of shingles using the 4-inch method

On hip roofs, lap shingles over the hips from both sides, as shown in Figure 4-19. Then cut the shingle edges of the upper layer in line with the centerline of the hip. See Figures 4-20 and 4-21.

■ **Dormers** If there's a dormer, shingle the top of it first. Then bring the shingles of the main roof up to and alongside the dormer, all the way to the ridge of the dormer. Extend one shingle course on the main roof on one side of the dormer to a distance at least one shingle beyond the ridge of the dormer roof. Notice the top shingle on the left side of the dormer in Figure 4-22.

Snap vertical chalk lines down from the ridge starting with the edge of the extended shingle, as shown in Figure 4-22. Use those chalk lines as guides to align the shingle courses as you install them on the right side of the dormer. Slip the last shingle course under the course that's in line with the ridge of the dormer. Aligning shingles on both sides of a dormer this way is called "tying in." You'll have to shingle the ridge of the dormer before you finish the main roof above the dormer. We'll describe that later in this chapter, on page 89 under the heading, "Ridge and Hip Units, Cap Shingles."

Figure 4-19 Lapping shingles over a hip

No matter where you begin shingling, roofing material manufacturers recommend you apply the shingles in the diagonal pattern described earlier. Then you'll be sure you've nailed every shingle properly because you can see each one until you cover it with the next course above.

Asphalt Shingles

Figure 4-20 Trimming the hip

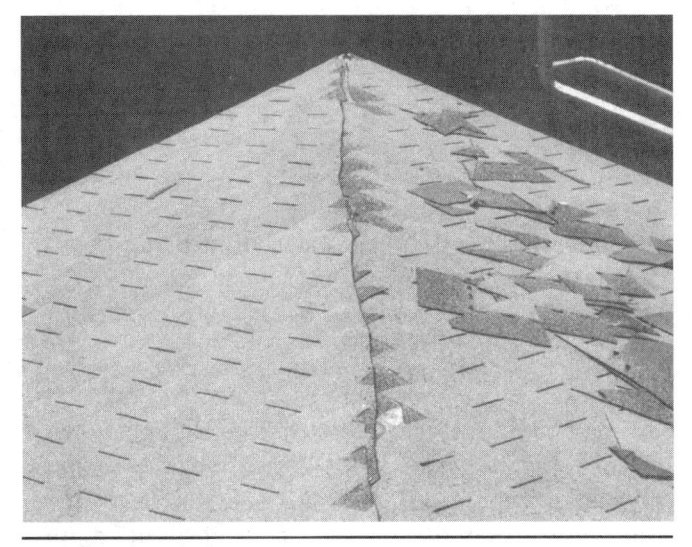

Figure 4-21 A trimmed hip

You can also use the straight-up (racking) method shown in Figure 4-23. But then you have to install some shingles under shingles you've already laid in the course above. That's shown in Figure 4-24. Since part of the underlying shingle is hidden, there's a possibility you could miss nailing that part of the shingle.

Some roofing contractors prefer the racking method because it's a more accurate way to align the shingles. You use the horizontal chalk lines and previously-laid shingle edges as guidelines. If you use this method, snap horizontal chalk lines on 5-inch centers starting at the eaves and allowing for an overhang. Then, snap two vertical chalk lines 6 inches apart. Install shingles up the roof offsetting every other course 6 inches, aligning them with the vertical chalk lines.

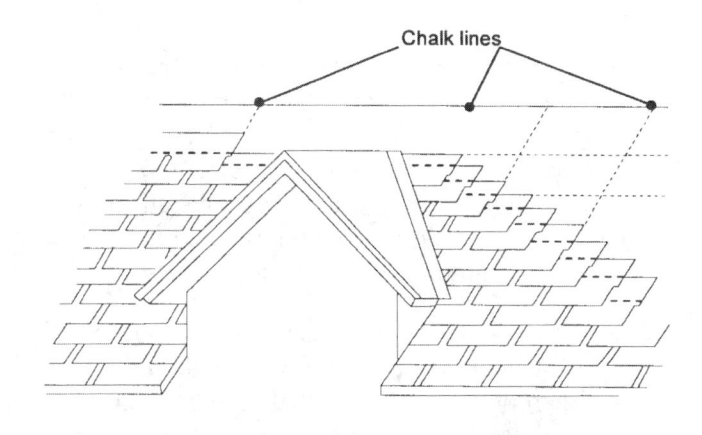

Courtesy of Asphalt Roofing Manufacturers Association (ARMA)

Figure 4-22 Tying in around a dormer

Figure 4-23 The straight-up (racking) method

Figure 4-24 Installing a shingle beneath one previously installed

83

That's shown in Figure 4-25. Then shingle the rest of the roof the same way, using the horizontal chalk lines and previously-installed shingle edges as guidelines.

■ **Patterning** Offset the joints in adjacent courses of 3-tab shingles to keep water from being channeled through the joints, where it can get under the shingles. Offset the joints of laminated shingles for the same reason.

To form a random pattern, start the first course at the rake with a full-length dimensional shingle. Then remove 4 inches from the first shingle of the second course, and 11 inches from the first shingle of the third course. Start the fourth course with a full-length shingle and repeat the pattern every third course. Finish the remainder of each course with full-length shingles.

By using this method, you won't get an obvious and unattractive repeated pattern throughout the roof. If you install dimensional shingles using the 6-inch pattern, you'll get repeated diagonal trails like the ones in Figure 4-26. If you install dimensional shingles using the racking method you'll get repeated vertical trails.

■ **Shading** Asphalt shingles sold as one color won't match perfectly. Some will look lighter or darker than others. This is called "shading" and it's due to the way they were made. It can also happen if the shingles have been stored too long, or in stacks so tall that backing material of one shingle rubs off onto the face of another. If you use the racking method, (straight-up application) you'll accent the shading. Use the diagonal method of application to help blend the shingles.

Figure 4-25 Vertical chalk lines used with the racking method

Figure 4-26 Repeated diagonal pattern

Low or High Slopes

In general, install asphalt strip shingles only on roof slopes of 4 in 12 and steeper. You can install square-tab shingles on slopes as low as 2 in 12 (but never lower) if you follow special application procedures. The primary requirement is that you install the proper underlayment and eaves flashing (if required) to prevent damage caused by ice dams. Refer back to Chapter 3 for details on ice dams. Also, for added wind resistance, use shingles with self-sealing factory-applied adhesive strips or apply a spot of roofing cement about the size of a quarter under every shingle tab. Use cement sparingly. Too much cement can cause blisters.

Asphalt Shingles

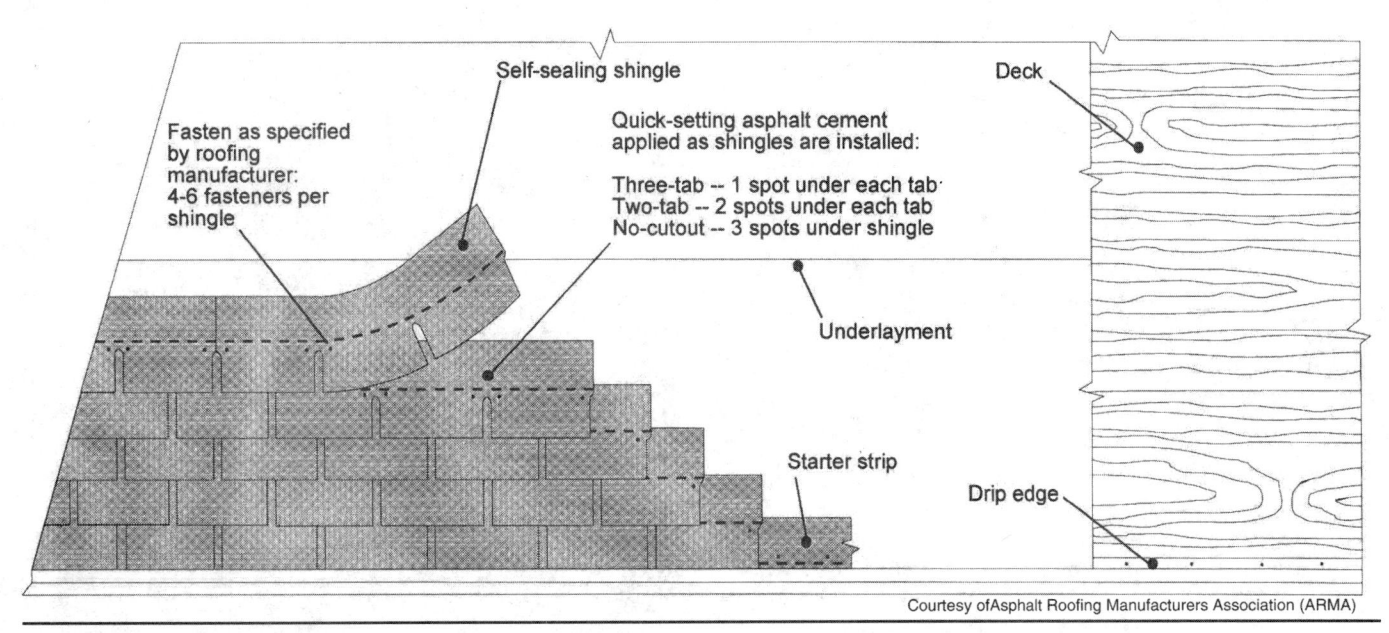

Figure 4-27 Application of shingles on steep slopes

Normally, you don't install asphalt strip shingles on roof slopes steeper than 21 in 12. The main problem is that the factory-applied self-sealing adhesive strip isn't very effective, especially on colder or shaded portions of the roof. However, you can install asphalt strip shingles on steeper slopes if you follow modified application procedures. Depending on the manufacturer's specifications, install each shingle with 4 to 6 fasteners.

Use roofing cement to attach shingle tabs to underlying shingles, Apply the cement in spots about the size of a quarter.

- For shingles with three or more tabs, apply a spot of cement under each tab.

- For 2-tab shingles, apply two spots of cement under each tab.

- For no-cutout shingles, apply three spots of cement under the exposed portion of each shingle.

Figure 4-27 shows this.

Installing Asphalt Strip Shingles in Valleys

The three main types of valley are:

1) open (Figure 4-28 on the following page)

2) closed-cut (half-lace) (look ahead to Figures 4-32 and 4-34)

3) woven (full-lace) (look ahead to Figures 4-38 and 4-39)

The valleys of aggregate-surfaced roofs are usually made of underlayment covered with aggregate embedded in bitumen. Turn back to Chapter 3 for information on valley flashing requirements. Never install a vent pipe or any other roof penetration in a valley.

■ **Open Valleys** Although Figure 4-28 shows it, I don't recommend you use open-valley construction on roofs with 3-tab asphalt shingles. Open valleys are more likely to leak than other types of valleys. The valley can get clogged by leaves, twigs, pine needles or other debris and cause a backup. Or water may be forced up under shingles adjacent to the valley during a heavy rain.

To construct an open valley, install shingles at the upper end of an open valley up to within 3 inches on each side of the centerline of the valley. Widen this distance by about $1/8$-inch per foot going down the valley. You need to make this area wider because, as a stream of water flows down a valley, the stream will get wider. This widening is helpful because it lets ice free itself and slide down the valley as it melts.

Trim 1 to 2 inches from the upper corner of the last shingle in each course in the valley at a 45-degree angle. This is to direct water into the valley and not between the shingle courses. That's called "dubbing," and it's shown in Figure 4-28. In addition, you should cement the end of the shingle to the valley flashing with a 3-inch width of roofing cement. Don't allow exposed nails along the valley flashing. I also recommend dubbing-off and cementing shingle corners in closed-out and woven valleys.

■ **Open Valleys at Dormer Roofs** Install dormer valley flashing after you've installed the shingles on the main roof deck up to a point just above the lower end of the dormer valley. Figure 4-29 shows this. Then install the valley flashing. Trim the lower part of the flashing so that it goes at least 2 inches below where the two roof decks meet. Also, trim the flashing so it overlaps the uppermost shingles (the ones you installed before) down to the top of the cutouts. In addition, cut a small arc in the flashing where the dormer and main roof decks meet, as shown in Figure 4-30.

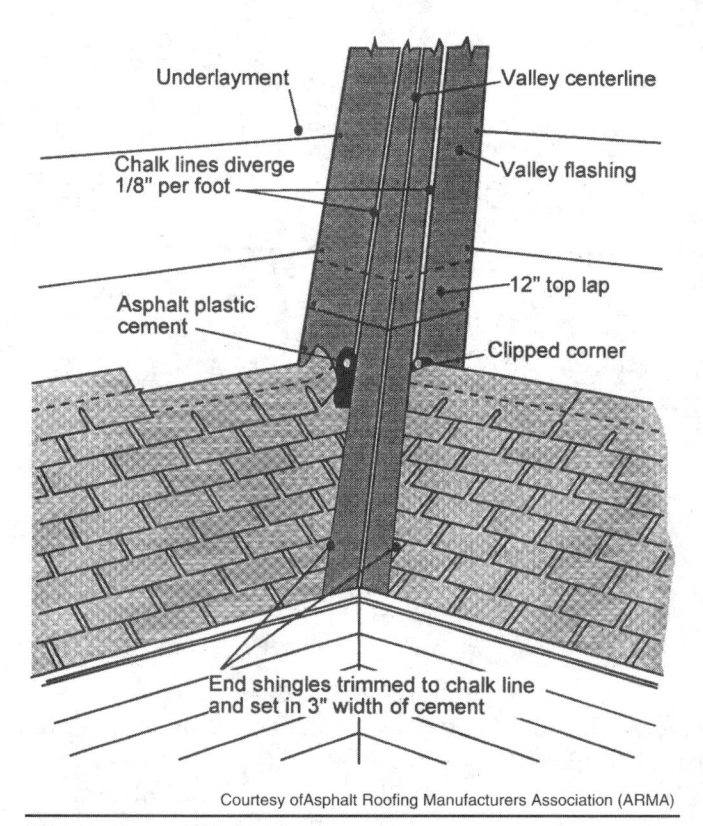

Figure 4-28 Application of shingles in an open valley

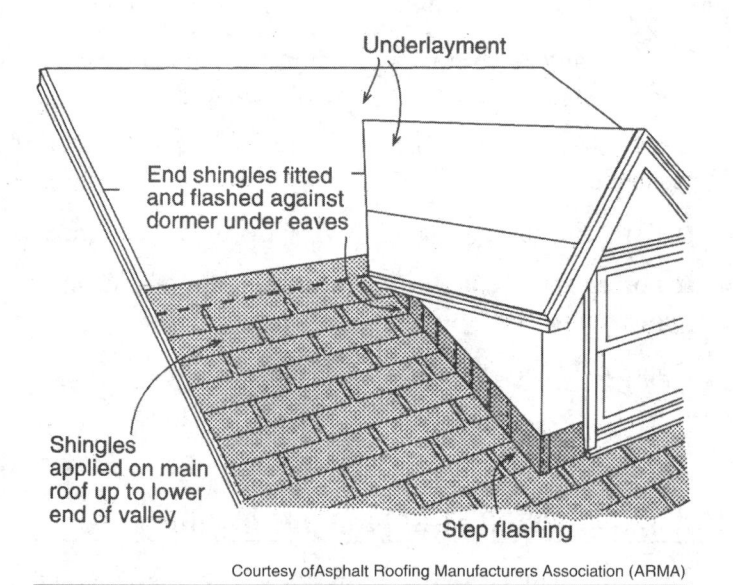

Figure 4-29 Point at which installation of open valley at dormer roof begins

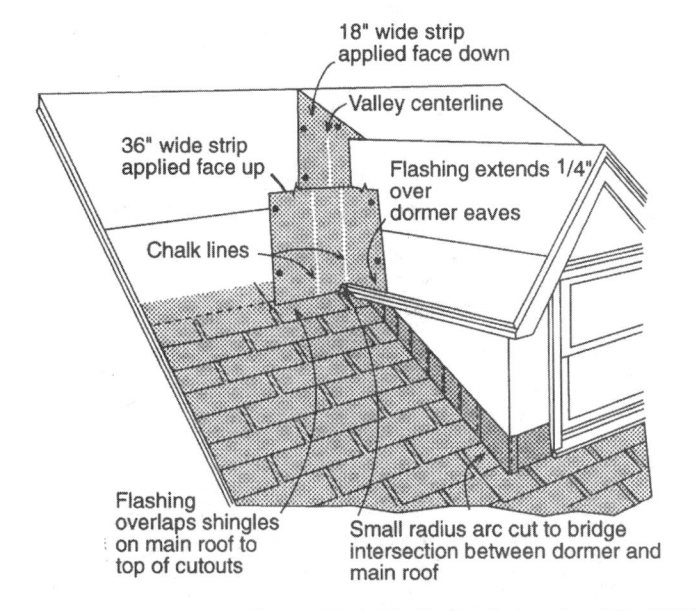

Asphalt Shingles

Figure 4-30 Application of roll roofing as flashing for an open valley at a dormer roof

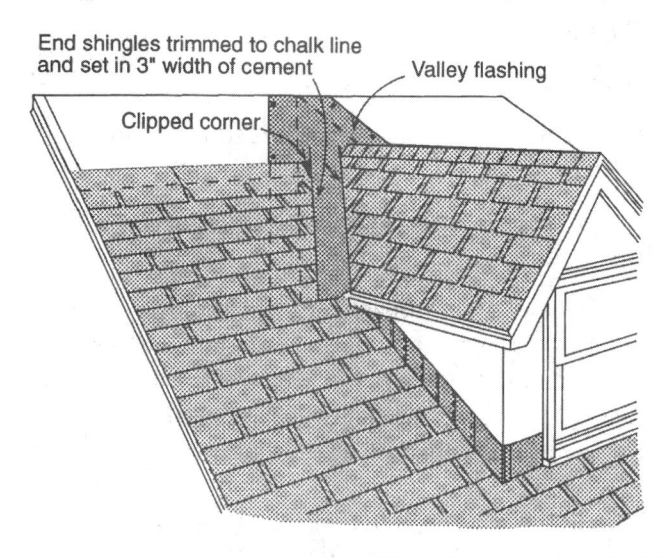

Figure 4-31 Application of shingles in open valley at dormer roof

Overlap, trim, cement, and nail down the upper part of the flashing above the dormer ridge. Then install shingles over the main roof deck and dormer roof as shown in Figure 4-31.

■ **Closed-Cut Valleys** I prefer the closed-cut (half-lace) valley construction shown in Figure 4-32 over woven (full-lace) valley construction because it looks neater and more professional. And, you can usually install this type of valley faster because you can shingle each side of the valley independently.

Install each shingle course along the eaves of one side of the valley and at least 12 inches across to the other side (Figures 4-33 and 4-34). Make sure the shingle end joints are at least 10 inches from the centerline of a closed-cut or woven valley. To keep a joint from ending up in a valley, insert an individual 12-inch-wide tab within a shingle course on either side of the valley. That's shown in Figures 4-35 and 4-36. Use two fasteners to secure the end of each shingle you install across the valley.

Figure 4-32 A closed-cut valley has a neat appearance

Next, apply shingles to the other side of the valley, extending them beyond the valley and over the shingles you just laid. Then, trim the overlying shingles back 2 inches from the centerline of the valley, as shown in Figure 4-37. Snap a chalk line and use it for a cutting guide. Also, trim 1 inch from

Roofing Construction & Estimating

Figure 4-33 Shingle at least 12 inches onto the adjoining roof plane. Note the roofing cement.

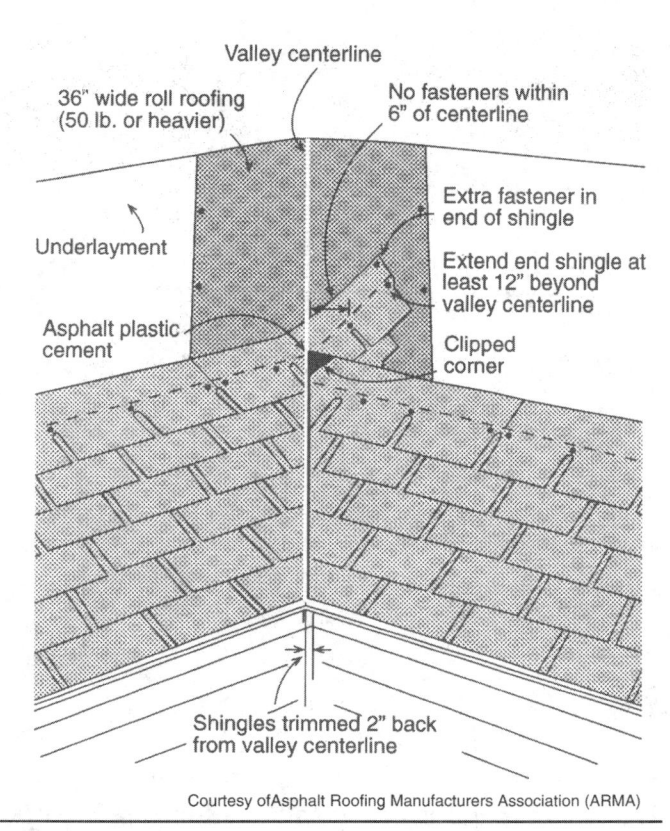

Figure 4-34 Application of shingles in a closed-cut valley

Courtesy of Asphalt Roofing Manufacturers Association (ARMA)

Figure 4-35 An inserted single tab prevents a joint from falling within the valley

Figure 4-36 Single 12-inch tab inserted within the shingle course

88

Asphalt Shingles

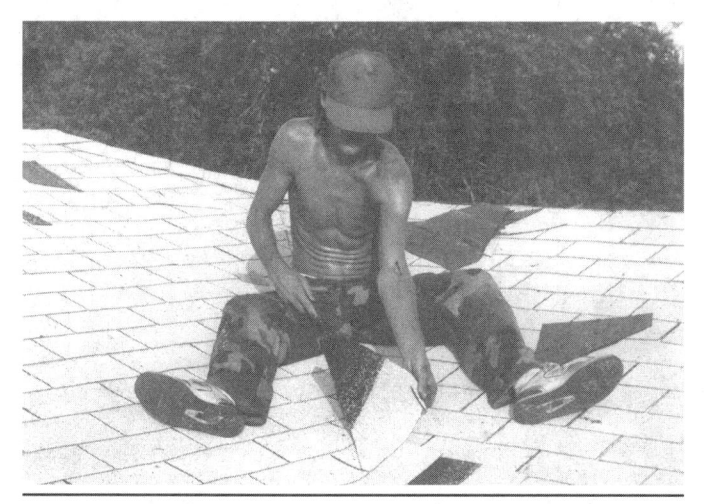

Figure 4-37 Trimming overlaying shingles in a closed-cut valley

the upper corner of the last shingle in each course at a 45-degree angle to direct water into the valley and not between the shingle courses. In addition, cement the end of the shingle with a 3-inch width of roofing cement. Refer back to Figure 4-33.

■ **Woven Valleys** To install shingles into a woven (full-lace) valley, apply them alternately to both sides of the valley. Extend the shingles across the valley, and at least 12 inches on each side. As with a closed-cut valley, be sure the shingle end joints are at least 10 inches from the centerline of the valley. Also, secure the end of each shingle that goes across the valley with two fasteners. See Figure 4-38.

It's best to use woven valleys only when the roof slope is 3 in 12 or steeper. Even though you don't have to trim the shingles when you make a woven valley, it'll still take you longer to install it. That's because you have to work both sides of the valley at the same time. I don't like this type of valley because it doesn't look clean and professional. See Figure 4-39. What do you think?

■ **Ridge and Hip Units (Cap Shingles)** Some asphalt shingles come with a prefabricated ridge roll or prefabricated individual 12" x 12" units. The advantage of the prefab units is that they save you time — all you have to do is install them. Sometimes, when you use laminated 4-tab and 6-tab shingles, you also need to use special hip and ridge shingles. You can field-fabricate the cap shingles from the same material as the rest of the roof, but as just stated, the factory-supplied units save time.

Courtesy of Asphalt Roofing Manufacturers Association (ARMA)

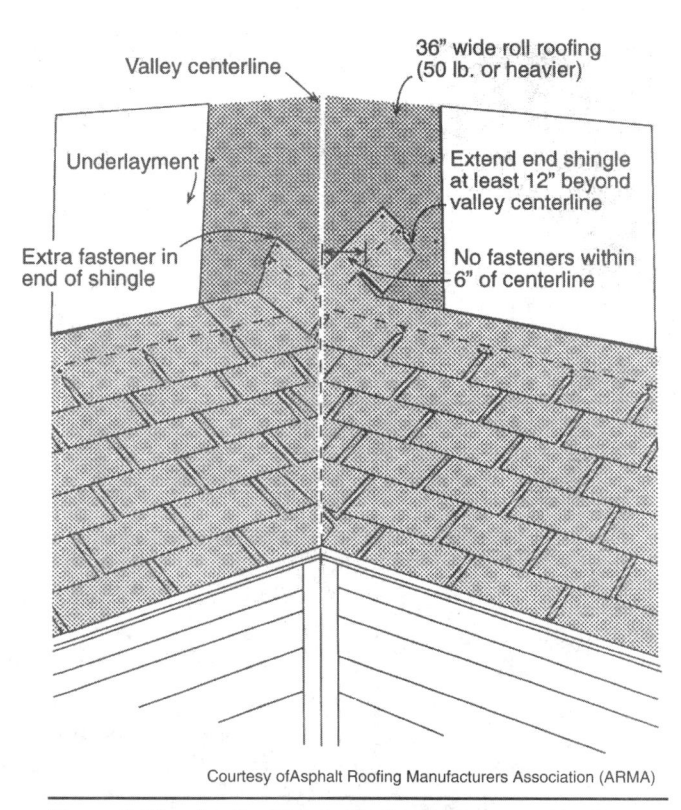

Figure 4-38 Application of shingles in a woven valley

Figure 4-39 A woven valley

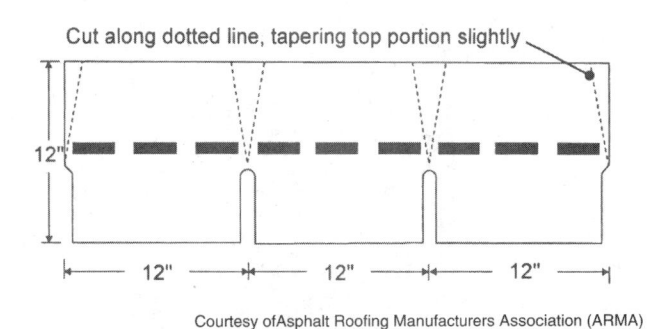

Figure 4-40 Fabrication of hip and ridge shingles from 3-tab strip shingles

With 2- and 3-tab shingles, or shingles with no tabs, you can cut hip and ridge units from standard shingles. Cut a 3-tab shingle down to three 12" x 12" units as in Figure 4-40. To get a neat, professional look, taper the lap portion of each unit so it's slightly narrower than the exposed part, as shown in Figure 4-41. To make cap shingles from 2-tab or no-tab shingles, trim units to a minimum of 9" x 12". Salvage parts of shingles left over from the rakes, hips, and valleys, and make them into cap shingles.

Install the hip units before you install the ridge units. Start shingling the hips at the eaves and work up slope toward the ridge. In high-wind areas, use roofing cement to secure the first hip unit. Trim the first hip unit so its edges overhang the eaves by $1/4$ to $3/8$ inch, depending on the overhang you allowed for the starter course. Then temporarily tack another hip unit at the top of the hip. Snap a chalk line down the hip aligned with one or both edges of the two units as a guide for intervening hip units. That's shown in Figure 4-42. Trim the top hip units where they meet at the ridge, as shown in Figure 4-43.

To cap the end of a ridge above the hips, nail down the end ridge shingle as shown in Figure 4-44 and cut about 6 inches through the center of the shingle tab. Then nail down one flap, as shown in Figure 4-45 and fold the opposite flap down into a bed of roofing cement to cover the nail and seal the hip-ridge junction. See Figure 4-46.

On a gable or hip roof, install ridge units at opposite ends and snap a chalk line along one or both edges to align the intervening ridge units. Install ridge shingles over a gable roof beginning at the end of the roof facing

Figure 4-41 Cap shingles with tapered lap portions

Figure 4-42 Snap a chalk line to line up hip units

Asphalt Shingles

Figure 4-43 Trim the uppermost hip units at the ridge

Figure 4-44 Install the first ridge unit above the hips

Figure 4-45 Cut the ridge unit and nail down one flap

Figure 4-46 Seal the hip-ridge juncture

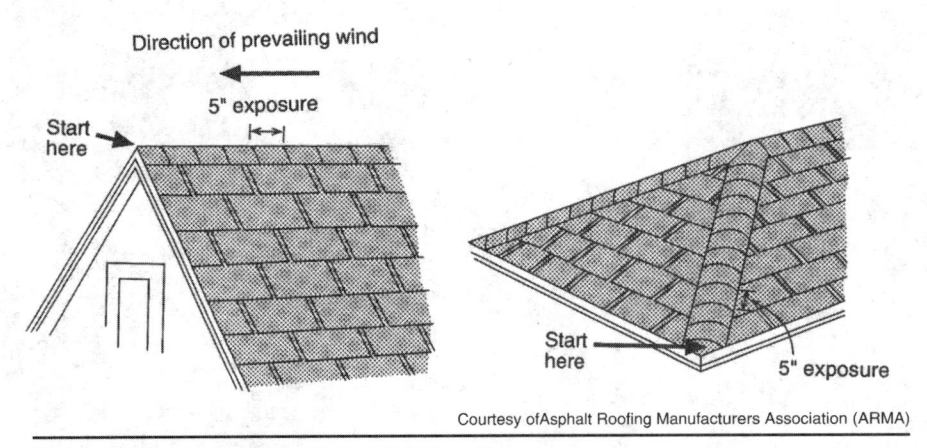

Figure 4-47 Application of hip and ridge units

into the wind, as shown in Figure 4-47. Install ridge shingles on a hip roof starting at both ends and working toward the center of the ridge. You can also follow this procedure on a gable roof. When you reach the center of the ridge, trim a shingle to use as a cap over the last ridge units. Nail the cap and cover the nails with roofing cement, as in Figure 4-48.

Install hip and ridge units at a 5-inch exposure. Secure each unit with two fasteners, one on each side. Drive the fasteners $5\frac{1}{2}$ inches back from the exposed end and 1 inch up from the edge of the shingle. See Figure 4-49.

On dormers, install the ridge units starting at the front of the dormer and working toward the main roof. Extend the last unit you install at least 4 inches onto the main roof. Split the part of the shingle that extends over the main roof down the center, and nail it into place as shown in Figure 4-50. Then cover this last dormer shingle with shingles you apply to the main

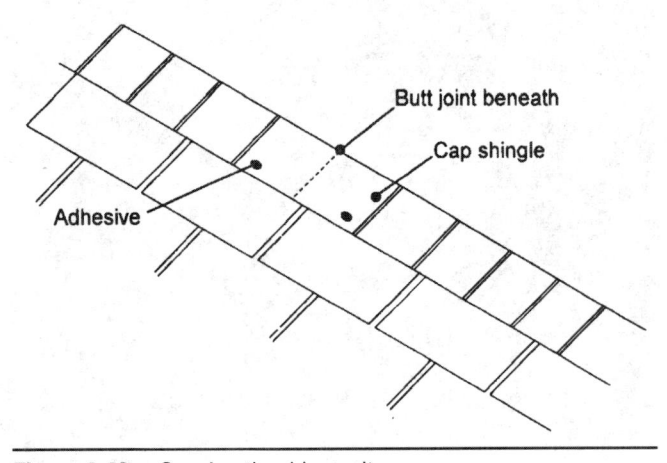

Figure 4-48 Capping the ridge units

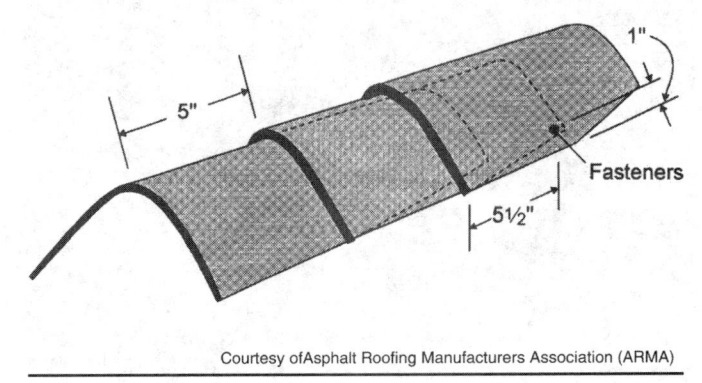

Figure 4-49 Nail location for hip and ridge shingles

Asphalt Shingles

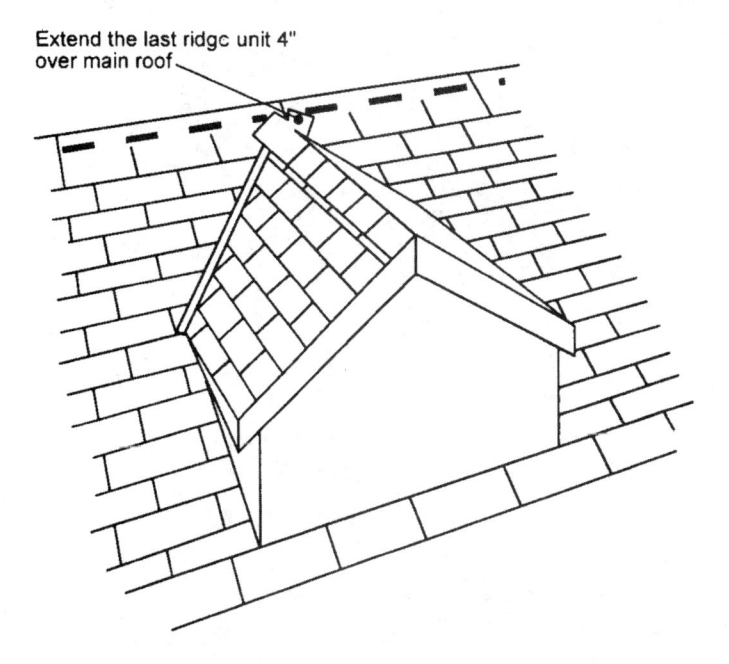

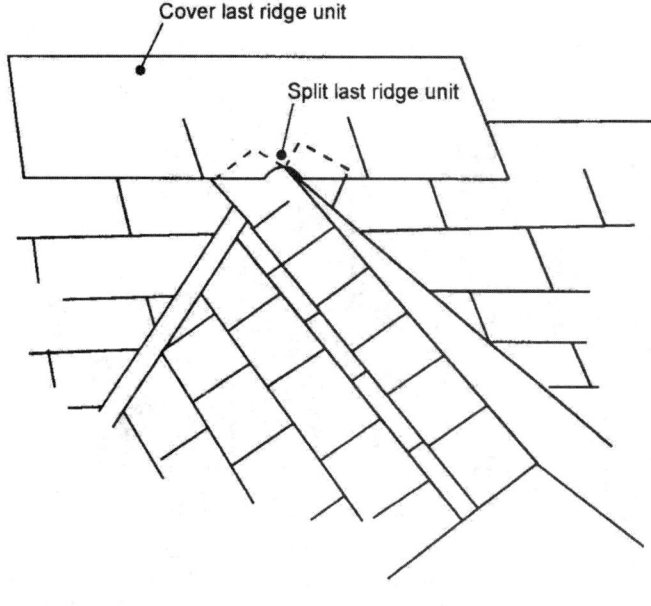

Figure 4-50 Application of ridge units over a dormer

Figure 4-51 Covering the last dormer ridge unit

roof, as shown in Figure 4-51. If a cutout in a main roof shingle falls over the dormer ridge shingle, coat the dormer shingle with roofing cement under the main roof shingle. To provide extra waterproofing at the place where the dormer ridge and the main roof meet, install 6-inch-wide strips of water shield material between the last dormer ridge unit and the shingle beneath it.

Flashing at Chimneys and Other Vertical Structures

Any flashing turned up on a vertical surface is called a *base flashing*. Flashing built into the vertical surface and bent down over the base flashing is called *counter-flashing* or *cap flashing*. All metal flashing must be corrosion-resistant and at least 0.019 inch thick.

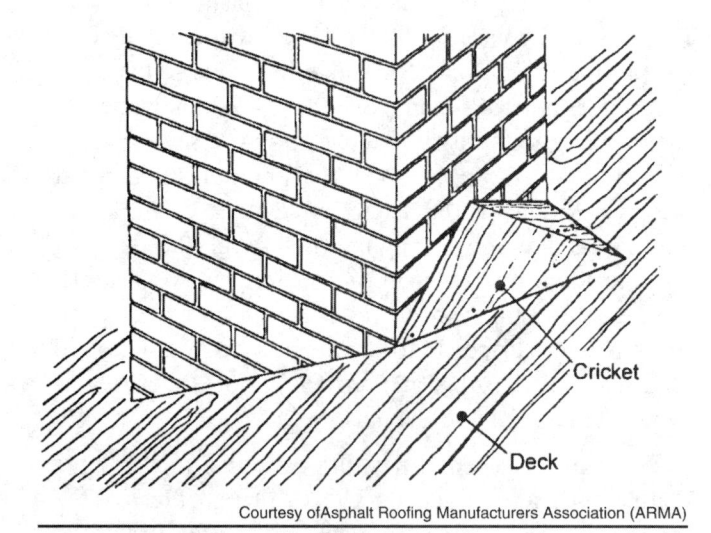

Courtesy of Asphalt Roofing Manufacturers Association (ARMA)

Figure 4-52 Location and configuration of chimney cricket

If the horizontal width of a chimney is greater than 2 feet, install a fabricated galvanized metal saddle flashing, or a wooden cricket, above the chimney, as in Figure 4-52. The cricket or saddle helps keep ice and snow from building up at the upper side of the chimney, and diverts rainwater around it. Build the saddle flashing or cricket with the same slope as the main roof.

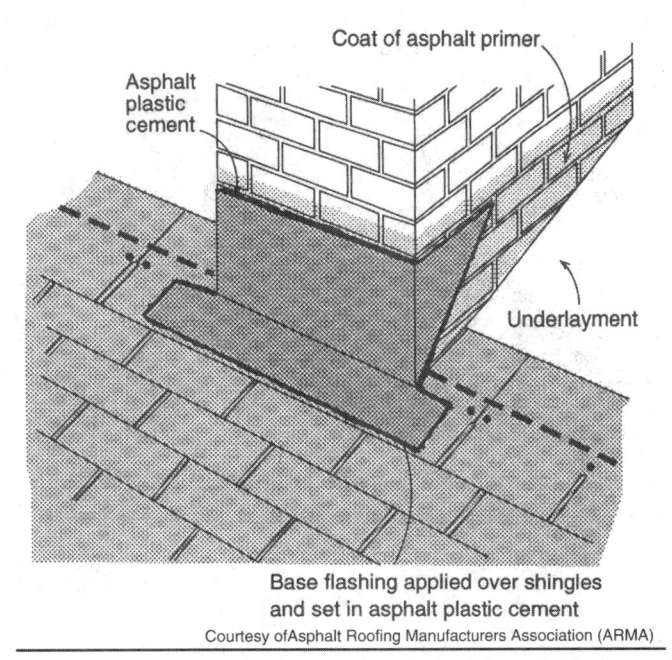

Figure 4-53 Application of base flashing at front of chimney

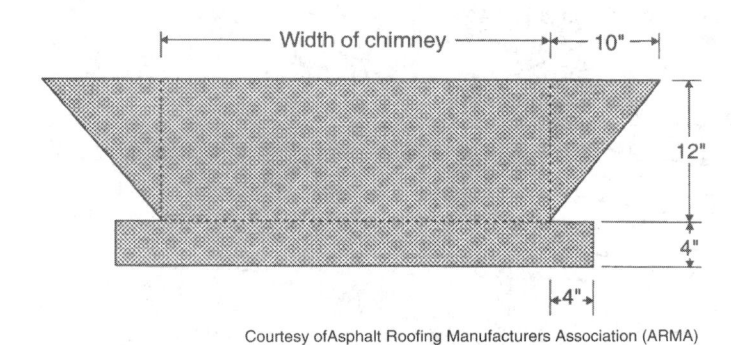

Figure 4-54 Pattern for cutting front base flashing

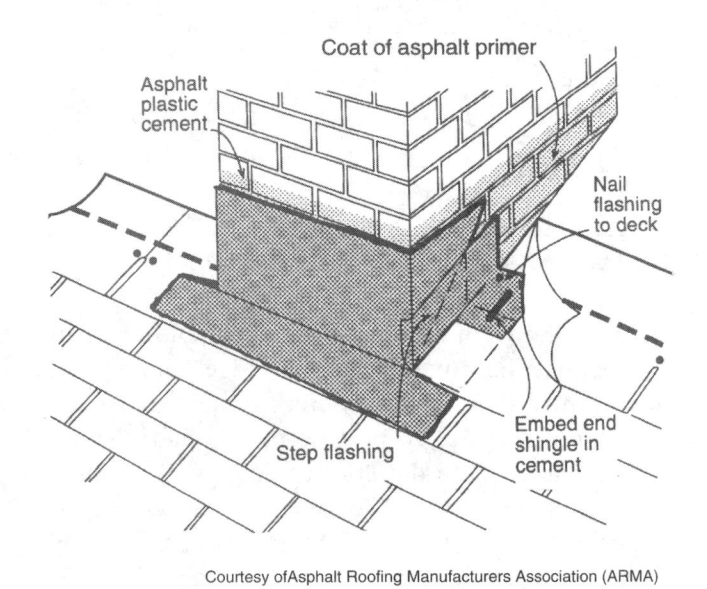

Figure 4-55 Application of base flashing at side of chimney

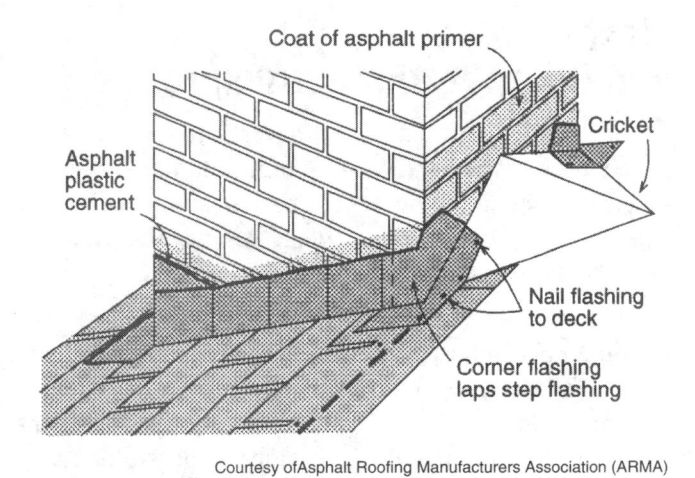

Figure 4-56 Application of corner base flashing rear of chimney

Apply asphalt shingles up to the lower edge of a chimney before you install any metal flashing material. Then install the base flashing on the down-slope face of the chimney (Figure 4-53). Make this piece so the lower part goes at least 4 inches over the shingles and the upper section goes at least 12 inches up the chimney face. See Figure 4-54. Apply a bed of asphalt plastic cement over the shingles and masonry and set the entire flashing in it. Drive only enough nails through the flashing into mortar joints to keep the flashing in place until the cement sets. Apply a coat of asphalt primer to any masonry surface before you apply roofing cement. That seals the masonry and provides good adhesion between the cement and the masonry.

You can buy special flashing cements you can use at all temperatures and on wet or dry surfaces. This cement comes in one-gallon cans or five-gallon pails.

Install metal step flashing (baby tins) and shingles at the sides of the chimney, as shown in Figure 4-55. Step flashing installation is discussed in greater detail shortly.

Install the base flashing at the rear of the chimney and over the cricket, as in Figures 4-56, 4-57, and 4-58. Extend the flashing at least 6 inches onto the roof sheathing and 6 inches up the chimney.

Install shingles over the cricket, or up to the cricket valleys. Install any shingles you apply over the cricket in a bed of asphalt plastic cement. You don't have to

Asphalt Shingles

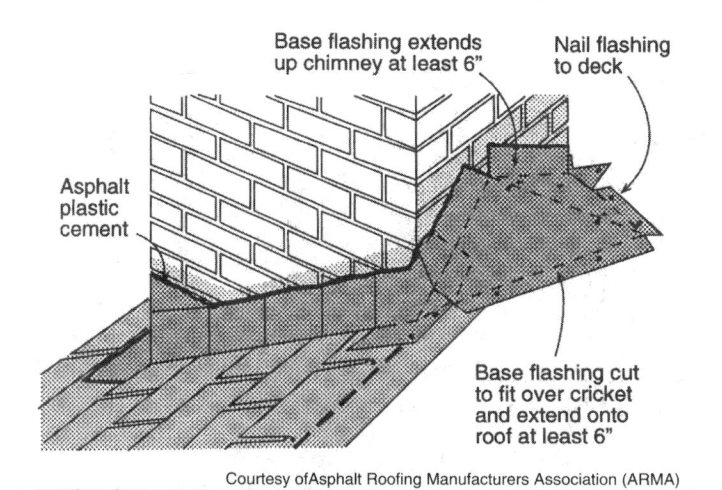

Figure 4-57 Application of base flashing over cricket

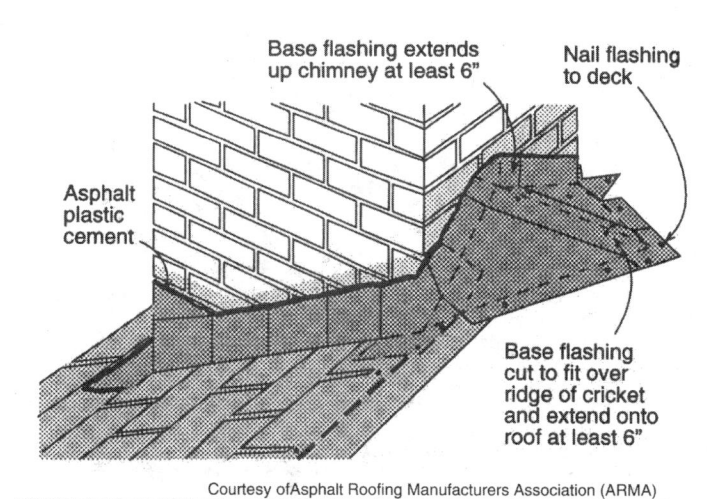

Figure 4-58 Application of base flashing over ridge of cricket

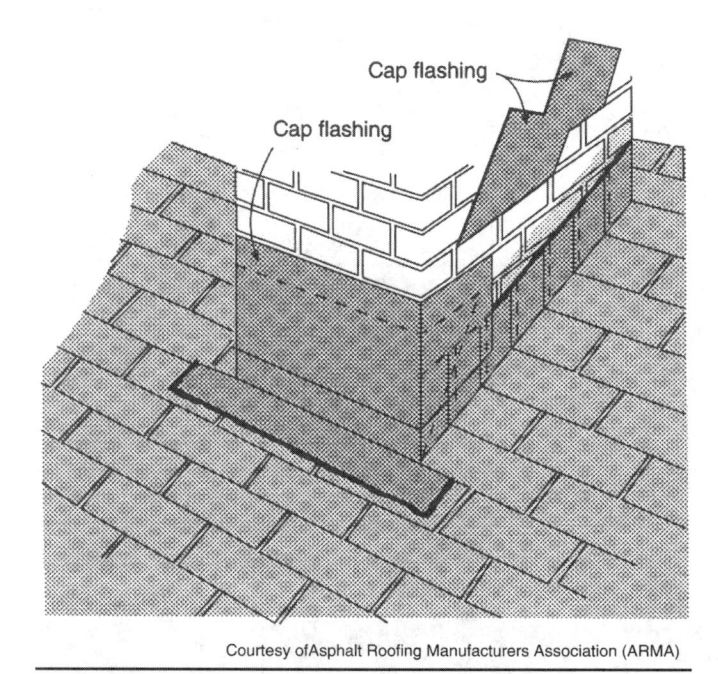

Figure 4-59 Application of cap flashing at front and side of chimney

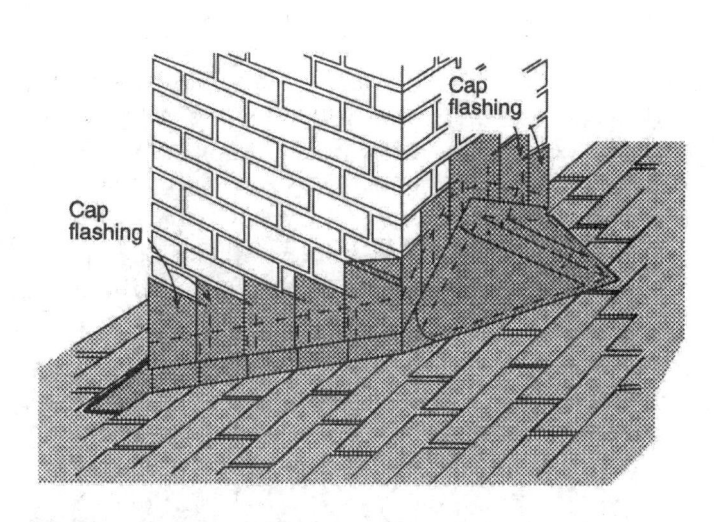

Figure 4-60 Application of cap flashing at side and rear of chimney

install shingles over a metal cricket if you can't see it from the ground or surrounding viewpoints.

Install metal cap flashing as shown in Figures 4-59, 4-60 and 4-61. The maximum height flashing may reach is to the top of the third brick. Don't install flashing that reaches above the third brick, as in Figure 4-59, except when it's over a cricket, as in Figure 4-60. Chisel and rake clean the mortar joints to a depth of 1½ inches before you install the cap flashing. Make some mortar that's 1 part portland cement and 3 parts fine mortar sand, and refill the joints with it. Wet the joints before you apply the fresh mortar.

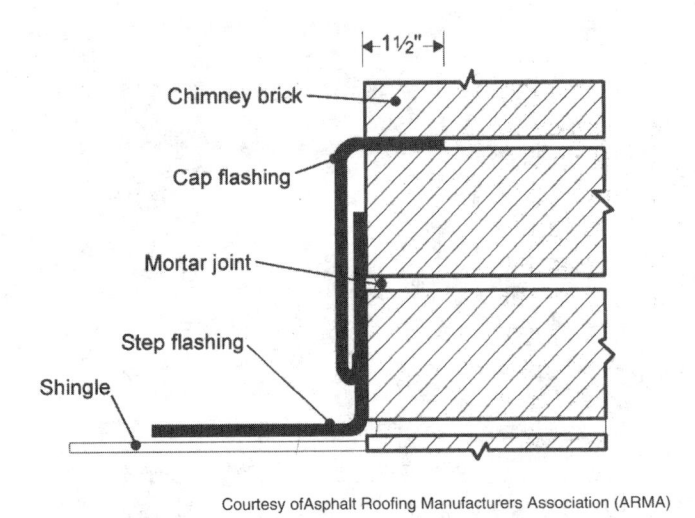

Figure 4-61 Application of cap flashing

Or, you can caulk the joints after you install the flashing pieces. Install the cap flashing at the front of the chimney in one piece, as shown in Figure 4-59. Install the cap flashing at the sides and back of the chimney as individual units, beginning at the lowest point. Install each piece into mortar joints so each cap flashing unit overlaps the base flashing by at least 3 inches, as in Figure 4-60. Bend the last piece of cap flashing around the upper corners of the chimney. Use a good grade of butyl rubber sealant to seal the flashing joints at the chimney corners.

Make all exposed flashing (such as cap flashing) with its bottom edge turned under $1/2$ inch, as in Figure 4-61. This adds stiffness against the wind and prevents snow from packing in under the flashing.

Use galvanized metal step flashing where a sloping roof meets a vertical surface such as the chimney shown in Figures 4-55 through 4-61, or the vertical side wall in Figure 4-62. The step flashing is later protected by cap flashing installed in the masonry, or by siding. The cap flashing makes a good water seal even when the roof and chimney (or wall) move independently due to expansion or settlement. You can see cap flashing in Figures 4-59 through 4-61. There is no cap flashing in Figure 4-62 because the siding takes the place of cap flashing. Overlap the step flashing joints at least 2 inches, and extend the metal under the shingles and up the chimney about 4 to 5 inches. See Figure 4-63. Lap the cap flashing down over the base flashing at least 3 inches and extend it down to within 1 inch of the finished roof. See Figure 4-61.

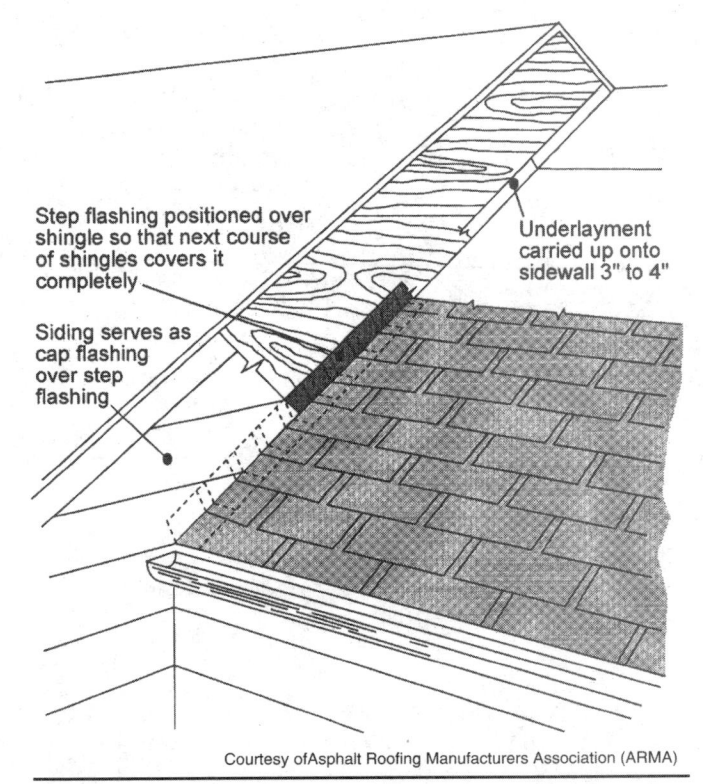

Figure 4-62 Application of step flashing against vertical side

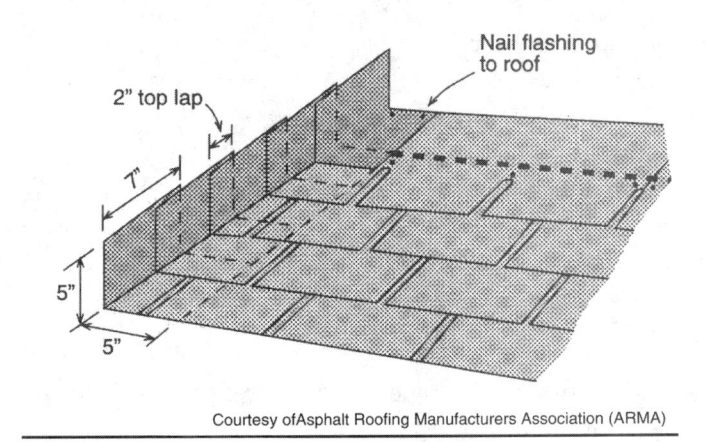

Figure 4-63 Application of step flashing

Asphalt Shingles

Step Flashing

Install step flashing as an individual piece for each course of shingles you lay by starting at the bottom and ending at the top of the chimney or side wall. Place the first piece of step flashing over the unexposed area of the shingle next to the lower edge of the chimney (refer back to Figure 4-55), or over the starter-course shingle at a side wall, as shown in Figure 4-62. Install a shingle over the step flashing so its butt is flush with the lower edge of the flashing. Install the next piece of step flashing over the shingle 5 inches above the butt. Install the next shingle so its butt is in line with the step flashing. Then, the horizontal leg of each piece of step flashing will cover the unexposed part of the underlying shingle. And, each piece of step flashing will be covered by the exposed part of the overlying shingle. See Figure 4-63. Continue this way until you've flashed and shingled the entire roof-wall intersection.

Use roofing cement to embed the end of each shingle which extends over the step flashing. Cut each piece of step flashing 10 inches wide and 2 inches longer than the shingle exposure. With 3-tab shingles, make each piece 7 inches long, providing a 2-inch lap.

Where a sloping roof and vertical wall meet, extend each piece of step flashing you install at least 5 inches up the wall and at least 5 inches under the shingles, as in Figure 4-63. Embed the horizontal leg of each piece of step flashing into roofing cement and secure it with two nails. Since the roof could eventually settle, don't nail the flashing to the wall. Cover the vertical leg of each piece of step flashing later with siding or cap flashing (in brick or stucco walls). In either case, extend the underlayment at least 3 inches up the wall, as in Figure 4-62.

Install continuous flashing like the one in Figure 4-64, where a sloping roof and a vertical side wall meet to form a horizontal line. One example of this is the intersection of a sloping roof and the front of a dormer. Another is a shed roof intersecting a wall, as in Figure 4-65.

Figure 4-64 Continuous flashing

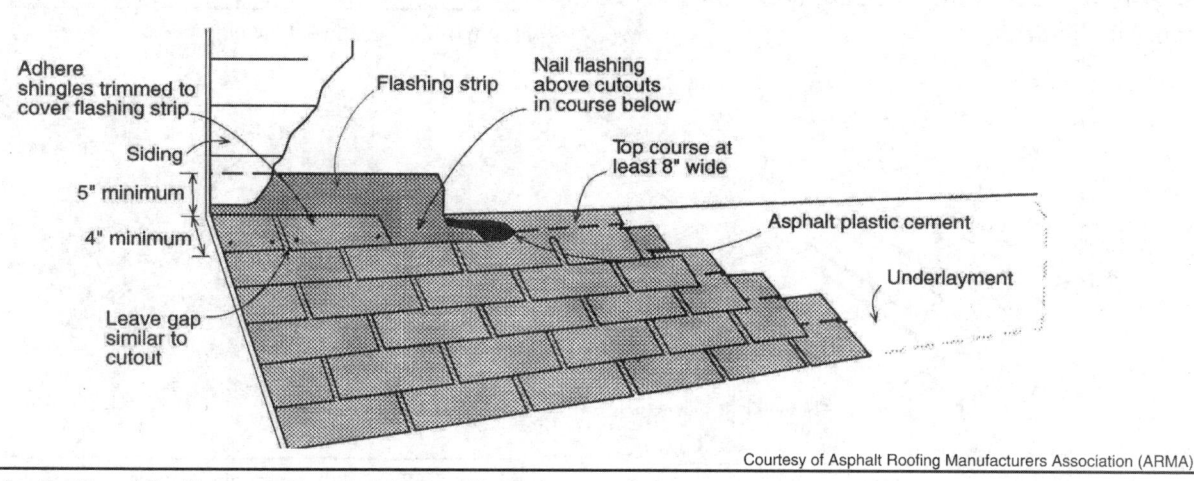

Figure 4-65 Application of flashing against vertical front wall

Courtesy of Asphalt Roofing Manufacturers Association (ARMA)

To install continuous flashing, embed it into roofing cement and nail it over the last course of shingles you apply to the roof deck. Nail it above the cutouts of the shingles below it, as shown in Figure 4-65. Don't nail the flashing to the wall. This way, the roof and wall can move independently. Instead, install the flashing before the siding. If the siding's already installed, pry up the lowest siding board enough to slip the flashing beneath it.

If the side wall is brick, use a masonry saw to remove $1\frac{1}{2}$ inches of mortar from a joint at a point about 5 inches above the roof-wall intersection. Bend the top of the flashing strip so you can insert it into the joint. Close the joint with mortar or caulking compound.

If the wall finish is stucco, saw out a joint and re-pack it with mortar or caulk. After you get the flashing in, cover it with one course of shingles trimmed to fit over the flashing. Nail the shingles into place and cover each nail head with roofing cement. Use 26-gauge galvanized metal flashing and extend it at least 5 inches up the wall and 4 inches over the last shingle course, as in Figure 4-65. Where front-wall flashing turns a corner (at a dormer, for example), extend the flashing at least 7 inches around the corner. From there on, install step flashing up the slope.

Flashing Soil Stacks and Vents

Normally, you flash vent pipes with a one-piece lead flange and sleeve (collar) as in Figure 4-66. Turn the top of the sleeve down into the stack. Look ahead to Figure 4-73. Or you can cut a lead sleeve flush with the top of the vent pipe and counterflash it with lead extending 4 inches down the outside of the pipe and 2 inches down the inside of the pipe, as shown in Figure 4-67.

Another popular type of vent flashing is a rubber flange, which you slip down over the pipe, as in Figure 4-68. This type of flashing is often installed on metal roof decks. Vent pipes are normally $1\frac{1}{2}$ to 3 inches in diameter.

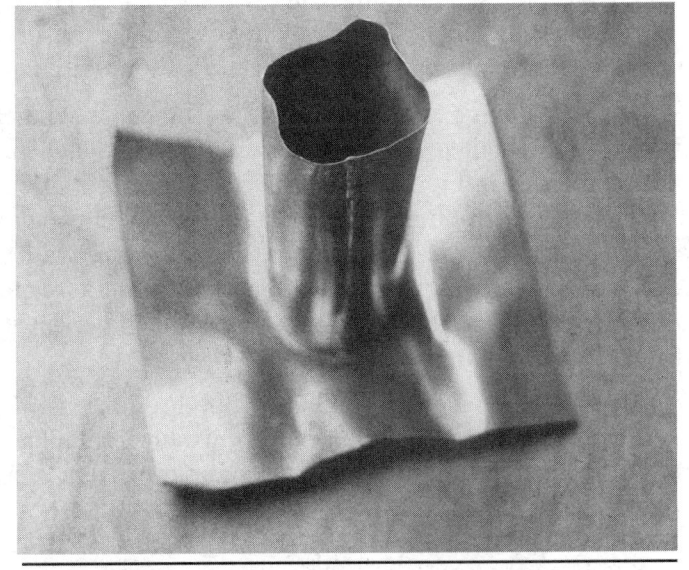

Figure 4-66 Lead vent flashing

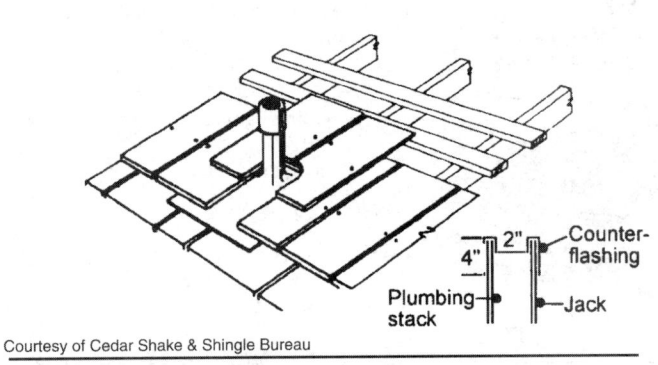

Figure 4-67 Counterflashed vent flashing

Figure 4-68 Rubber vent pipe flange

Asphalt Shingles

Figure 4-69 Shingle up to the bottom of vent pipe

When you come to a vent pipe, install shingles up to the bottom edge of the pipe, as in Figure 4-69. If the top edge of a shingle hits the pipe, notch it so it fits around the pipe, as shown in the figure. If a shingle ends up over the pipe, cut a hole in the shingle and slip it over the pipe, as in Figure 4-70.

Slip the flange over the pipe and underlying shingle, as shown in Figure 4-71. Embed the flange and overlying shingle in roofing cement. See Figures 4-72 and 4-73. Always install a full-width shingle over the pipe (Figure 4-74).

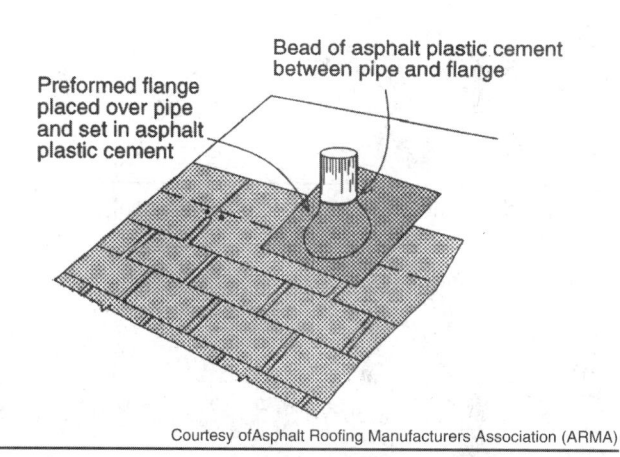

Courtesy of Asphalt Roofing Manufacturers Association (ARMA)

Figure 4-70 Application of shingle over vent pipe

Figure 4-71 Slip the flange over the vent pipe and underlying shingle

99

Roofing Construction & Estimating

Figure 4-72 Embed the flange in roofing cement

Figure 4-73 Embed the overlying shingle in roofing cement

Asphalt Shingles

Figure 4-74 Arrange to install a full shingle over the vent pipe

Insert a single tab along the shingle course to rearrange and offset the joints so there won't be a joint above the pipe.

Continue shingling around and above the pipe, trimming successive shingle courses to fit around the pipe. Nail shingles over the vent pipe so the nails don't go into the metal flashing flange. Allow a 1/2-inch space between the vent sleeve and the overlying shingle so debris won't get caught between the shingle and the vent stack. Most of the debris there will be dislodged mineral granules which won't wash away easily if there's no gap. You don't need to cover the down-slope part of the flange with shingles. In fact, I recommend that you leave the bottom third of any vent flashing flange exposed, as shown in Figure 4-75. An exposed flange isn't as likely to trap debris, but it doesn't look as neat. Let your customer have the final word on this.

For added leak protection on low-sloped roofs, embed a strip of mineral-surfaced roll roofing into roofing cement under the vent pipe flange.

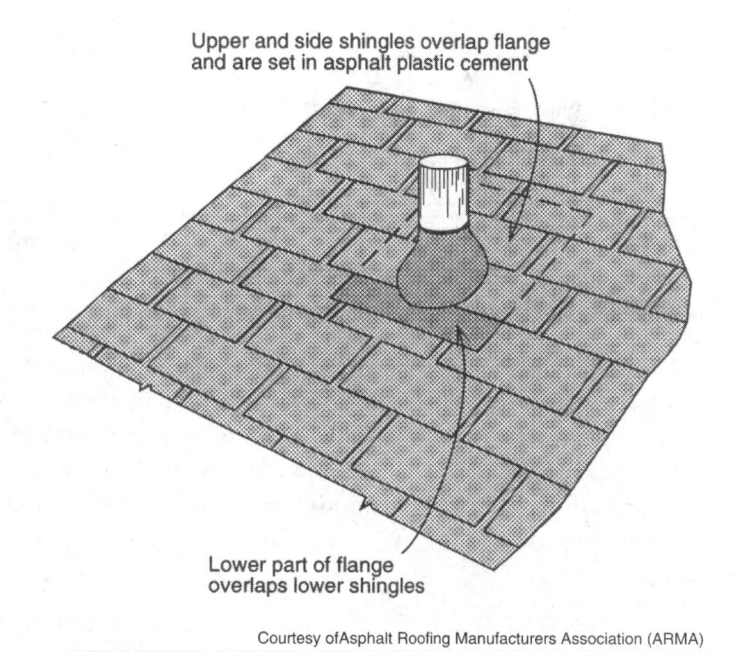

Courtesy of Asphalt Roofing Manufacturers Association (ARMA)

Figure 4-75 Bottom portion of vent flange exposed

Figure 4-76 Applying roofing cement over the flange of a heater vent

Figure 4-77 Installing a course above the vent

When you come to a larger roof penetration such as a heater vent, nail down the flange of the vent and install shingles up to the lower edge of the vent. Apply roofing cement over the flange, as in Figure 4-76. Cut shingles to fit around and above the vent as in Figure 4-77. Don't drive shingle nails through the vent flange.

Application	Nail length (inches)
Roll roofing on new deck	1
Strip or individual shingles on new deck	1¼
Roofing over old asphalt roofing	1½ to 2
Roofing over old wood shingles	2

Figure 4-78 Recommended nail lengths

Fasteners

Install asphalt roofing materials over solid roof sheathing. Use 11- or 12- gauge hot-dipped galvanized or aluminum roofing nails with heads that are at least ³⁄₈ inch in diameter and barbed or deformed shanks that are 1 to 2 inches long. Recommended lengths are shown in Figure 4-78. When re-roofing with asphalt shingles, use nails that are long enough to go at least ³⁄₄ inch into the sheathing. You can make sure the nails you use are long enough by checking the underside of the sheathing to see if they come through it. Allow for about 2½ pounds of nails per square when you install asphalt shingles.

When you install shingles across a roof, start nailing from the end nearest the shingle you just laid and proceed across. This will prevent buckling. Drive nails straight so the nail head doesn't damage the surface of the shingle. Don't drive nails into knotholes or cracks in the sheathing. If you have to remove a nail, seal the hole with roofing cement or remove and replace the entire shingle.

Asphalt Shingles

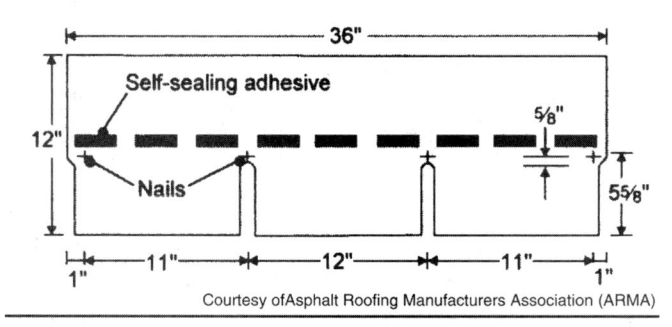

Figure 4-79 Nail locations for 3-tab strip shingle

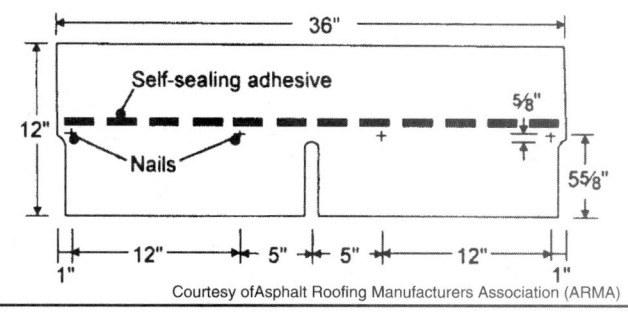

Figure 4-80 Nail locations for 2-tab strip shingle

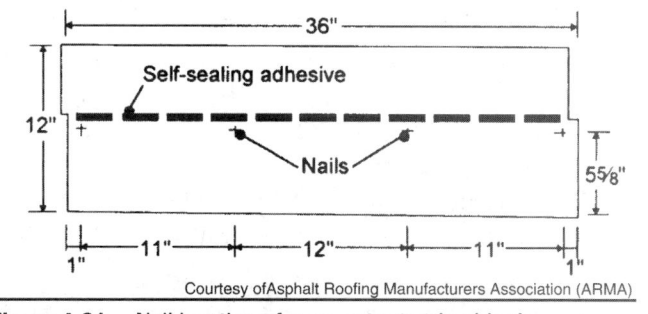

Figure 4-81 Nail locations for no-cutout strip shingle

Drive nails into the shingle along a line just below the factory-applied adhesive strip. Depending on what part of the country you're building in, use at least four nails for each 3-tab shingle. When you're laying a shingle at a 5-inch exposure, drive the nails along a line $5^{5}/_{8}$ inches above the butt edge of the shingle.

Earlier you recall we said not to nail in a straight line in order to avoid splitting the sheathing boards. But in this case, nail placement is very important so the adhesive strip will do its job. That's not a problem with starter-course material, so stagger-nail when you can. But in this case, nail in a straight line. Drive the two outermost nails 1 inch from each end of the shingle. Center the innermost nails over each cutout, as in Figure 4-79.

Two-tab shingles also need at least four nails. When you're laying a shingle at a 5-inch exposure, drive the nails along a line $5^{5}/_{8}$ inches above the butt edge of the shingle, and at 1 and 13 inches from each end of the shingle, as in Figure 4-80.

Shingles with no cutouts also require at least four nails. When you're laying a shingle at a 5-inch exposure, drive nails $5^{5}/_{8}$ inches above the butt edge of the shingle, and 1 and 12 inches from each end, as in Figure 4-81.

Never use fewer than four nails to install each strip shingle. Some roofing contractors don't drive the fourth nail because it's hidden under the overlapping shingle above. This is called "three-nailing" and I don't recommend it.

Many building codes, especially in high wind and hurricane areas, or where the eaves are 20 feet or more above grade, require six nails. Check with your building inspector beforehand, and make the necessary material and labor cost adjustments to your estimate.

Stapling

I also don't recommend you use staples because they tend to come loose eventually. Roofing contractors who hand-nail shingles (and explain to their clients why they do so) have more work than they can handle, at a price

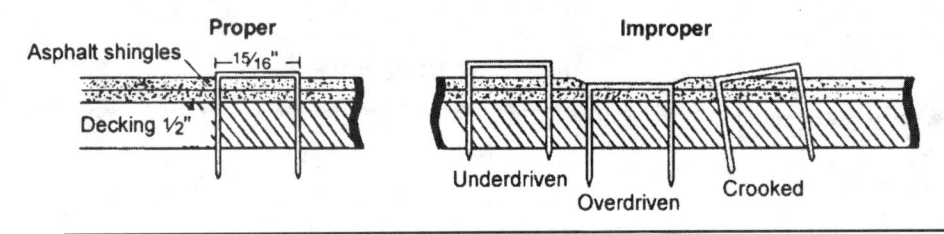

Figure 4-82 Driving staples

that yields high profit. Many have a waiting list of customers which includes general contractors. That's because the quality and durability of their work is consistently above that of their competitors.

If you decide to use staples, use them only on new construction to fasten wind-resistant asphalt shingles with factory-applied adhesives. If the old roofing has been removed, you can use staples for re-roofing. Use galvanized staples that are at least 16 gauge with a minimum crown of $^{15}/_{16}$ inch. Make sure the staples are long enough to go at least $^{3}/_{4}$ inch into the sheathing. Locate staples the same way as roofing nails.

It's very important to hold the staple gun so the staples go in at the correct angle so the crown is practically flush with the shingle surface. And be sure to adjust the air gun so the staples go far enough into the sheathing. Figure 4-82 shows good and bad stapling.

Be aware that a pneumatic stapler has no "feel" to it. It won't be obvious when you're stapling into a joint or knothole. In warm weather, it's easy to drive a staple all the way through a soft asphalt shingle. Wind will also tear off a shingle unless you drive the crown of the staple parallel to the long shingle edge.

Number of Shingles Required per Square

Asphalt strip shingles are usually 3' x 1' and come in 3-bundle squares (for lighter-weight shingles), or 4-bundle squares (for heavier shingles). That means you need three or four bundles of asphalt strip shingles (whichever the case), laid at the recommended exposure (usually 5 inches) to cover a square of roofing surface.

Whatever the size or exposure a strip shingle is, you can figure out how many shingles you need for each square of roof surface by:

$$\text{Shingles/Square} = \frac{100 \text{ SF}}{\text{Shingle Length (in.) x Exposure (in.)}} \times 144 \text{ sq. in./SF}$$

Equation 4-1

▼ **Example 4-1**: Assume you're using 3' x 1' asphalt strip shingles at a 5-inch exposure. Find the number of shingles you need to cover one square of roof area.

$$\text{Shingles/Square} = \frac{100 \text{ SF}}{36 \text{ in.} \times 5 \text{ in.}} \times 144 \text{ sq. in./SF}$$
$$= 80 \text{ shingles per square}$$

You shouldn't "stretch" the exposure, but you can install shingles at an exposure *less* than what's recommended. In this case, the extra shingles you need per square, in terms of a percentage-of-increase factor, are:

$$\text{Percentage-of-Increase Factor} = \frac{\text{Recommended Exposure}}{\text{Actual Exposure}}$$

Equation 4-2

▼ **Example 4-2**: An area of 20 squares is to be covered with 3-tab strip shingles. Assume the recommended exposure is 5 inches, then find the number of shingles required to install shingles at:

a) 4½ inches, or

b) 4 inches

Solution:

a) Percentage-of-Increase Factor, 4½" exposure $= \dfrac{5 \text{ in.}}{4½ \text{ in.}}$
   $= 1/10$

Thus, you increase the shingle quantity by 10 percent:
20 squares x 1.1 = 22 squares

b) Percentage-of-Increase Factor, 4" exposure $= \dfrac{5 \text{ in.}}{4 \text{ in.}}$
   $= 1.20$

Thus, you increase the shingle quantity by 20 percent:
20 squares x 1.2 = 24 squares

Number of Shingle Courses

You can determine the number of shingle courses required to cover a wall or roof by:

$$\text{Courses} = \frac{\text{Dimension of Structure}}{\text{Exposure}}$$

Equation 4-3

where the Dimension of Structure is the wall height, or the width of a roof section, measured from the eaves to the ridge, along the top of a common rafter.

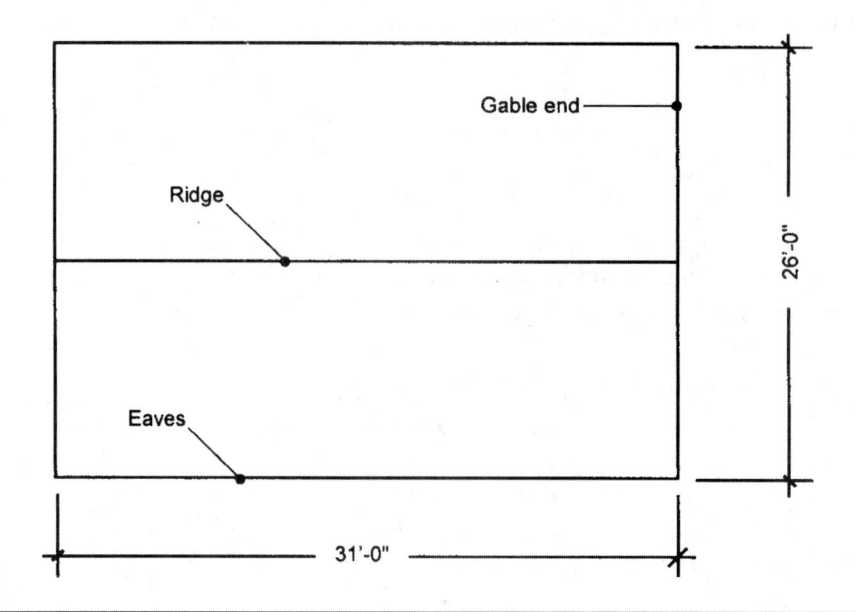

Figure 4-83 Gable roof example

You can also use Equation 4-3 to determine the number of hip and ridge units required (if you know the lengths of the hips and ridge).

▼ **Example 4-3:** Assume an exposure of 5 inches, then find the number of courses of 12-inch-wide asphalt shingles required to cover the roof of the building diagrammed in Figure 4-83. The roof slope is 5 in 12.

Solution: From Column 2 of Appendix A, the actual width of each side of the roof, measured along any rafter from eaves to ridge is:

Length = 13' x 1.083 = 14.1 linear feet

From Equation 4-3, the number of shingle courses required on one side of the roof is:

$$\text{Courses} = \frac{14.1 \text{ ft.} \times 12 \text{ in.} / \text{LF}}{5 \text{ in.}}$$
$$= 33.84, \text{ rounded to 34 courses.}$$

Double that for a total of 68 courses for both sides of the roof. Remember, this formula produces the number of courses for a *section* of roof. In this case, you multiply by 2 because both sides of the roof are the same.

Notice that the calculated answer in the previous example didn't come out to an even number of courses. Since it's impractical to install 0.84 of a course, we changed the answer to an even 34 courses, which will result in slightly less exposure throughout. We also could have used 33 courses, with slightly more exposure. Either way, use the following equation to find the exact exposure for this example.

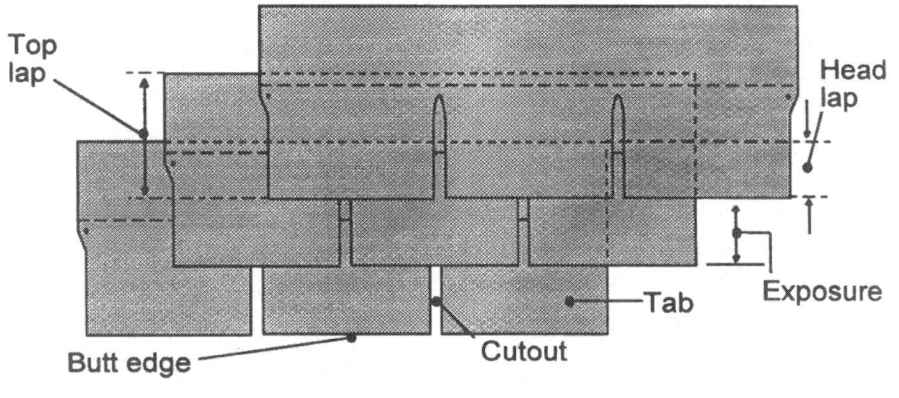

Asphalt Shingles

Figure 4-84 Exposure diagram

$$\text{Exposure} = \frac{\text{Dimension of Structure}}{\text{Number of Courses}}$$

Equation 4-4

▼**Example 4-4:** Determine the consistent exposure required to install:

a) 33 courses, or

b) 34 courses of shingles on the roof described in Example 4-3.

Solution: The consistent exposures are:

a) For 33 courses:

$$\text{Exposure} = \frac{14.1 \text{ ft.} \times 12 \text{ in./LF}}{33}$$
$$= 5.127 \text{ inches, or } 5^{1}/_{8} \text{ inches (rounded off)}$$

Since you shouldn't "stretch" the exposure, use 34 courses in this case, with a decreased exposure.

b) For 34 courses:

$$\text{Exposure} = \frac{14.1 \text{ ft.} \times 12 \text{ in./LF}}{34}$$
$$= 4.98 \text{ inches, which rounds to 5 inches}$$

There's another way you can install shingles on roofs or walls whose dimensions aren't evenly divisible by the shingle exposure. To do this, apply courses at the recommended exposure and decrease the exposure near the ridge so you finish with a full shingle width at the ridge. It's better to shorten than to stretch the exposure of the shingle near the ridge. This method for shortening an exposure is called "stacking" the shingles.

107

Head Lap, Top Lap and Exposure

Head lap is where shingles (or other roof coverings) are three layers thick. Top lap is where shingles (or other roof coverings) are at least two layers thick. See Figures 4-84 and 4-85. Thus, the 7-inch lap on a 12-inch-wide shingle, where the shingles are two layers thick, is the top lap. The 2-inch lap where the shingles are three layers thick is the head lap, as shown in Figure 4-84. Exposure is the part of the shingle not covered by the next course of shingles. The relationships between head lap, top lap and exposure are given in the following equations, where TL = top lap, W = width of shingle, E = exposure, and HL = head lap.

Top Lap = W - E, or   **Equation 4-5**

Top Lap = E + HL   **Equation 4-6**

Head Lap = TL - E, or   **Equation 4-7**

Head Lap = W - 2E   **Equation 4-8**

Exposure = $\dfrac{W - HL}{2}$ **Equation 4-9**

E = exposure
HL = head lap
TL = top lap
W = width of shingle

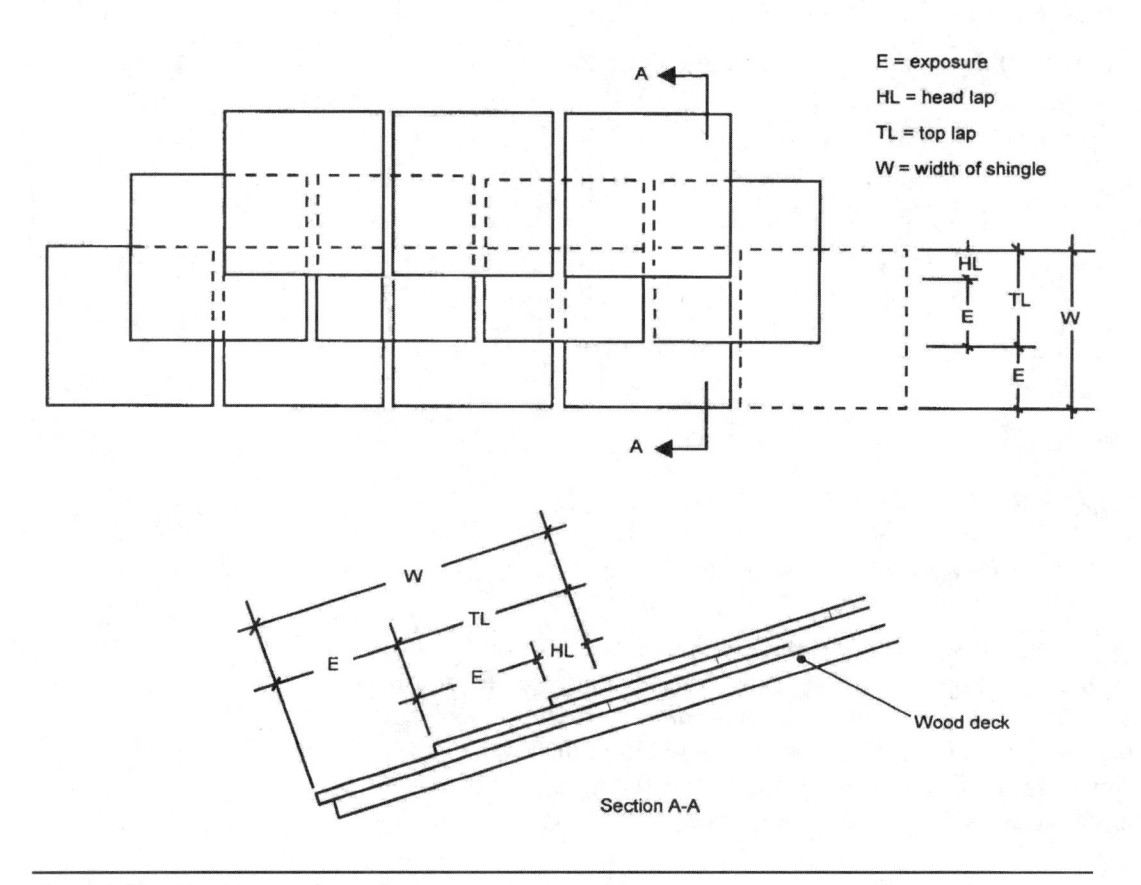

Figure 4-85 Head lap, top lap and exposure

Coverage Based on Number of Shingle Plies

Shingle coverage is often designated as single, double or triple, depending on the number of plies or layers of shingles. The number of plies is defined as the number of shingle layers at the head lap. Figure 4-85 shows 3-ply construction.

If the number of plies isn't the same throughout the roof, coverage is generally considered as the number of layers installed over a majority of the roof area. Most strip shingles are designed for double coverage. To get two layers of shingles over an entire roof, overlap shingles by a bit more than half. Overlap by slightly more than two-thirds to get triple coverage.

Sometimes roof coverage specifications call for the acceptable minimum number of shingle plies.

▼**Example 4-5:** Assume that 3-ply coverage is specified, then find the maximum exposure allowed to apply 12-inch-wide asphalt shingles with a minimum of a 2-inch head lap.

Solution: From Equation 4-9, the maximum allowable exposure is:

$$E = \frac{12 \text{ in.} - 2 \text{ in.}}{2}$$

$$= 5 \text{ inches}$$

Estimating Asphalt Strip Shingle Quantities

First, take off quantities for asphalt strip shingles in square feet. Then, convert these quantities into the number of bundles of shingles required. The total quantity of shingles required must include material for the starter course, hip and ridge units, cutting waste at the rakes, hips and valleys, and waste due to crew errors.

Starter Course

You can find the amount of field-fabricated starter-course material required by:

Starter Course (SF) = Eaves (LF) x Exposed Area (SF/LF) | Equation 4-10 |

You can find the coverage per linear foot of shingle, by:

$$\text{Area (SF/LF)} = \frac{\text{Exposure (in.)}}{12}$$ | Equation 4-11 |

▼**Example 4-6:** Assume a 5-inch exposure, then find the coverage of starter-course material in square feet per linear foot of eaves.

Solution: The coverage is:

$$\text{Area} = \frac{5}{12} = 0.42 \text{ square feet per linear foot}$$

As a convenience, use Figure 4-86 to quickly determine the area-per-linear-foot values for shingles installed at various exposures along the eaves.

Sometimes you need to know the *number* of starter-course units. The number of eaves shingles required at any given eaves is:

Number of Eaves Shingles $= \dfrac{\text{Eaves Length (ft.)}}{\text{Shingle Length (ft.)}}$ | **Equation 4-12** |

Estimate eaves units carefully. section by section. The high cost of some materials, and the inconvenience and delay caused when you're short of materials, makes underestimating eaves materials very expensive.

▼ **Example 4-7:** Assume application of 10-inch-long shingles, then find the number of eaves shingles required for the roof of the building diagrammed in Figure 4-83.

Solution: The number of eaves shingles required along each eaves is:

$$\text{Number of Eaves Shingles (each eaves)} = \frac{31 \text{ feet}}{.83 \text{ feet}} = 38 \text{ (rounded up)}$$

Because there are two eaves in this example, you need to order 76 shingle units.

Asphalt shingle manufacturers recommend that the starter course and first course overhang the eaves and rake by $1/4$ to $3/8$ inch. However, you don't have to increase your order to account for that since the added area is so small.

Cutting Waste at Rakes, Hips and Valleys

Don't forget to allow for shingles lost due to cutting waste at rakes, hips and valleys. Use Figure 4-87 to account for wasted material. The table doesn't include material wasted due to crew error. The table assumes 3' x 1' strip shingles laid at a 5-inch exposure.

Ridge and Hip Units

Using a 12-inch-wide shingle, you'll need 1 square foot of shingles for each linear foot of hip and ridge. Assuming a conscientious and prudent crew that uses every salvageable single-tab shingle available from material cut at the rakes, hips and valleys, you can use Figure 4-88 to determine the quantity of shingles salvaged. The table assumes 3' x 1' 3-tab shingles laid at a 5-inch exposure.

The formula for allowance in hip and ridge units is the difference between the square feet required and the square feet salvaged:

Net Allowance (ridge and hip units) = SF Required - SF Salvaged | **Equation 4-13** |

Asphalt Shingles

Exposure (inches)	Area covered (SF) per LF of shingle	Exposure (inches)	Area covered (SF) per LF of shingle
3	0.25	9	0.75
3½	0.29	9½	0.79
3¾	0.31	10	0.84*
4	0.34*	10½	0.88
4¼	0.35	11	0.92
4½	0.38	11½	0.96
4¾	0.40	12	1.00
5	0.42	12½	1.04
5⅛	0.43	13	1.08
5¼	0.44	13½	1.13
5½	0.46	14	1.17
5⅝	0.47	14½	1.21
5¾	0.48	15	1.25
6	0.50	15½	1.29
6½	0.54	16	1.34*
7	0.58	16½	1.38
7½	0.63	18	1.50
8	0.67	20	1.67
8½	0.71	22	1.84*

*For the sake of cautious estimating, I round up the answer for 4, 10, 16 and 22 inches of exposure, respectively. This results in quantities that are a bit long, but safe.

Figure 4-86 Area of coverage per linear foot of eaves

	Waste (SF per linear foot)				
Shingle type	Rake	Hip	Open Valley	Closed-cut (half-lace) valley	Woven (full-lace) valley
3-tab	0.25	0.64	1.41	2.12	2.83
Other than 3-tab	negligible	0.30	0.30	1.00	1.71

Figure 4-87 Asphalt strip shingle cutting waste

	Shingles salvaged (SF per linear foot)				
Shingle type	Rake	Hip	Open Valley	Closed-cut (half-lace) valley	Woven (full-lace) valley
3-tab	1.00	0.50	2.00	1.00	0

Figure 4-88 Asphalt hip and ridge shingles salvaged from cutting waste

On hip roofs, you'll require more units than you've salvaged. On gable roofs, you'll salvage more units than you need. Unless you can use the excess on another roof, the salvaged units will be wasted.

A "Shortcut" Method for Determining Asphalt Strip Shingle Waste

As a "rule of thumb," some contractors add 10 percent waste on small-to average-sized gable roofs, and 15 percent on hip roofs (3-tab shingles). They add 2 percent on gable roofs and 3 percent on hip roofs for laminated shingles with prefabricated ridge and hip units. This allows for additional material required for the starter course and site-fabricated ridge and hip units (in the case of 3-tab strip shingles), cutting waste at rakes, hips, and valleys, and waste due to crew error. They add a larger percentage for more complex roofs.

I don't rely on this method of estimating waste. Remember that exclusive of crew-error waste (which varies from crew to crew and from job to job), actual waste depends on the roof type, the ratio of roof length to roof width, the roof slope, the shingle exposure, the type of shingle installed, and whether the hip and ridge units are prefabricated or site-constructed.

Waste on asphalt strip-shingle roofs varies from 1 to 8 percent on average-sized gable roofs, and from 3 to 18 percent on hip roofs. You can use Figure 4-89 and 4-90 to quickly get a "ball-park" percentage of asphalt strip shingle waste. Waste due to crew error isn't included in these figures.

Figure 4-89 lists the total percentage you add to net roof area of a *gable roof* for the starter course, cutting waste at rakes, and site-fabricated ridge units (if applicable).

This table assumes 3' x 1' strip shingles laid at a 5-inch exposure.

Figure 4-90 shows the total percentage you add to net roof area of a *hip roof*, including the starter course, cutting waste at hips, and site-fabricated ridge units (if applicable). The table assumes 3' x 1' strip shingles laid at a 5-inch axposure. The table doesn't include crew-error waste.

▼ **Example 4-8:** Assume a roof slope of 5 in 12 and 3-bundle squares installed at a 5-inch exposure. How many bundles of 3' x 1' 3-tab shingles are required for the roof diagrammed in Figure 4-83? Assume also that you'll use field-fabricated 3-tab shingles for the starter course at the eaves and 12" x 12" tabs salvaged at the rakes for ridge units.

Solution: First, remember the formula for net roof area to be covered (from Chapter 1):

Actual (Net) Roof Area = Roof Plan Area x Roof-Slope Factor

Thus, Net Roof Area = 31' x 26' x 1.083
  (from Column 2 of Appendix A)
  = 873 square feet ÷ 100
  = 8.73 squares (use this quantity to estimate labor costs)

Asphalt Shingles

Cutting waste and overruns (gable roof)			
Building dimensions (LxW)	Roof slope		
	3 in 12	6 in 12	12 in 12
30 x 20	8[1](4)[2]	8 (4)	8 (3)
40 x 30	6 (3)	6 (3)	6 (2)
45 x 30	5 (3)	6 (3)	5 (2)
50 x 30	5 (3)	5 (3)	5 (2)
60 x 30	4 (3)	4 (3)	4 (2)
70 x 30	4 (2)	4 (3)	4 (2)
80 x 30	5 (2)	5 (3)	3 (2)
50 x 40	5 (2)	5 (2)	5 (2)
60 x 40	4 (2)	4 (2)	5 (2)
70 x 40	4 (2)	4 (2)	4 (2)
80 x 40	3 (2)	3 (2)	4 (2)
90 x 40	3 (2)	3 (2)	4 (2)
60 x 50	4 (2)	5 (2)	4 (1)
70 x 50	4 (2)	4 (2)	3 (1)
80 x 50	4 (2)	4 (2)	3 (1)
90 x 50	3 (2)	3 (2)	3 (1)

(1) 3-tab shingles using site-fabricated hip and ridge units.
(2) Laminated strip shingles using a prefabricated ridge and hip roll. The roll must be taken off separately.

Figure 4-89 Cutting waste and overruns (gable roof)

Cutting waste and overruns (hip roof)			
Building dimensions (LxW)	Roof slope		
	3 in 12	6 in 12	12 in 12
30 x 20	18[1](10)[2]	17 (9)	14 (8)
40 x 30	14 (7)	13 (6)	12 (6)
45 x 30	13 (7)	12 (6)	10 (5)
50 x 30	12 (6)	11 (6)	9 (5)
60 x 30	11 (6)	11 (5)	9 (4)
70 x 30	11 (5)	11 (5)	9 (4)
80 x 30	10 (5)	9 (4)	8 (4)
50 x 40	11 (6)	9 (4)	9 (5)
60 x 40	9 (5)	9 (4)	9 (4)
70 x 40	9 (4)	9 (4)	8 (3)
80 x 40	9 (4)	8 (4)	6 (3)
90 x 40	8 (4)	8 (4)	6 (3)
60 x 50	9 (5)	9 (4)	7 (3)
70 x 50	9 (4)	8 (4)	7 (3)
80 x 50	9 (4)	6 (4)	6 (3)
90 x 50	7 (4)	6 (4)	6 (3)

(1) 3-tab shingles using site-fabricated hip and ridge units.
(2) Laminated strip shingles using a prefabricated ridge and hip roll. The roll must be taken off separately.

Figure 4-90 Cutting waste and overruns (hip roof)

The total eaves length is 2 x 31, or 62 linear feet.

The total area of starter-course material required is:

Area (Starter Course) = 62' x 0.42 SF/LF
 (Equation 4-10 or Figure 4-86)
 = 26 square feet

The total rake length is:

LF (Rake) = 4 ea. x 13' x 1.083 (Column 2, Appendix A)
 = 56 linear feet

Cutting waste at the rakes is:

Waste (Rake) = 56' x 0.25 SF/LF (from Figure 4-87)
 = 14 square feet

Remember, you need 1 square foot of single-tab shingles per linear foot of ridge, or 31 square feet.

From Figure 4-88, the amount of salvaged tabs at the rakes is 56 square feet. That's more single-tab units than needed, so from Equation 4-13, the net allowance for ridge units is:

Net Allowance (Ridge Units) = 56 SF - 31 SF
= 25 square feet

Thus, the gross roof area = 873 SF + 26 SF + 14 SF + 25 SF
= 938 square feet ÷ 100
= 9.38 squares

From Chapter 1, the waste factor (excluding crew-error waste) is:

$$\text{Waste Factor} = \frac{\text{Area Covered (including waste)}}{\text{Net Roof Area}}$$

Therefore,

$$\text{Waste Factor} = \frac{9.38 \text{ Squares}}{8.73 \text{ Squares}}$$
$$= 1.08$$

As you can see, that answer agrees with the information from Figure 4-89. In that table, the nearest size to our example is a 30' x 20' roof with a 6 in 12 slope, and the table shows a waste factor of 8 percent for 3-tab shingles.

Total material required: 9.38 squares x 3 bundles/square = 29 bundles.

▼ **Example 4-9:** Work the same problem assuming use of 3' x 1' laminated strip shingles with field-fabricated shingles used for the starter course at the eaves. Also, assume using 3-bundle squares and a prefabricated ridge roll.

Solution: First, you need 31 linear feet of ridge roll to cover the length of the ridge. We already know the net roof area is 8.73 squares (from the last example). We also know there's 26 square feet of starter course material required. From Figure 4-87 we see that cutting waste at the rakes is negligible for this type roof.

Thus, the gross roof area = 873 SF + 26 SF
= 899 square feet ÷ 100, or 8.99 squares

The waste factor is 3 percent (8.99 ÷ 8.73), close enough to the 4 percent waste predicted in Figure 4-89. At 3 bundles per square, this job requires 27 bundles.

The following examples show you how to estimate roofs of any size, with any slope. If you calculate the waste factors for each example, you'll see that you can use the tables (Figures 4-89 and 4-90) with confidence.

Asphalt Shingles

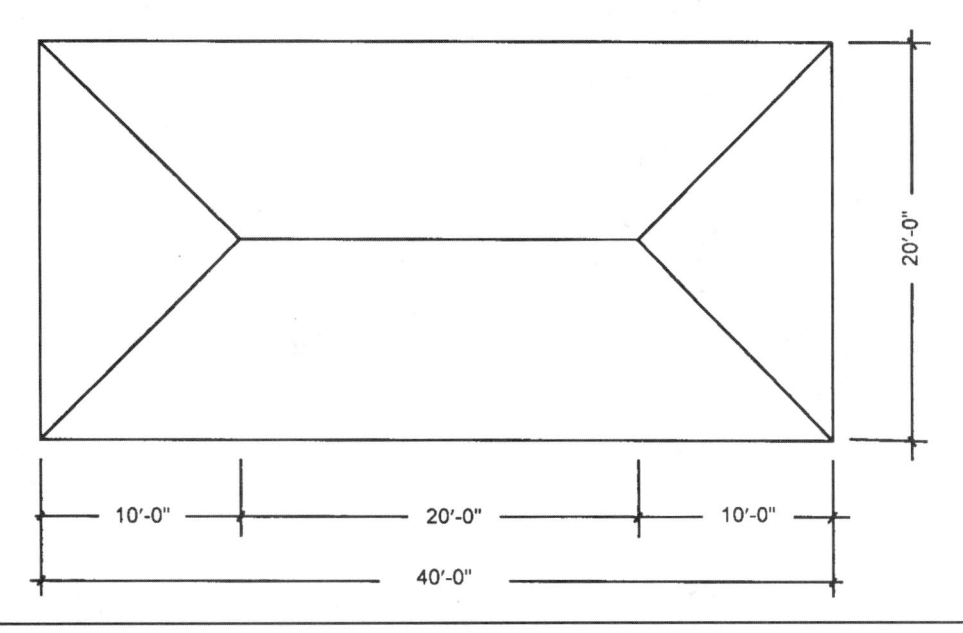

Figure 4-91 Hip roof example

▼**Example 4-10:** Assume a roof slope of 5 in 12 and the use of 3-bundle squares installed at a 5-inch exposure, then find how many bundles of 3' x 1' 3-tab shingles are required for the roof of the building diagrammed in Figure 4-91. Also, assume using field-fabricated 3-tab shingles for the starter course at the eaves and 12" x 12" tabs salvaged at the hips for hip and ridge units.

Solution: The net roof area to be covered is:

Actual (Net) Roof Area = Roof Plan Area x Roof-Slope Factor

Net Roof Area = 40' x 20' x 1.083 (from Column 2 of Appendix A)
= 866 square feet ÷ 100
= 8.66 squares (use this quantity to determine labor costs)

The total eaves length is:

Perimeter = 2(L + W)

LF (Eaves) = 2 x (40' + 20')
= 120 linear feet

The total area of starter-course material required is:

Area (Starter Course) = 120' x 0.42 SF/LF
(from Equation 4-10, or Figure 4-86)
= 50 square feet

From Column 3 of the Slope Factor table, each hip length is:

Length (Hip) = 10' x 1.474
= 14.74 linear feet

The total hip length is 4 x 14.74, or 59 linear feet.

From Figure 4-87, calculate cutting waste at the hips:

Waste (Hips) = 59' x 0.64 SF/LF
 = 38 square feet

You need 1 square foot of single-tab shingles for each linear foot of ridge and hips, therefore:

Hips and Ridge = (59' + 20') x 1 SF/LF
 = 79 square feet

From Figure 4-88, calculate the number of tabs salvaged at the hips for use on ridge and hips:

59' x 0.5 SF/LF = 30 square feet

You need more shingles than you salvaged, so you have to allow for additional shingles:

New Waste (Hip and Ridge Units) = 79 SF - 30 SF
 = 49 square feet

The total of the above calculations produces the gross roof area to be covered:

Gross Roof Area = 866 SF + 50 SF + 38 SF + 49 SF
 = 1,003 square feet ÷ 100
 = 10.03 squares

The waste factor (excluding crew-error waste) is:

$$\text{Waste Factor} = \frac{\text{Area Covered (including waste)}}{\text{Net Roof Area}}$$

$$= \frac{10.03 \text{ Squares}}{8.66 \text{ Squares}}$$
$$= 1.158$$

This job requires 31 bundles of shingles:

Bundles = 10.03 Squares x 3 Bundles/Square
 = 31 bundles (10.33 squares)

▼ **Example 4-11:** Using the same dimensions as those in Example 4-10 (Figure 4-91), assume the application of 3' x 1' laminated strip shingles using field-fabricated shingles for the starter course at the eaves. Also, assume 3-bundle squares and a prefabricated hip and ridge roll. Now calculate how many bundles of 3' x 1' laminated strip shingles you need to roof the building.

Solution: The total hip and ridge length is the same as in the previous example, 79 linear feet. The net roof area is also the same, 8.66 squares, as is the total area for the starter course, 50 square feet.

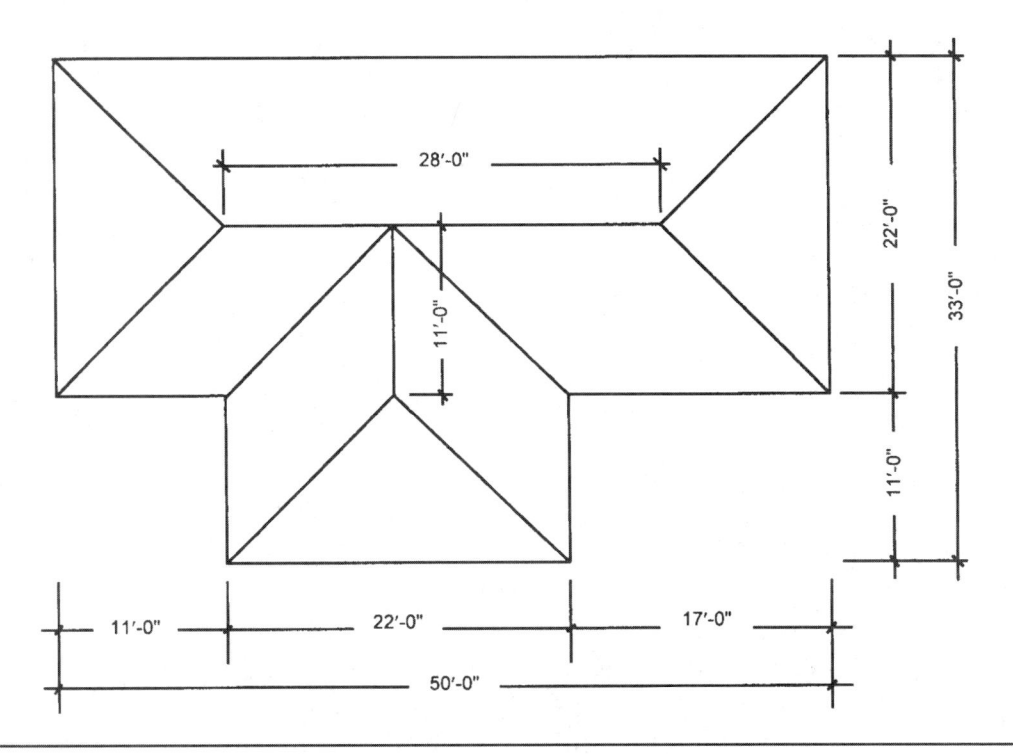

Figure 4-92 Hip and valley roof example

From Figure 4-87 cutting waste at the hips is:

Waste (Hips) = 59' x 0.30 SF/LF
= 18 square feet

The gross roof area in this case is 866 SF + 50 SF + 18 SF

= 934 square feet ÷ 100
= 9.34 squares

This job requires 29 bundles of shingles:

Bundles = 9.34 Squares x 3 Bundles/Square
= 29 bundles (9.67 squares)

▼ **Example 4-12:** Assuming a roof slope of 4 in 12 and the use of 3-bundle squares installed at a 5-inch exposure, determine the number of bundles of 3' x 1' 3-tab shingles required for the roof of the building diagrammed in Figure 4-92. Also assume using field-fabricated 3-tab shingles for a starter course at the eaves and 12" x 12" tabs salvaged at the hips and valleys for ridge and hip units. Also assume the closed-cut valley method of construction.

Solution: The formula for net roof area is:

Actual (Net) Roof Area = Roof Plan Area x Roof-Slope Factor

Net Roof Area = [(50' x 22') + (22' x 11')] x 1.054
(Column 2, Appendix A)
= 1,415 square feet ÷ 100
= 14.15 squares (Use this figure to estimate labor costs.)

The total length of eaves is:

Perimeter = 2(L + W)

LF (Eaves) = 2 x (50' + 33')
= 166 linear feet

From Equation 4-10 or Figure 4-86, the total area of starter-course material required is:

Area (Starter Course) = 166' x 0.42 SF/LF
= 69 square feet

From Column 3 of the Slope Factor table in Appendix A, the length of each hip or valley is:

Length (Hip or Valley) = 11' x 1.453
= 16 linear feet

There are 6 hips and 2 valleys. Use Figure 4-87 to calculate the waste allowance:

Waste (Hips) = (6 ea. x 16' x 0.64 SF/LF)

Waste (Valleys) = (2 ea. x 16' x 2.12 SF/LF)

Add the two together:

= (96' x 0.64) + (32' x 2.12)
= 129 square feet

Now, calculate the total length of ridge and hips to find the number of single-tab shingles required:

LF (Ridge and Hips) = 39' + 96'
= 135 linear feet
= 135 square feet (at 1 SF/LF)

From Figure 4-88, the quantity of tabs salvaged at the hips and valleys for hip and ridge units is:

SF salvaged at hips = (96' x 0.50 SF/LF)

SF salvaged at valleys = (32' x 1.00 SF/LF) (closed-cut valley)
= 80 square feet

Use Equation 4-13 to calculate the net waste allowance. Since we can expect to salvage 80 square feet, and need 135 square feet, we need an additional 55 square feet of single-tab shingles for the hips and ridges.

Thus, Gross Roof Area = 1,415 SF + 69 SF + 129 SF + 55 SF
= 1,668 square feet ÷ 100
= 16.68 squares

At 3 bundles per square, this job requires 50 bundles of shingles.

▼ **Example 4-13:** Work the last example, using 3-bundle, 3' x 1' laminated strip shingles with field-fabricated shingles used for the starter course at the eaves. Assume closed-cut valley construction, and a prefabricated hip and ridge roll.

Solution: The total hip and ridge length is the same as the previous example, 135 linear feet. The net roof area is also the same, 1,415 square feet, as is the total area of starter-course material required, 69 square feet.

The total length of hips is 96 linear feet, so from Figure 4-87, calculate the waste allowance:

Waste (Hips) = 96' x 0.30 SF/LF
             = 29 square feet

The total length of valleys is 32 linear feet, so from Figure 4-87, calculate the waste allowance:

Waste (Valleys)   = 32' x 1 SF/LF
                    = 32 square feet

Thus, Gross Roof Area = 1,415 SF + 69 SF + 29 SF + 32 SF
                       = 1,545 square feet ÷ 100
                       = 15.45 squares

This job requires a total of 47 bundles (15.45 x 3).

Notice that if you calculate the waste factors for the previous two examples, you'll see that the rule of thumb for estimating waste we discussed on page 112 ("shortcut") isn't accurate. The calculated waste factors are 18 percent (16.68 ÷ 14.15) and 9 percent (15.45 ÷ 14.15). The rule of thumb method would have produced waste factors of 15 percent and 18 percent, respectively. That's probably not close enough on a large job in a competitive market.

Estimating Asphalt Shingle Roofing Costs

Let's say you've got to cover the roof shown in Figure 4-91 under the conditions given in Example 4-10. At the time of this writing, Craftsman Book Company's *National Construction Estimator* gave the national average cost of a square of shingles, including felt, flashing, fasteners, vents and sales tax, at $176.13. Let's use that number. From Example 4-10, we must purchase 31 bundles of shingles, or 10.33 squares (31 ÷ 3 = 10.33). So the total material cost is 10.33 squares x $176.13/square = $1,819.43 (rounded off).

Additionally, 79 linear feet of hip/ridge shingles are purchased at $2.15 per linear foot.

79 x $2.15 = $169.85

Our Total Material Cost = $1,819.43 + $169.85 = $1,989.28

We have a roof deck area of 8.66 squares and a total of 79 linear feet of hips and ridge. According to Craftsman's *National Construction Estimator*, an R1 crew consisting of one roofer and one laborer can install composition shingles at the rate of 1 square per 1.83 manhours, and hip and ridge units at the rate of 1 linear foot per 0.028 manhours. Assuming the roofer (costing $41.66/hour) and the laborer (costing $30.31/hour) cost the employer $35.99 per manhour, the average cost per manhour is: ($41.66 + $30.31) = $71.97 ÷ 2 = 35.99

So the labor cost will be:

Labor (installing shingles) = 8.66 squares x 1.83 manhours/square x $35.99 /manhour = $570.36

Labor (installing hip and ridge units) = 79 LF x 0.028 manhours/LF x $35.99 = $79.61

The total labor cost is $570.36 + $79.61 = $649.97.

The total cost is $1,989.28 + $649.97 = $2,639.25.

Now that you know how to install asphalt shingles and estimate their quantities, let's move on to another asphalt roof covering material, mineral-surfaced roll roofing. That's the topic of the next chapter.

5 Mineral-Surfaced Roll Roofing

▶ The asphalt roofing industry has been around since roll roofing products were developed about a hundred years ago. Technically, roll roofing includes all materials such as asphalt-saturated felts, base sheets, cap sheets, and smooth-coated rolls. I've discussed asphalt-saturated felts in Chapter 3 on underlayment, and Chapter 10 covers roll roofing materials used in built-up roofing. In this chapter, I'll just talk about mineral-surfaced roll roofing — roll roofing whose exposed face is surfaced with granular material.

Roll-Roofing Materials Described

Mineral-surfaced roll roofing is inexpensive. It's also quick and easy to install. But it probably won't last more than 10 years. That's because a maximum of two layers is all that's ever installed, and you use fewer nails than for other types of roofing materials. It also blows off easily. Any weakness, even a small tear, will affect a large deck area, so it's necessary to quickly replace damaged roll roofing to protect the deck.

Mineral-surfaced roll roofing has also been used as underlayment for long-life roof coverings such as tile. But I don't recommend it. It's not intended for use as underlayment and there are far better underlayment materials available that are designed for the purpose.

Mineral-surfaced roll roofing is made of the same materials as asphalt shingles. Use it as a single- or double-coverage roofing material, or for valley flashing as described in Chapter 3.

Single-coverage rolls are usually 3 feet wide and 36 to 38 feet long (one factory square per roll). Rolls weigh from 40 to 90 pounds per square. The 90-pound roll is most common. For use as a single-coverage material, the exposed face is entirely covered with crushed slate embedded in the asphalt surface coating.

As a double-coverage material, 17 inches of the exposed face of the 36-inch-wide roll is surfaced with granular material. This is called *selvage roll roofing* or *split-sheet roofing*. The remaining 19 inches (selvage edge) may have any of a variety of finishes. That finish is usually granule-free and saturated, or saturated and coated with asphalt. Selvage roll roofing usually comes in 36-inch-wide rolls, and in lengths of 18 or 36 feet. Selvage rolls weigh from 55 to 70 pounds per square.

Pattern-edge roll roofing comes in 36-inch widths 42 feet long, or in 32-inch widths 48 feet long. Rolls weigh about 105 pounds per square. One side of the roll is mineral-surfaced, except for a 4-inch-wide center strip. You cut the roll in half down the center, so you get two 18-inch-wide strips with a 16-inch exposure and a 2-inch top lap. In the same way, you get two 16-inch strips with a 14-inch exposure and a 2-inch top lap from the 32-inch roll. Including normal waste, one factory square will cover 100 square feet of roof area.

Storing Mineral-Surfaced Roll Roofing

Store mineral-surfaced rolls upright in a dry, cool area. If you stack the rolls, put plywood sheets between the tiers so the roll ends don't get damaged. If you store rolls outdoors, put them on a platform raised above the ground. Cover the material to protect it from the weather. Don't store rolls in the hot sun.

Modified Bitumen Asphalt (MBA) Roofing

Asphalt mineral-surfaced roll roofing is now largely replaced with a roll product called MBA. It's made of a rubberized asphalt mat reinforced with fiberglass and surfaced on one side with mineral granules. Each roll weighs about 110 pounds. You install the rolls over a 45-pound base sheet and roll on cold-applied bitumen. Figure 5-1 shows this.

You can also *torch on* this material. Use a blow torch to heat the base sheet as you "walk" the MBA roll over and into the base sheet. The torch heats the asphalt in the base sheet and makes the base sheet stick to the MBA.

Figure 5-1 Cold-applied bitumen for roll roofing

Figure 5-2 Installing roll roofing

You can use MBA on flat roofs. Lap the sides about 3 to 4 inches and the ends about 12 inches. Adhere the laps with cold-applied bitumen. Always walk the roll down into the cement as shown in Figure 5-2. If you're installing MBA roofing over insulation, always first install a cover board on top of the insulation.

Installing Mineral-Surfaced Roll Roofing

The biggest problem you'll have with mineral-surfaced roll roofing is its tendency to buckle. Try to install rolls when the weather is above 45° F. In cold weather, cut the rolls into shorter lengths (12 to 18 feet long) and lay them out on the ground to let them warm in the sun.

No matter what method you use to install mineral-surfaced roll roofing, these rules apply:

- Install sheets parallel or perpendicular to the eaves, allowing a $1/4$- to $3/8$-inch overhang at the eaves and rakes.

Direction of rolls	Nailing Method	Minimum roof slope allowed
Parallel to rake	Exposed	4/12
Parallel to rake	Concealed	3/12
Parallel to rakes	Exposed	2/12
Parallel to rakes	Concealed	1/12

Figure 5-3   Methods of applying single-coverage mineral-surfaced roll roofing

- Cut the sheets flush with the rake edges, or run the rolls 3 to 4 inches beyond the roof edge and cut off the excess with a hook blade.
- Install mineral-surfaced roll roofing over a solid deck — plywood is best.
- Be sure the drip edge extends at least 3 inches onto the roof.
- Flash valleys with a 36-inch-wide strip of mineral-surfaced roll roofing before you install the main roof.
- Don't try to bend roll roofing at a 90-degree angle. It'll crack.
- Don't cement mineral-surfaced roll roofing to the sheathing. If the roof deck shifts, the sheets may split. To prevent this when you hot-mop the sheets, nail down a base sheet before you install the mineral-surfaced roll roofing.

Valley Flashing

Starting at the low point, nail valley flashing on 6-inch centers along rows $3/4$ inch from each edge. If a flashing strip is too short to cover the full length of the valley, lap the upper strip over the lower by at least 6 inches. Nail the under layer down and embed the overlap into a bed of roofing cement. Apply only enough cement to securely fasten the sheets. Too much cement will cause blistering.

Mineral-Surfaced Roll Roofing (Single Coverage)

There are four methods of applying single-coverage mineral-surfaced roll roofing. They are described in Figure 5-3. The concealed-nail methods result in a more durable roof.

Exposed-Nail Method

Use this method only for temporary roofing, or on buildings such as storage sheds, because it doesn't last very long. You can install rolls parallel to the eaves, as in Figure 5-4, or parallel to the rake, like Figure 5-5.

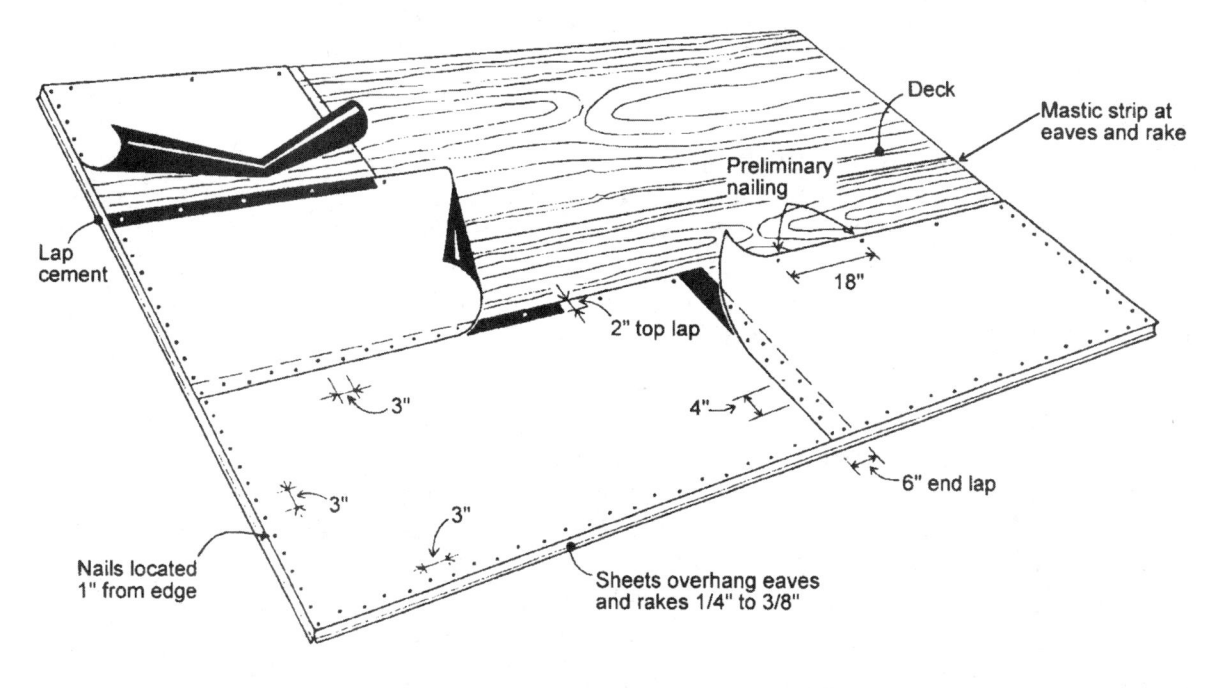

Mineral-Surfaced Roll Roofing

Figure 5-4 Exposed-nail method of applying roll roofing parallel to the eaves

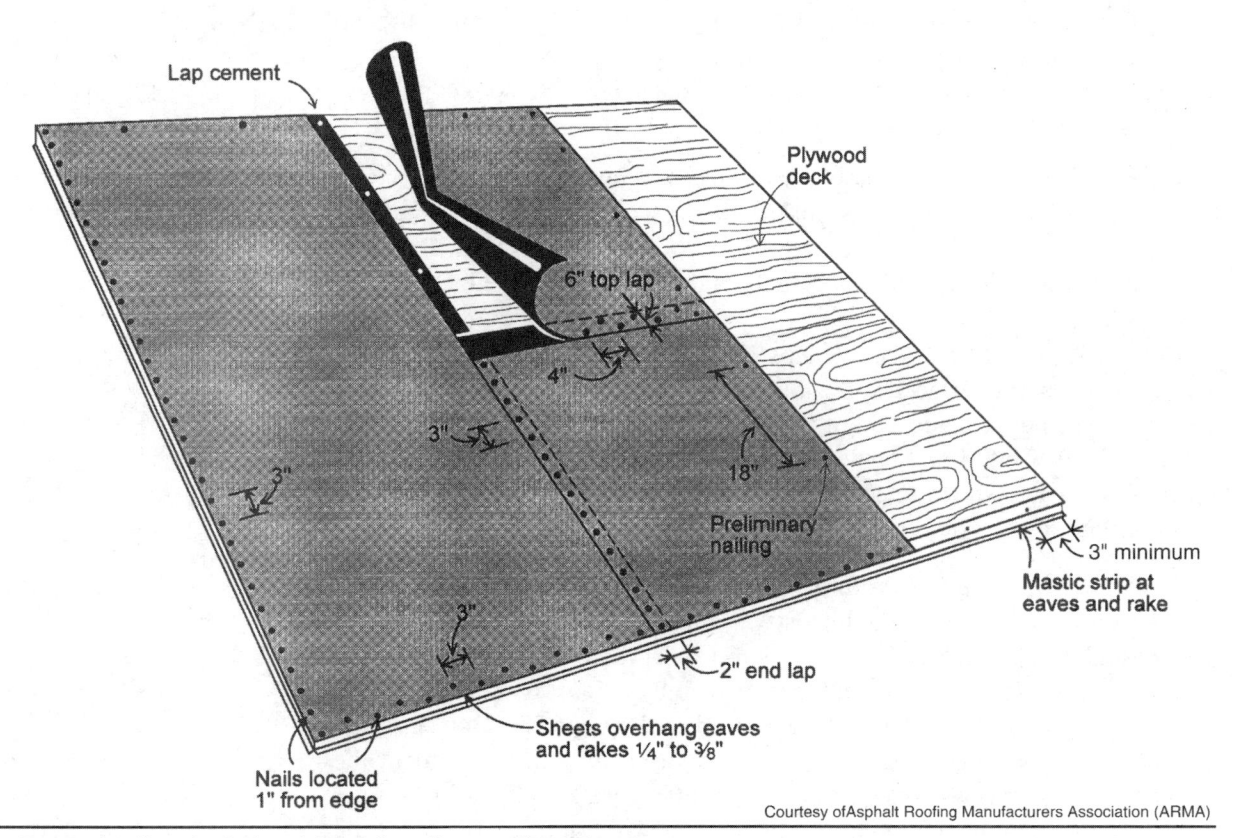

Figure 5-5 Exposed-nail method of applying roll roofing parallel to the rakes

125

When you install mineral-surfaced roll roofing using the exposed-nail method, nail a drip edge along the eaves and rakes. Then flash the valleys as just described.

■ **Exposed-Nail Method (Rolls Applied Parallel to Eaves)** To position the first course of roll roofing, snap a chalk line at 35$\frac{3}{4}$ inches above the eaves (assuming a $\frac{1}{4}$-inch overhang beyond the eaves). Nail the top edge of the first sheet at $\frac{1}{2}$ to $\frac{3}{4}$ inch from the edge on 18- to 20-inch centers. Apply a 2-inch-wide band of roofing cement along the edges of the roof, and nail the sheet on 3-inch centers along the eaves and rake.

Drive nails along the eaves and rake 1 inch from the edges of the sheet. Stagger the nails along the eaves to prevent splitting a solid deck. If you need more than one sheet to complete a course, lap the next sheet by 6 inches. Then stagger-nail the underlay on 4-inch centers in rows 1 and 5 inches from the end of the sheet. Embed the overlap in roofing cement and nail it in place. You can get special lap cement to secure laps of roll roofing. It comes in one-gallon cans or five-gallon pails. Be sure to stagger the end laps in succeeding courses.

Snap a chalk line 2 inches down from the upper edge of the first course. Position the second course on this line so it overlaps the first by 2 inches. Nail the top of the second strip on 18-inch centers as before, to temporarily secure the sheet to the sheathing. Spread a 2-inch-wide strip of roofing cement along the top edge of the underlying sheet and stagger-nail the lap on 3-inch centers at about $\frac{3}{4}$ inch from the edge of the sheet. (Staggering the nails just a little will prevent splitting the deck.) Attach the second sheet at the rakes as described previously.

Continue up the roof as shown in Figure 5-4 until you've covered the entire roof deck. Trim, butt and nail the sheets where they meet at the hips and ridge. If it's not windy, you can save time by nailing down the top edges of all the strips before you cement and nail the bottom edges.

■ **Hip and Ridge Units** To cover hips and ridges, make 12-inch-wide strips by cutting mineral-surfaced roll roofing material lengthwise. Snap a chalk line at 5$\frac{1}{2}$ inches on each side of the centerline of the hips or ridge. Apply a 2-inch-wide band of roofing cement inside each chalk line. Bend the strips lengthwise and lay them over the hips and ridge. In cold weather, heat the roofing with a torch before you bend it.

Install all hip coverings before you lay the ridge coverings. Starting at the low end of the hip, nail the hip cap on 3-inch centers along rows $\frac{3}{4}$ inch from each edge. If a single cap strip is too short to cover the length of the hip, overlap the lower strip with the next strip by at least 6 inches. Nail the under layer down and embed the overlap in roofing cement. Don't use too much cement as it may cause blistering. Neatly trim, nail and cement the end of the cap strip at the ridge, as shown in Figure 5-6. Install the ridge cap the same way as the hip caps. To prevent wind uplift, work with the prevailing wind at your back.

Mineral-Surfaced Roll Roofing

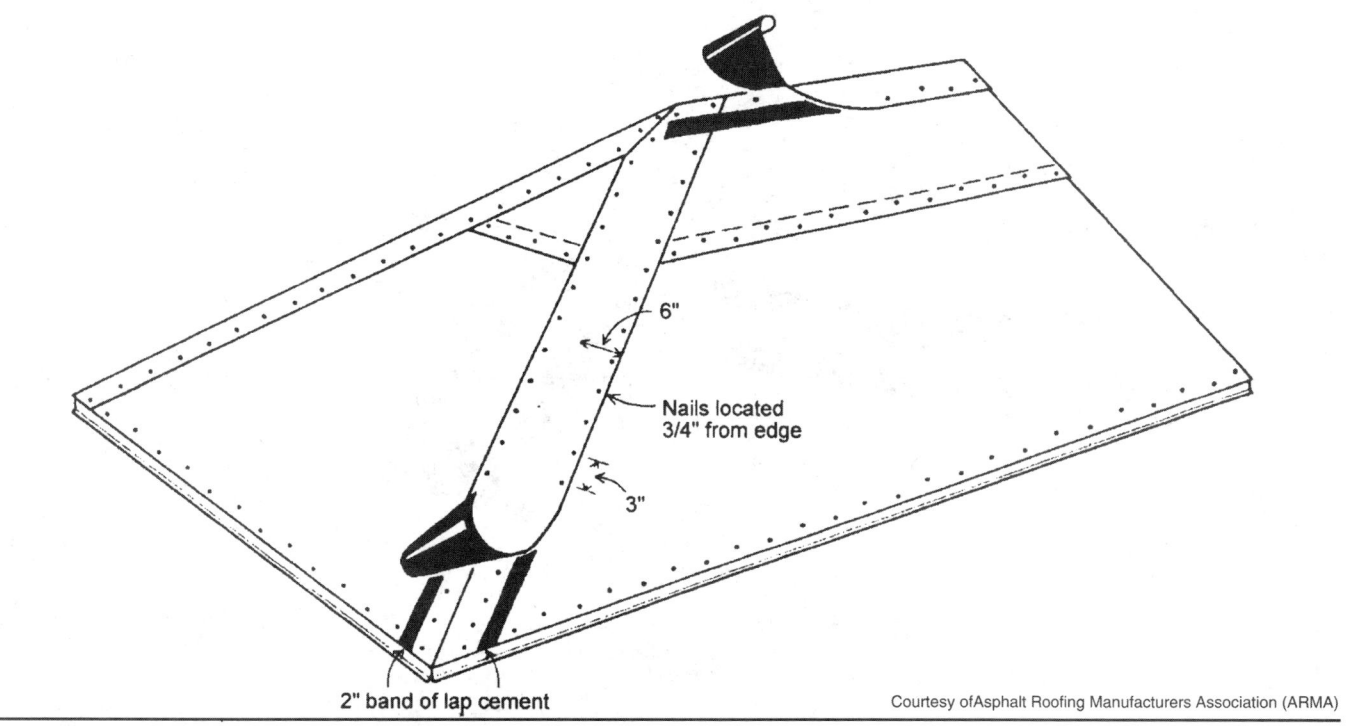

Figure 5-6 Exposed-nail method of applying roll roofing to hips and ridges

■ **Exposed Nail Method (Rolls Applied Parallel to Rake)** Apply the sheets vertically from the ridge. Fasten them with three or four nails. Then unroll toward the eaves. Now, you can use the same application methods as for horizontal installation, as far as laps, cementing, nailing and finishing of hips and ridges are concerned. See Figure 5-5.

Concealed-Nail Method

You can install mineral-surfaced roll roofing using the concealed-nail method on slopes as low as 1 in 12 as long as you install the rolls with a 3-inch top lap.

To use the concealed-nail (blind-nail) method, nail a drip edge to the deck along the eaves and rakes, then flash the valleys as described earlier in this chapter. Before you install the first full sheet, nail down a 9-inch-wide strip of mineral-surfaced roll roofing (edge strip) along the eaves and rakes on 4-inch centers at 1 inch from both edges of the strip. Allow a $1/4$- to $3/8$-inch overhang. See Figure 5-7.

■ **Concealed-Nail Method (Rolls Applied Parallel to Eaves)** Install the first roll with its lower edge and ends flush with the edge strips at the eaves and rakes. To position the first course of mineral-surfaced roll roofing, snap a chalk line $35^3/4$ inches above the eaves (assuming a $1/4$-inch overhang).

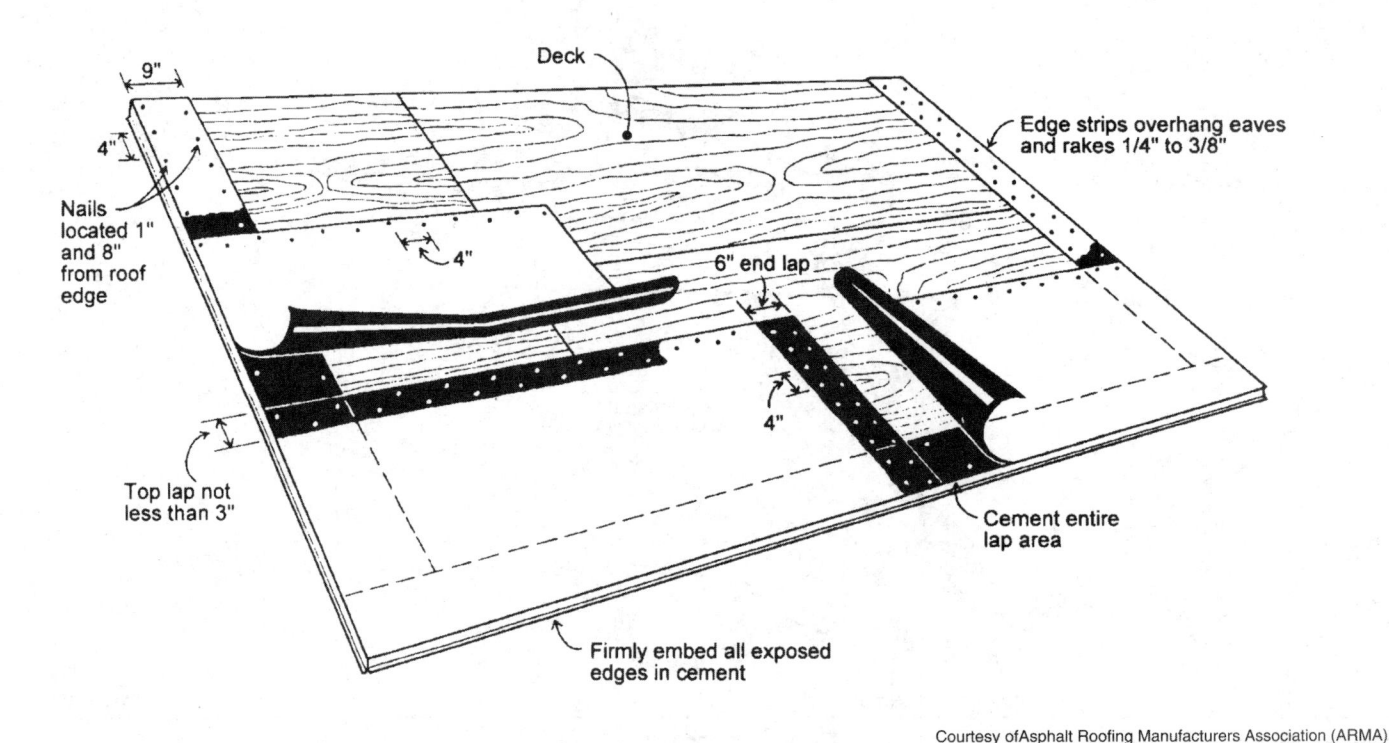

Figure 5-7 Concealed-nail method of applying roll roofing parallel to the eaves

Stagger-nail the top edge of the first sheet on 4-inch centers in rows $3/4$ inch and 2 inches from the edge of the sheet. Locate nails so that the next course overlaps them by at least 1 inch.

Apply roofing cement over the 9-inch strip along the edges of the roof. Embed (don't nail) the edges of the roll along the eaves and rakes into the roofing cement applied over the edge strip. If you need more than one sheet to finish a course, lap the next sheet by at least 6 inches and stagger-nail the under layer on 4-inch centers along rows 1 and 5 inches from the end of the sheet. Embed the overlap in a 6-inch-wide bed of cement.

Snap a chalk line 3 inches down from the upper edge of the first course. Position the second course at the chalk line so it overlaps the first by 3 inches. Nail the upper edge of the second course as you did the first course. Also, embed the edges along the rakes into cement and install overlaps as you did when you installed the first course. Be sure you stagger the end laps in succeeding courses.

Spread a 3-inch-wide strip of cement along the upper edge of the underlying sheet and embed the overlap into the cement (don't nail). Continue up the roof this way until you've covered the entire roof deck. Figure 5-7 shows this process. Trim, butt and nail the sheets as they meet at the hips and ridge. If it isn't too windy, you can save time by nailing down the top edges of all the strips before you cement the bottom edges.

Mineral-Surfaced Roll Roofing

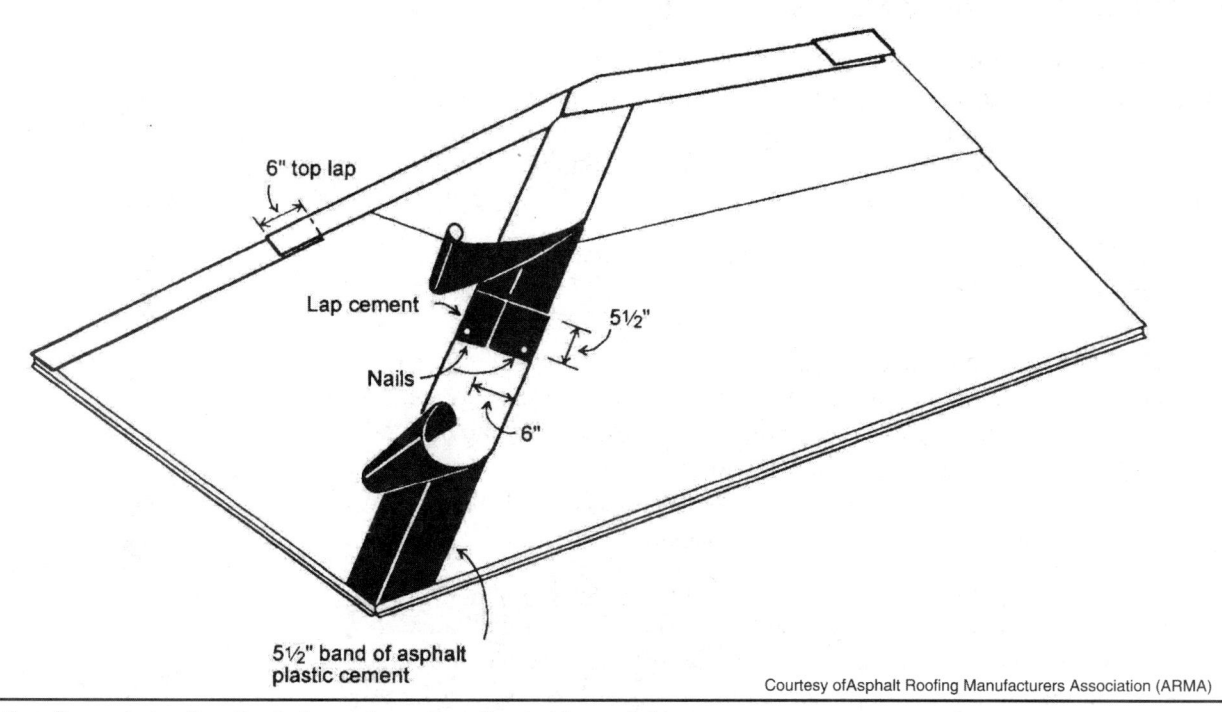

Figure 5-8 Concealed-nail method of applying roll roofing to hips and ridges

■ **Hip and Ridge Units** To make hip and ridge units, cut the roofing material across the roll into 12" x 36" strips. To cover hips and ridges, snap a chalk line 5$\frac{1}{2}$ inches on each side of the centerline of the hip or ridge. Apply a 5$\frac{1}{2}$-inch-wide band of roofing cement within the chalk lines. Bend the strips lengthwise and lay them over the hips and ridge. In cold weather, be sure to heat the roofing before you bend it.

As with the exposed nail method, install the hip units before the ridge units. Starting at the low end of the hip, embed a hip unit into the cement and secure the top with two nails 5$\frac{1}{2}$ inches from the top of the strip. Overlap the lower strip with the next strip by at least 6 inches. Embed the overlap into roofing cement, as shown in Figure 5-8. Use cement sparingly or it may cause blistering. Neatly trim, nail and cement the end of the cap strip at the ridge. Install the ridge caps the same way as the hip caps. To prevent wind uplift, work with the wind at your back.

■ **Concealed-Nail Method (Rolls Applied Parallel to Rake)** Apply the sheets vertically from the ridge. Fasten them with three or four nails. Then unroll toward the eaves. Now, you can use the horizontal application methods as far as edge strips, laps, cementing, nailing and finishing of hips and ridges are concerned.

Double-Coverage Mineral-Surfaced Roll Roofing

Double-coverage mineral-surfaced roll roofing is also called *19-inch selvage double-coverage roll roofing*, or *split-sheet roofing*. You can install this type of roofing on roof slopes as low as 1 in 12. You can install rolls parallel to the eaves as in Figure 5-9, or parallel to the rake, as in Figure 5-10.

To install double-coverage mineral-surfaced roll roofing, nail the drip edge along the eaves, then flash the valleys as previously described.

■ **Roofing Rolls Applied Parallel to Eaves** Before you install the first full sheet, snap a chalk line $18^3/_4$ inches above the eaves (assuming a $^1/_4$-inch overhang). Nail a 19-inch-wide selvage piece (the part without the mineral surface) of the roll along the eaves, allowing a $^1/_4$-inch overhang. Drive staggered nails on 12-inch centers along two rows $4^3/_4$ inches from the top edge, and 1 inch from the bottom edge of the strip.

To locate the first full-width course of roll roofing, snap a chalk line at $35^3/_4$ inches above the eaves. Stagger-nail the top edge on 12-inch centers in rows $4^3/_4$ and $13^1/_4$ inches from the top edge of the sheet. If a course requires more than one sheet, lap succeeding sheets at least 6 inches, and secure it the same as for single-coverage roll roofing.

Apply roofing cement at the rate of $1^1/_2$ gallons per 100 square feet over the 19-inch selvage strip. Embed (don't nail) the exposure part of the first full-width roll into the cement. Don't apply more cement than necessary because it will cause blistering.

Lay the lower edge of the second course in line with the top edge of the exposed part of the underlying sheet. Nail the top of this sheet and embed the bottom of the sheet in cement, as you did the first course. Be sure you stagger the end laps in succeeding courses. Continue up the roof this way until you've covered the entire roof deck, as shown in Figure 5-9. Trim, butt and nail sheets where they meet at the hips and ridge.

Use the mineral-surfaced part of the roll you trimmed from the starter strip at the eaves to finish at the ridge.

■ **Hip and Ridge Units** Cut the roofing material across the roll into 12" x 36" strips. Snap a chalk line at $5^1/_2$ inches on each side of the centerline of the hip or ridge. In cold weather, heat the roofing before you bend it lengthwise to lay over the hips and ridge. Install roofing on the hips before you do the ridge.

Starting at the low end of a hip, nail a 12-inch-wide selvage starter strip 19 inches long along the lower end of each hip. Drive the nails on 4-inch centers 1 inch from the edges of the starter strip. Apply a $5^1/_2$-inch-wide band of roofing cement over the starter strip. Embed the exposed part of the next hip strip into the cement. Secure the selvage part with nails on 4-inch centers 1 inch from the edges of the strip. Continue this up to the

Mineral-Surfaced Roll Roofing

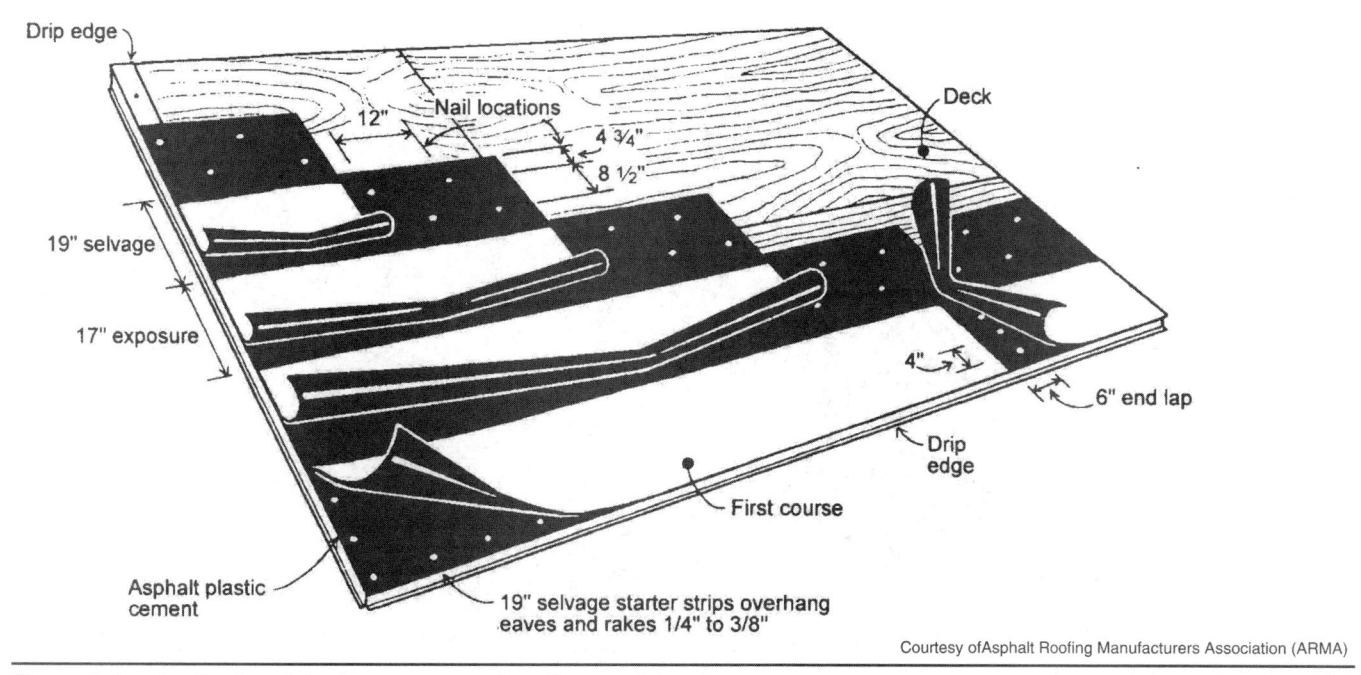

Figure 5-9 Application of double-coverage roll roofing parallel to the eaves

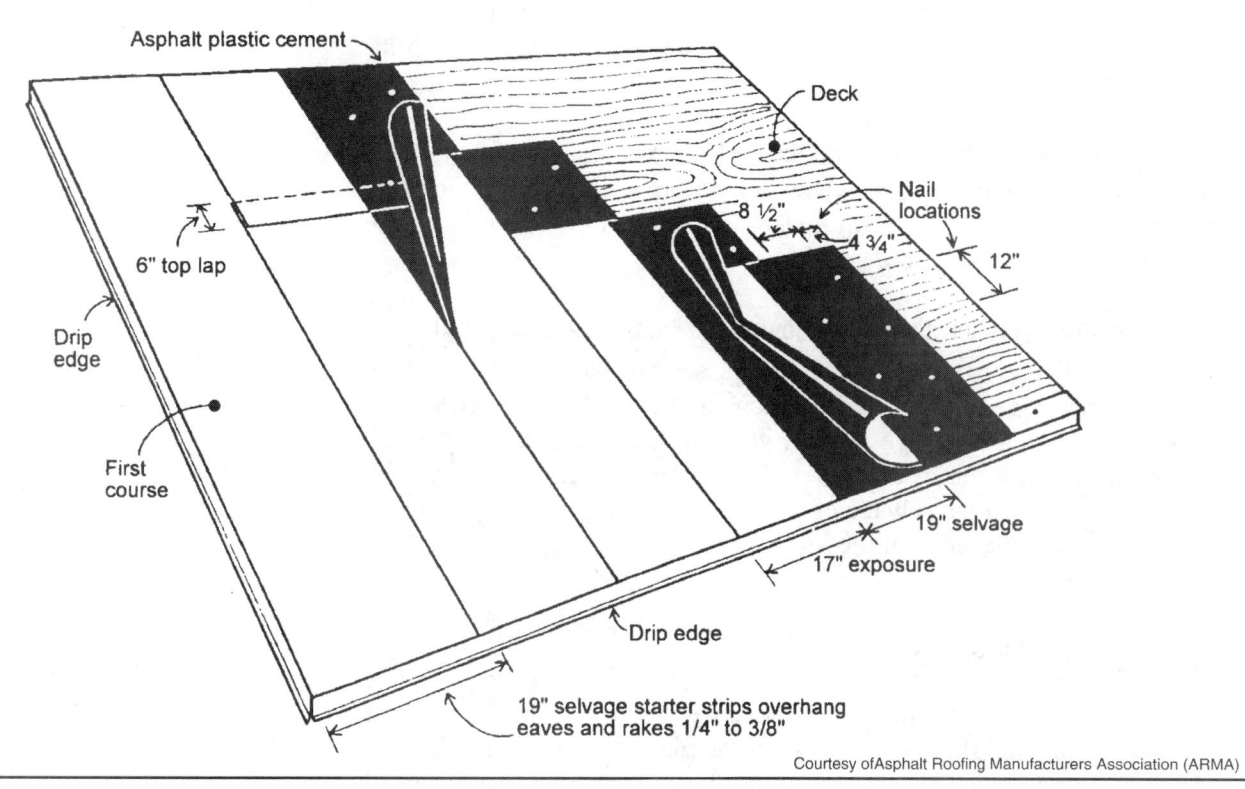

Figure 5-10 Application of double-coverage roll roofing parallel to the rake

131

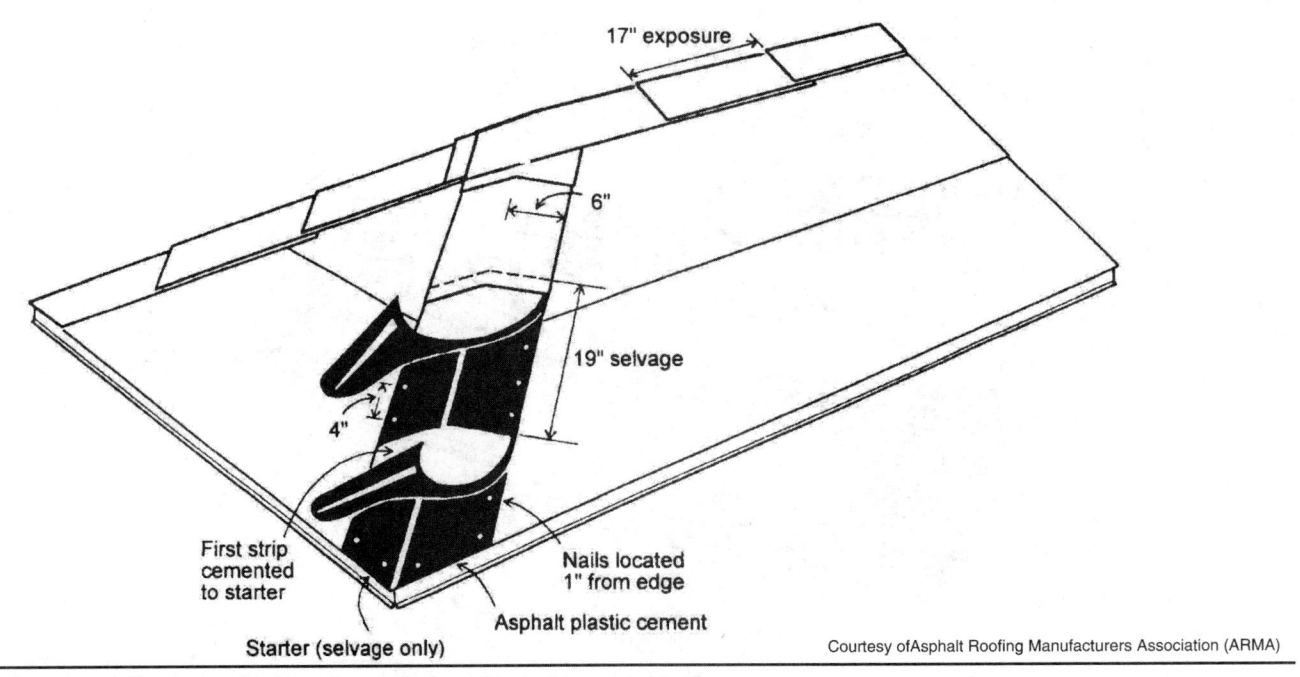

Figure 5-11 Application of double-coverage roll roofing to hips and ridges

ridge, as shown in Figure 5-11. Neatly trim the last hip unit at the ridge. Install the ridge units the same way as the hip units. To prevent wind uplift, work with the wind at your back.

■ **Roofing Rolls Applied Parallel to Rake** Use the installation method just described, except apply the sheets vertically from the ridge. Fasten them with three or four nails and unroll toward the eaves. Look back at Figure 5-10 to see vertical installation.

■ **Finishing Shed Roofs with Double-Coverage Mineral-Surfaced Roll Roofing** When a roof has no ridge, nail the selvage part of the last course along the upper edge of the roof. Then cement a granular-surfaced part of another sheet over the selvage strip. If you're not concerned about looks, use the full width of surfaced material to save trimming. Otherwise, trim the sheet to cover just the selvage of the course below. Then cement metal flashing over the upper edge of the roof.

Vent Pipes and Flashing

To work around a vent pipe with single-coverage roll roofing, cut a hole in the roll roofing and slip the roofing over the pipe. Then slip vent flashing over the pipe and embed the flashing in roofing cement. Nail the edges of the flashing on 2-inch centers. Cover the flashing with roofing cement.

Mineral-Surfaced Roll Roofing

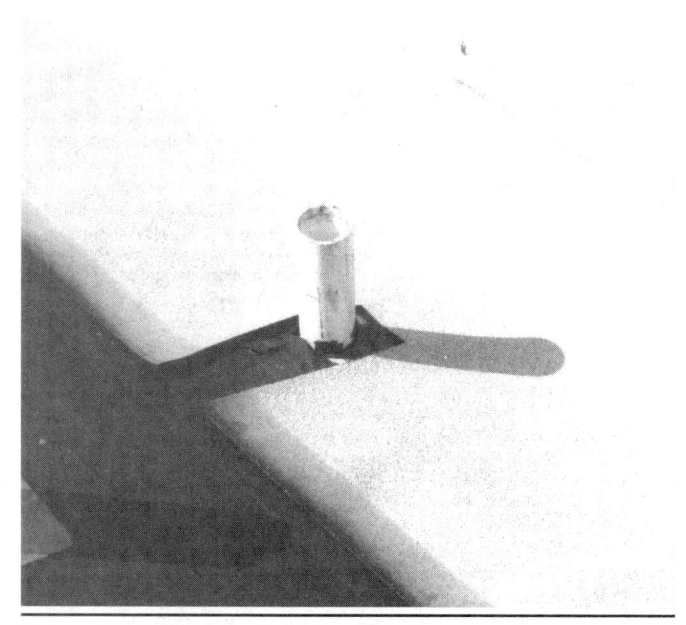

Figure 5-12 Cut the roll to fit around the vent pipe

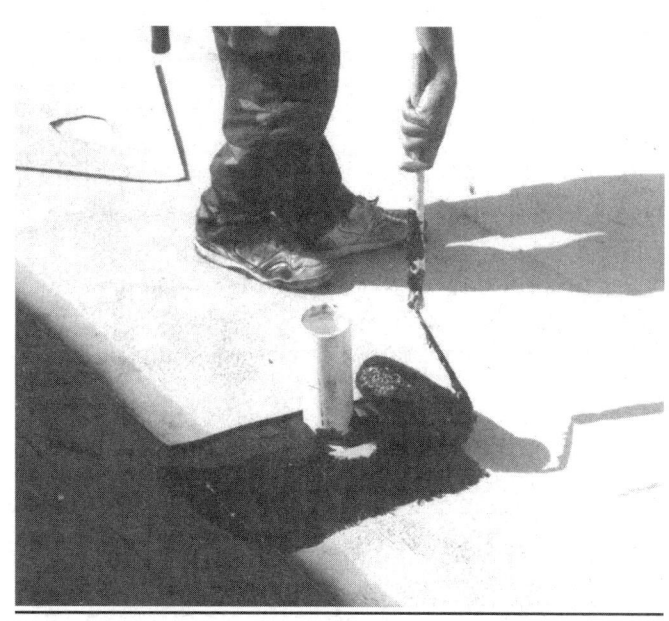

Figure 5-13 Apply roofing cement around the vent pipe

With double-coverage roll roofing, cut a piece out of the underlying roll to fit around the pipe, as shown in Figure 5-12. Next, cut a hole in a sheet of mineral-surfaced roll roofing to use as additional weather protection (shown in the upper left corner of Figure 5-13). Embed the roll roofing flashing in roofing cement. See Figure 5-14. Now, slip the vent flashing over the pipe and embed that in roofing cement. Then cover the vent flashing with roofing cement and install the overlying roll roofing with a hole cut to neatly fit over the vent pipe. You'll have five layers:

1) the deck

2) the bottom layer of mineral-surfaced roll roofing with a slot cut out

3) the embedded piece of coated roofing material with a hole cut to fit the vent

4) an embedded metal flashing covered with roofing cement

5) the top layer of roofing material.

If a roof intersects a vertical surface such as a wall or chimney, install continuous flashing, not step flashing. Refer back to Chapter 4 for details on flashing junctions between roofs and vertical structures.

Figure 5-14 Install roll roofing flashing for added leak protection

Cement Smudges

It's almost impossible to keep roofing cement from squeezing out from between laps and onto the roof surface. To make matters worse, someone always steps in it and tracks it across the roof. It doesn't detract from the roof's performance, but it's unsightly. The easiest way to fix it is to sprinkle out loose granules and rub them into the cement. You can buy tubes of granules or collect your own by rubbing two granule-surfaced shingles or pieces of surfaced roll material together.

Coverage classification	Coverage (FS/Sq.)	Coverage (SF/FS)
Single	1	100
Double	2	50

Figure 5-15 Covering capacity of roll roofing

Estimating Mineral-Surfaced Roll Roofing

First you estimate mineral-surfaced roll roofing quantities (single- or double-coverage) by the square foot. Then you change those measurements to rolls, since that's the way this material is sold. Remember, you'll lose some material due to laps. The amount depends on the installation method you use (exposed or concealed nails). Use Figure 5-15 to roughly change from square feet to rolls.

Figure 5-15 doesn't allow for additional material required for overlaps, starter strips, flashings, concealed nailing, roof layout, or waste due to crew errors.

Additional Material Waste

The type and amount of additional material waste depends on:

- single- or double-coverage installation
- applying the material parallel or perpendicular to the eaves
- exposed or concealed nailing
- the type of roof (gable or hip).

■ **Over-Cutting at the Edge of the Roof** When you apply mineral-surfaced roll roofing (single- or double-coverage; exposed- or concealed-nail method) perpendicular to the edge of a roof, you usually cut the material off beyond the edge, and trim it flush with the edge later. So, let's assume an average of 4 inches of waste beyond:

1) the gable ends of a gable roof when you apply the rolls parallel to the eaves, or

2) the eaves of a gable or hip roof when you apply the rolls perpendicular to the eaves.

This is 0.34 square feet per linear foot of roof edge. Here are some other allowances you'll have to make:

Starter-Strip Allowance: When you use the concealed-nail method on a gable or hip roof, either parallel or perpendicular to the eaves, you must also allow for additional material required for a 9-inch-wide starter strip at the edges of the roof. You'll need 0.75 square foot of material per linear foot of roof edge.

Ridge and Hip Units: No matter which application method you use, add 1 square foot of area to be covered per linear foot of ridge and hips. This assumes you install these units with the same exposure as the rest of the roof.

Selvage-Strip Waste: You have to allow for waste due to the 19-inch-wide selvage strip when you install mineral-surfaced roll roofing for double-coverage.

Whether you install the 19-inch selvage strip at the eaves (rolls parallel to the eaves), or at the rake (rolls perpendicular to the eaves) of a gable roof, you can install the remaining surfaced part of the roll next to the ridge, or at the opposite rake. Therefore, selvage waste on a gable roof is minimal.

When you install rolls parallel to the eaves on a hip roof, you can only use part of the surfaced section remaining from the starter strip roll at the ridge. (You'll also have some cutting waste at the hips.)

Here's a formula for the starter-strip waste at the ridge:

Waste (Ridge) = 4 x E x (L - R - E) | Equation 5-1

where: E = exposure, L = roof length, and R = ridge length

For double coverage, with a 17-inch exposure, the equation becomes:

Waste (Ridge) = 5.67' x (L - R - 1.417') | Equation 5-2

The waste at the hips is:

Waste (Hips) = 4 x E x (Run - E) x Roof-Slope Factor | Equation 5-3

where the Roof-slope factor comes from Column 4 of Appendix A.

Substitute 17 inches for the exposure, and the equation becomes:

Waste (Hips) = 5.67' x (Run - 1.417') x Roof-Slope Factor | Equation 5-4

Find the starter-strip and hip-cutting waste for double-coverage with mineral-surfaced roll roofing *parallel* to the eaves on a hip roof by combining Equations 5-1 and 5-3 as follows:

Waste (At Starter Strip & Hips)
         = 4 x E x [(L - R - E) + (Run - E)] x Roof-Slope Factor | Equation 5-5

where the Roof-slope factor comes from Column 4 of Appendix A.

Again, substituting 17 inches for the exposure, the equation is:

= 5.67' x [(L - R - 1.417') + (Run - 1,417')] x Roof-Slope Factor   **Equation 5-6**

When you apply roll roofing (double coverage) *perpendicular* to the eaves on a hip roof, the equation for total waste (starter-strip and hip-cutting waste) is:

= 4 x E x (Run - E) x Roof-Slope Factor   **Equation 5-7**

where Roof-slope factor comes from Column 4 of Appendix A.

A "Shortcut" for Estimating Mineral-Surfaced Roll Roofing Waste

Use Figures 5-16 through 5-21 to estimate material overrun and waste on various sizes and slopes of roofs, depending on the coverage, roof type, direction of application and nailing method. Notice that these tables don't include waste due to roof layout (where dimensions aren't evenly divisible by roofing material exposure). You'll find the formulas for non-conforming roof layout waste following these tables.

Waste from Non-conforming Roof Layout

Regardless of roof type, coverage, direction of application, or nailing method, you may have extra material waste due to the length of:

1) The rake (material installed parallel to the eaves), or

2) The eaves (material installed perpendicular to the eaves).

For example, you're using double coverage (17-inch exposure) installed perpendicular to the eaves. If the length of the eaves is evenly divisible by the exposure, there won't be any waste. Here are some more cases:

Eaves length	Courses
28 feet, 4 inches	20
29 feet, 9 inches	21
31 feet, 2 inches	22

Mineral-Surfaced Roll Roofing

Gable roof material overrun (Roll roofing, single coverage, exposed nail method)			
Total percentage to be added to net roof area including 4-inch cutting waste at rake (roofing applied parallel to eaves), or eaves (roofing applied perpendicular to eaves). Figures include allowance for material required to make ridge units			
Building Dimensions	Roof Slope		
(L x W)	3/12	6/12	12/12
30 x 20	5 (7)	5 (7)	4 (5)
40 x 30	5 (6)	5 (6)	4 (4)
45 x 30	5 (6)	4 (6)	4 (4)
50 x 30	4 (6)	4 (6)	3 (4)
60 x 30	4 (6)	4 (6)	3 (4)
70 x 30	4 (6)	4 (6)	3 (4)
80 x 30	4 (6)	4 (6)	3 (4)
50 x 40	3 (4)	3 (4)	3 (4)
60 x 40	3 (4)	3 (4)	3 (4)
70 x 40	3 (4)	3 (4)	3 (4)
80 x 40	3 (4)	3 (4)	3 (4)
90 x 40	3 (4)	3 (4)	3 (4)
60 x 50	3 (4)	3 (4)	2 (2)
70 x 50	3 (4)	3 (4)	2 (2)
80 x 50	3 (4)	3 (4)	2 (2)
90 x 50	3 (4)	3 (4)	2 (2)
Roofing applied parallel to eaves. (Figures in parentheses are for roofing applied perpendicular to eaves.) Waste due to crew error and non-conforming roof layout not included.			

Figure 5-16   Gable roof material overrun (Roll roofing, single coverage, exposed nail method)

Hip roof material overrun (Roll roofing, single coverage, exposed nail method)			
Total percentage to be added to net roof area. Allowance is included for ridge and hip units and 4-inch waste at the eaves when roofing is applied perpendicular to eaves.			
Building Dimensions	Roof Slope		
(L x W)	3/12	6/12	12/12
30 x 20	11 (16)	11 (16)	9 (13)
40 x 30	8 (12)	8 (12)	7 (10)
45 x 30	7 (11)	7 (10)	6 (9)
50 x 30	7 (10)	6 (9)	6 (9)
60 x 30	7 (10)	7 (10)	5 (7)
70 x 30	6 (9)	6 (9)	5 (7)
80 x 30	6 (9)	5 (8)	4 (6)
50 x 40	7 (10)	6 (9)	5 (7)
60 x 40	6 (9)	6 (9)	5 (7)
70 x 40	5 (8)	5 (7)	5 (7)
80 x 40	5 (7)	4 (6)	4 (6)
90 x 40	4 (6)	4 (6)	4 (6)
60 x 50	3 (4)	5 (7)	4 (6)
70 x 50	5 (7)	5 (7)	4 (6)
80 x 50	5 (7)	4 (6)	4 (6)
90 x 50	4 (6)	4 (6)	4 (6)
Roofing applied parallel to eaves. (Figures in parentheses are for roofing applied perpendicular to eaves.) Waste due to crew error and non-conforming roof layout not included.			

Figure 5-17   Hip roof material overrun (Roll roofing, single coverage, exposed nail method)

Gable roof material overrun (Roll roofing, single coverage, concealed nail method)			
Total percentage to be added to net roof area including 4-inch cutting waste at rake (roofing applied parallel to eaves), or eaves (roofing applied perpendicular to eaves). Also included are ridge units and a 9-inch wide starter strip.			
Building Dimensions	**Roof Slope**		
(L x W)	3/12	6/12	12/12
30 x 20	18 (20)	17 (19)	14 (15)
40 x 30	14 (15)	13 (14)	11 (11)
45 x 30	13 (12)	13 (14)	11 (11)
50 x 30	12 (14)	12 (14)	10 (11)
60 x 30	11 (13)	11 (13)	9 (10)
70 x 30	11 (13)	11 (13)	9 (10)
80 x 30	11 (13)	10 (12)	8 (9)
50 x 40	10 (11)	9 (10)	9 (10)
60 x 40	9 (10)	9 (10)	8 (9)
70 x 40	9 (10)	9 (10)	8 (9)
80 x 40	9 (10)	8 (9)	8 (9)
90 x 40	8 (9)	8 (9)	7 (8)
60 x 50	8 (9)	8 (9)	7 (7)
70 x 50	8 (9)	8 (9)	6 (6)
80 x 50	8 (9)	8 (9)	6 (6)
90 x 50	8 (9)	7 (9)	6 (6)
Roofing applied parallel to eaves. (Figures in parentheses are for roofing applied perpendicular to eaves.) Waste due to crew error and non-conforming roof layout not included.			

Figure 5-18 Gable roof material overrun (Roll roofing, single coverage, concealed nail method)

Hip roof material overrun (Roll roofing, single coverage, exposed nail method)			
Total percentage to be added to net roof area. Allowance is included for ridge and hip units and 4-inch waste at the eaves when roofing is applied perpendicular to eaves.			
Building Dimensions	**Roof Slope**		
(L x W)	3/12	6/12	12/12
30 x 20	23 (28)	22 (27)	18 (22)
40 x 30	17 (21)	16 (20)	13 (16)
45 x 30	15 (19)	15 (18)	12 (15)
50 x 30	15 (18)	13 (16)	12 (15)
60 x 30	14 (17)	14 (17)	10 (12)
70 x 30	13 (17)	12 (15)	10 (12)
80 x 30	13 (17)	11 (14)	9 (11)
50 x 40	14 (17)	12 (15)	10 (12)
60 x 40	12 (15)	12 (15)	9 (11)
70 x 40	11 (14)	10 (12)	9 (11)
80 x 40	11 (13)	9 (11)	8 (10)
90 x 40	9 (11)	9 (11)	8 (10)
60 x 50	10 (12)	10 (12)	8 (10)
70 x 50	10 (12)	10 (12)	8 (10)
80 x 50	10 (12)	8 (10)	8 (10)
90 x 50	8 (9)	8 (10)	7 (9)
Roofing applied parallel to eaves. (Figures in parentheses are for roofing applied perpendicular to eaves.) Waste due to crew error and non-sconforming roof layout not included.			

Figure 5-19 Hip roof material overrun (Roll roofing, single coverage, concealed nail method)

Mineral-Surfaced Roll Roofing

Gable roof material overrun (Roll roofing, double coverage)			
Total percentage to be added to net roof area including 4-inch cutting waste at the rake (roofing applied parallel to eaves), or eaves (roofing applied perpendicular to eaves). Also included are ridge units.			
Building Dimensions	Roof Slope		
(L x W)	3/12	6/12	12/12
30 x 20	5 (7)	5 (7)	4 (5)
40 x 30	5 (6)	5 (6)	4 (4)
45 x 30	5 (6)	4 (6)	4 (4)
50 x 30	4 (6)	4 (6)	3 (4)
60 x 30	4 (6)	4 (6)	3 (4)
70 x 30	4 (6)	4 (6)	3 (4)
80 x 30	4 (6)	4 (6)	3 (4)
50 x 40	3 (4)	3 (4)	3 (4)
60 x 40	3 (4)	3 (4)	3 (4)
70 x 40	3 (4)	3 (4)	3 (4)
80 x 40	3 (4)	3 (4)	3 (4)
90 x 40	3 (4)	3 (4)	3 (4)
60 x 50	3 (4)	3 (4)	2 (2)
70 x 50	3 (4)	3 (4)	2 (2)
80 x 50	3 (4)	3 (4)	2 (2)
90 x 50	3 (4)	3 (4)	2 (2)
Roofing applied parallel to eaves. (Figures in parentheses are for roofing applied perpendicular to eaves.) Waste due to crew error and non-conforming roof layout not included.			

Figure 5-20 Gable roof material overrun (Roll roofing, double coverage)

Hip roof material overrun (Roll roofing, double coverage)			
Total percentage to be added to net roof area including cutting waste at starter strip and hip, plus a 4-inch waste allowance at the eaves (roofing applied perpendicular to eaves) Also included are hip and ridge units.			
Building Dimensions	Roof Slope		
(L x W)	3/12	6/12	12/12
30 x 20	35 (23)	33 (23)	29 (20)
40 x 30	27 (18)	26 (18)	22 (19)
45 x 30	24 (17)	23 (15)	19 (14)
50 x 30	23 (15)	22 (15)	18 (14)
60 x 30	19 (13)	20 (13)	15 (11)
70 x 30	17 (13)	16 (13)	14 (10)
80 x 30	16 (12)	14 (11)	13 (10)
50 x 40	22 (14)	21 (14)	17 (12)
60 x 40	19 (13)	17 (12)	15 (11)
70 x 40	16 (12)	16 (11)	13 (9)
80 x 40	15 (10)	14 (10)	12 (9)
90 x 40	14 (10)	12 (9)	11 (9)
60 x 50	18 (11)	17 (11)	14 (10)
70 x 50	16 (11)	15 (10)	13 (9)
80 x 50	14 (9)	13 (9)	12 (9)
90 x 50	13 (9)	12 (9)	10 (8)
Roofing applied parallel to eaves. (Figures in parentheses are for roofing applied perpendicular to eaves.) Waste due to crew error and non-conforming roof layout not included.			

Figure 5-21 Hip roof material overrun (Roll roofing, double coverage)

But when the eaves length isn't evenly divisible by the exposure, you have to trim and throw away part of the last strip. Or you can increase the size of the top lap of each roll (and decrease the exposure) so the edge of the last roll is flush with the edge of the rake. In either case, you'll need extra material.

The waste due to non-conforming roof layout can be so small you don't have to think about it. That's true of most of the example problems that follow. However, in Example 5-3 you can see that roof layout waste can be major, even with a short ridge. So, you should always calculate waste due to non-conforming roof layout.

To find waste due to non-conforming roof layout, first find the width of the last strip installed. For rolls parallel to the eaves (gable or hip roofs), the width of the last strip is:

Last strip width (LSW) = E x the decimal part (numbers to the right of the decimal point) of the expression:

$$\frac{W/2 \times \text{Roof–Slope Factor}}{E}$$

Equation 5-8

where W = roof width, E = exposure length, and the roof-slope factor comes from Column 2 of Appendix A.

Here are the exposure widths for various mineral-surfaced roll roofing applications:

Exposure for double coverage: 17 inches, or 1.417 feet.

Exposure for exposed-nail, single coverage: 34 inches, or 2.83 feet.

Exposure for single coverage, concealed-nail method: 33 inches, or 2.75 feet.

For rolls perpendicular to the eaves of a hip roof, single or double coverage, the non-conforming roof layout waste is negligible. For rolls perpendicular to the eaves of a gable roof, the last strip width is:

LSW = E x the decimal part of the expression: $\dfrac{L}{E}$

Equation 5-9

Where L is the roof length and E is the exposure.

For single-coverage roll roofing:

1) If LSW is less than or equal to half the exposure, the waste in square feet per linear foot of ridge (rolls parallel to the eaves on gable or hips roofs) or per linear foot of a single rake (rolls perpendicular to the eaves on gable roofs), is:

Waste (SF/LF) = E - (2 X LSW)

Equation 5-10

Mineral-Surfaced Roll Roofing

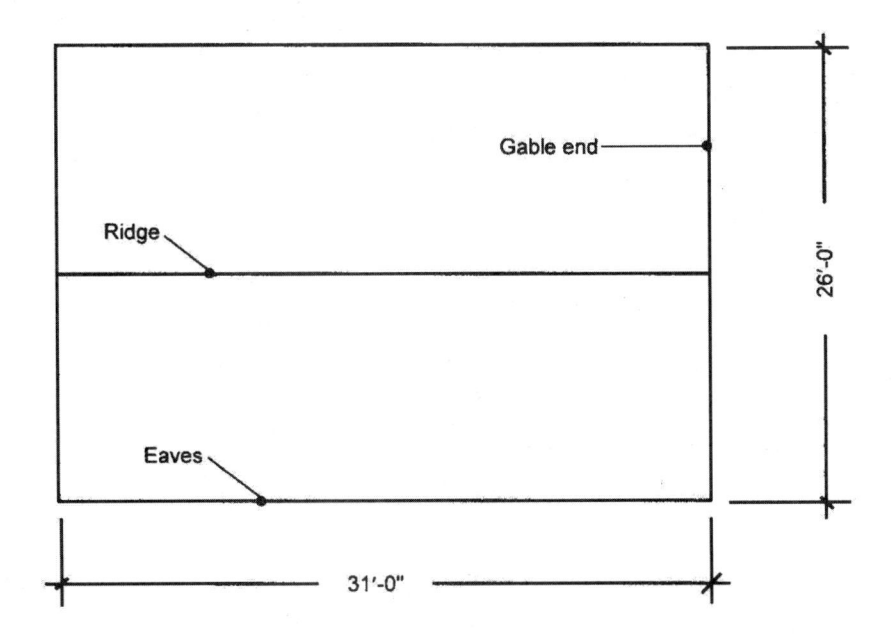

Figure 5-22 Gable roof example

2) If LSW is greater than half the exposure, the waste is:

Waste (SF/LF) = 2 x (E - LSW) | **Equation 5-11**

This equation is also true for double-coverage roofing. Waste is in square feet per linear foot of ridge (rolls applied parallel to eaves), or per linear foot of a single rake (rolls applied perpendicular to eaves).

Examples

▼ **Example 5-1:** Assume a 5-in-12 roof slope and single-coverage roll roofing (exposed-nail method) applied parallel to the eaves. Find the number of rolls required to cover the roof of the building in Figure 5-22.

First, remember from Chapter 1 that **Net Roof Area = Roof Plan Area x Roof-Slope Factor.** The roof-slope factor is from Column 2 of Appendix A.

Net Roof Area = 31' x 26' x 1.083
             = 873 square feet ÷ 1.083
             = 8.73 squares (Use this quantity to determine labor costs)

The total length of rake is 56 linear feet (4 x 13' x 1.083).

Cutting waste at the rakes is 19 square feet (that's 56 feet x 0.34 SF/LF where 0.34 SF/LF is the 4-inch waste for over-cutting at the eaves or rake).

The total length of ridge is 31 feet. Since you must allow 1 square foot per linear foot of ridge, allow 31 square feet for the ridge.

In Chapter 1, we defined the waste factor with the equation:

Waste Factor = Total Area Covered ÷ Net Roof Area

Therefore, waste so far, excluding waste due to non-conforming roof layout, is:

$$\text{Waste Factor} = \frac{873 \text{ SF} + 19 \text{ SF} + 31 \text{ SF}}{873 \text{ SF}}$$

$$= \frac{923 \text{ SF}}{873 \text{ SF}}$$

$$= 1.06$$

This is consistent with Figure 5-16 which predicts about 5 percent waste for a building of this approximate size and roof slope.

From Equation 5-8, we know that

Last strip width (LSW) = E x the decimal part of the expression:

$$\frac{\textbf{W/2 x Roof-Slope Factor}}{\textbf{E}}$$

where E is the exposure and W is the width of the roof

The expression $\frac{W/2 \times \text{Roof-slope factor}}{E}$ evaluates to $\frac{(26'/2) \times 1.083}{2.83'}$

$$= (13' \times 1.083) \div 2.83'$$
$$= 4.97$$

Now, multiply only the decimal part of 4.97 by the exposure:

$$= 0.97 \times 2.83'$$
$$= 2.77 \text{ feet}$$

Now, using Equation 5-11, you can calculate waste per linear foot of ridge:

Waste (SF/LF) = 2 x (E - LSW)

$$= 2 \times (2.83' - 2.77')$$
$$= 0.12 \text{ square feet per linear foot}$$

Waste due to non-conforming roof layout is:

Waste (Roof Layout) = 31' x 0.12 SF/LF
= 4 square feet (rounded up from 3.72)

So, total roof area to cover is:

Gross Roof Area = 873 SF + 19 SF + 31 SF + 4 SF
= 927 square feet ÷ 100
= 9.27 squares

Since this is a single-coverage job, Figure 5-15 shows we need one roll of material per square:

Rolls = 9.27 squares x 1/roll/square
= 10 rolls

▼ **Example 5-2:** Here's how to calculate this for roofing applied perpendicular to the eaves:

The net roof area from Example 5-1 is 8.73 squares. (Use this quantity to figure labor cost.)

In Figure 5-22 you see the total eaves length is 62 linear feet (2 x 31'). Cutting waste at the eaves is 21 square feet (62 feet x 0.34 SF/LF). Allowance for the ridge is 31 square feet (allowing 1 SF/LF).

When you calculate the waste factor (Waste factor = Total area covered ÷ net roof area), you get 1.06. This is consistent with Figure 5-16 which predicts a waste factor of 6 to 7 percent.

Now, from Equation 5-9, calculate the last strip width:

LSW = E x the decimal part of the expression L ÷ E

First, calculate L ÷ E

31 ÷ 2.83 = 10.95

Now, multiply only the decimal part of that number by the exposure:

= 0.95 x 2.83'
= 2.69 feet

Since LSW is more than half the exposure, waste (from Equation 5-11) per linear foot of a single rake is:

Waste (SF/LF) = 2 x (2.83' - 2.69')
= 0.14 square feet per linear foot

The length of a single rake is:

LF (One Rake) = 13' x 1.083 (length times roof-slope factor
from Appendix A)
= 14.1 linear feet

Waste due to non-conforming roof layout is:

Waste (Roof Layout) = 14.1' x 0.14 SF/LF
= 2 square feet

So, the total material required is:

Gross Roof Area = 873 SF + 21 SF + 31 SF + 2 SF
= 927 square feet ÷ 100
= 9.27 squares

You should order 10 squares of rolled roofing (1 roll per square, single coverage).

▼ **Example 5-3:** Now let's assume we're using the concealed-nail method with rolls parallel to the eaves for the same building as the last two examples (Figure 5-22).

You already know the net roof area is 8.73 squares (that's the amount you use to figure labor costs). Cutting waste at the rakes is 19 square feet and the material allowance at the ridge is 31 square feet.

The total length of 9-inch-wide starter strip required is:

LF (Starter Strip) = (2 x 31') + (2 x 26' x 1.083)
= 118.3 linear feet
= 89 square feet (118.3 x 0.75 SF/LF)

To check the waste factor:

$$\text{Waste factor} = \frac{873 \text{ SF} + 19 \text{ SF} + 31 \text{ SF} + 89 \text{ SF}}{873 \text{ SF}}$$
= 1012 SF ÷ 873 SF
= 1.16

This agrees with Figure 5-18, which predicts a range from 13 to 17 percent.

Now, find the last strip width:

First, $\frac{26'}{2}$ x 1.083 = 14.08

Then, 14.08 ÷ 2.75' = 5.12

Multiply the decimal part of that answer by the exposure:

0.12 x 2.75' = 0.33

Now, because the LSW is less than half the exposure, use Equation 5-10 to calculate the waste due to non-conforming roof layout this way:

Waste (SF/LF) = E - (2 X LSW)
Waste (SF/LF) = 2.75' - (2 X 0.33')
= 2.09 square feet per linear foot of ridge

Mineral-Surfaced Roll Roofing

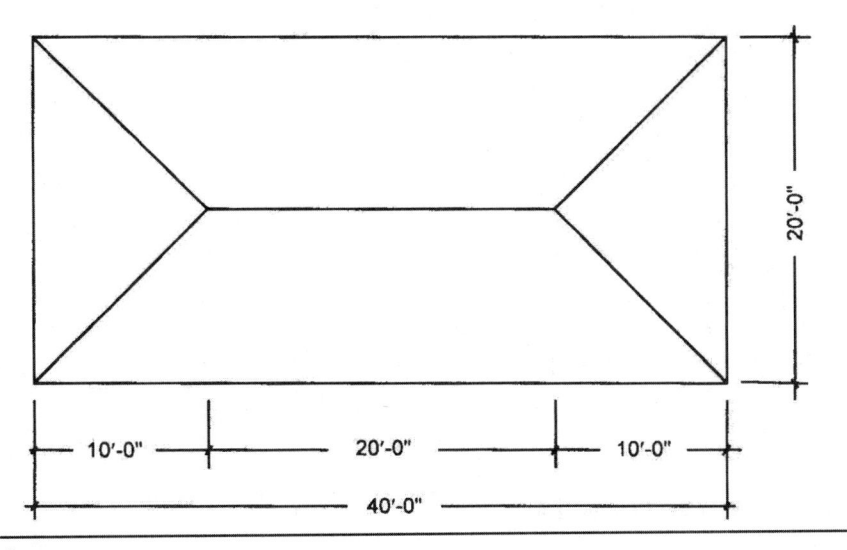

Figure 5-23 Hip roof example

Waste due to non-conforming roof layout is:

Waste (Roof Layout) = 31' x 2.09 SF/LF
= 65 square feet

You should order a total of 1012 SF + 65 SF, or 1077 SF. For single coverage, you need 1 roll per square, so this job requires 11 rolls of material.

▼ **Example 5-4:** Now, let's try a couple of examples for a hip roof. Assume a roof slope of 6 in 12 and the installation of single-coverage roll roofing (exposed nail method) applied parallel to the eaves, and find the number of rolls required to cover the roof of the building in Figure 5-23.

You calculate the net roof area from the equation:

Net Roof Area = Roof Plan Area x Roof-Slope Factor

Net Roof Area = 40' x 20' x 1.118 (from Column 2, Appendix A)
= 894 square feet ÷ 100
= 8.94 squares (use this quantity to estimate labor costs)

The total ridge length is **Length - Width**, or 20 linear feet.

From Column 3 of Appendix A, you calculate the hip length as

Run x Roof-Slope Factor

From the drawing in Figure 5-23 you see the hip is 10' x 1.5, or 15 linear feet.

There are four hips, so the total hip length is 60 linear feet. Total length of hips and ridge is 80 linear feet.

Fabrication allowance at hips and ridge is 1 SF/LF, or 80 square feet.

So far, the waste excluding allowance for non-conforming roof layout is:

$$\text{Waste Factor} = \frac{894 \text{ SF} + 80 \text{ SF}}{894 \text{ SF}}$$
$$= \frac{974 \text{ SF}}{894 \text{ SF}}$$
$$= 1.09, \text{ or } 9 \text{ percent}$$

Figure 5-17 predicts a range of 8 to 11 percent waste for a building about this size and roof slope.

The last strip width is 2.69 feet (from Equation 5-8).

From Equation 5-11, the waste per linear foot of ridge is 0.28 square feet per linear foot.

Now you can calculate waste due to non-conforming roof layout as 0.28 SF/LF x 20', or 6 square feet (rounded up from 5.6)

Thus, the total roof area is 894 SF + 80 SF + 6 SF = 980 square feet, or 9.8 squares. Since this is a single-coverage job, you need 1 roll per square (Figure 5-15), or 10 rolls of material.

▼ **Example 5-5:** This time, figure on double-coverage applied parallel to the eaves.

From the above example, we already know the net roof area (8.94 squares) and the waste for ridge and hip units (80 square feet).

From Equation 5-6, cutting waste at the 19-inch-wide selvage strip and along the eaves is:

Waste (Starter Strip and Hips)
    = 5.67' x [(40' - 20' - 1.417') + (10' - 1.417')] x 1.061
    = 163 square feet

Waste excluding that for non-conforming roof layout calculates to 27 percent (1137 ÷ 894). This agrees with the range of 26 to 33 percent in Figure 5-21.

From Equation 5-8 the last strip width is 1.27 feet.

The ridge length is 20 feet, the waste per linear foot of ridge (from Equation 5-11) is 0.29 square feet per linear foot. So the waste due to non-conforming roof layout is 6 square feet (rounded up from 5.8).

The gross roof area   = 894 SF + 80 SF + 163 SF + 6 SF
                           = 1,143 square feet ÷ 100
                           = 11.43 squares

From Figure 5-15, the material requirement is 2 rolls per square, or 23 rolls.

Estimating Mineral-Surfaced Roll Roofing Costs

Let's say you've got to cover the roof shown in Figure 5-22 with single-coverage mineral-surfaced roll roofing using the concealed-nail method with rolls parallel to the eaves. Using prices from the *National Construction Estimator* at the time of this writing, each roll of mineral-surfaced roll roofing costs $60.34, including sales tax.

From Example 5-3, you must purchase 11 rolls, at a cost of:

11 rolls x $60.34/roll = $663.74

The roof deck area is 8.73 squares. According to the *National Construction Estimator*, an R1 crew consisting of one roofer and one laborer can install mineral-surfaced roll roofing at the rate of 1 square per 1.03 manhours, and a roofer is costing the employer $41.66 an hour and the laborer $30.31, including labor burden. Average cost per manhour is $35.99:

$41.66 + $30.31 = $71.97 ÷ 2 = $35.99

So the labor cost will be:

8.73 squares x 0.20 manhours/square x $35.99/manhour = $62.84

Total job cost will be:

$663.74 + $62.84 = $726.58

If all this is clear to you, you're ready to move on to the next chapter — on the application of wood shingles on sloping roofs.

6 Wood Shingles and Shakes

▶ You can expect a wood shingle or shake roof to last from 25 to 30 years. Some last up to 50 years. They're more expensive than asphalt shingles, but they're popular because they have a charming, rustic appearance that becomes even more attractive after a few years of weathering.

Although they can be made somewhat fire-resistant by pressure-treating with fire-retardant (Certi-guard), they're still more fire-prone than other roof coverings and in many areas of the country subject to forest fires, like a number of jurisdictions in California, you can no longer install them.

Other areas permit wood roofs provided they be the fire-resistant type and installed over a solid deck of at least 1/2-inch-thick plywood or equivalent material that will earn a Class "B" fire-rating. Getting fire insurance on a house with a wood shingle roof could also be a problem.

Don't install untreated wood shingles and shakes where heat and humidity are severe because they'll be very susceptible to decay from moss, mildew and fungus. Figure 6-1 shows humid locations in the United

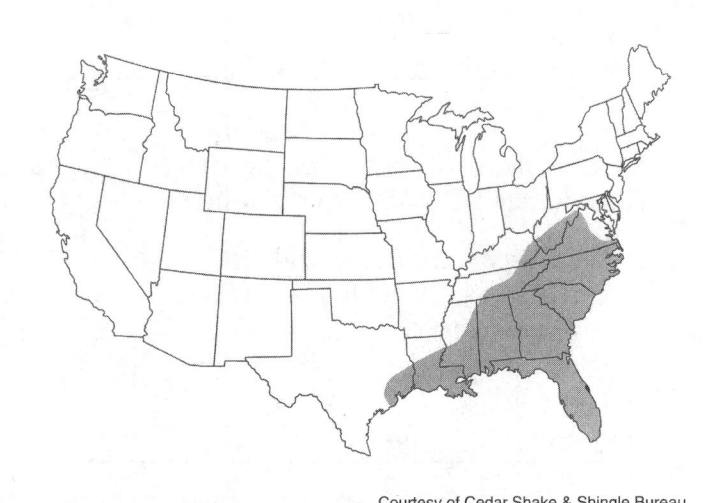

Courtesy of Cedar Shake & Shingle Bureau

Figure 6-1 Locations of severe humidity

149

States. Use treated (Certi-last) shingles on low-slope roofs, too. That's because a low-slope roof doesn't shed water as well as roofs with a higher pitch. And on roofs shadowed by overhanging trees use treated shingles because the trees will keep the roof wet longer.

You can get wood shingles and shakes that carry a 30-year warranty if they're pressure-treated with a wood preservative at the factory. If the shingles haven't been factory-treated, you can apply a fungicidal wood preservative such as copper or zinc napthenate after the roof has weathered for a year. This waiting time lets the natural oils in the cedar leach out, so the wood is porous and soaks up more of the preservative.

In low-humidity climates, you can apply a commercial oil-based preservative to help prevent excessive dryness, which leads to curling or cracking.

Site-applied surface treatments are only temporary. Following initial application, you should re-apply a preservative about every five years. You can also extend the life of any roof by periodically removing leaves and other debris which hold moisture. This reduces the potential for growth of moss and fungi.

Wood shingles and shakes are rigid, so they're extremely resistant to wind uplift (up to 130 mph). Cedar shingles have twice the insulating value of standard asphalt shingles, four times the value of cement shingles, and five times that of slate. And they increase the overall strength of the building. Depressions from hailstones disappear after a short time, because the wood fibers recover their original shape.

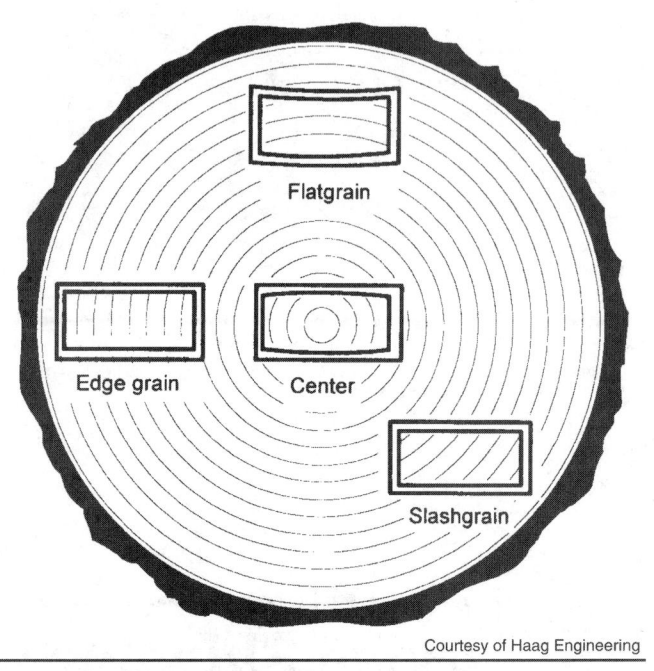

Figure 6-2 Wood shingle grain

How Wood Shingles and Shakes Are Made

Shingles are sawn, while shakes are split, or split and sawn. Shingles have a relatively smooth surface; shakes have at least one textured, natural-grain face. Wood shingles are usually pine, redwood or cedar.

Logs for wood shingles and shakes are cut into 16-, 18- and 24-inch lengths and shaped into blocks. Manufacturers try to produce blocks with an edge-grain face. (See Figure 6-2.)

Red Cedar Shingles

Red cedar trees produce fine, even-grained wood with a uniform texture and very few knots. Shingles are sawn from cedar blocks and have a straight, flat and uniform face. The five basic grades of red cedar shingles are shown in Figure 6-3. Western red cedar shingles are the most popular shingles.

Grade	Length	Thickness (at butt)	# courses per bdl.	# bdls. or cartons per square
No. 1 Blue Label® - the premium grade of shingles for roofs and sidewalls. These top-grade shingles are 100% heartwood, 100% clear and 100% edge-grain. 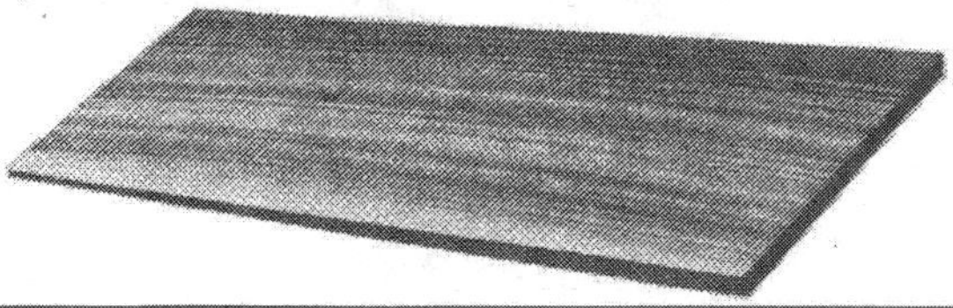	16" (Fivex) 18" (Perfections) 24" (Royals)	.40" .45" .50"	20/20 18/18 13/14	4 bdls. 4 bdls. 4 bdls.
No. 2 Red Label - a good grade for many applications. Not less than 10" clear on 16" shingles, 11" clear on 18" shingles and 16" clear on 24" shingles. Flat grain and limited sapwood are permitted in this grade. 	16" (Fivex) 18" (Perfections) 24" (Royals)	.40" .45" .50"	20/20 18/18 13/14	4 bdls. 4 bdls. 4 bdls.
No. 3 Black Label - a utility grade for economy applications and secondary buildings. Not less than 6" clear on 16" and 18" shingles, 10" clear on 24" shingles. 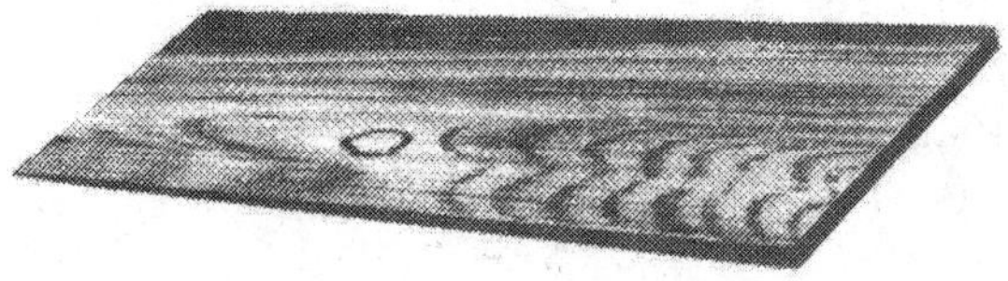	16" (Fivex) 18" (Perfections) 24" (Royals)	.40" .45" .50"	20/20 18/18 13/14	4 bdls. 4 bdls. 4 bdls.
No. 4 Undercoursing - a utility grade for starter course undercoursing. 	16" (Fivex) 18" (Perfections)	.40" .50"	14/14 or 20/20 14/14 or 18/18	2 bdls. 2 bdls.
No. 1 or No. 2 Rebutted and Rejointed - same specifications as above for No. 1 and No. 2 grades but machine trimmed for parallel edges with butts sawn at right angles. For sidewall applications where tightly fitting joints are desired. Also available with smooth sanded face. 	16" (Fivex) 18" (Perfections) 24" (Royals)	.40" .45" .50"	33/33 28/28 13/14	1 carton 1 carton 4 bdls.

Courtesy of Cedar Shake & Shingle Bureau

Figure 6-3 Certigrade® red cedar shingles

Red cedar shingles have several advantages, including:

- They weigh less than shakes, since they are thinner
- They have a high crushing strength.
- They don't expand or contract significantly when the humidity changes.
- They don't absorb a lot of water.
- They have good thermal and acoustical properties.

I recommend Number 1 Grade (Blue Label) cedar shingles because they're naturally decay-resistant. Also, you'll waste less material during installation because these shingles don't break as easily as the lower grades.

Number 2 Grade (Red Label) cedar shingles are cut from wood that's free of knots and checks for the lower two-thirds of the wood from the butt. Normally, these are flatgrain and slashgrain shingles, cut from sections of the log, as shown in Figure 6-2.

Grade 3 (Black Label; utility or economy grade) shingles are clear in the lower third of the wood from the butt.

Ideally, shakes are 100 percent edge grain. But with today's grading standards, one bundle of Grade 1 (Blue Label) cedar shingles is about 73 percent clear vertical-grain (edge grain) heartwood, 20 percent flatgrain and 7 percent Number 2 Grade. Number 2 and 3 grades are OK for siding; but you'll break a lot of them because they're weak around the knots.

Don't install wood shingles on slopes less than 4 in 12 unless you use the special low-slope application described later in this chapter, beginning on page 160. You can install wood shingles on slopes between 3 in 12 and 4 in 12, provided you reduce the recommended exposure by $1\frac{1}{2}$ inches.

Cedar Shakes

Figure 6-4 shows the five basic types of cedar shakes. Handsplit and resawn shakes are made with rough split faces and sawn backs. Tapersplit shakes are split along both faces so they have a rough face on both sides. Both the handsplit/resawn and tapersplit shingles have thick butts that taper to a thin head, as you can see in Figure 6-4. Straightsplit shakes are the same thickness throughout. Use them only over solidly framed roofs because they're very heavy, especially when they're wet. One square of 16-inch shingles weighs about 144 pounds. Each square of 18-inch shingles weighs 158 pounds. The 24-inch shingles weigh 192 pounds per square. Shakes weigh about 200 to 450 pounds per square, depending on the shake length, butt thickness, and the exposure. (The shorter the exposure, the more shakes it takes to cover a square.)

Shakes have a rougher face than wood shingles so they don't shed water as easily. You don't need underlayment or interlayment with wood shingles, but you do with shakes.

Grade	Length and thickness	18" pack # course per bdl.	# bdls. per sq.
No. 1 Handsplit & Resawn - these shakes have split faces and sawn backs. Cedar logs are first cut into desired lengths. Blanks or boards of proper thickness are split and then run diagonally through a band saw to produce two tapered shakes from each blank.	15" Starter-finish 18" x ½" Mediums 18" x ¾" Heavies 24" x ⅜" 24" x ½" Mediums 24" x ¾" Heavies	9/9 9/9 9/9 9/9 9/9 9/9	5 5 5 5 5 5
No. 1 Certi-sawn® (Tapersawn) - these shakes are sawn both sides. No. 2 and 3 are also available. For details contact the Bureau.	24" x ⅝" 18" x ⅝"	9/9 9/9	5 5
No. 1 Tapersplit - Produced largely by hand, using a sharp-bladed steel froe and a mallet. The natural shingle-like taper is achieved by reversing the block, end-for-end, with each split. 	24" x ½"	9/9	5

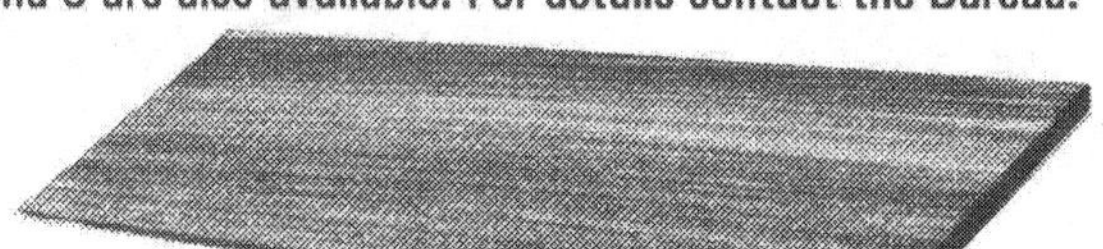

Grade	Length and thickness	20" pack	# bdls. per sq.
No. 1 Straightsplit - produced by machine or in the same manner as tapersplit shakes except that by splitting from the same end of the block, the shakes acquire the same thickness throughout.	18" x ⅜" True-edge 18" x ⅜" 24" x ⅜"	14 Straight 19 Straight 16 Straight	4 5 5
No. 1 Machine Grooved - machine-grooved shakes are manufactured from shingles and have striated faces and parallel edges. Used double-coursed on exterior sidewalls.	16" x .40" 18" x .45" 24" x .50"	16/17 14/14 12/12	2 ctns. 2 ctns. 2 ctns.

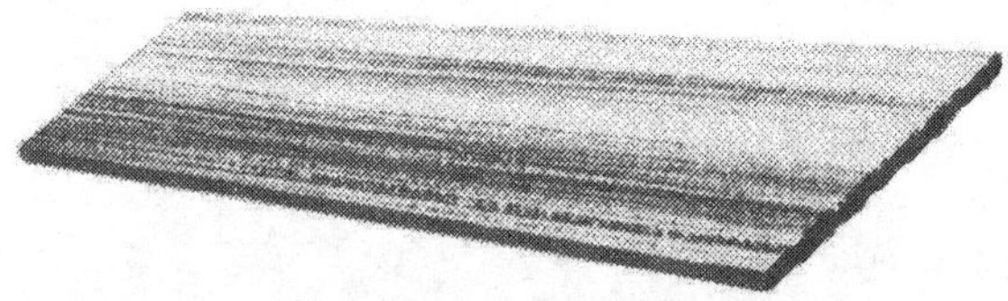

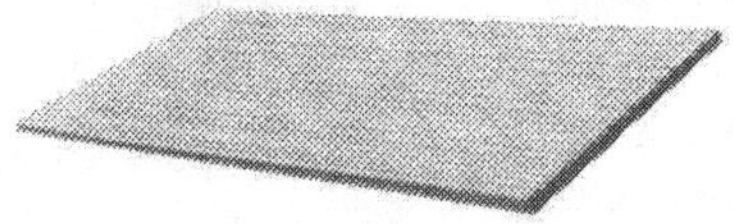

Courtesy of Cedar Shake & Shingle Bureau

Figure 6-4 Certi-split® red cedar shakes

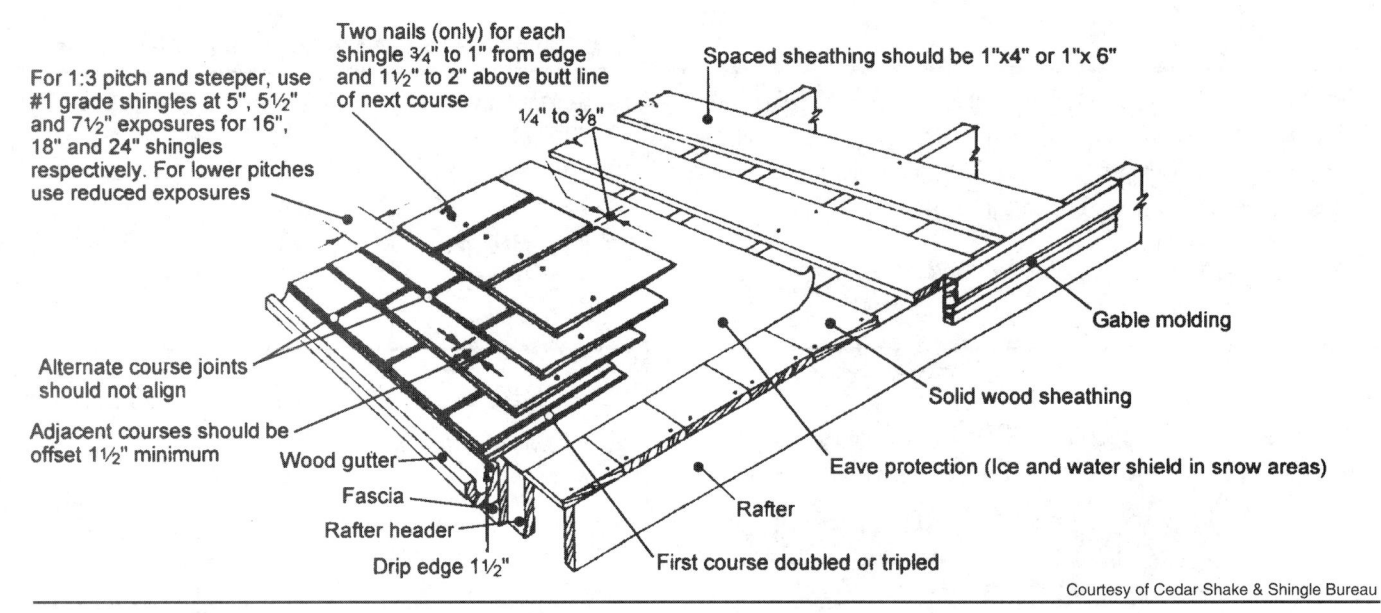

Figure 6-5 Shingles applied over spaced sheathing

Shakes have a higher insulating value, they're larger, and can be laid with a longer exposure than wood shingles, So you can cover an area faster with shakes than shingles.

In dry climates you can install hand-split shakes on a roof as flat as 4 in 12, provided you use the low-slope application method. In wet climates, I recommend a minimum slope of 6 in 12 so water will run off quickly. For more information on wood shingles or shakes, contact the Cedar Shake & Shingle Bureau, 355 Lexington Avenue, 15th Floor, New York, NY 10017. Their website is www.cedarbureau.org

Installing Wood Shingles and Shakes

You can install wood shingles and shakes over solid or spaced sheathing, as discussed in Chapter 2. Figures 6-5 and 6-6 show installation details. In wet climates, it's better to use spaced sheathing because the increased air circulation reduces dampness. But always check the local codes, which may say you have to install solid sheathing for structural purposes, or in areas subject to wind-driven rain or snow.

After you've got the underlayment (or interlayment), drip edge and valley flashing material in place, install a starter course at the eaves of the roof. Allow a 1½-inch overhang beyond the eaves fascia and a ½- to ¾-inch overhang beyond the rake fascia. (Without a starter course, you'd only have single coverage at the eaves.) For 24-inch shakes laid at a 10-inch exposure, install 1x4s on 10-inch centers with an additional 1x4 placed between the on-center boards, or use 1x6s on 10-inch centers. For 18- and 24-inch shakes laid at less than a 10-inch exposure, use 1x4 spaced sheathing.

Wood Shingles and Shakes

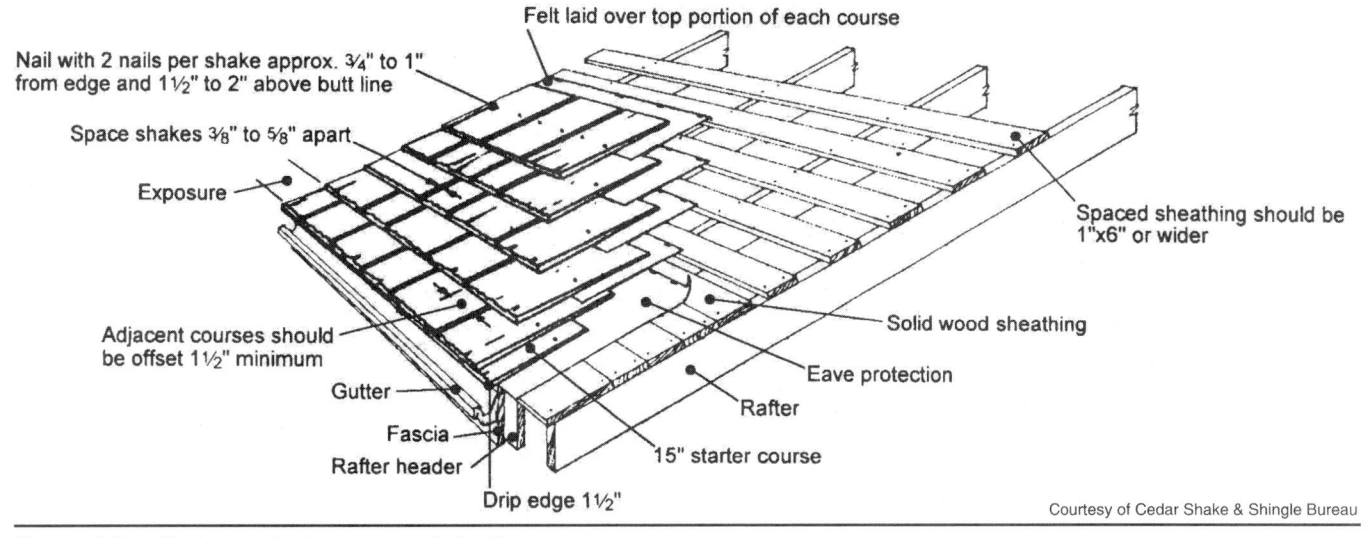

Figure 6-6 Shakes applied over spaced sheathing

Install the first course of wood shingles or shakes on top of the starter course. In heavy-snow regions, overlay the starter course with two layers of wood shingles or shakes. On shake roofs, use 15-inch wood shingles for the starter course and overlay them with a course of shakes. If you can't find 15-inch shakes to use as a starter course, you can cut the tops off longer shakes to make them fit.

To align the starter course and first course, nail down a shingle with the correct overhang on each end of the eaves. Then drive a nail into the butt of each shingle and stretch and tie a string between the nails. Align the butts of the intervening shingles along the string, as in Figure 6-7. Every third or fourth course, measure from the eaves up to the butts of end shingles and snap a chalk line to align the butts of the next course of shingles.

To allow for expansion when wet, space shakes $^3/_8$ to $^5/_8$ of an inch apart. Allow $^1/_4$ to $^3/_8$ inch for wood shingles. Offset adjacent shake or shingle courses by at least 1$^1/_2$ inches. Don't let two joints line up directly in any three courses. In heavy-snow regions, increase the offset of adjacent courses to 2 inches. Treat knots and similar defects of a wood shingle at the edge of the shingle and align the joint in the course above at least 1$^1/_2$ inches from the edge of the defect. That's shown in Figure 6-8.

Use only wood shingles that are from 3 to 8 inches wide. If a wood shingle or shake is too wide, split it with a roofer's hatchet. Saw-cutting takes too long.

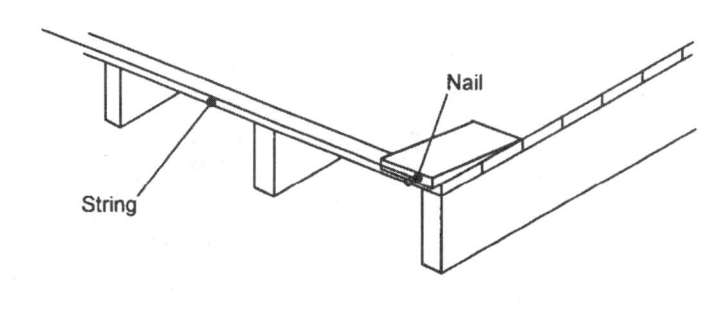

Figure 6-7 Lining up the first course of wood shingles or shakes

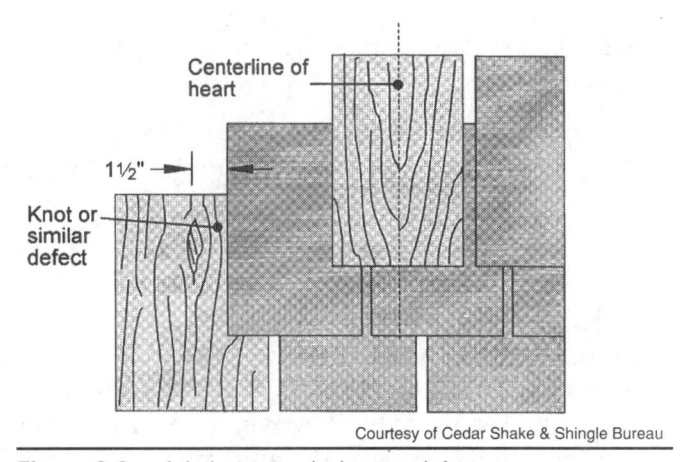

Figure 6-8 Jointing around a knot or defect

Shake exposures vary, as you'll see later in Figure 6-23. Common exposures for 2-ply coverage are $7^1/_2$ inches for 18-inch shakes and 10 inches for 24-inch shakes. For 3-ply coverage, use exposures of $5^1/_2$ inches for 18-inch shakes and $7^1/_2$ inches for 24-inch shakes. Install 3-ply roofing in heavy-snow regions.

To avoid having to cut wood shingles or shakes at the ridge, shorten the exposures of the last few courses installed below the ridge. Try to reduce shake exposure approaching the ridge so you can use a 15-inch starter-finish shake immediately below the ridge. You'll have to cut these to fit if you can't get stock units.

Valleys

Valley flashing should be corrosion-resistant metal at least 0.017-inch thick, and it should extend no less than 8 inches on each side of the valley centerline. I prefer to shingle away from both sides of valleys and hips. It saves time because this way, you can put all shingles adjacent to valleys and hips using the same pattern. Then you split shingles to fit where each course meets at mid-roof. Save wider shakes for finishing the valley edges.

However, if you're right-handed, you may be more comfortable working from left to right all the time. Shingle into the left side of a valley, then cut the last shingle or shake individually at the end of each course. Place a 1 x 4 in the valley as a cutting guide. Place the end shingle over the board, eyeball, and score a line, then cut the shingle with a portable power saw.

Now, shingle out from the right side of the valley. Since this is the starting point for all shingle courses, position the 1 x 4 and cut one shingle or shake and use it for a pattern to cut the rest of the starting shingles or shakes. Just be sure your starting shingles are different widths, or the joints will fall in a line. Reverse the working direction if you're left-handed.

Use the same basic principles when you shingle into a hip. When you begin shingling at a hip, use one shingle as a pattern for cutting the others. When you end a shingle course at a hip, you must cut each end shingle individually. Use the triangular pieces cut from shingles used in valleys to finish the hip edges.

Figure 6-9 shows yet another way to shingle valleys. Shingle into the valley from both sides, stopping the course before you get into the valley. Then install a pre-cut shingle at the end of the course. Finish the course by selecting (or cutting) a shingle or shake to fit between the pre-cut piece and the last full shingle or shake.

Install wood shingles or shakes to within 2 to 4 inches on each side of the centerline of a valley. Never allow joints between wood shingles or shakes to break into a valley; that is, be sure all joints have a solid shingle beneath and on top of them. And never lay shingles or shakes with the grain parallel with the centerline of a valley. Install wood shingles or shakes so they lap at least 7 inches over each side of the valley flashing. Drive nails

Wood Shingles and Shakes

Keep nails well away from the center of valley

Order of applying shingles or shakes at valley

1. Stop course line here
2. Place pre-cut piece so that cut-angle is positioned on chalk line with tip on course line
3. Select a shingle or shake of the required width to complete the course

Courtesy of Cedar Shake & Shingle Bureau

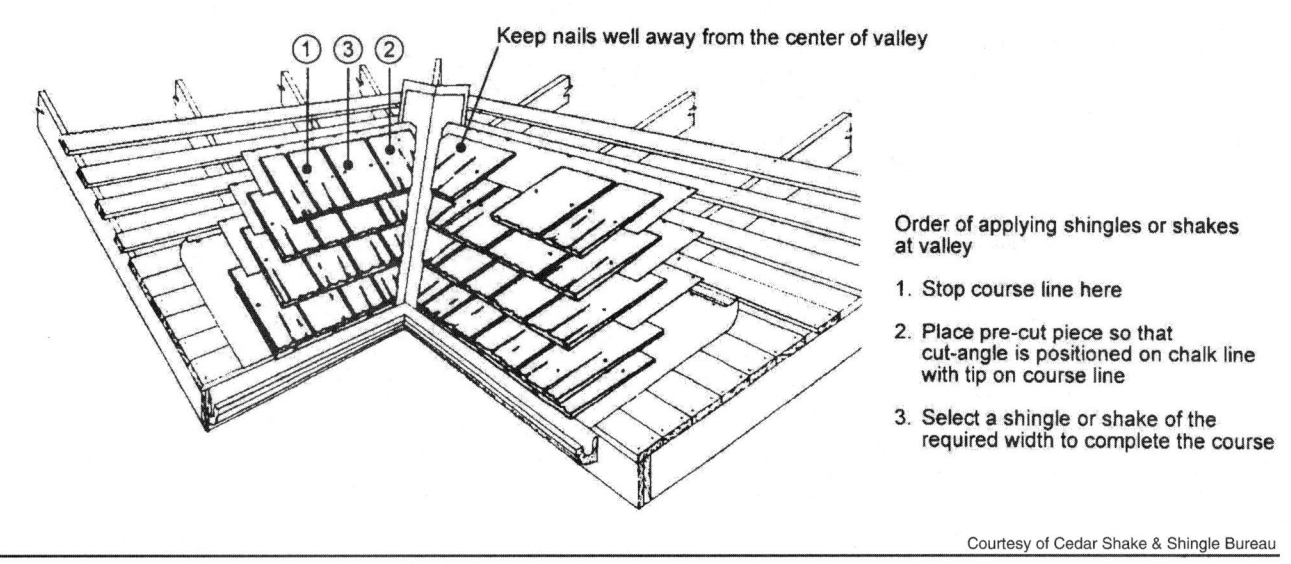

Figure 6-9 Installing wood shingles or shakes in a valley

at least 2 inches from the edge of the metal. Install wood shingles or shakes at the top of the valley at least 4 inches from each side of the centerline. Increase this distance about $1/8$ inch per foot down the valley. Snap chalk lines down both sides of the valley for guidelines.

Nailing Shingles and Shakes

Regardless of the width of a wood shingle or shake, drive two nails about $3/4$ to 1 inch from each side of the unit, and $1\frac{1}{2}$ to 2 inches above the butt of the succeeding course. If the shake or shingle splits, drive two nails into each piece. Use stainless steel, hot-dipped galvanized nails. Bright, blue, or steel wire nails don't mix with wood preservatives. They'll quickly dissolve. And aluminum nail heads break off. Figure 6-10 shows the nail types and minimum lengths you should use to install wood shingles and shakes.

Use 5d or 6d nails for wood shingle re-roofs. Use 7d or 8d nails on shake re-roofs. When in doubt, use nails long enough to go $1/2$ inch into the deck. But check your local code. Some building codes require $3/4$ inch. Use ring-shank nails when plywood sheathing is less than a half-inch thick.

Nails are better than staples because staples make twice as many punctures. Also, if you don't hold the staple gun correctly of if it's out of adjustment, you can damage the shingles. If you use staples, make sure they're 16-gauge aluminum or stainless steel with a $7/16$-inch (minimum) crown. Don't use blue-steeled fasteners. Make sure the staple is long enough to go at least $1/2$ inch into the sheathing, and that its crown is flush with the surface of the wood shingle or shake.

Type of shingle or shake	Nail type and minimum length	
Shingles - New Roof	Type	(in.)
16" and 18" shingles	3d box	1¼
24" shingles	4d box	1½
Shakes - New Roof	Type	(in.)
18" Straight-split	5d box	1¾
18" and 24" handsplit-and-resawn	6d box	2
24" tapersplit	5d box	1¾
18" and 24" taper-sawn	6d box	2

Courtesy of Cedar Shake & Shingle Bureau

Figure 6-10 Nails used to install wood shingles and shakes

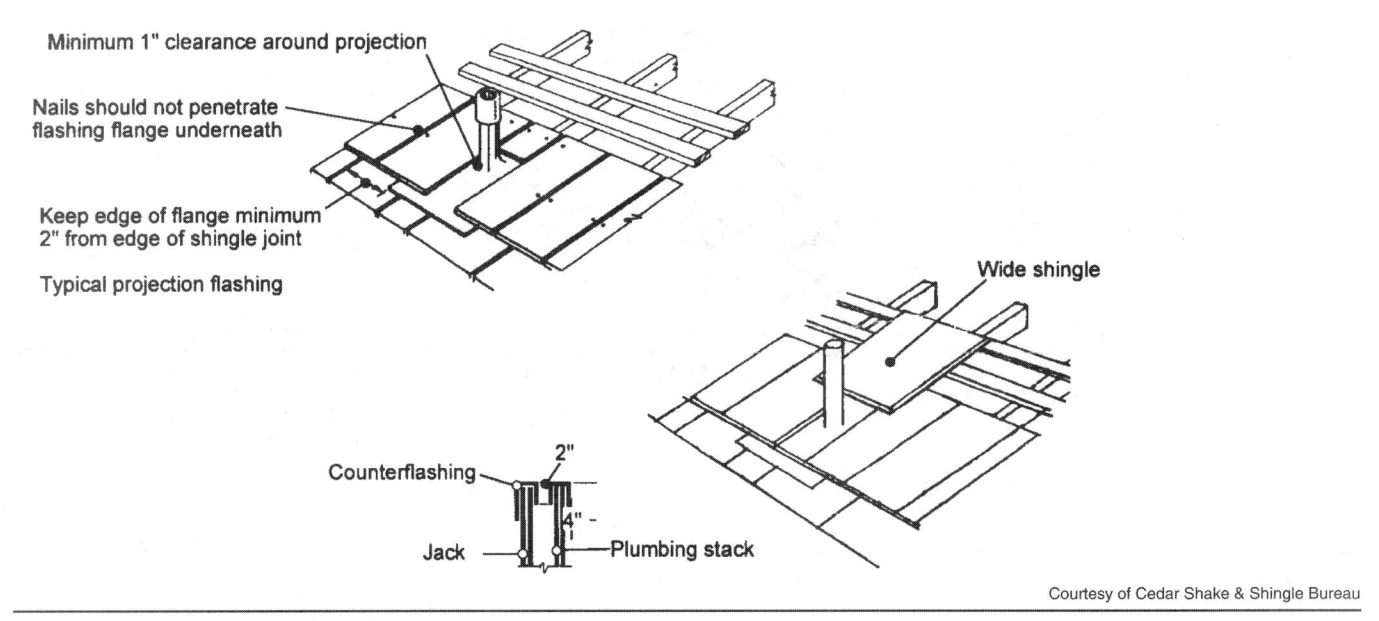

Figure 6-11 Installing wood shingles at a vent pipe

Shingling Around Vents

When you come to a vent pipe, use a saber saw or keyhole saw to notch the wood shingles or shakes that meet the pipe so they'll fit closely around it. Or, you can install shingles individually to fit around the pipe. Slip the vent flashing over the pipe and the adjacent shingles. In either case, leave the flashing area open below the pipe so that debris doesn't collect. That's shown in Figure 6-11. As added protection, install a felt strip overlaid with a wide shingle immediately above the pipe. Drop it below the butt line of other shingles in the same course if aligning it with the rest of the course would leave a gap above the pipe.

Installing Shingles or Shakes at Chimneys

Figure 6-12 illustrates flashings around chimneys. Saddle flashing goes upslope of the chimney, apron flashing goes on the downslope side. You can use a cricket flashing instead of saddle flashing. Look back at Figure 4-52 in Chapter 4 for a picture of a cricket.

Extend the apron flashing at least 3 inches up the vertical surface. It should also go at least $1^{1}/_{2}$ times the shingle or shake exposure (6 inches minimum) over the roof slope. Carry cricket flashing at least 10 inches under the shingles or shakes. Extend step flashing over the roof at least 3 inches and up the chimney. It should be covered by at least 4 inches of the counterflashing. Lap each step flashing over the next piece by at least 3 inches. Install counterflashing so it extends down within 1 inch of the finished roof surface.

Wood Shingles and Shakes

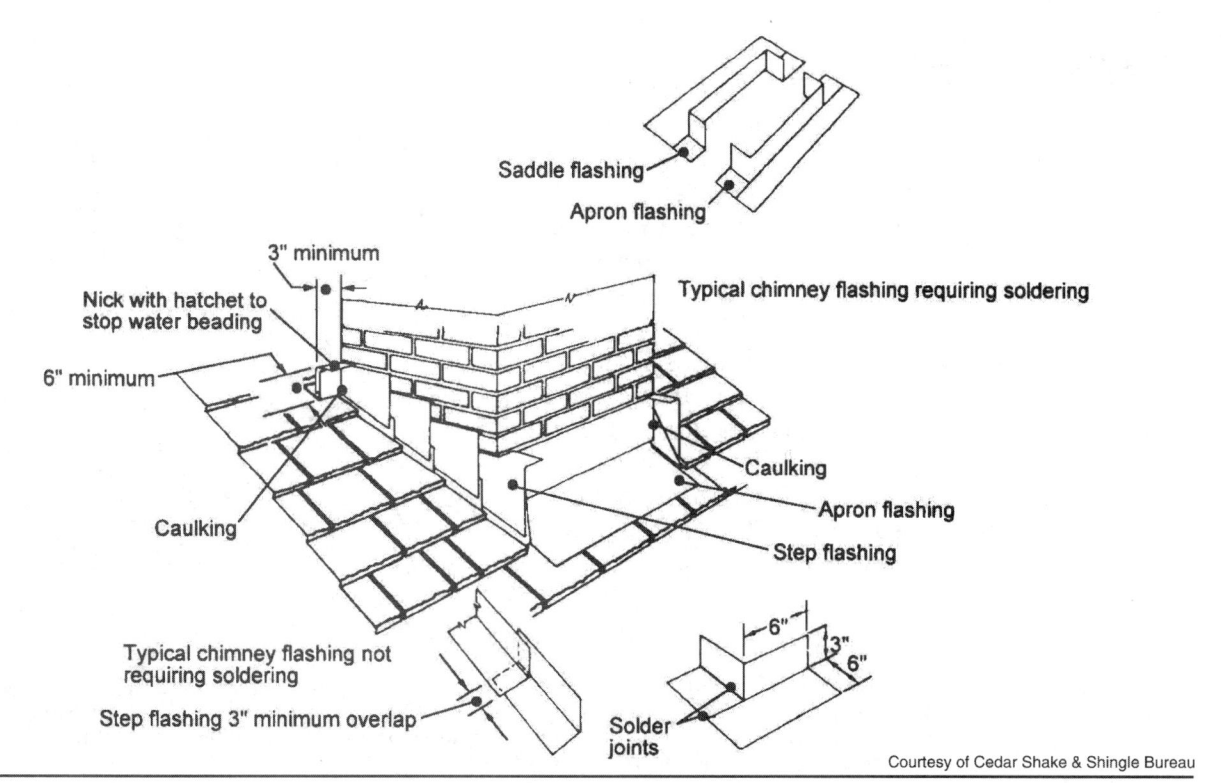

Figure 6-12 Typical chimney flashing

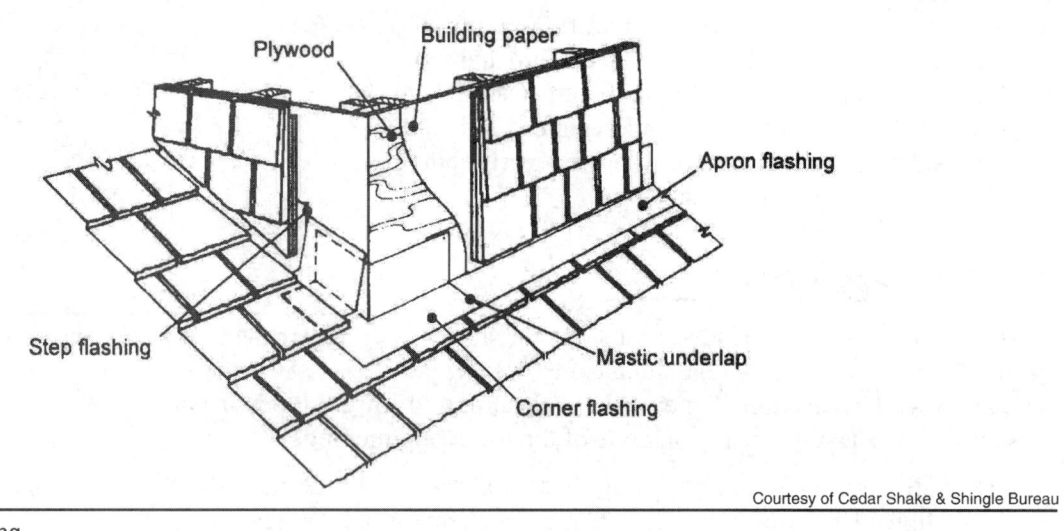

Figure 6-13 Dormer flashing

Dormers

At dormers, extend the apron flashing up the walls at least 3 inches under sheathing paper and at least 3 inches over the roof slope. Install step flashing as prescribed earlier, using shingles or siding for counterflashing. You can see that in Figure 6-13.

Hips and Ridge Units

You can field-fabricate hip and ridge units, but it's faster to use prefabricated units which come with alternating mitered joints and concealed nailing (staples, usually two, which hold the units together). Install the hip units before the ridge units. Install double starter units over the first course of shingles or shakes at the low end of each hip, as in Figure 6-14. Then temporarily install a hip unit at the top of the hip. Snap chalk lines on each side of the hip along the edges of the two hip units. The chalk lines serve as guides for positioning the rest of the hip units. Install hip units starting at the low end of each hip, alternating the direction of the miter joints. Trim the top hip units so they meet at the ridge with their edges flush. Use the same exposure you used for the shingles to install hip and ridge units.

Install ridge units starting at both ends of the ridge, working toward the center of the roof. Create a saddle to cover where both courses meet.

Use two 8d or 10d galvanized or aluminum nails to install each hip and ridge unit. When in doubt, use nails that are long enough to go at least $1/2$ inch into the deck.

I recommend you apply a strip of underlayment; synthetic or 30-pound felt, under hip and ridge units. It's also a good idea to install concealed metal hip and ridge flashing, as in Figure 6-15. A wood roof lasts a long time, but it's only as good as its underlayment. I consider these steps a good precaution to prevent early failure of a wood roof, especially a shake roof.

Figure 6-14 Hip and ridge application

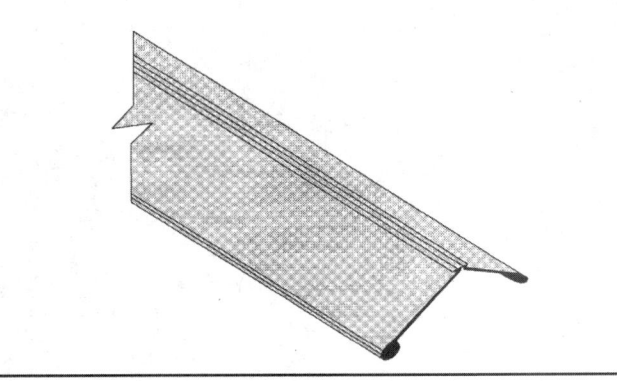

Figure 6-15 Hip and ridge flashing

Low-Slope Applications

The minimum recommended roof slope for shakes is 4 in 12, and 3 in 12 for wood shingles. However, you can apply shakes and shingles over solid sheathing to lower slopes if you first cover the sheathing using any one of the following methods:

1) Nail down a 43-pound organic sheet or a 28-pound fiberglass sheet followed by a cold-applied surface coat of asphalt or ice shield material.

2) Nail down two plies of 30-pound felt followed by a cold-applied surface coat of asphalt or ice shield material.

3) Hot-mop two plies of 15-pound felt followed by a cold-applied surface coat of asphalt or ice-shield material.

4) Nail down a 30-pound felt followed by hot-mopping 90-pound mineral-surfaced roll roofing.

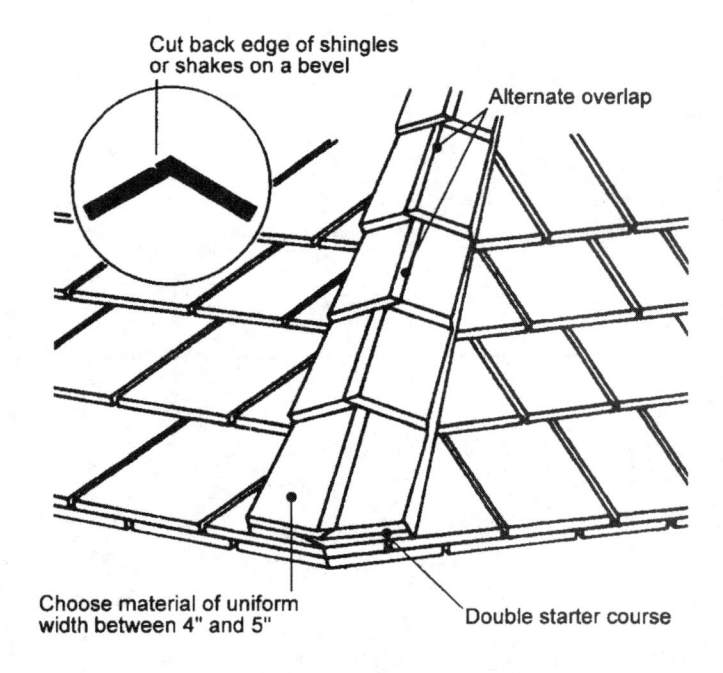

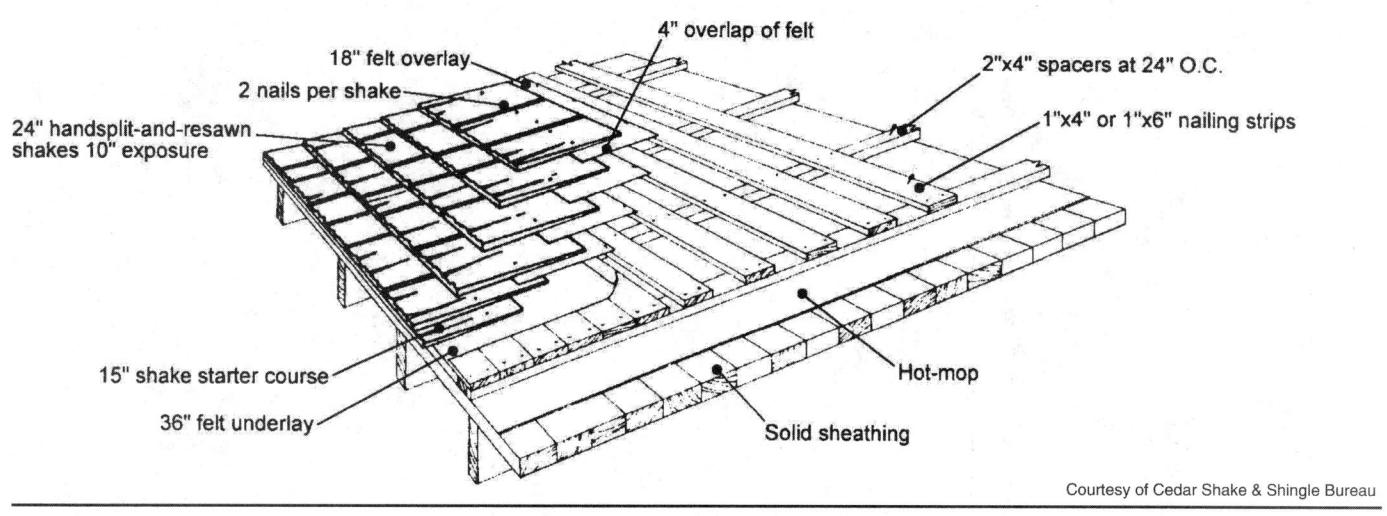

Figure 6-16 Application of wood shingles or shakes over low-slope roofs

When you use methods 1, 2 or 3, embed treated 2 x 4 spacers into the hot asphalt surface coat on 24-inch centers extending from the eaves to the ridge. If you use method 4, you can embed the 2 x 4s into cold-applied roofing cement.

Nail the 2 x 4s at the eaves and ridge only. This will be enough to hold the spacers in place. Also, avoid driving nails into the built-up roof because every puncture provides a path for water to leak through. If you *have* to put in additional nailing, seal the penetrations well with roofing cement.

Next, install 1 x 4 or 1 x 6 nailing strips, spaced according to the weather exposure selected (see Chapter 2, page 29 for spacing requirements). Nail the strips across the spacers to form a lattice-like nailing base. Finally, install wood shingles or shakes in the normal manner. This is shown in Figure 6-16.

Steep-Slope Applications

Wood shingles or shakes are often applied to mansard roofs over panel sheathing or spaced battens. Construction details for typical mansard roofs are shown in Figure 6-17.

Swept or Bell Eaves

You may have to soak (usually overnight) or steam the shakes or shingles you install over swept or bell eaves so you can bend them. See Figure 6-18. Install the roof in the usual manner above the sweep.

Installing Wood Shingles or Shakes over Rigid Insulation

Don't nail wood shingles or shakes through rigid insulation into the deck below. You'll create several problems:

- Since you'll have to use longer nails, they'll have thicker shanks. They'll split shingles and shakes you drive them through.

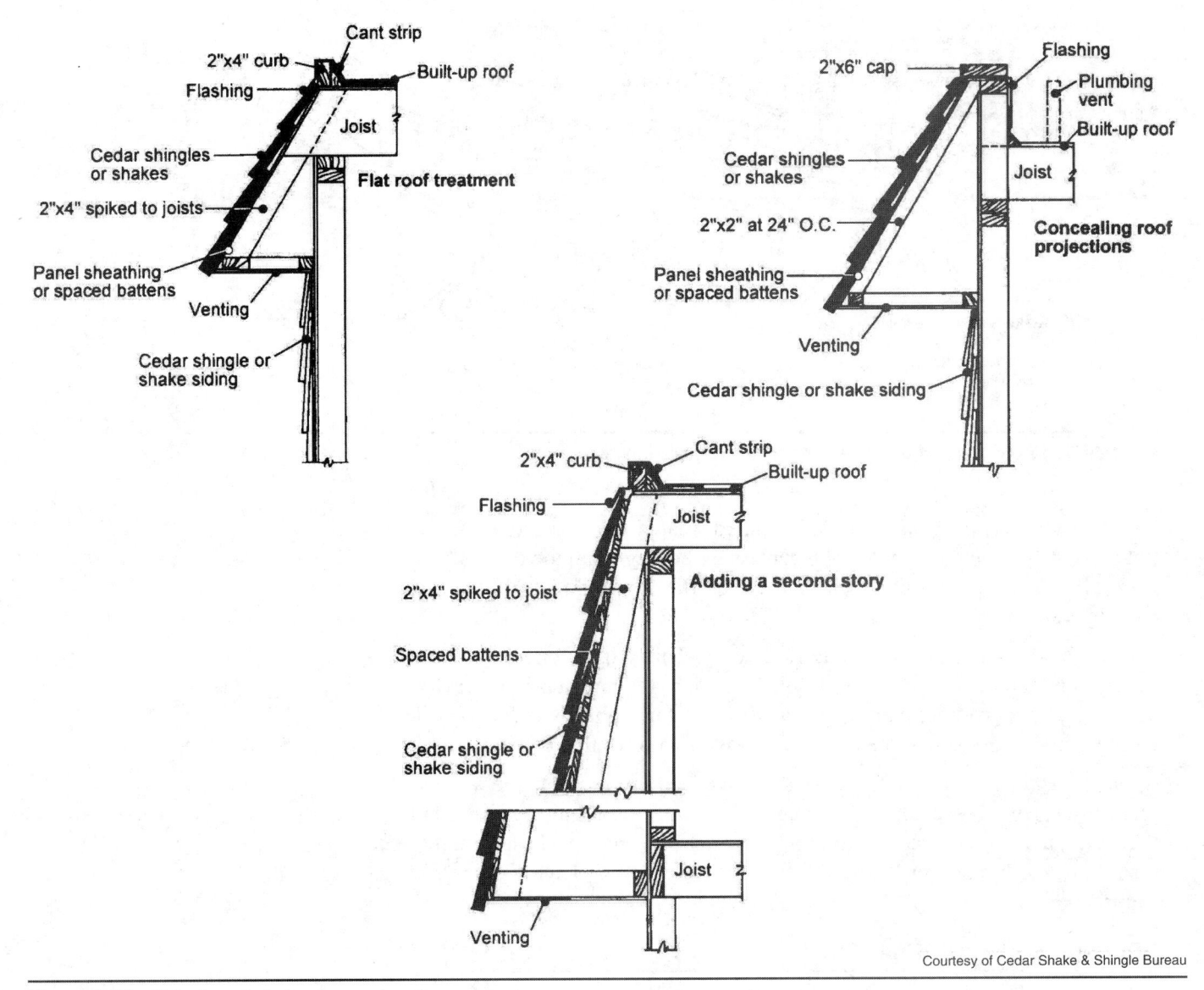

Figure 6-17 Shingle and shake mansard roofs

- The nails expand and contract, enlarging the holes in the insulation and reducing its efficiency.
- Nails allow the shingles and underlayment to move, causing leaks.

To eliminate these problems, install a plywood or *false deck* (shown in Figure 6-19) or horizontal strapping, as in Figure 6-20, immediately over the insulation. Horizontal strapping is 1 x 4s or 1 x 6s installed on centers as prescribed for spaced sheathing. Use the false deck or strapping as a nailing surface for the wood shingles or shakes, and as an air space beneath the shingles or shakes for good air circulation.

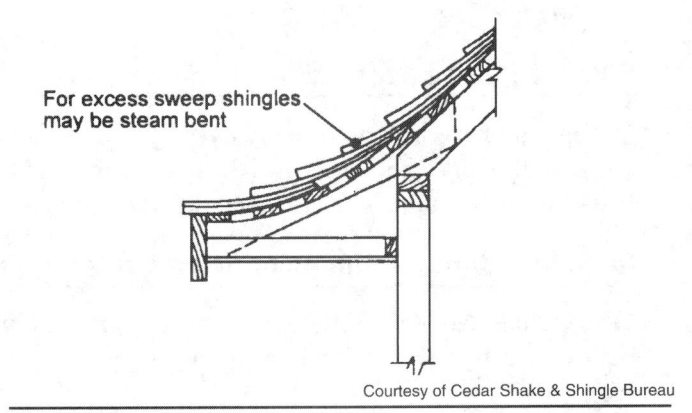

Figure 6-18 Swept or bell eaves

Wood Shingles and Shakes

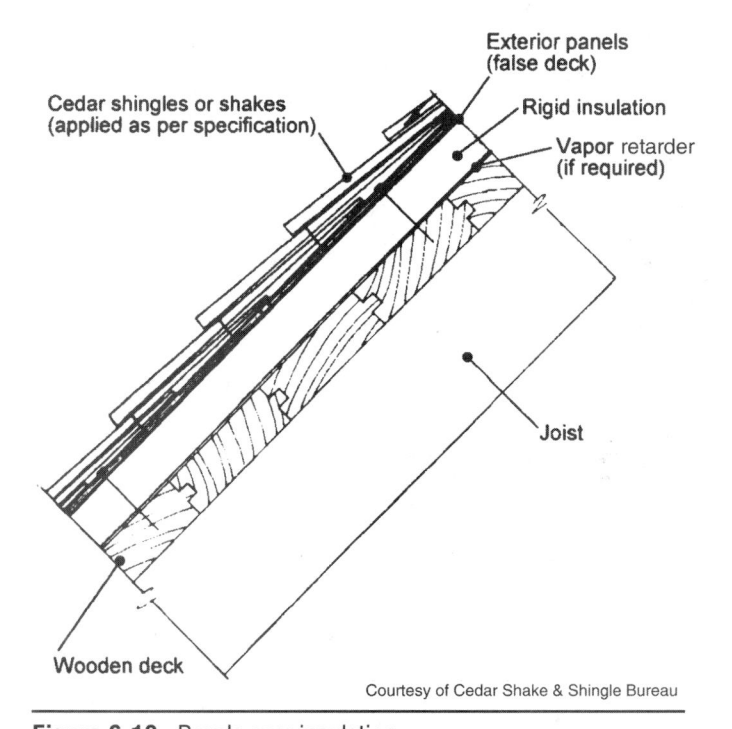

Figure 6-19 Panels over insulation

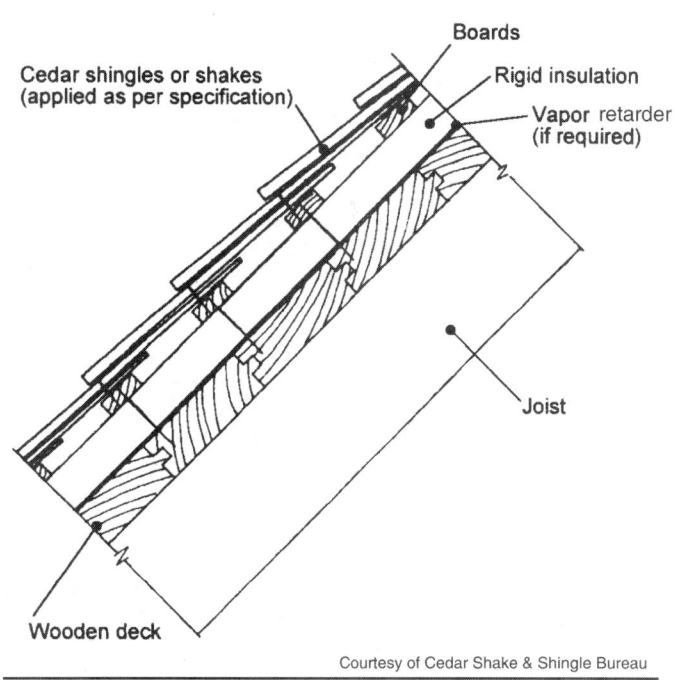

Figure 6-20 Strapping over insulation

Although you nail the false deck and strapping through the insulation into the sheathing beneath, far fewer nails penetrate the insulation. To eliminate nails penetrating the insulation or to avoid the possibility of a steep roof creeping downward under a heavy snow load, install 2 x 4s on edge from eaves to ridge between insulation panels. You can then install the false deck or strapping to the 2 x 4s, as in Figure 6-21.

Installing Wood Shingles or Shakes over a Metal Deck

Over a metal deck, fasten wood shingles or shakes to a false deck or horizontal strapping as just described. Fasten the false deck or strapping to 2-inch-wide lumber you install from eaves to ridge, as shown in Figure 6-22. You can fasten the lumber to the metal deck with screws or bolts. On decks with shallow corrugations, you might have to use clip angles (small "L" brackets) to attach the lumber to the deck. If so, nail or screw the clips to the lumber and screw or bolt the clips to the deck.

If you install a vapor retarder, put it *under* the rigid insulation. Never *choose* to install lumber between two vapor retarders but you may have to do it to comply with an architect's design. If that's the case, use preservative-treated lumber.

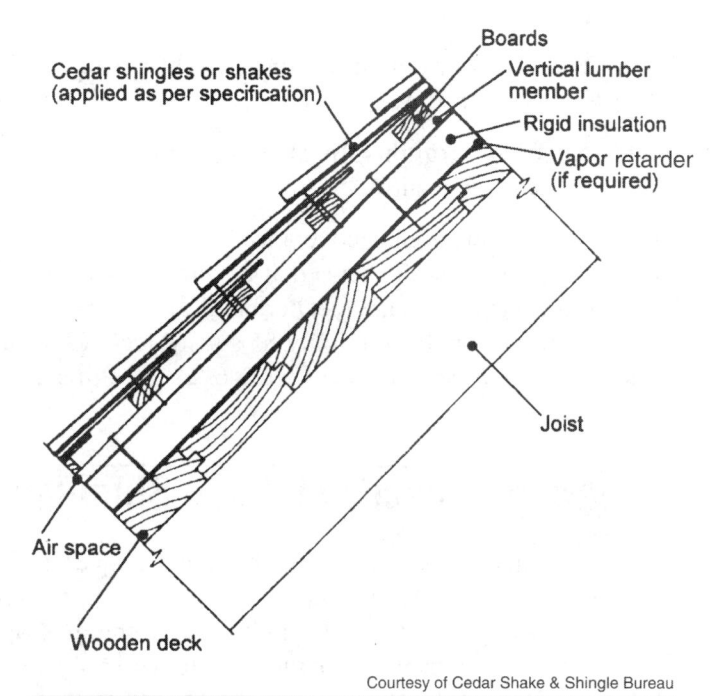

Figure 6-21 Strapping over 2 x 4s turned edgewise

163

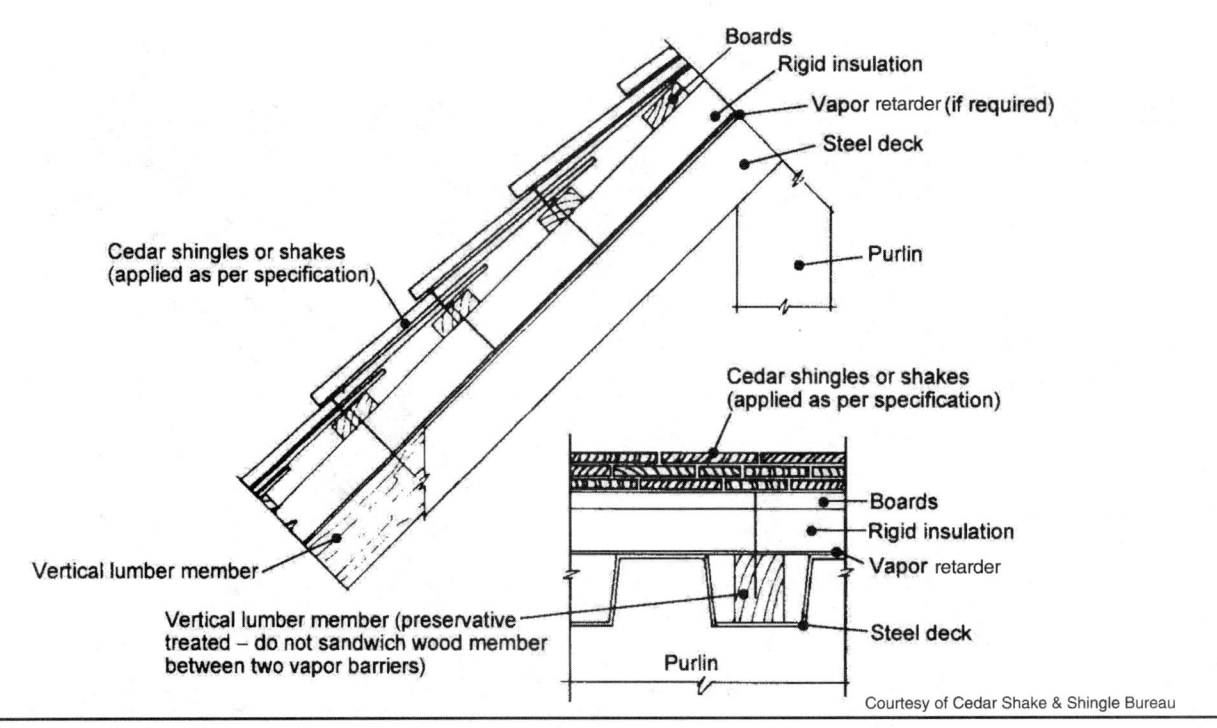

Figure 6-22 Installation over steel deck

Covering Capacity of Shakes

Shakes usually come in 5-bundle squares. Five bundles installed with a 10-inch exposure will cover 100 square feet of roof area (Figure 6-23). Exceptions are:

- True-edge straightsplit shakes, which come in 4-bundle squares (14-inch exposure)

- 18" x 3/8" straightsplit shakes, which come in 5-bundle squares (8½-inch exposure).

Shakes provide double coverage if you install them at the recommended exposures. Figure 6-23 shows that for most shakes, five bundles will cover 100 square feet (20 square feet per bundle) when laid at a 10-inch exposure. But five bundles will cover only 85 square feet (17 square feet per bundle) at an 8½-inch exposure. Use Figure 6-24 to estimate quantities.

Covering Capacity of Wood Shingles

Wood shingles generally come in 4-bundle squares. (Turn back to Figure 6-3 for exceptions.) Figure 6-25 shows the maximum recommended exposures for various wood shingle grades. Wood shingles, applied at the correct exposure, provide triple coverage. Widths will vary in a particular bundle. Lower-grade shingles are typically narrower than the higher grade ones and cost more to install.

Wood Shingles and Shakes

Grade	Length & Thickness	\multicolumn{10}{c	}{Approximate coverage (square feet) of 5 bundles (4 bundles in the case of true-edge straight split) of shakes when applied with ½-inch spacing, at the following exposures (in inches). }
		5½	7½	8	8½	10	11½	12	14	16	18
No. 1 Handsplit and Resawn	18" x ½"	55[a]	75[bcd]	80	85[e]	100	115	120	140[f]		
	18" x ¾"	55[a]	75[bcd]	80	85[e]	100	115	120	140[f]		
	24" x ⅜"		75[ac]	80	85	100[bd]	115[e]	120	140	160	180[f]
	24" x ½"		75[a]	80	85	100[bcd]	115[e]	120	140	160	180[f]
	24" x ¾"		75[a]	80	85	100[bcd]	115[e]	120	140	160	180[f]
No. 1 Tapersawn	18" x ⅝"	55[a]	75[bcd]	80	85[e]	100	115	120	140[f]		
	24" x ⅝"		75[a]	80	85	100[bcd]	115[e]	120	140	160	180[f]
No. 1 Tapersplit	24" x ½"		75[a]	80	85	100[bcd]	115[e]	120	140	160	180[f]
No. 1 Straight-split	18" x ⅜"	65[a]	88[bcd]	94	100[e]	118	135	141	165	188	
	24" x ⅜"		75[a]	80	85	100[bcd]	115[e]	120	140	160	180[f]
True-edge	18" x ⅜"	39	53	56	60[e]	70	81	84	98	112[f]	
15" Starter-finish	\multicolumn{11}{c	}{Use supplementary with shakes applied not over 10" exposure}

[a] Maximum recommended weather exposure for 3-ply roof construction.
[b] Maximum recommended weather exposure for 2-ply roof construction.
[c] Maximum recommended weather exposure for application on roof slopes between 4 in 12 and 8 in 12.
[d] Maximum recommended weather exposure for application on roof slopes of 8 in 12 and steeper.
[e] Maximum recommended weather exposure for single-coursed wall construction.
[f] Maximum recommended weather exposure for double-coursed wall construction. Use a 24-inch cedar shingle for underlay.

Figure 6-23 Covering capacity of shakes

Grade	\multicolumn{11}{c	}{Bundles of shakes required per square when applied with ½-inch spacing, at the following exposures (in inches).}
	5½	7½	8	8½	10	11½	12	14	16	18	20	22
No. 1 Handsplit and Resawn, No. 1 Tapersawn, No. 1 Tapersplit, and No. 1 Straight-split (24")	9.09	6.67	6.25	5.88	5.00	4.35	4.17	3.57	3.13	2.78	2.50	2.27
No. 1 Straight-split (True edge)	10.26	7.55	7.14	6.67	5.71	4.94	4.76	4.08	3.57			
No. 1 Straight-split (18" long)	7.69	5.68	5.32	5.00	4.24	3.70	3.55	3.03	2.66			

Figure 6-24 Bundles of shakes per square

Roof Slope	\multicolumn{3}{c	}{Number 1 Blue Label}	\multicolumn{3}{c	}{Number 2 Red Label}	\multicolumn{3}{c	}{Number 3 Black Label}
	16"	18"	24"	16"	18"	24"	16"	18"	24"
3/12 – 4/12	3¾"	4¼"	5¾"	3½"	4"	5½"	3"	3½"	5"
4/12 and Steeper	5"	5½"	7½"	4"	4½"	6½"	3½"	4"	5½"

Figure 6-25 Maximum recommended exposures for wood shingles

Roofing Construction & Estimating

Length and Thickness	Approximate coverage (square feet) of 4 bundles of wood shingles when applied at the following exposures (in inches)											
	3	3½	3¾	4	4¼	4½	5	5½	5¾	6	6½	7	7½
16" x 5/2"[i]	60[a]	70[bd]	75[c]	80[e]	85	90	100[f]	110	115	120	130	140	150[g]
18" x 5/2¼"		63.6[a]	68.2	72.7[bd]	77.3[c]	81.8[e]	90.9	100[f]	104.5	109.1	118.2	127.3	136.4
24" x 4/2"						66.7[a]	73.3[bd]	76.7[c]	80	86.7[e]	93.3	100[f]
Length and Thickness	8	8½	9	9½	10	10½	11	11½					
16" x 5/2"	160	170	180	190	200	210	2.20	2.30					
18" x 5/2¼"	145.5	154.5[g]	163.6	172.7	181.8	190.9	200	209.1					
24" x 4/2"	106.7	113.3	120	126.7	133.3	140	146.7	153.3[g]					
Length and Thickness	12	12½	13	13½	14	14½	15	15½	16				
16" x 5/2"	240[h]												
18" x 5/2¼"	218.2	227.3	236.4	245.5	254.5								
24" x 4/2"	160	166.7	173.3	180	186.7	193.3	200	206.7	213.3[h]				

[a] Maximum recommended weather exposure for No. 3 grades applied on roof slopes between 3 in 12 and 4 in 12.
[b] Maximum recommended weather exposure for No. 2 grades applied on roof slopes between 3 in 12 and 4 in 12.
[c] Maximum recommended weather exposure for No. 1 grades applied on roof slopes between 3 in 12 and 4 in 12.
[d] Maximum recommended weather exposure for No. 3 grades applied on roof slopes 4 in 12 and steeper.
[e] Maximum recommended weather exposure for No. 2 grades applied on roof slopes 4 in 12 and steeper.
[f] Maximum recommended weather exposure for No. 1 grades applied on roof slopes 4 in 12 and steeper.
[g] Maximum recommended weather exposure for single-coursing No. 1 grades on sidewalls. Reduce exposure for No. 2 grades.
[h] Maximum recommended weather exposure for double-coursing No. 1 grades on sidewalls.
[i] Sum of thickness; e.g., 5/2" means that 5 butts stacked have a total thickness of 2 inches.

Figure 6-26 Covering capacity of wood shingles

Referring to Figure 6-26, 16-inch wood shingles are packaged so 4 bundles will cover 100 square feet if applied at a 5-inch exposure. However, 4 bundles will cover only 90 square feet if applied at a 4½-inch exposure. The 18- and 24-inch shingles will cover 100 square feet if applied at 5½- and 7½-inch exposures, respectively. Use Figure 6-27 to estimate quantities.

Estimating Wood Shingle and Shake Quantities

In the following examples, the equations are the same as those for asphalt shingles, but the material dimensions are different.

First, you take off shingle quantities by the square foot. Then convert that figure to bundles required. The total required must include material for the starter course, hip and ridge units, cutting waste at the rakes, hips and valleys, and crew-error waste.

Wood Shingles and Shakes

Length and Thickness	Bundles of wood shingles required per square when applied at the following exposures (in inches)												
	3	3½	3¾	4	4¼	4½	5	5½	5¾	6	6½	7	7½
16" x 5/2"[a]	6.67	5.71	5.33	5.00	4.71	4.44	4.00	3.64	3.48	3.33	3.08	2.86	2.67
18" x 5/2¼"		6.29	5.87	5.50	5.18	4.89	4.40	4.00	3.83	3.67	3.38	3.14	2.93
24" x 4/2"							6.00	5.46	5.22	5.00	4.61	4.29	4.00
Length and Thickness	8	8½	9	9½	10	10½	11	11½					
16" x 5/2"	2.50	2.35	2.22	2.11	2.00	1.91	1.82	1.74					
18" x 5/2¼"	2.75	2.59	2.45	2.32	2.20	2.10	2.00	1.91					
24" x 4/2"	3.75	3.53	3.33	3.16	3.00	2.86	2.73	2.61					
Length and Thickness	12	12½	13	13½	14	14½	15	15½	16				
16" x 5/2"	1.67												
18" x 5/2¼"	1.83	1.76	1.69	1.63	1.57								
24" x 4/2"	2.50	2.40	2.31	2.22	2.14	2.07	2.00	1.94	1.88				

[a] Sum of thicknesses; e.g., 5/2" means that 5 butts stacked have a total thickness of 2 inches.

Figure 6-27 Bundles of wood shingles per square

Starter Course

Look back at Figures 6-5 and 6-6 and you see that shakes and shingles are both installed with a double starter course. From Chapter 4, on asphalt shingles, we know from Equation 4-10 that:

Starter Course (SF) = Eaves (LF) x Exposed Area (SF/LF)

The starter course and first course overhang the eaves and rake to help shed water from the roof. Add a half inch (0.04 square feet per linear foot) to the actual roof size along each rake for the additional material you need for the overhang. Then add 1½ inches (0.125 square feet per linear foot) to the roof size along each eave.

Cutting Waste at Rakes, Hips and Valleys

Add 1 square foot of area to be covered per linear foot of hip or valley and 0.34 square foot of area per linear foot of rake for cutting waste and breakage. You may need to allow more than that if it's a complex roof or if you're using lower-grade shingles (lower grades break more easily). And if you're using a less experienced roofing crew, add a little more.

Ridge and Hip Units

Use the following equation to determine the number of prefabricated hip and ridge units required:

$$\text{Hip and Ridge Units} = \frac{\text{Total LF Ridge and Hips}}{\text{Exposure (ft.)}}$$

Equation 6-1

You must also add one unit at each hip for the starter course:

$$\text{Total Units} = \frac{\text{Total LF Ridge and Hips}}{\text{Exposure (ft.)}} + \text{Number of Hips}$$

Equation 6-2

Prefabricated wood shingle and shake ridge and hip units are sold by the bundle. Each bundle contains 40 units. If the units are field-fabricated, add 1 square foot for each linear foot of ridge and hips.

A Shortcut Method for Determining Wood Shingle and Shake Waste

As a rule of thumb, some contractors add:

- 10 percent waste to the net shingle or shake quantities on small- to average-size gable roofs
- 15 percent on hip roofs with field-fabricated hip and ridge units
- 5 percent on gable roofs with prefabricated hip and ridge units
- 10 percent on hip roofs with prefabricated hip and ridge units

This includes material required for the starter course and site-fabricated ridge and hip units (if used), cutting waste at rakes, hips and valleys, and waste due to crew errors. Sometimes you'll have to increase the allowance for more complex roofs.

I prefer a more accurate, realistic way to estimate waste which depends on:

- roof type
- ratio of roof length to roof width
- roof slope
- shingle exposure
- type of shingle installed (lower grades break more easily)
- whether hip and ridge units are prefabricated or site-constructed

In Figures 6-28 and 6-29 you see that the waste on shake or wood shingle roofs, exclusive of crew errors, is from 3 to 17 percent on average-sized gable roofs, and from 6 to 33 percent on hip roofs. Use Figures 6-28 and 6-29 to get a "ball-park" percentage for waste.

Wood Shingles and Shakes

Total percentages to be added to net roof area, including double starter course, cutting waste at rakes and additional material required for overhang.

Building dimension (L X W)	3 in 12 Roof slope Exposure 5"		7.5"		10"		6 in 12 Roof slope Exposure 5"		7.5"		10"		12 in 12 Roof slope Exposure 5"		7.5"		10"	
30 x 20	8*	(13**)	10*	(15**)	12*	(17**)	8*	(13**)	9*	(15**)	11*	(16**)	7*	(11**)	8*	(12**)	10*	(14**)
40 x 30	6	(9)	8	(11)	9	(12)	6	(9)	8	(11)	9	(12)	5	(7)	6	(8)	7	(9)
45 x 30	6	(9)	8	(11)	9	(12)	6	(9)	8	(11)	9	(12)	5	(7)	6	(8)	7	(9)
50 x 30	5	(8)	7	(10)	8	(11)	5	(8)	7	(10)	8	(11)	4	(6)	5	(7)	6	(8)
60 x 30	5	(8)	7	(10)	8	(11)	5	(8)	7	(10)	8	(11)	4	(6)	5	(7)	6	(8)
70 x 30	5	(8)	7	(10)	8	(11)	5	(8)	7	(10)	8	(11)	4	(6)	5	(7)	6	(8)
80 x 30	5	(8)	7	(10)	8	(11)	5	(8)	7	(10)	8	(11)	4	(6)	5	(7)	6	(8)
50 x 40	4	(6)	5	(7)	6	(8)	4	(6)	5	(7)	6	(8)	4	(6)	5	(7)	6	(8)
60 x 40	4	(6)	5	(7)	6	(8)	4	(6)	5	(7)	6	(8)	4	(6)	5	(7)	6	(8)
70 x 40	4	(6)	5	(7)	6	(8)	4	(6)	5	(7)	6	(8)	4	(6)	5	(7)	6	(8)
80 x 40	4	(6)	5	(7)	6	(8)	4	(6)	5	(7)	6	(8)	4	(6)	5	(7)	6	(8)
90 x 40	4	(6)	5	(7)	6	(8)	4	(6)	5	(7)	6	(8)	4	(6)	5	(7)	6	(8)
60 x 50	4	(6)	5	(6)	6	(8)	4	(6)	5	(7)	6	(8)	3	(4)	4	(5)	4	(5)
70 x 50	4	(6)	5	(7)	6	(8)	4	(6)	5	(7)	6	(8)	3	(4)	4	(5)	4	(5)
80 x 50	4	(6)	5	(7)	6	(8)	4	(6)	5	(7)	6	(8)	3	(4)	4	(5)	4	(5)
90 x 50	4	(6)	5	(7)	6	(8)	4	(6)	5	(7)	6	(8)	3	(4)	4	(5)	4	(5)

* Indicates the use of prefabricated ridge units. These units must be taken off separately.
** Parentheses indicate the use of site-fabricated ridge units. These unit are included in the data given in this table.

Figure 6-28 Cutting waste and overruns (gable roof)

Building dimension (L X W)	3 in 12 Roof slope Exposure						6 in 12 Roof slope Exposure						12 in 12 Roof slope Exposure					
	5"		7.5"		10"		5"		7.5"		10"		5"		7.5"		10"	
30 x 20	18*	(28**)	21*	(31**)	23*	(33**)	17*	(27**)	20*	(30**)	23*	(33**)	15*	(23**)	17*	(26**)	20*	(29**)
40 x 30	14	(22)	16	(24)	19	(27)	13	(21)	15	(23)	17	(25)	12	(19)	14	(21)	16	(23)
45 x 30	12	(19)	14	(21)	17	(24)	11	(18)	13	(20)	15	(22)	9	(15)	11	(17)	12	18)
50 x 30	11	(18)	13	(20)	15	(22)	10	(17)	12	(19)	14	(21)	9	(15)	11	(17)	12	18)
60 x 30	10	(16)	12	(18)	14	(20)	10	(16)	12	(18)	14	(20)	8	(13)	10	(15)	11	(16)
70 x 30	9	(15)	11	(17)	13	(19)	9	(15)	11	(17)	13	(19)	8	(13)	10	(15)	11	(16)
80 x 30	9	(15)	11	(17)	13	(19)	7	(12)	9	(14)	10	(15)	7	(12)	9	(14)	10	(15)
50 x 40	11	(17)	13	(19)	15	(21)	9	(15)	11	(17)	12	(18)	9	(14)	11	(16)	12	(17)
60 x 40	9	(15)	11	(17)	12	(18)	9	(14)	11	(16)	12	(17)	8	(13)	10	(15)	11	(16)
70 x 40	8	(13)	10	(15)	11	(16)	8	(13)	10	(15)	11	(16)	7	(11)	8	(12)	9	(13)
80 x 40	8	(13)	10	(15)	11	(16)	7	(12)	9	(14)	10	(15)	6	(10)	7	(11)	8	(12)
90 x 40	7	(12)	9	(14)	10	(15)	7	(11)	9	(13)	10	(14)	6	(10)	7	(11)	8	(12)
60 x 50	9	(14)	11	(16)	12	(17)	9	(14)	11	(16)	12	(17)	8	(11)	8	(12)	9	(13)
70 x 50	8	(13)	10	(15)	11	(16)	8	(12)	10	(14)	11	(15)	7	(11)	8	(12)	9	(13)
80 x 50	8	(12)	10	(14)	11	(15)	6	(10)	7	(11)	8	(12)	6	(10)	7	(11)	8	(12)
90 x 50	7	(11)	9	(13)	10	(14)	6	(10)	7	(11)	8	(12)	6	(9)	7	(10)	8	(11)

Total percentages to be added to net roof area, including double starter course, cutting waste at hips and additional material required for overhang.

* Indicates the use of prefabricated ridge units. These units must be taken off separately.
** Parentheses indicate the use of site-fabricated ridge units. These unit are included in the data given in this table.

Figure 6-29 Cutting waste and overruns (hip roof)

Wood Shingles and Shakes

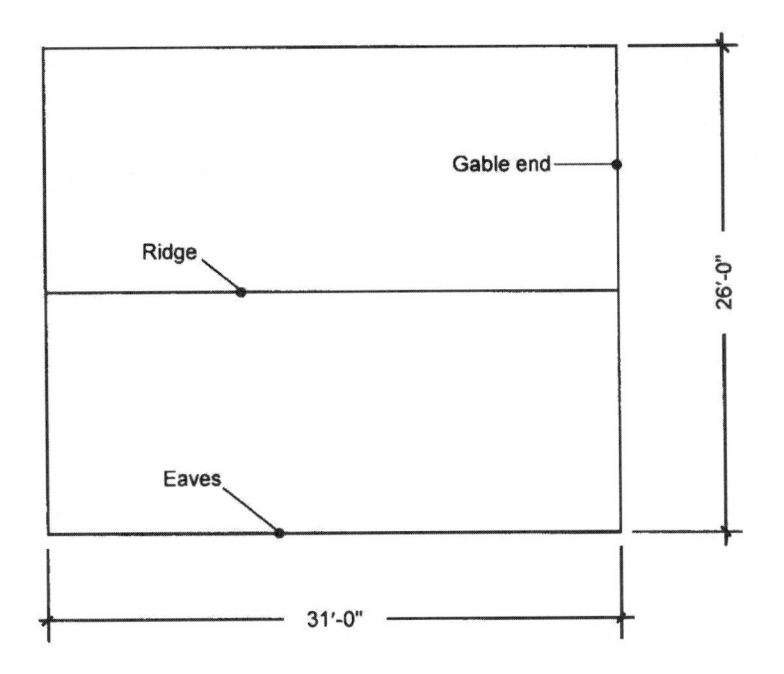

Figure 6-30 Gable roof example

▼**Example 6-1:** Assume a roof slope of 5 in 12, and find the number of bundles of No. 1 grade (16" Fivex) wood shingles required to cover the roof of the gable roof in Figure 6-30. Also assume a double starter course at the eaves and the use of prefabricated ridge units:

From Figure 6-25 you see that the maximum recommended exposure for this 5 in 12 roof is 5 inches. Therefore, use the following equation (from Chapter 1) to find the net roof area:

Actual (Net) Roof Area = Roof Plan Area x Roof-Slope Factor

Net Roof Area
= (31' x 26' x 1.083) (from Column 2 of Appendix A)
= 873 square feet ÷ 100
= 8.73 squares (use this figure to estimate labor costs)

The total eaves length is 2 x 31, or 62 linear feet.

The total area of starter-course material required is 62' x 0.42 SF/LF (5-inch exposure divided by 12), or 26 square feet from the equation:

Starter Course (SF) = Eaves (LF) x Exposure Area (SF/LF)

The job requires 4 x 13' x 1.083, or 56 linear feet of material for the rakes (4 rakes at 13 feet each, times the roof-slope factor from Appendix A).

Figure waste for breakage at the rakes at 19 square feet (56 LF times 0.34 SF/LF).

Waste allowance for the overhang at the rakes is 56' x 0.04 = 2.24 square feet ($1/2$ inch per actual linear foot of rake).

Waste allowance for the overhang at the eaves is 62' x 0.125 = 7.75 square feet ($1\,1/2$ inches per linear foot of eaves).

Therefore, total waste for rakes and eaves is 10 square feet.

Add these to get the gross roof area:

= 873 SF + 26 SF + 19 SF + 10 SF
= 928 square feet ÷ 100
= 9.28 squares

Divide the net roof area by that answer to confirm the waste allowance:

Waste Factor $= \dfrac{9.28 \text{ Squares}}{8.73 \text{ Squares}}$
= 1.06

Figure 6-28 predicts a range from 6 to 8 percent waste, so this is correct.

Figure 6-27 says this material comes in 4 bundles per square, so the job requires 38 bundles of shingles (9.28 x 4).

The ridge length is 31 linear feet, so from Equation 6-1, here's how to figure how many prefabricated ridge units to order:

Ridge Units $= \dfrac{31'}{0.42' \text{ (5–inch exposure divided by 12)}}$
= 74 each

Since these come 40 to a bundle, the job requires 2 bundles of ridge units.

If you had used field-fabricated ridge units, you'd need an additional 31 square feet (1 SF/LF of ridge), or 959 square feet. That's 9.59 squares, or 39 bundles (at 4 bundles per square).

Now, here's an example showing how this works for a hip roof:

▼ **Example 6-2:** Assuming a roof slope of 6 in 12, find the number of bundles of No. 2 grade (18" Perfections) wood shingles required for the hip roof in Figure 6-31. Assume a double starter course at the eaves and lower ends of the hips, and the use of prefabricated ridge and hip units.

Figure 6-25 says that the maximum recommended exposure for this type of shingle on a 6 in 12 roof is $4\,1/2$ inches.

The net roof area is 40' x 20' x 1.118 (Column 2 of Appendix A), or 894 square feet. Base your labor cost on 8.94 squares of material.

The total eaves length is 2 x (40' + 20'), or 120 linear feet.

The total area of starter-course material required is 120' x 0.38 ($4\,1/2$-inch exposure divided by 12), or 46 square feet.

172

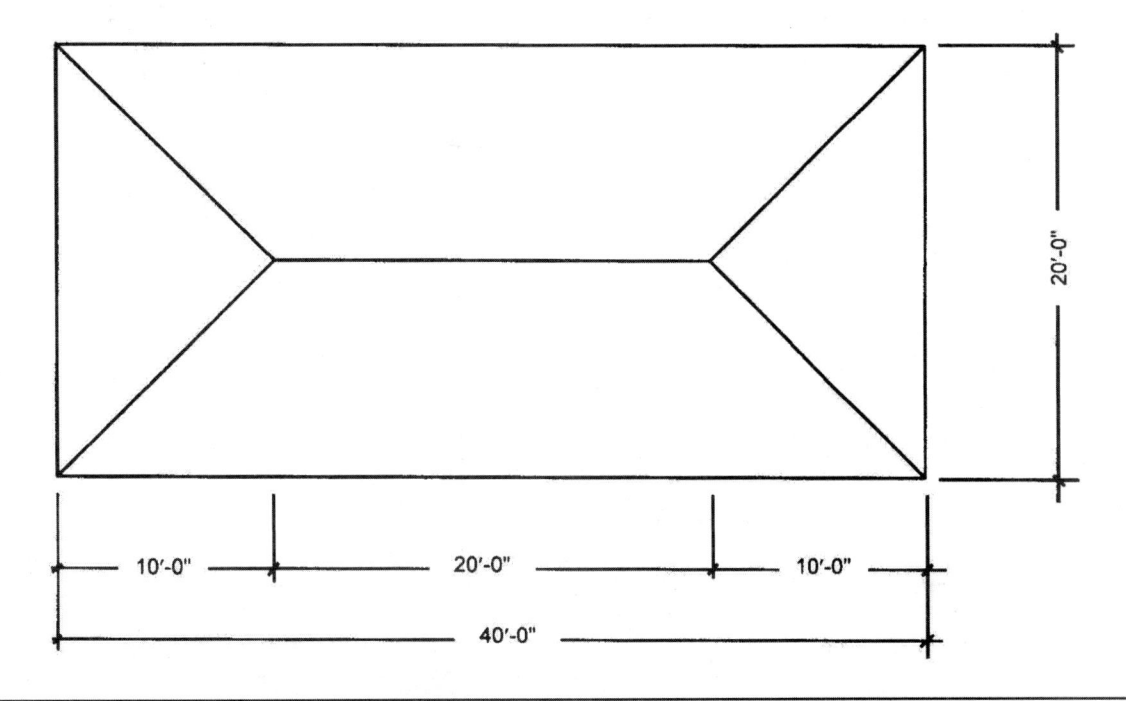

Figure 6-31 Hip roof example

The length of each hip is 10 x 15 (from Column 3 of Appendix A) or 15 linear feet. So there's a total of 60 linear feet of hip.

The cutting waste at the hips is 1 square foot per linear foot, or 60 square feet.

The additional material required for the 1$\frac{1}{2}$-inch overhang at the eaves is 15 square feet (120' x 0.125 square feet per linear foot).

Therefore, the gross roof area is 894 SF + 46 SF + 60 SF + 15 SF, which equals 1,015 square feet, or 10.15 squares of material.

Figure 6-29 predicts a range of waste from 13 to 17 percent. This example calculates to 14 percent (1015 ÷ 894).

From Figure 6-27, you see that these shingles come 4.89 bundles per square, so you have to order 50 bundles (10.15 squares x 4.89).

The total length of ridge and hips is 80 linear feet, so from Equation 6-2 the job requires 215 units (80 divided by 0.38, plus 4 units).

There are 40 units per bundle, so the job requires 6 bundles (215 ÷ 40).

For field-fabricated hip and ridge units, add another 80 square feet based on 1 square foot per linear foot of ridge and hip.

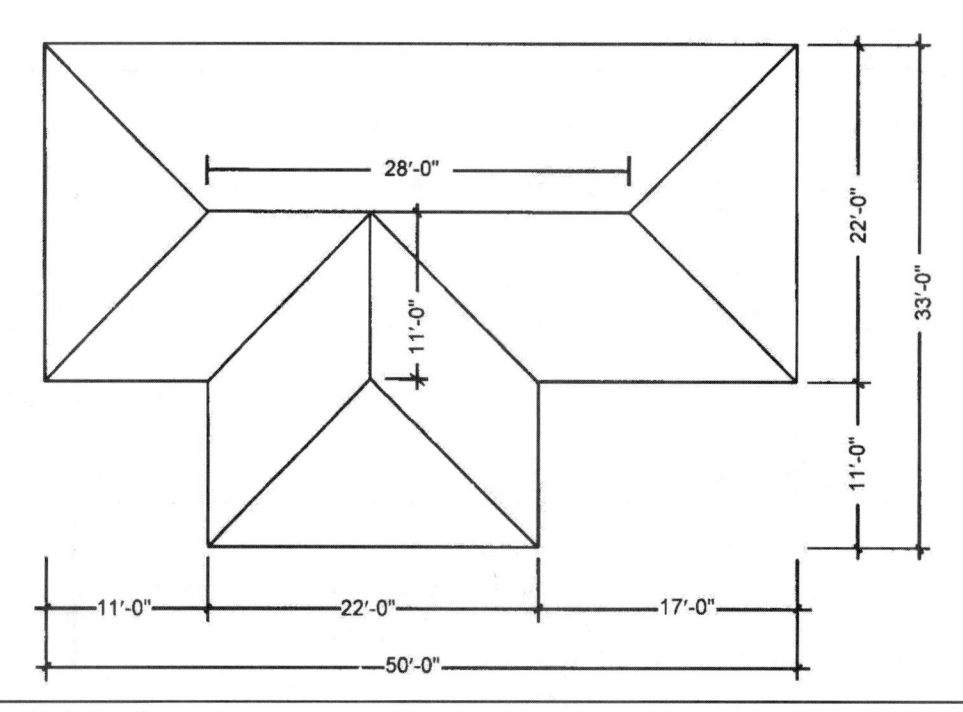

Figure 6-32 Hip and valley roof example

Here's an example for a hip and valley roof:

▼ **Example 6-3:** Assume a roof slope of 6 in 12, and find the number of bundles of No. 3 grade (24" Royals) wood shingles required for the roof in Figure 6-32. Figure on a double starter course at the eaves and lower ends of the hips, and the use of prefabricated ridge and hip units.

Figure 6-25 calls for a 5½-inch exposure for this job.

The net roof area is [(50' x 22') + (22' x 11')] x 1.118 (the roof-slope factor), or 1,415 square feet for 14.15 squares of shingles (use this figure to estimate labor costs).

Eaves length is 166 linear feet (2 x (50' + 33'))

The total area of starter-course material required is 76 square feet (166' x 0.46 SF/LF), where 0.46 is the 5½-inch exposure divided by 12.

The length of any hip or valley from Column 3 of Appendix A is 16.5 linear feet, so the total length of hips and valleys is 132 linear feet (8 ea. x 16.5').

Cutting waste at the hips and valleys is 1 square foot per linear foot, or 132 square feet.

Waste at the eaves overhang is 21 square feet (166' x 0.125 SF/LF).

Therefore, the gross roof area is:

Gross Roof Area = 1,415 SF + 76 SF + 132 SF + 21 SF
= 1,644 square feet ÷ 100
= 16.44 squares

The waste factor, excluding crew-error waste and breakage, is 16 percent (1644 ÷ 1415).

Figure 6-27 shows that we require 5.46 bundles per square for these shingles, so this job requires 90 bundles (16.44 x 5.46).

The total ridge length is 39 linear feet (28' + 11') and the total hip length is 99 linear feet (6 x 16.5') for a total of 138 linear feet.

The number of prefabricated ridge and hip units required is 306 units (138 ÷ 0.46, plus 6 units).

At 40 units per bundle, the job requires 8 bundles (306 ÷ 40).

Staggered Patterns

Staggered (serrated) wood shingle and shake patterns are popular. You can calculate material quantities required for these patterns the same way as for conventional patterns in the previous examples. This is true because the only difference is that, beginning with the second row of shingles, every other shingle is gauged up or down with an offset of around 1 inch, producing two different exposures in each row.

Dutch Weave Patterns

Another popular wood shingle application is the Dutch weave pattern (also referred to as a *thatch*, or *shake-look* roof). Increase the quantity (field shingles only) for a conventional wood shingle roof by 25 percent for this pattern. This is because you need an additional shingle under every fourth shingle in each course. Offset the double-thick set of shingles from other shingles by no more than 1 inch.

Sidewall Shakes and Wood Shingles

Shake and wood shingle siding is durable, attractive and requires no paint. Wood sidewall shingles are manufactured from red cedar or white cedar and sold in Number 1, 2 and 3 grades. A wide variety of cut patterns is available. They come in lengths of 16, 18 and 24 inches.

All wood shingles and shakes you use in roofing can also be used on sidewalls. But there are some shingles and shakes which are specifically manufactured for sidewalls. They include No. 4 undercoursing and rebutted

Roofing Construction & Estimating

| Diagonal | Half cove | Diamond | Round | Hexagonal | Octagonal | Arrow | Square | Fish-scale |

Courtesy of Cedar Shake & Shingle Bureau

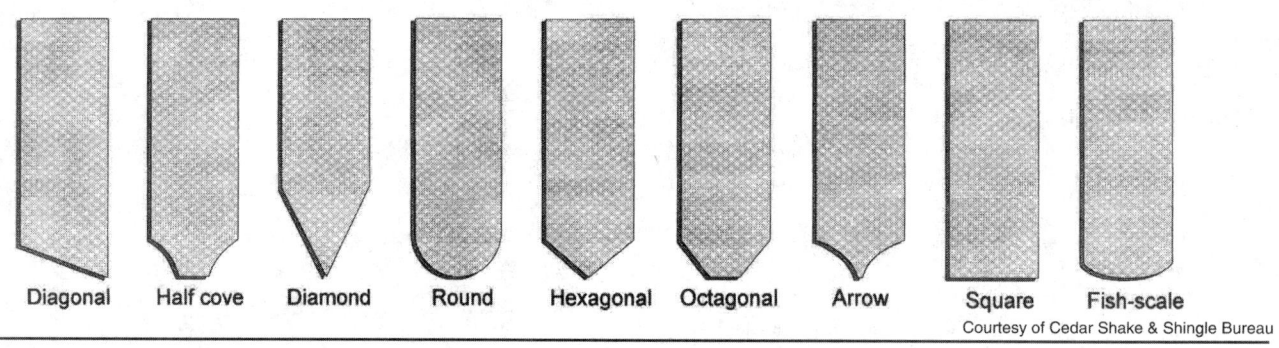

Figure 6-33 Fancy-butt red cedar shingles

and rejointed shingles which are shown in Figure 6-3. For double-coursing on sidewalls, use No. 1 machine-grooved shakes which are described in Figure 6-4.

Fancy-butt red cedar shingles are 5 inches wide and 16 or 18 inches long. A 96-piece carton will cover 25 square feet when laid at a $7^{1}/_{2}$-inch exposure. Fancy-butt shingles can also be custom-made. Figure 6-33 shows some example of fancy-butt shingles.

Installing Sidewall Shingles and Shakes (New Construction)

I always applied 15- or 30-pound felt to walls on new construction before installing wood shingles or shakes. Certain synthetic underlayments may also be applicable, but because of moisture issues, which will be different from those on a roof, be sure to check with your local officials beforehand.

To apply felt, start at the base of the wall and use a 2-inch side lap and a 6-inch end lap. Wrap the felt 4 inches each way beyond both inside and outside corners. Install metal flashing above window and door openings and caulk around the openings. I recommend that you also install metal flashing over all inside corners. Be sure all door and window casings are in place and caulked before you start the shingle or shake application. That's shown in Figure 6-34.

Find the number of courses required by measuring the height of the wall from a point 1 inch below the top of the foundation to the top of the wall. Divide the height into equal parts corresponding closely to the recommended exposure. Don't exceed the maximum recommended exposure. Transfer the measurement

Furring when required

1. Replace old casing with brick or shingle molding or other trim to suit, or apply new casing over the old.
2. Caulk.

Courtesy of Cedar Shake & Shingle Bureau

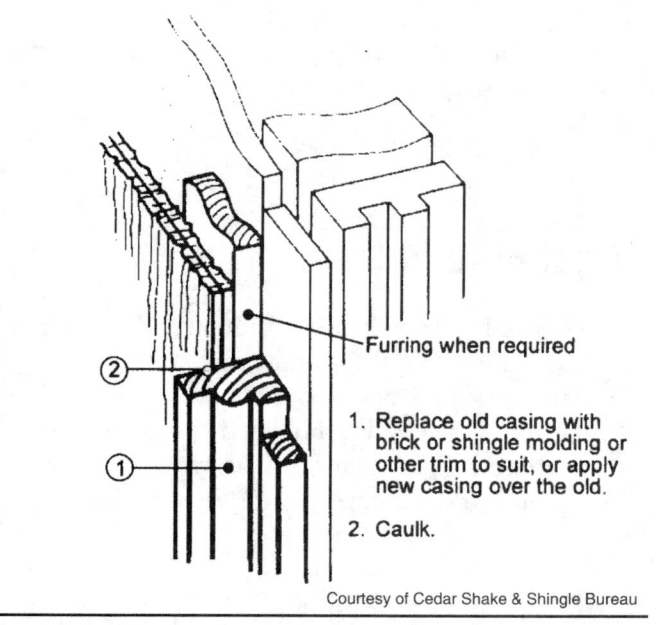

Figure 6-34 Typical window and door casing detail

Courtesy of Cedar Shake & Shingle Bureau

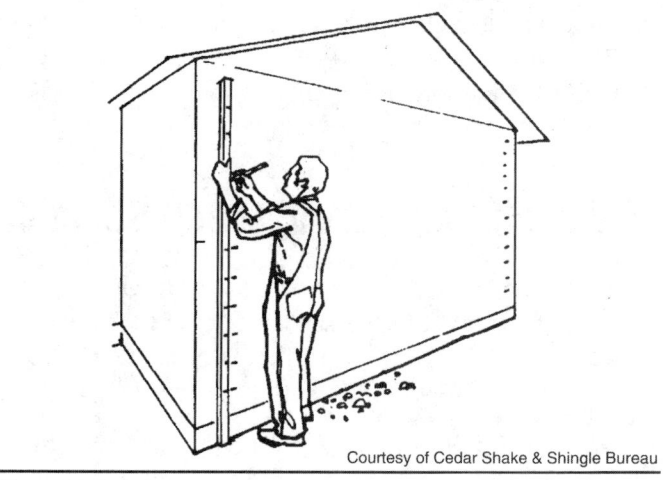

Figure 6-35 Story pole

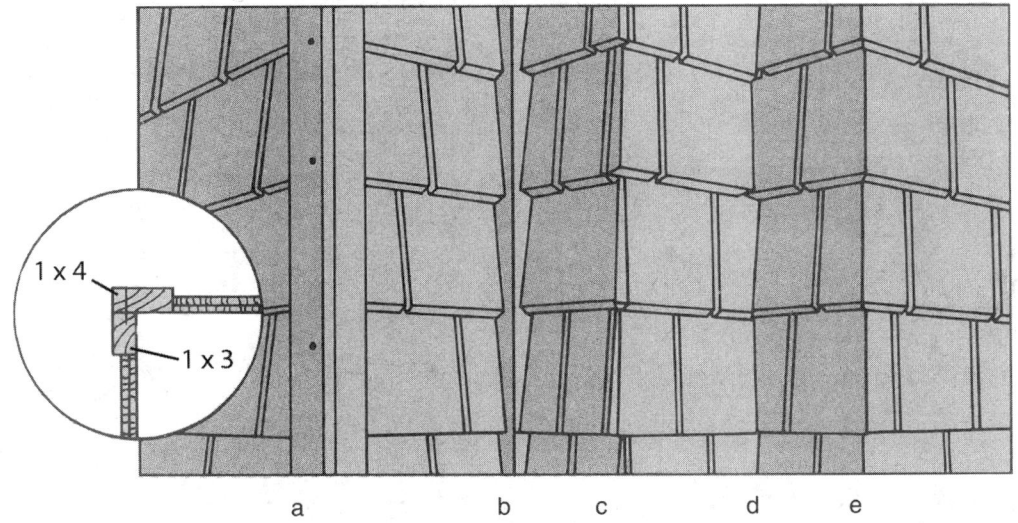

a) Shingles butted against corner boards
b) Shingles butted against square wood strip on inside corner, flashing behind
c) Laced outside corner
d) Laced inside corner with flashing behind
e) Mitered cornier

Courtesy of Cedar Shake & Shingle Bureau

Figure 6-36 Corner details

and the number of courses to a story pole (a straight 1x3 or 1x4) to lay out the courses on the walls, as in Figure 6-35.

Whenever possible, align the butt lines with the tops and bottoms of windows and other openings. Then you won't have to cut so many shingles to fit. Also try to line up the courses so the exposure of the final course matches those below. Use the story pole to make marks at opposite ends of each wall. Place the marks on corner boards, or on the felt. Figure 6-36 shows alternative methods for installing shingle siding at corners.

To align the first course of shingles, shakes, or undercoursing, nail down a shingle at the correct location at each end of the wall. Tack a nail into the butt of each shingle and stretch and tie a string line between the two shingles. Align the butts of the other shingles along the string. You can align succeeding courses the same way, or snap a chalk line over the preceding course to line up the butts of the next course.

When you install shingles at corners, miter, weave, or butt them against the trim boards, as shown in Figure 6-36. Mitered corners provide the least protection against wind-driven rain, and it'll take a long time to install them. Woven corners provide more protection and are quicker to install. The fastest method and most weather-resistant corner treatment is to butt shingles up to trim boards. Use 2 x 2s at inside corners and a prefabricated outside corner made up of a 1 x 3 and a 1 x 4. For added weather protection, install a bead of caulk between the shingles and corner boards.

Shingle away from corners, especially if you install mitered or laced corners, and trim shingles or shakes to fit around door and window openings. Allow a $1/8$- to $1/4$-inch joint space between shingles or shakes. You can install units with tight-fitting joints if you prime and stain them first. Offset the joints of adjacent courses at least $1 1/2$ inches, as shown in Figure 6-37.

Sidewall Shingling Methods

The method you use to install shingles on sidewalls depends on the coursing style you're using, and the material under the shingles.

To install shingle coursing over nailable sheathing, start with a double starter course. For economy, you can use undercourse-grade shingles overlaid with a higher-grade outer course. Drive nails about 1 inch above the butt line of the succeeding course and $3/4$ inch from each edge, as in Figure 6-38. If a shingle or shake is wider than 8 inches, drive two additional nails about 1 inch apart near the center. This rule applies no matter which coursing style you use. Use galvanized ring-shank nails to install sidewall shingles or shakes.

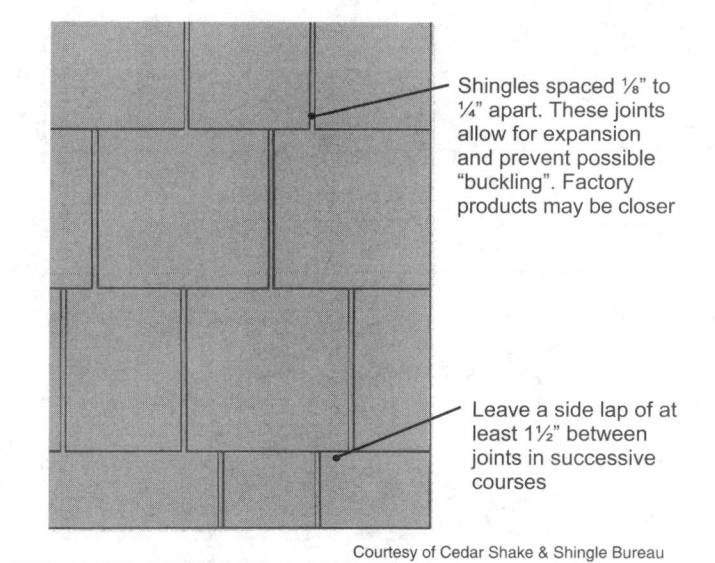

Figure 6-37 Spacing detail

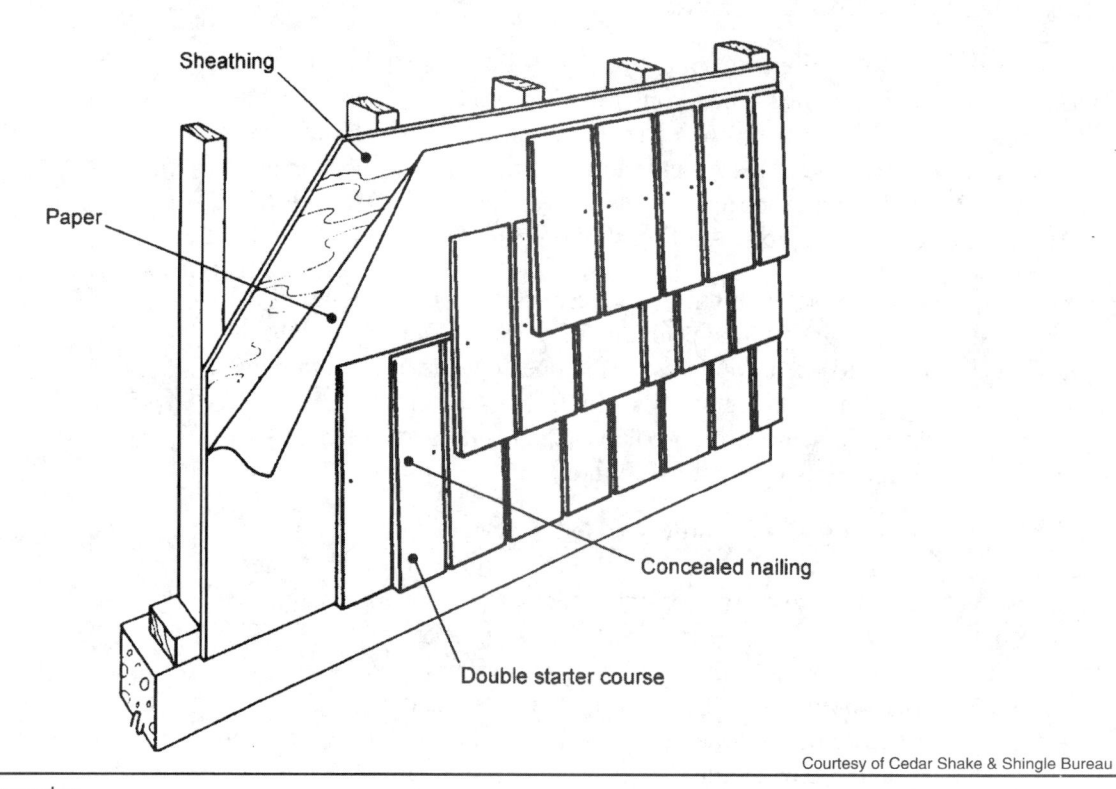

Figure 6-38 Single coursing

Wood Shingles and Shakes

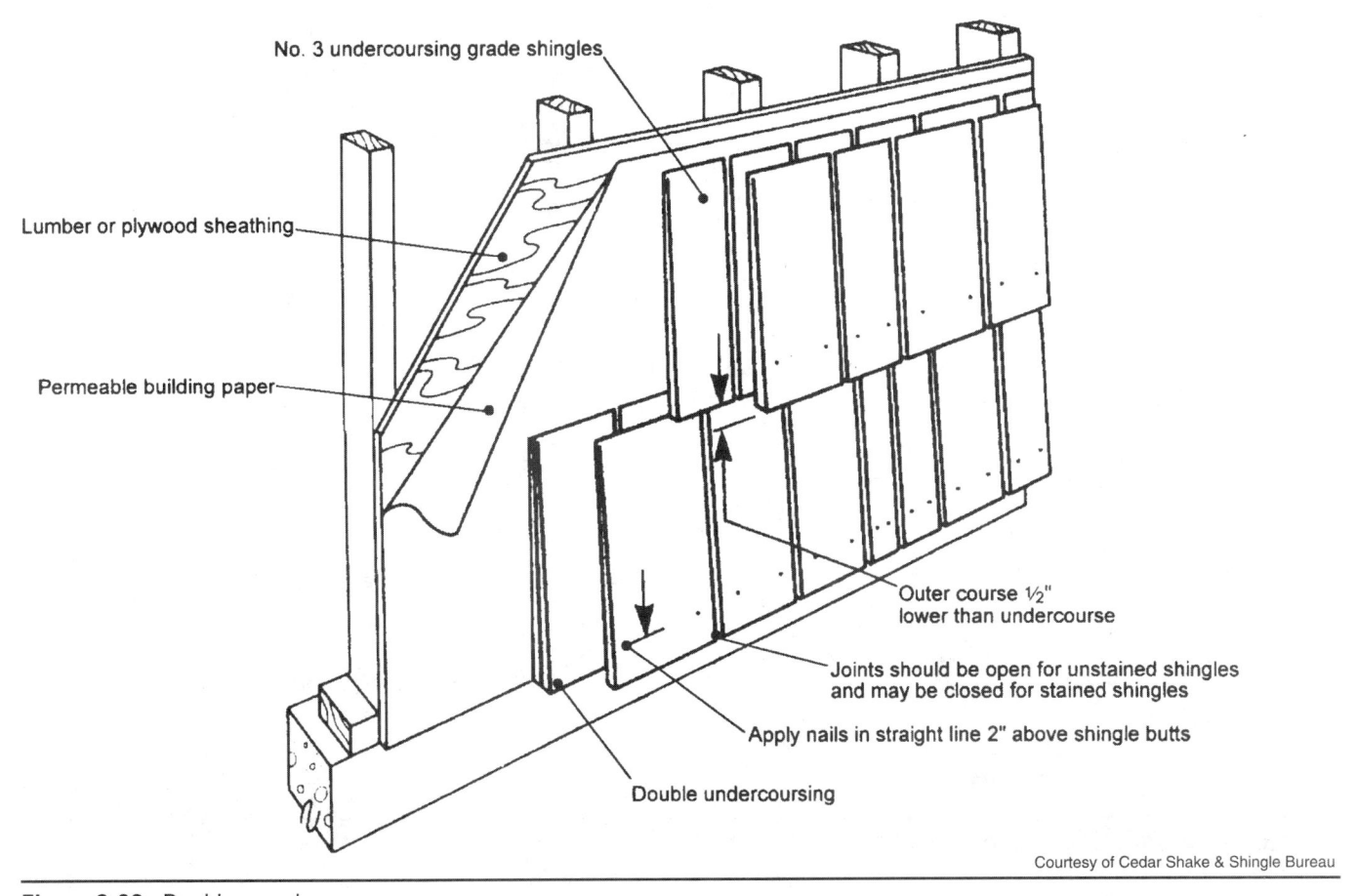

Figure 6-39 Double coursing

Labels in figure:
- No. 3 undercoursing grade shingles
- Lumber or plywood sheathing
- Permeable building paper
- Outer course ½" lower than undercourse
- Joints should be open for unstained shingles and may be closed for stained shingles
- Apply nails in straight line 2" above shingle butts
- Double undercoursing

Courtesy of Cedar Shake & Shingle Bureau

The double-coursing method provides a wide exposure and deep shadow lines. It's an economical application because the wide exposure requires fewer shingles. You'll need undercoursing, but you can use lesser-grade inexpensive shingles.

To install double coursing over nailable sheathing, install a triple starter course using two undercourses and one outer course. Apply all outer courses ½ inch lower than the undercourse. Install each undercourse unit with one nail or staple driven into the center of the unit. Face-nail each exposed shingle or shake to the undercourse units with two casing nails driven about 2 inches above the butt line and ¾ inch from each edge, as shown in Figure 6-39.

To install double coursing over non-nailable sheathing, first fasten lath strips to the studs with 7d or 8d nails. Space the lath strips vertically so the distance between their centers is the same as the shingle exposure. Then rest the butts of the undercoursing on the lath strips you nailed to the studs.

For the first course, use two undercourses resting on double lath strips overlaid with an outer course. Nail or staple the undercourse units to the sheathing using two fasteners per unit. Apply all outer courses ½ inch below each lath and face-nail each outer course to the laths with two small-headed 5d nails (Figure 6-40).

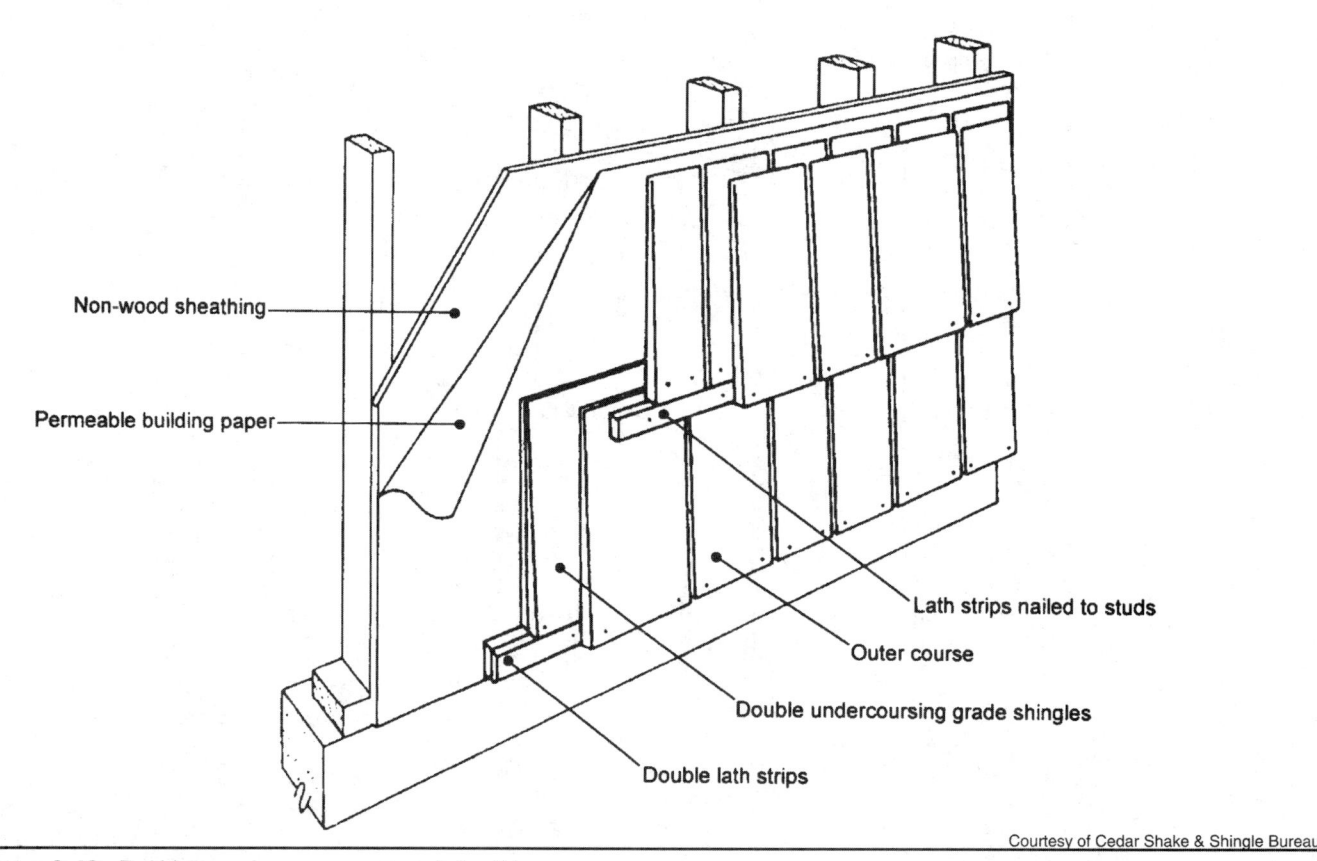

Figure 6-40 Double coursing over non-wood sheathing

To install staggered single coursing over nailable sheathing, set the butts of the shingles or shakes below a horizontal line you set with a story pole. Be sure you don't go above the horizontal line, or the exposure of the underlying shingle will be too great. Don't vary the distance from the butts to the horizontal by more than 1 inch for 16- and 18-inch shingles, or more than $1\frac{1}{2}$ inches for 24-inch shingles. That's shown in Figure 6-41.

Ribbon coursing gives a double shadow-line effect. You can install ribbon coursing over nailable sheathing by raising the outer course units about 1 inch above the undercoursing. Use number 1 grade units for the undercoursing. You can see this installation in Figure 6-42.

Installing Sidewall Shingles and Shakes over Existing Walls

If you remove exterior sheathing and felt from an existing wall, be sure to take out all nails and anything else sticking out before you install new shakes or shingles. Apply new 15- or 30-pound felt, or other underlayment approved for this purpose, over the old sheathing. This is called *re-walling*.

Usually, you can install new shakes or shingles over an existing exterior finish (*over-walling*). This saves time because you don't have to get rid of the old wall covering. But you still have to remove old door and window casings and install new ones. Figure 6-34 in the section on new construction shows this.

Wood Shingles and Shakes

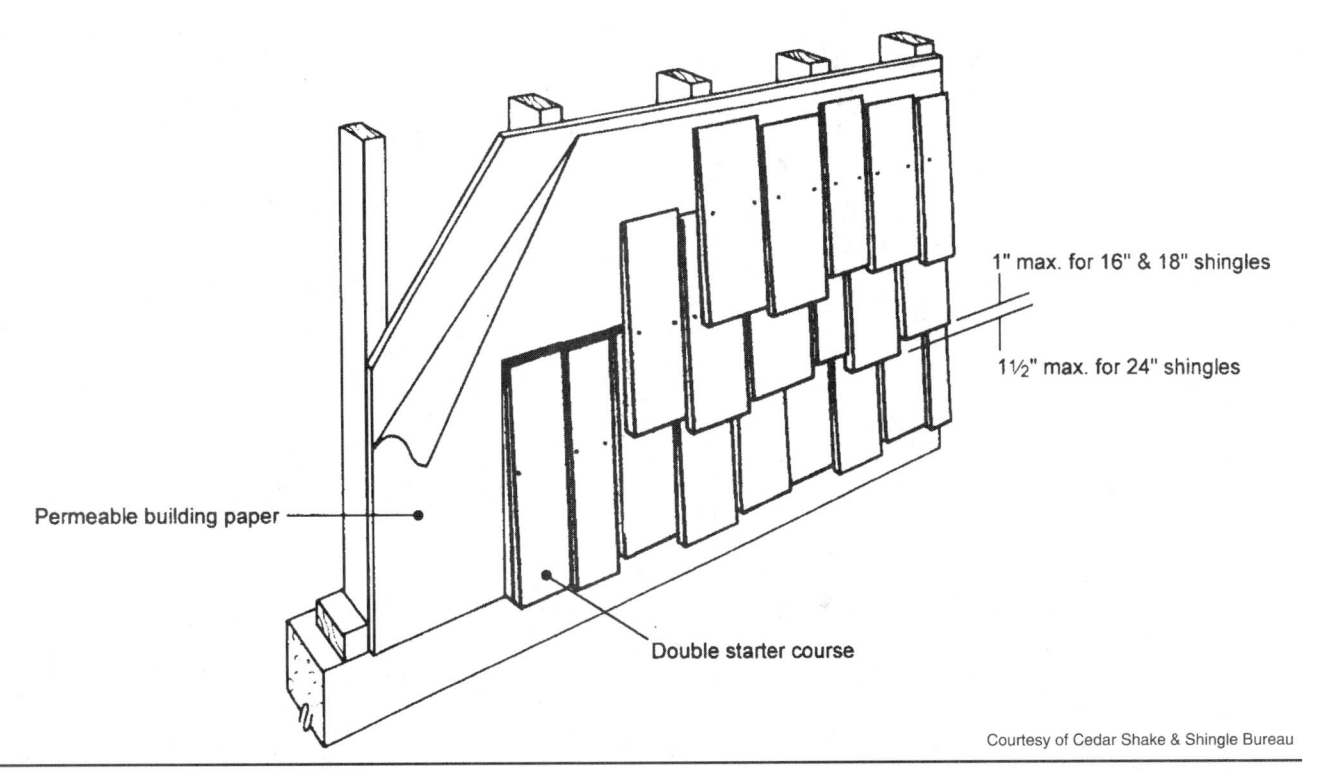

Figure 6-41 Staggered coursing

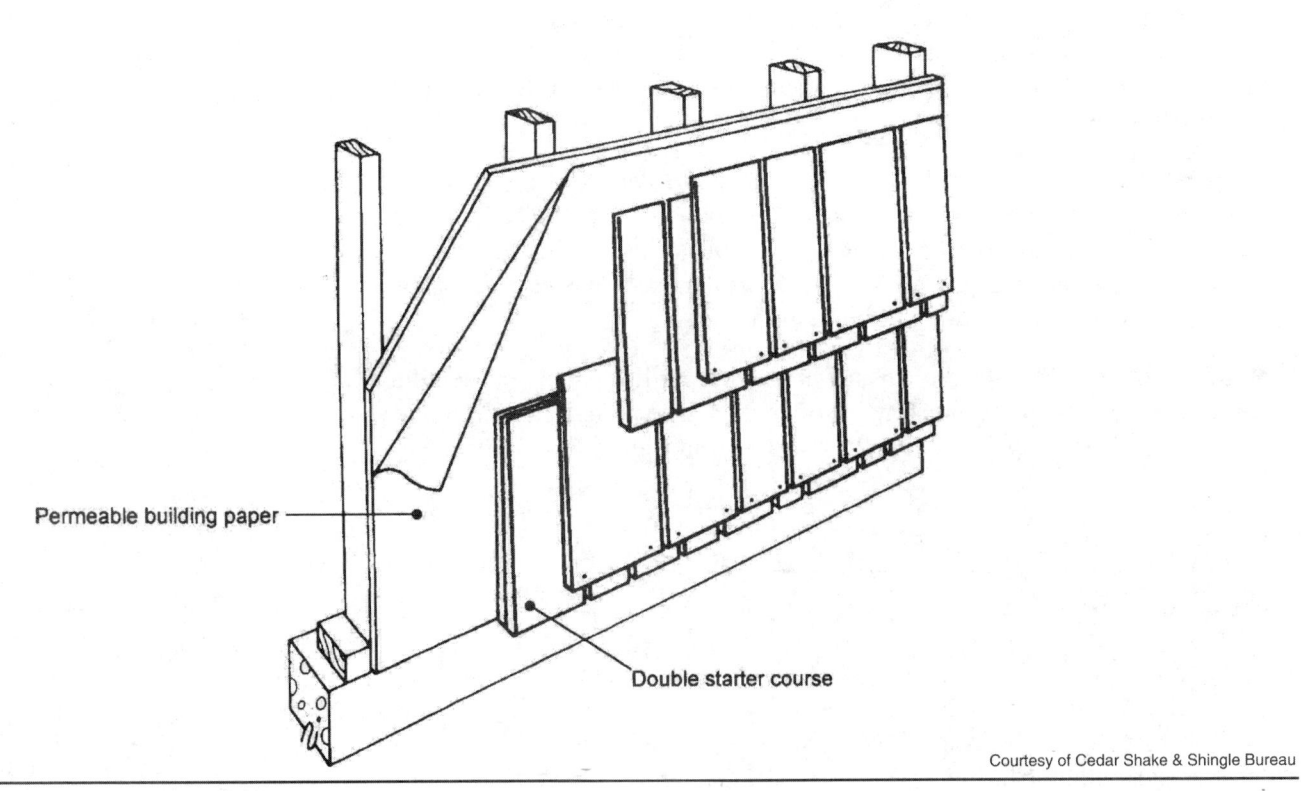

Figure 6-42 Ribbon coursing

181

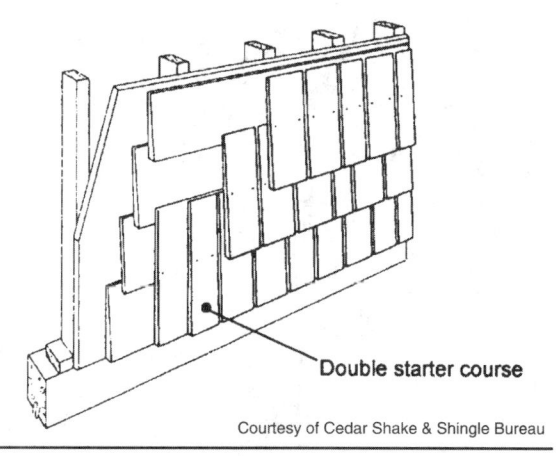

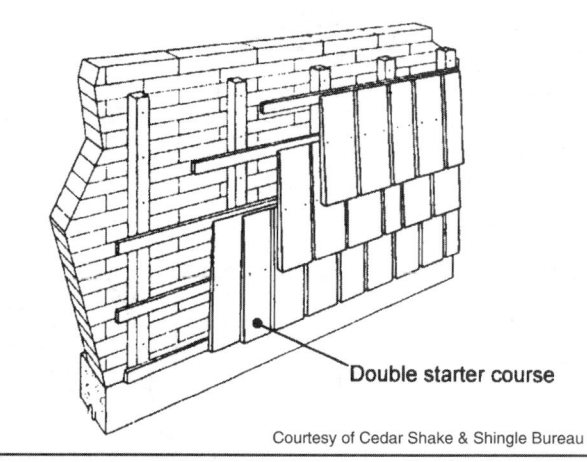

Figure 6-43 Over beveled siding detail

Figure 6-44 Over masonry detail

■ **Over Siding** You can install new shakes or shingles directly over old smooth wood siding the same way you would over any nailable sheathing. In this case, you don't have to apply roofing felt first, assuming there's underlayment beneath the existing siding.

To install shakes or shingles over beveled siding, fill in the low points of the existing wall with lumber or plywood strips (called *horse feathers*). This gives you more area to use as a nailing surface. Then install the shakes or shingles the same as you would over sheathing.

You can also nail the shingles or shakes to the high points of the bevels of each course of the old wall, as in Figure 6-43. Or you can nail the units to alternate siding courses, as long as you don't go over the recommended shingle exposure.

■ **Over Masonry** To install shingles or shakes over existing masonry walls, first use masonry nails at the masonry joints to install 2 x 4 furring strips vertically. Then install 1 x 3 or 1 x 4 nailing strips horizontally across the furring strips spaced at centers the same as the exposure distance of the shingles or shakes, as shown in Figure 6-44.

■ **Over Stucco** Over stucco, install horizontal nailing strips on centers equal to the exposure of the shingles or shakes. Use nails long enough to go into the underlying sheathing. That's shown in Figure 6-45.

Roof Junctures

All roof junctures must be weathertight. For flashing, use at least 26-gauge galvanized steel painted on both sides with metal or bituminous paint. If you have to bend the flashing strips at a sharp angle (greater than 15 degrees), paint the flashing after you make the bends. (Sheet metal shops have special machines to bend flashing materials.)

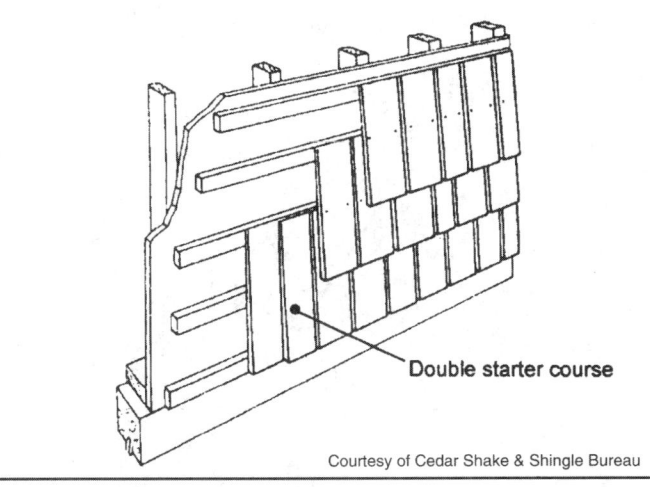

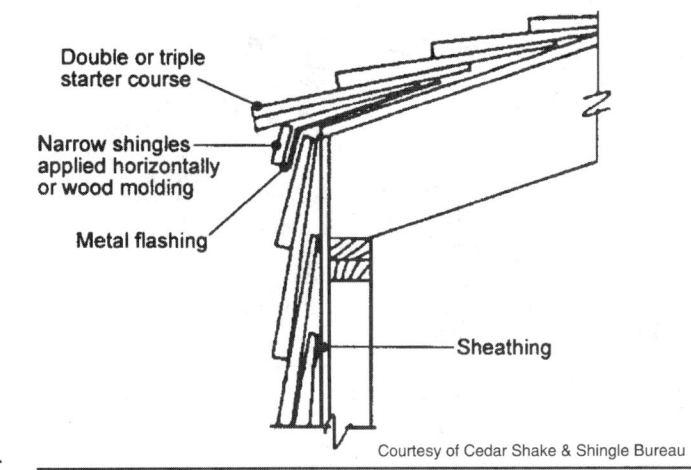

Figure 6-45 Over stucco detail

Figure 6-46 Convex roof juncture

■ **Convex Juncture**   For a convex juncture, cover the top 4 inches of the final course of the wall and 8 inches of the roof with flashing, as shown in Figure 6-46. Install a narrow strip of shingles or shakes, or a wood molding strip over the downward leg of the flashing. Then install a double or triple starter course on the roof with a 1$\frac{1}{2}$-inch overhang at the eaves. Complete the roof in the normal way.

■ **Concave Juncture**   For a concave juncture, cover the top of the roof slope and the bottom 4 inches of the wall with flashing, as shown in Figure 6-47. The flashing can be either under or over the last course(s). After you've finished the roof part of the juncture, install a double starter course at the bottom of the wall. Then finish the remaining wall as described above.

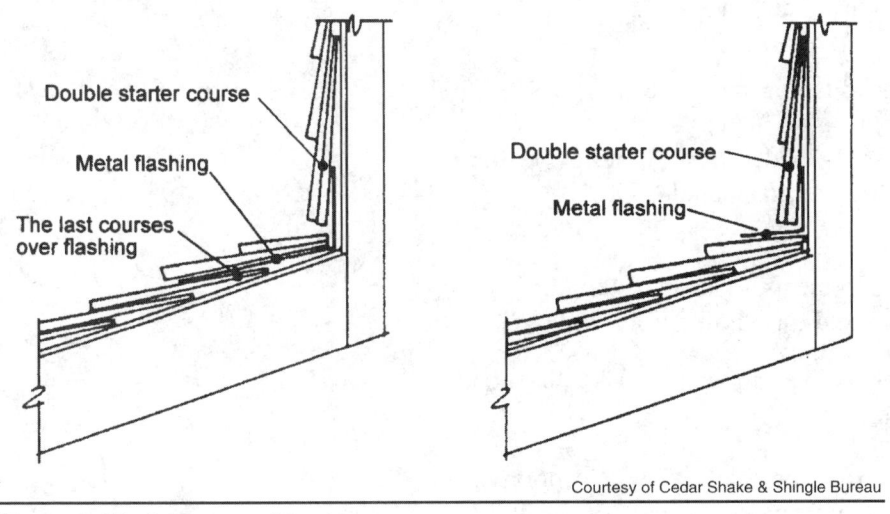

Figure 6-47 Concave roof juncture

Roofing Construction & Estimating

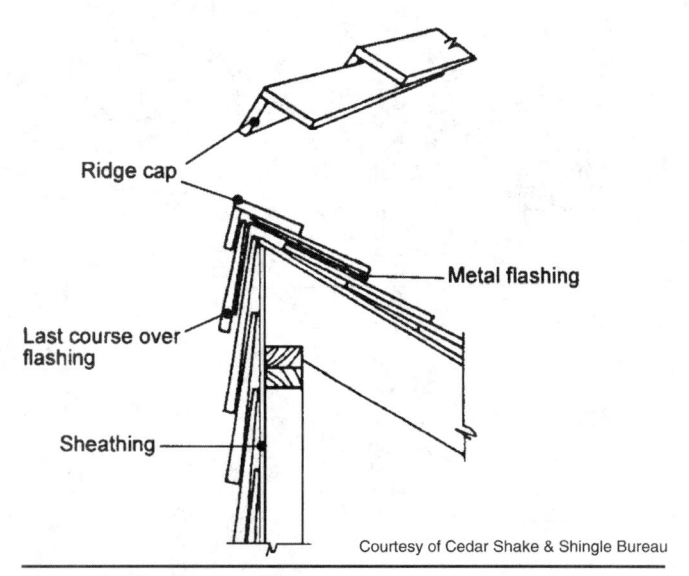

Figure 6-48 Apex roof juncture

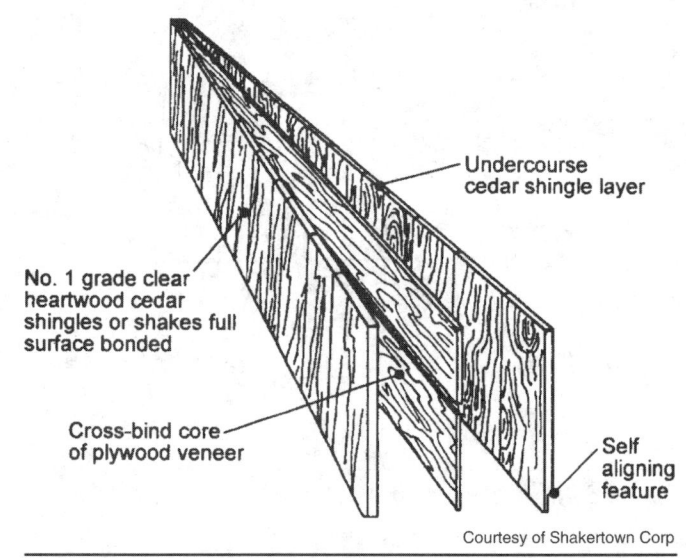

Figure 6-49 Sidewall and mansard panels

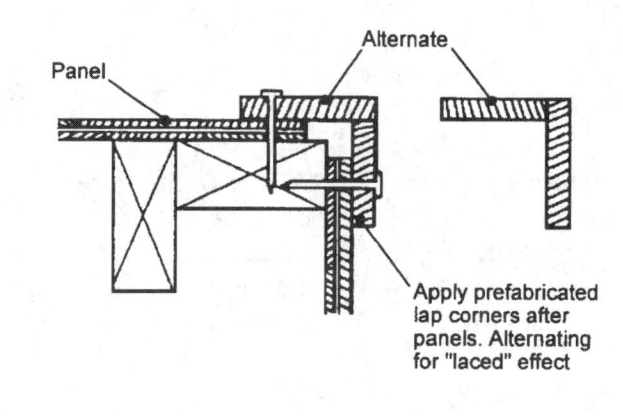

Figure 6-50 Lap corners

■ **Apex Juncture** Cover the top 8 inches of the roof and the top 4 inches of the wall with flashing, as in Figure 6-48. Finish the wall before you begin to shingle the roof. Then cover the juncture with prefabricated ridge units. The staples joining the prefab ridge units are flexible, so you can bend them to fit any angle.

■ **Shingle Sidewall and Mansard Panels** To save time, install prefabricated panels faced with cedar shingles bonded to various backing materials as shown in Figure 6-49. These panels are self-aligning. They're usually 8 feet long and 18 inches wide (for 14-inch exposure) or 9 inches wide (for 7-inch exposure). Panel faces come in a variety of styles and textures, and in a straight- or staggered-butt design.

You can also get prefabricated outside lap corners (Figure 6-50) and flush outside corners (Figure 6-51). The 18-inch and 9-inch panels come in 4-, and 8-panel bundles, respectively. Either panel size covers 37 square feet per bundle. This equals 2.7 bundles per square.

Depending on local building code requirements, you can install the panels over sheathing, or nail them directly to studs, as shown in Figure 6-52.

Figure 6-53 shows how to apply panels over a mansard. Use 30-pound roofing felt under all panel installations.

You can get additional information from the Shakertown Corporation. They're at P.O. Box 400, 1200 Kerron Street, Winlock, WA, 98596 telephone (360) 785-3501.

Wood Shingles and Shakes

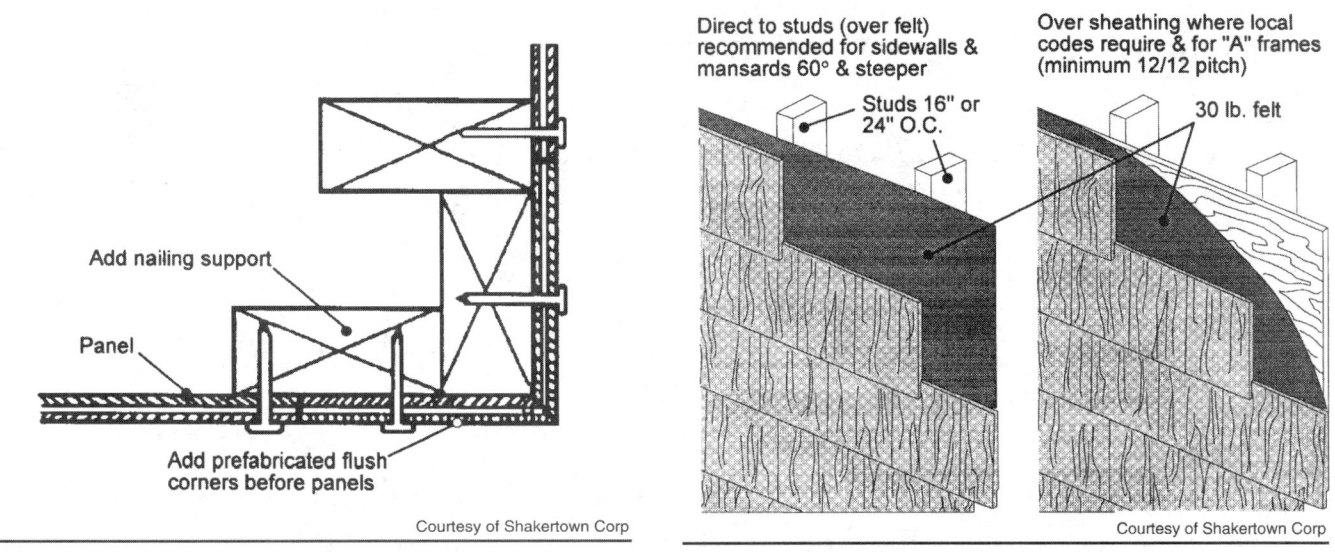

Figure 6-51 Flush corners

- Add nailing support
- Panel
- Add prefabricated flush corners before panels

Courtesy of Shakertown Corp

Figure 6-52 Siding panel applications

- Direct to studs (over felt) recommended for sidewalls & mansards 60° & steeper
- Over sheathing where local codes require & for "A" frames (minimum 12/12 pitch)
- Studs 16" or 24" O.C.
- 30 lb. felt

Courtesy of Shakertown Corp

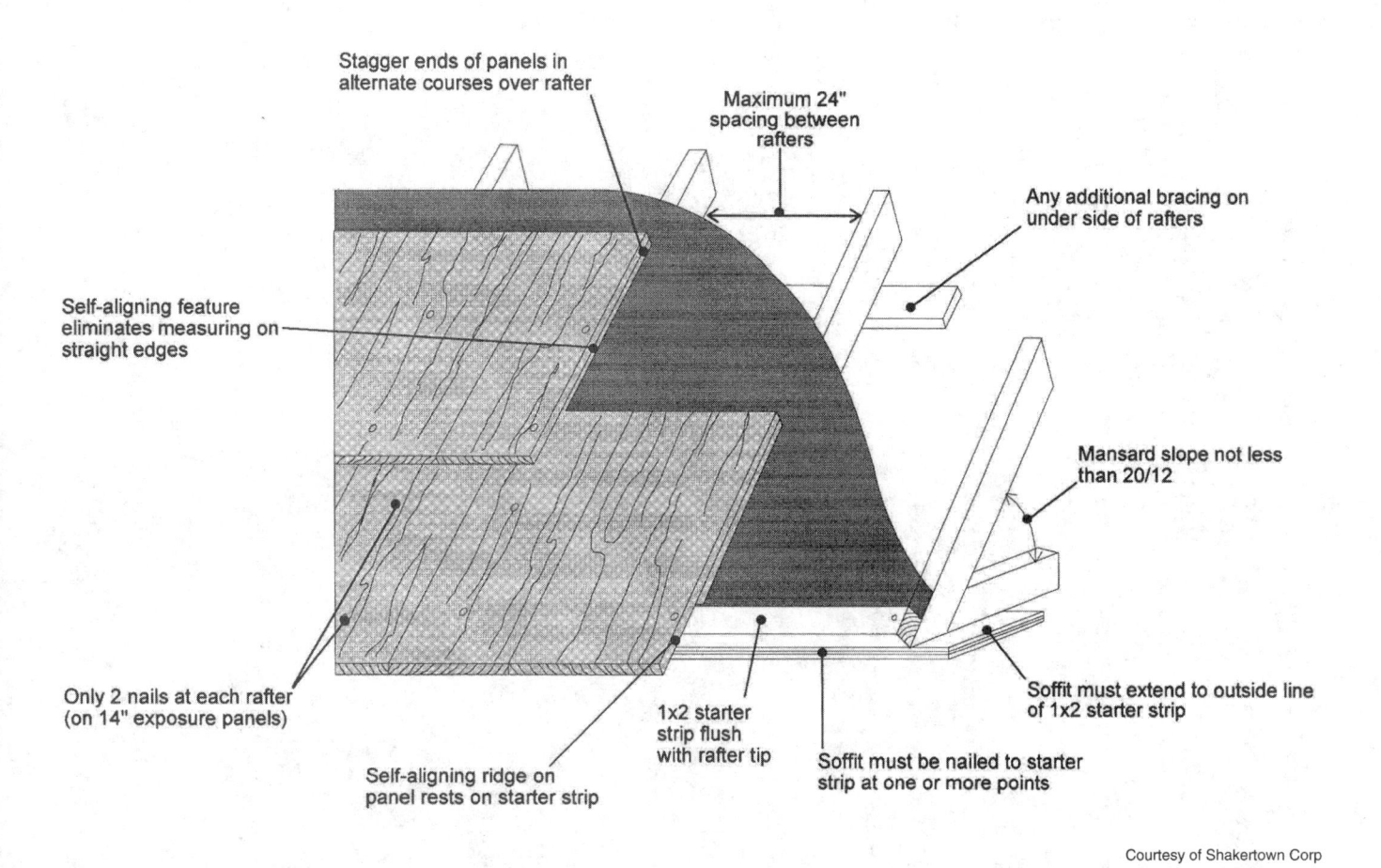

Figure 6-53 Mansard application

- Stagger ends of panels in alternate courses over rafter
- Maximum 24" spacing between rafters
- Any additional bracing on under side of rafters
- Self-aligning feature eliminates measuring on straight edges
- Mansard slope not less than 20/12
- Only 2 nails at each rafter (on 14" exposure panels)
- 1x2 starter strip flush with rafter tip
- Self-aligning ridge on panel rests on starter strip
- Soffit must be nailed to starter strip at one or more points
- Soffit must extend to outside line of 1x2 starter strip

Courtesy of Shakertown Corp

185

Estimating Wood Shingle Roofing Costs

Let's say you've got to cover the roof shown in Figure 6-30 under the conditions given in Example 6-1. The published national average cost at the time of this writing for a square of wood shingles, including felt, flashing, fasteners, vents and sales tax was $414.95. We'll use this cost in our example. From Example 6-1, you need 9.3 squares of shingles. So the total material cost is:

9.3 squares x $414.95/square = $3,859.04

You'll also need two bundles of ridge units, which we'll price at $91.81 per bundle, for a total cost of $183.62. This brings the total material cost to $4,042.66.

There's a roof deck area of 8.73 squares and 31 feet of ridge units. According to the *National Construction Estimator*, an R1 crew consisting of one roofer and one laborer can install wood shingles at the rate of one square per 3.52 manhours, and ridge units at the rate of one linear foot per 0.059 manhours. Assuming the roofer (costing $41.66/hour) and the laborer (costing $30.31/hour) cost the employer $35.99 per manhour, the average cost per manhour is:

$41.66 + $30.31 = $71.97 ÷ 2 = $35.99

So the labor cost will be:

Labor (installing shingles) = 8.73 squares x 3.52 manhours/square x $35.99/manhour = $1,105.96

Labor (installing ridge units) = 31 LF x 0.059 manhours/LF x $35.99/manhour = $65.83

The total labor cost is:

$1,105.96 + $65.83 = $1,171.79

The total cost is:

$4,042.66 + $1,171.79 = $5,214.45

7 Tile Roofing

A well-constructed clay or concrete tile roof should last more than 50 years. It costs more to install than most other types of roofing because the materials are more expensive and require a strong frame to support the heavy load of the tiles. But it needs very little maintenance and no preservatives. Also, tile is fireproof, so fire insurance costs less for the entire life of the building.

Clay and concrete tiles come in a variety of shapes, colors and sizes, and have similar properties and installation methods. Figure 7-1 shows a selection. Concrete tiles are usually 16-1/2 inches long and 13 inches wide, whereas clay tiles come in many different sizes, as shown in Figure 7-2. Clay tiles cost more, not only because they're more expensive to make, but because they can only be made near a clay quarry. If you're not close to one, freight costs to your building site are going to be high. Concrete tile is not only cheaper to make, but they can be made anywhere.

Concrete tiles weigh about 900 pounds per square when installed with a 3-inch top lap. It takes about 90 concrete field tiles laid with a 3-inch top lap to cover a square of roof area.

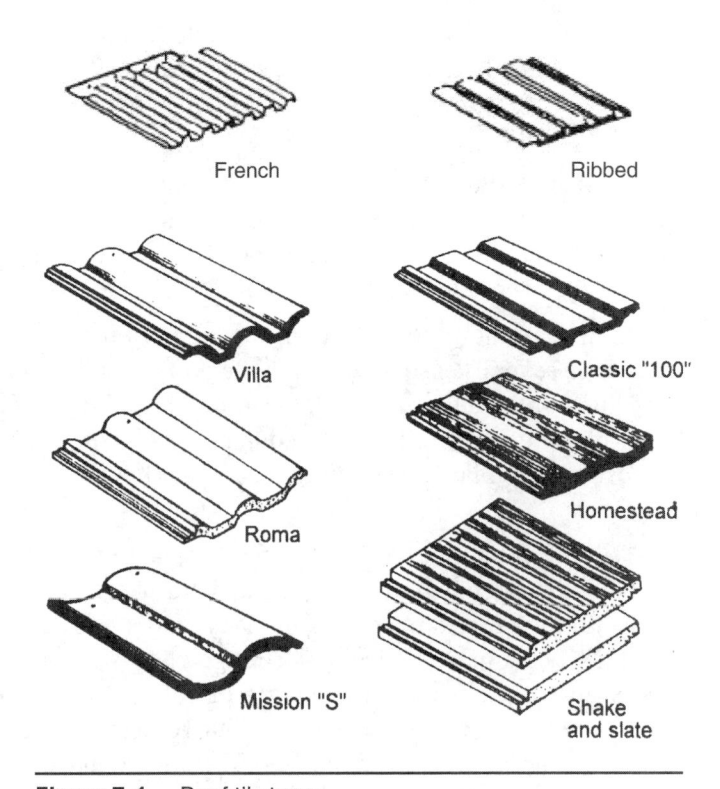

Figure 7-1   Roof tile types

Dimensions (inches)		
Tile type	Width	Length
Flat	5, 7, or 8	12, 15 or 24
English	8	13¼
French	9	16
Spanish	9¼	13¼
Roman	6 to 8	13
American	9	14
Lanai	8	14
Mission	8	14½ or 18
Norman	7	15

Figure 7-2 Common sizes of clay roofing tiles

Tile Type	Approximate weight (lb./sq.)
Flat	800-1600
French	900-1000
Spanish	850-900
Mission	1250-1350
Roman	1100-1200
Greek	1250
English	800-900
American	800
Lanai	800
Norman	1600

Figure 7-3 Clay tile weights

Clay tiles can take from 85 to 330 field tiles per square. Clay tiles weigh about 800 to 1,600 pounds per square, depending on the style, thickness and exposure. The weights of clay tile are shown in Figure 7-3, but this is just the tile. Add 1,000 lbs per square for the mortar to calculate the weight you'll be putting on the roof.

A concrete tile has baffles (weather checks) on its underside and water locks on its sides to keep wind-driven moisture and water runoff from getting under the tile. The weather check nearest the butt of the tile is called a *nose lug*. These are shown in Figure 7-4.

Some clay tiles interlock while others are joined with mortar. Concrete tiles have several sets of interlocking lugs. The top lap and exposure depend on which of the lugs you interlock with the head lug of the underlying tile. That makes it easy to keep the courses evenly spaced. If you shorten the exposure of the last few courses, you don't have to trim the top course at the ridge.

Clay tile comes glazed or unglazed. Concrete tile is colored either on its surface with a cementitious pigmented surface coating, or throughout its body. Surface-coated tile comes in many colors and deep shades. Body-colored tile has a more limited color range and its colors are more subdued. Color-coated tile is more resistant to discoloration from moss and fungus. I don't recommend color-coated tile where temperatures often range from mild days to below-freezing nights. The coating is susceptible to spalling. Body-colored tile, which is colored throughout, withstands short freeze/thaw cycles better.

Underlayment Under Tile Roof Coverings

There are three basic underlayment systems for tile roofs:

1) One-ply nonsealed system

2) One-ply sealed system

3) Two-ply sealed system

These underlayment systems are discussed in detail in Chapter 3.

One-ply underlayment systems are normally used over slopes of 4 in 12 and steeper. You must seal a 1-ply system (1-ply sealed system) over slopes between 4 in 12 and 6 in 12. On slopes steeper than 6 in 12 you can use a 1-ply nonsealed system.

It's possible to install a 1-ply sealed underlayment system over slopes as low as $2\frac{1}{2}$ in 12, if you put a counter batten system over the underlayment. I'll discuss the counter batten system in detail in the following section.

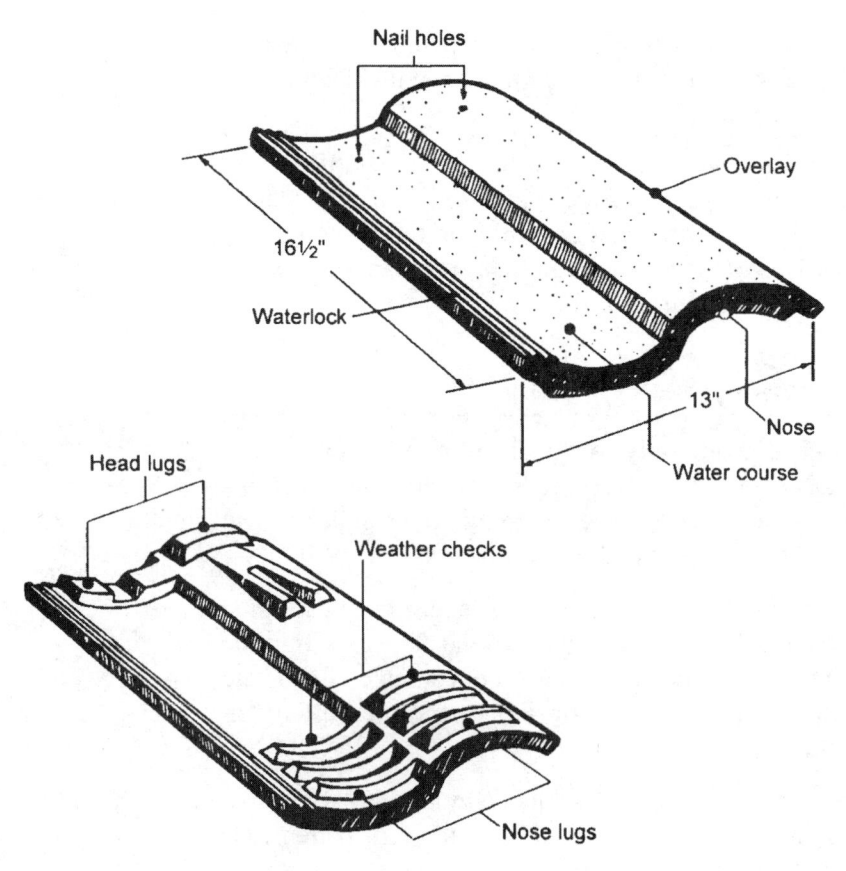

All measurements are approximate

Courtesy of Monier Roof Tile

Figure 7-4 Views of a concrete tile

You can use the 2-ply sealed underlayment system over roof slopes of 2½ in 12 and steeper. You *must* use this system when you embed the tiles in mortar.

You can direct-nail tiles over any of the underlayment systems just discussed. But you should use roofing cement to seal all nail penetrations into a sealed underlayment system. You don't have to seal nail penetrations into an unsealed underlayment system.

You can install tiles with a 2-inch top lap over any sealed underlayment system, but you have to use a 3-inch top lap over an unsealed system.

Installing Roof Tiles

Install roofing tile over solid sheathing. Some building codes specify plywood instead of solid board sheathing. After you've got the underlayment, drip edge and valley flashing material in, install a starter course at the eaves of the roof. Allow from ¾ inch to 2 inches overhang at the eaves, depending on the type of tile and whether there'll be gutters. You use a shorter overhang with gutters, because otherwise, a heavy runoff might pour right over, instead of into, the gutter.

Hold field tiles back 1 to 4 inches from the rake, depending on the type of rake tiles you'll use. We'll discuss that in detail later in the chapter.

To align the first course, nail down a tile with the correct overhang at each end of the eaves. Stretch a string between the top edges of the end tiles and snap a chalk line. Align the top edges of the rest of the tiles along the chalk line. Check alignment about every third or fourth course by measuring up from the eaves to the butts of end tiles and snapping a chalk line to align the butts of the next course of tiles.

Battens and Counter Battens

You can fasten roofing tiles directly to the sheathing or to pressure-treated battens you install over the underlayment. You can install battens on roof slopes of 2½ in 12 and steeper. They're required on slopes of 7 in 12 and steeper. Figure 7-5 shows tile installed over battens.

Beginning at the ridge, install battens at least 1½ inches from each side of the ridge, as in Figure 7-6. Work your way down the slope, spacing the battens according to the tile coursing requirements. For example, if concrete tiles are 16½ inches long with a 3-inch top lap, space the battens on 13½-inch centers.

When you install battens on roof slopes less than 6 in 12, use 4-foot-long battens with ½-inch drain slots between the ends of adjacent battens. Or shim the battens every 4 feet with ¼-inch moisture-resistant lath strips or a decay-resistant material such as an asphalt cap sheet or asphalt shingle, as in Figure 7-7.

Tile Roofing

Figure 7-5 Flat tiles installed over batten

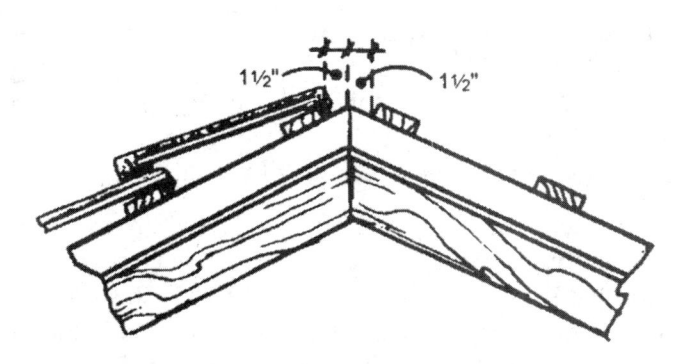

Note: Measurement should be made from the apex of the roof to the face edge of the first batten. Heads of the two tiles should not be more than 2" apart.

Courtesy of Monier Roof Tile

Figure 7-6 Battens at the ridge

Alternate to 4' slotted batten method: Shimming with moisture resistant ¼" nominal lath or strips of decay resistant material such as asphalt cap sheet or asphalt shingle.

½" drain slots every 4'
Metal closure
Underlayment
1" x 2" battens
Plywood deck

Courtesy of Monier Roof Tile

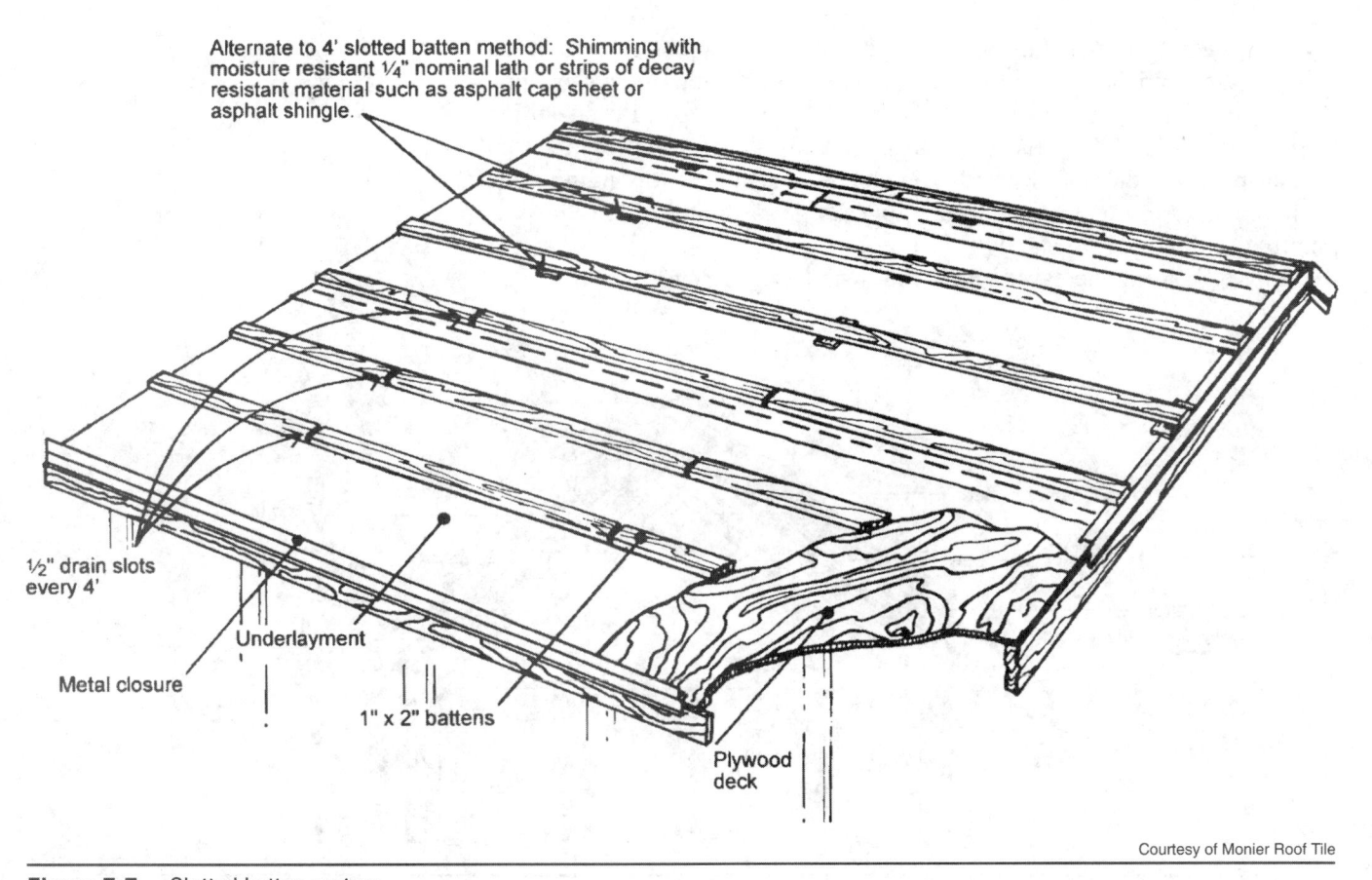

Figure 7-7 Slotted batten system

191

For battens over a sealed underlayment system, seal all nail penetrations with roofing cement. Use pressure-treated lumber and secure the battens to the sheathing using 6d corrosion-resistant nails you drive at 24-inch centers. Use nails at least ¾ inch long, or at least long enough to go into the underside of the sheathing.

A counter batten system of 1x2s (minimum) mounted vertically from ridge to eaves on 16-inch centers is required over single-ply underlayment systems on roof slopes less than 4 in 12, as shown in Figure 7-8. You install the 1x4s horizontally across the counter battens, according to the tile coursing requirements. For example, use counter battens on roofs in heavy-snow areas or over cathedral ceilings, as shown in Figures 7-9 and 7-10. A counter batten system is also recommended over roofs in damp climates where good ventilation is especially important.

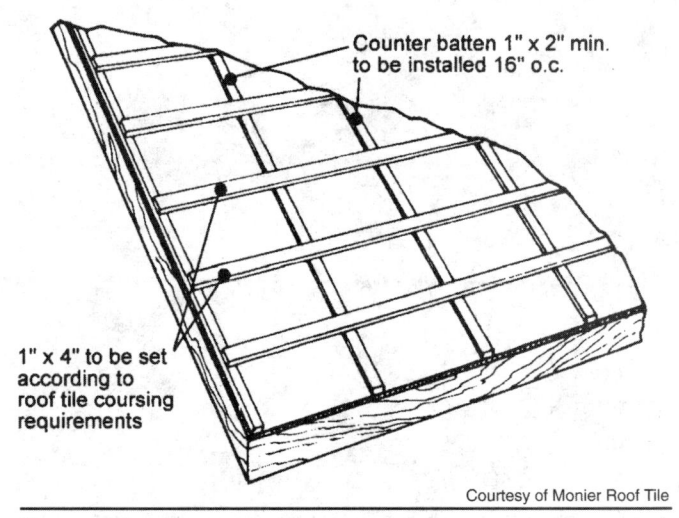

Figure 7-8 Counter batten system

The Starter Course

Raise the starter course to keep water runoff away from the eaves fascia. To do this, extend the fascia board 1½ inches above the top of the sheathing, as in Figure 7-11. Or install a 2x2 wood starter strip on top of the sheathing, as in Figure 7-12. In either case, you should install an 8-inch-wide wood cant strip behind the raised wood. You can use a strip of antiponding metal flashing nailed to the deck along the upper edge of the flashing to provide positive drainage. That's shown in Figures 7-11 and 7-12. You can omit the drip edge along the eaves when you install antiponding flashing.

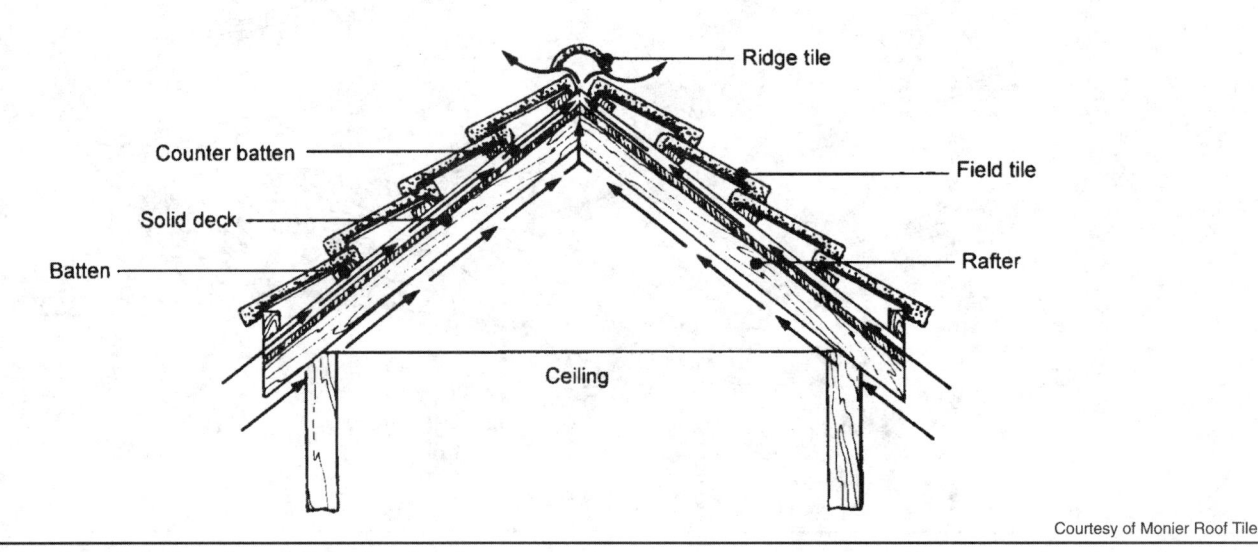

Figure 7-9 Counter battens for a well-ventilated roof

Tile Roofing

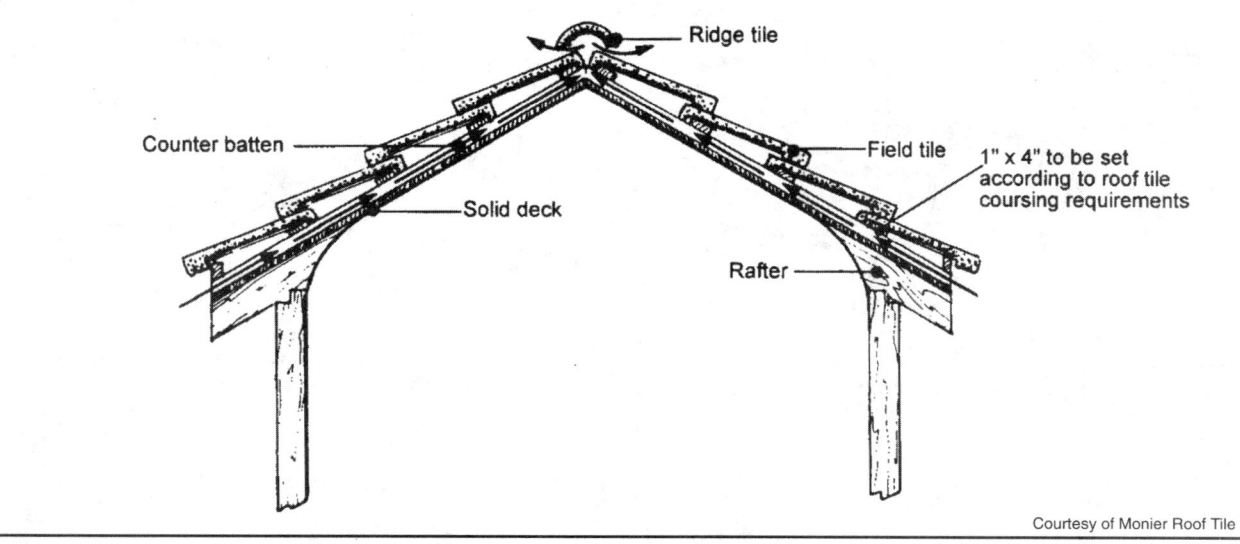

Figure 7-10 Counter battens over a cathedral ceiling

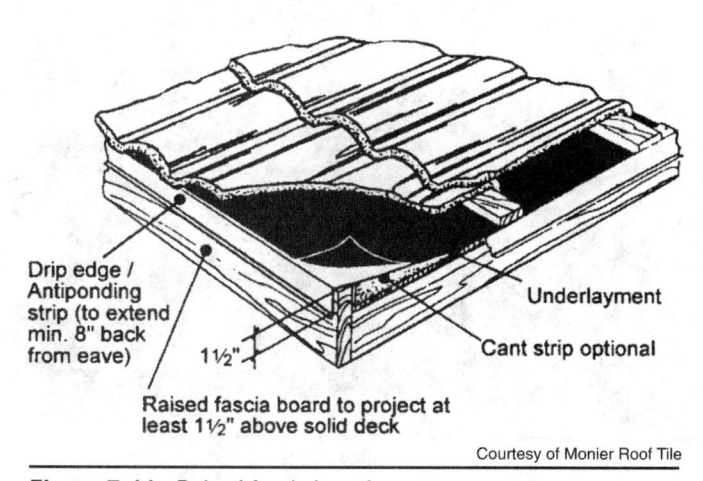

Figure 7-11 Raised fascia board

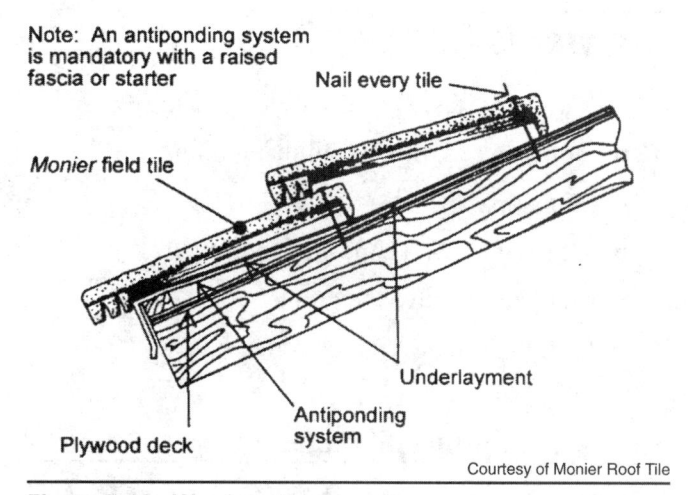

Figure 7-12 Wood starter strip

You can also install a 26-gauge metal eaves closure like the one in Figure 7-13, or a prefabricated synthetic (EPDM) rubber eaves closure strip, as shown in Figure 7-14.

Flat tiles are often elevated by under-eaves tiles installed along the eaves, as in Figure 7-15. That's to divert water away from the eaves, which helps preserve the fascia board.

You can also raise tiles by installing them in a bed of mortar as shown in Figure 7-16. Pack the spaces under the front of the eaves tiles with color-matched Type M mortar. Point (smooth) the mortar to form an

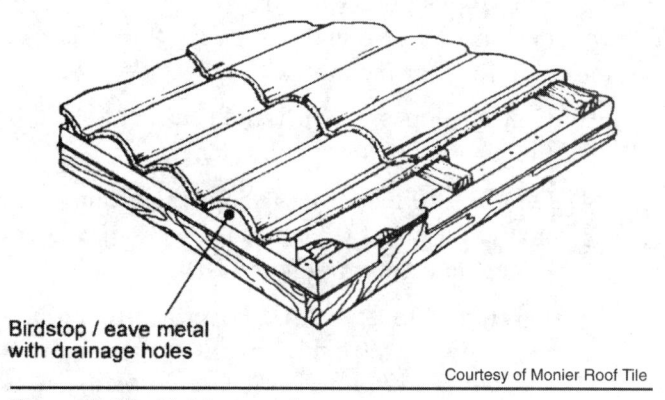

Figure 7-13 Metal eaves closure

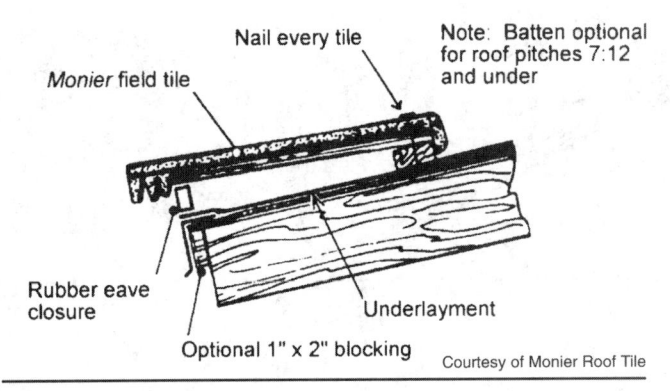

Figure 7-14 Rubber eaves closure

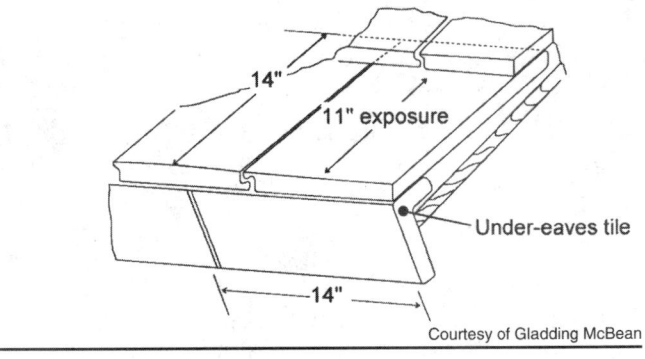

Figure 7-15 Under-eaves tiles

even, flat surface, as in Figure 7-17. Before the mortar sets up, punch holes ("bird holes") for air circulation and water drainage. That's shown in Figure 7-18.

Fastening Roofing Tiles

The way you fasten roofing tile depends on:

- mortar-set vs. direct-nail installation
- roof slope
- anticipated wind exposure and speed
- roof height above the ground
- local building code requirements

Using the Mortar-Set Method

Figure 7-16 Setting eaves tiles in mortar

For the mortar-set method, follow these rules:

1) On roof slopes from 4 in 12 to 6 in 12, you must nail each eaves tile with one nail in addition to using mortar.

2) On slopes from 6 in 12 to 7 in 12, install the eaves tiles as described above and nail every third tile in every fifth course in addition to using mortar.

3) On roof slopes of 7 in 12 and steeper, install battens and nail every tile in addition to using mortar.

4) On roofs 40 feet or less above the ground (measured to the eaves) in areas of wind velocities up to 80 miles per hour, follow the fastening guidelines in Figure 7-19.

5) On roofs 40 feet or less above the ground in areas of wind velocities between 80 and 110 miles per hour, follow the fastening guidelines in Figure 7-20. Use corrosion-resistant hot-dipped galvanized nails long enough to penetrate $3/4$ inch into the roof deck.

Tile Roofing

Figure 7-17 Pointing mortar at the eaves

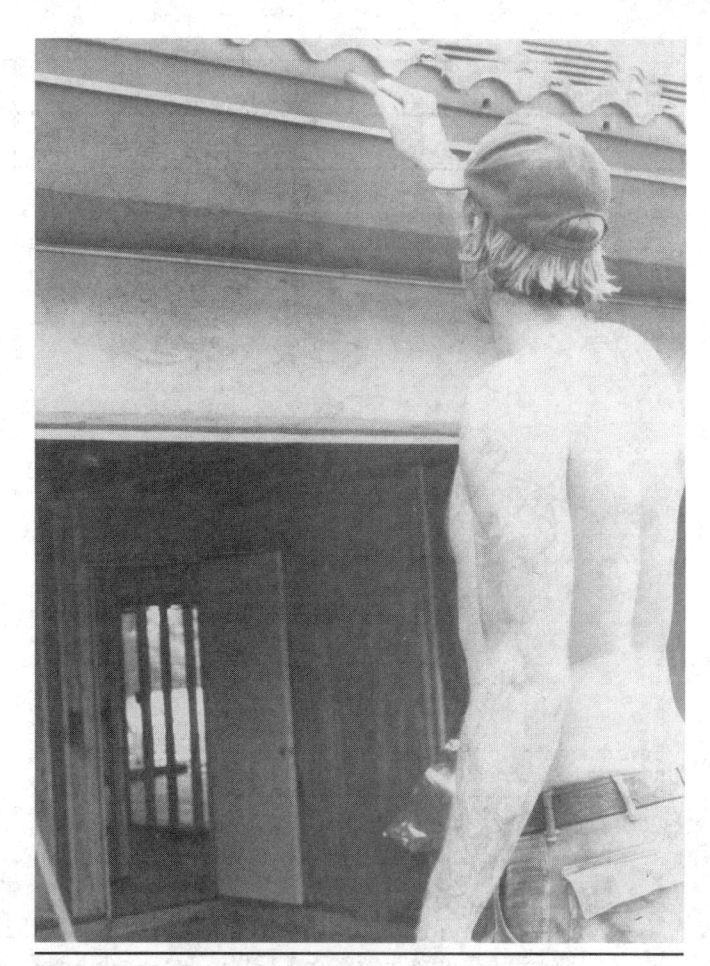

Figure 7-18 Forming "bird holes"

Roof Slope	Field tiles	Eaves course	Perimeter tiles[2]
3 in 12 through 12 in 12	Nail every tile	Nail and clip[1]	Nail every tile
Over 12 in 12	Nail and clip[1]	Nail and clip[1]	Nail and clip[1]

Use 10d nails

Courtesy of Monier Roof Tile

[1] You can use two nails per tile instead if clips.
[2] Perimeter nailing areas include the distance equal to three tiles (but not less than 36") from the edges of hips, ridges, eaves, rakes, and major roof penetrations.

Figure 7-19 Fastening requirements for tiles in wind velocities up to 80 MPH (roof height not exceeding 40 feet)

Roof Slope	Field tiles	Eaves course	Perimeter tiles [2]
All slopes	Nail and clip[1]	Nail and clip[1]	Nail and clip[1]

Use one 12d nail or two 10d nails.

Courtesy of Monier Roof Tile

[1] You can use two nails per tile inlieu of clips.
[2] Perimeter nailing areas include the distance equal to three tiles (but not less than 36") from the edges of hips, ridges, eaves, rakes, and major roof penetrations.

Figure 7-20 Fastening requirements for tiles in wind velocities between 80 MPH and 110 MPH (roof height not exceeding 40 feet)

Roof Slope	Field tiles	Eaves course	Perimeter tiles [2]
All slopes	2 nails and clip	2 nails and clip	2 nails and clip

Use two 10 d nails.

Courtesy of Monier Roof Tile

[1] Perimeter nailing areas include a distance of three tile widths (but not less than 36") from the edges of hips, ridges, eaves, rakes and major roof penetrations

Figure 7-21 Fastening requirements for tiles in wind velocities between 80 MPH and 120 MPH (roof height exceeds 40 feet)

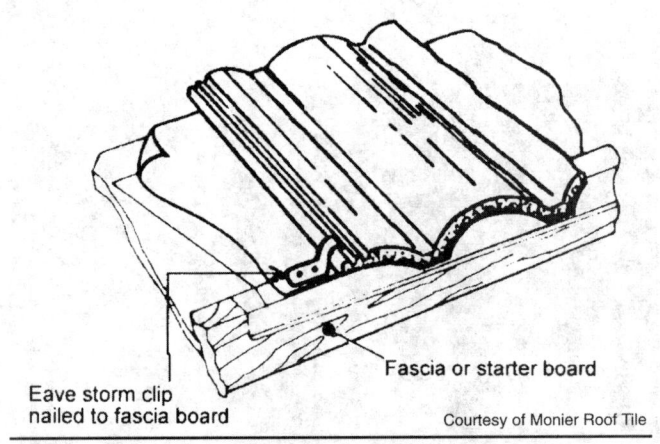

Figure 7-22 Eaves storm clip nailed to raised wooden fascia

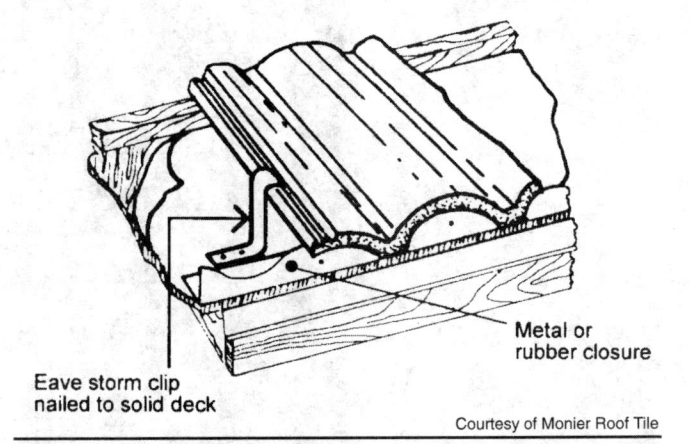

Figure 7-23 Eaves storm clip nailed to sheathing

6) On roofs more than 40 feet above the ground in areas of wind velocities between 80 and 120 miles per hour, follow the fastening guidelines in Figure 7-21.

The type of storm clip you use depends on the brand of tile you use, the location of the tile, the type of fascia installed and the fastening system.

- For a raised fascia, nail eaves-tile storm clips to the fascia board, as in Figure 7-22.

- For a metal or rubber eaves closure, nail eaves-tile storm clips to the sheathing, as in Figure 7-23.

- For battens, nail field-tile storm clips to the back of each batten, as in Figure 7-24.

- If you don't use battens, nail field-tile storm clips to the sheathing, as in Figure 7-25.

With the mortar-set method, wet each tile before you lay it, so the tile doesn't draw water out of the mortar, weakening the bond. Embed each tile in mortar applied with a filled 10-inch masonry trowel. Place the mortar (Type M) under the pan or flat part of each tile (Figure 7-26). Don't place the mortar under the lugs or heads of the tiles since this can result in tilted or crooked tile. Apply mortar from the head of the tile below; up to 2 to 4 inches from the head of the tile you're setting. Be sure the mortar meets the head of the lower course of tile and the underside of the tile you're setting.

Tile Roofing

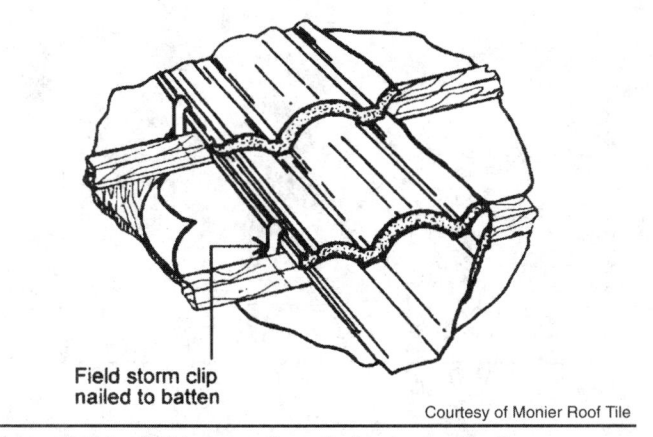

Figure 7-24 Field storm clip nailed to back of batten

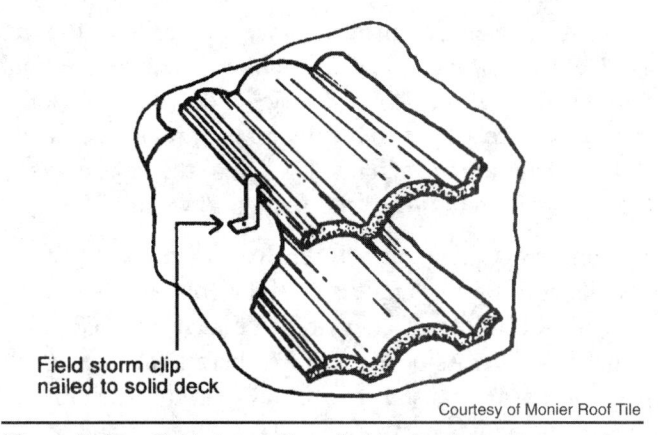

Figure 7-25 Field storm clip nailed to sheathing

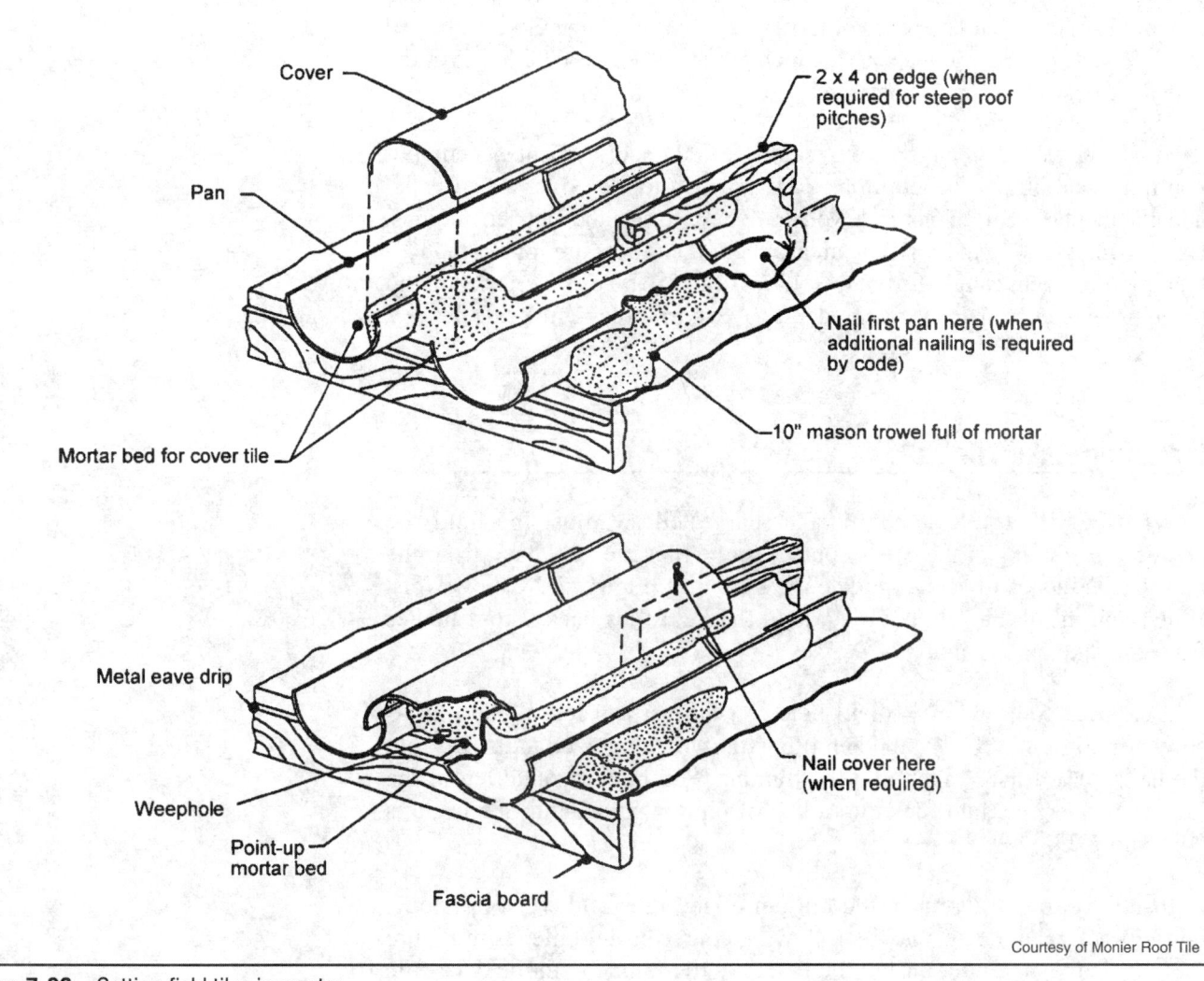

Figure 7-26 Setting field tiles in mortar

197

Install two-piece barrel tile like the ones in Figure 7-27 with mortar applied under the center of the pan with the narrow end of the pan facing down slope. Bond the pan you're setting to the pan in the previous course. Place mortar along each inside edge of the pans and set the covers with the wide end facing down slope.

Concrete tile manufacturers recommend that you install roofing tile using the straight-bond method, that is, with the joints of every other course aligned. The staggered- or cross-bond method where joints are offset laterally by a set distance is usually used to install flat tiles. But where heavy snow loads are likely, you should use the straight-bond method for flat tile also. That provides the maximum distance between joints from one course to the next and cuts down the likelihood of leaks where joints fall close together.

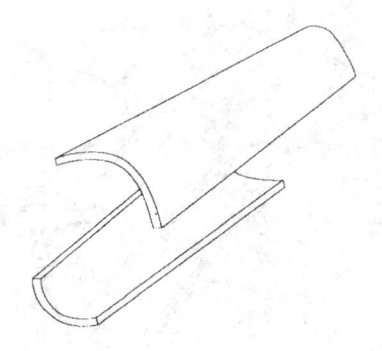

Figure 7-27 Barrel tile

To prevent tiles, which are rigid, from cracking, it's important that tiles can move independently of each other, especially if you've nailed or clipped them down. So, allow at least a 1 1/16-inch gap between vertical joints at the side interlock of each tile.

You can leave a 2-inch top lap over any sealed underlayment system, but you must seal all nail penetrations. Use a 3-inch top lap over an unsealed underlayment system. In this case, you don't have to seal the nail penetrations. Use a 4-inch top lap in hurricane areas where the rafters are longer than 20 feet on roof slopes ranging from 4 in 12 to 7 in 12. You can increase the top lap to provide even spacing so you don't have to trim tiles at the ridge.

Installing Rake Tiles

When you'll install standard rake (barge) tiles, adjust the field tile position (beginning at the eaves course) so the tiles are held back from the rake 1 to 2 inches to accommodate the rake tiles. Figure 7-28 shows this. When you install barrel rake tiles, hold the field tiles back 2 to 4 inches from the rake, as in Figure 7-29.

Rake tiles come with two nail holes on the vertical leg of the tile, as shown in Figure 7-30. Install each tile with two galvanized nails. Be sure the nails go at least 3/4 inch into the framing. Set the horizontal leg of the rake tiles over the field tiles into a bed of mortar or sealant. See Figures 7-28 and 7-29.

Install the rake tiles from the bottom of the rake and work up-slope, butting each rake tile against the following course of field tile, as in Figure 7-31. If there's no gutter, install the first rake tile in line with the eaves tile course.

Tile Roofing

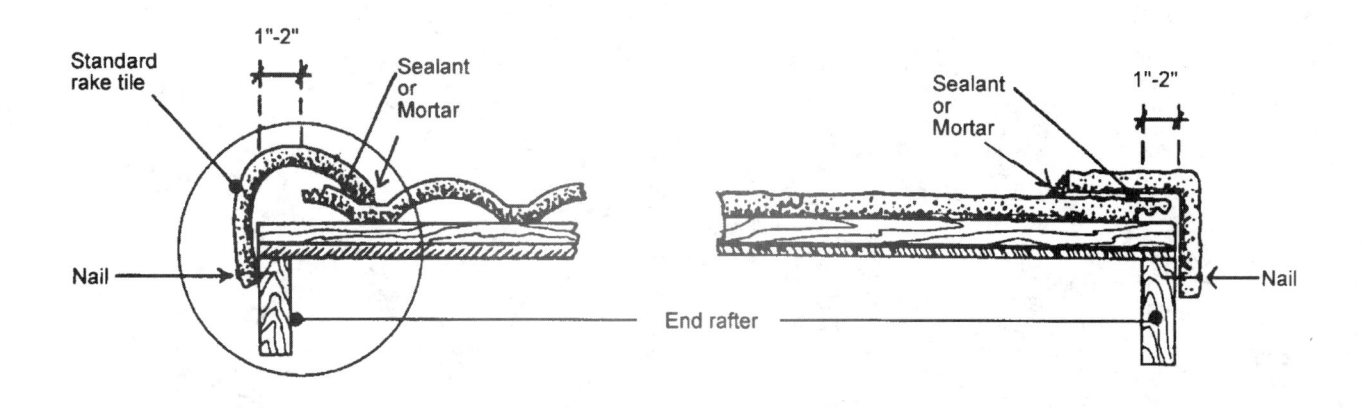

Figure 7-28 Standard rake tiles

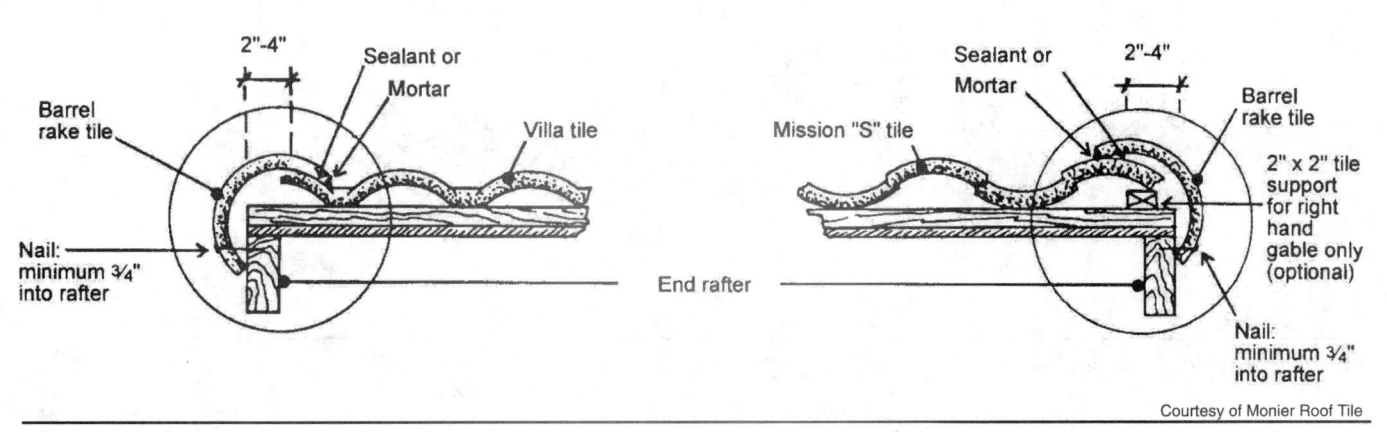

Figure 7-29 Barrel rake tiles

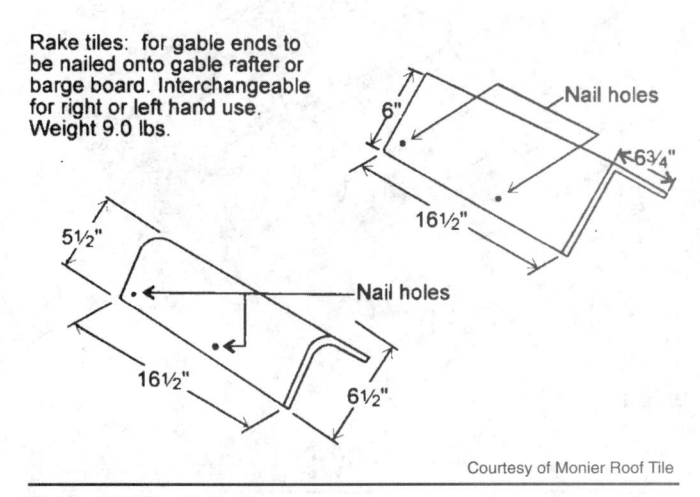

Figure 7-30 Rake tiles

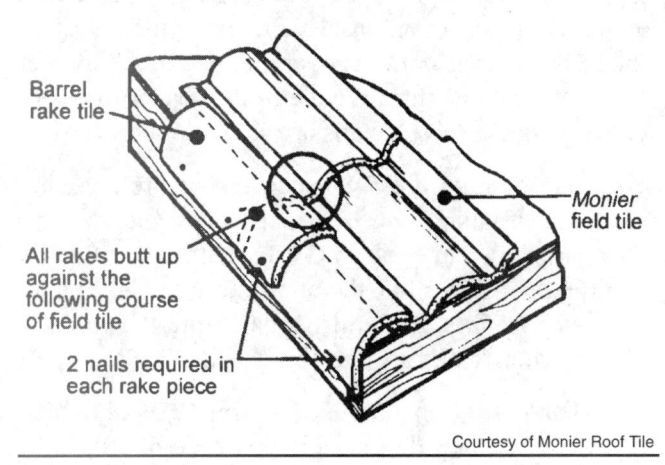

Figure 7-31 Installing rake tiles

199

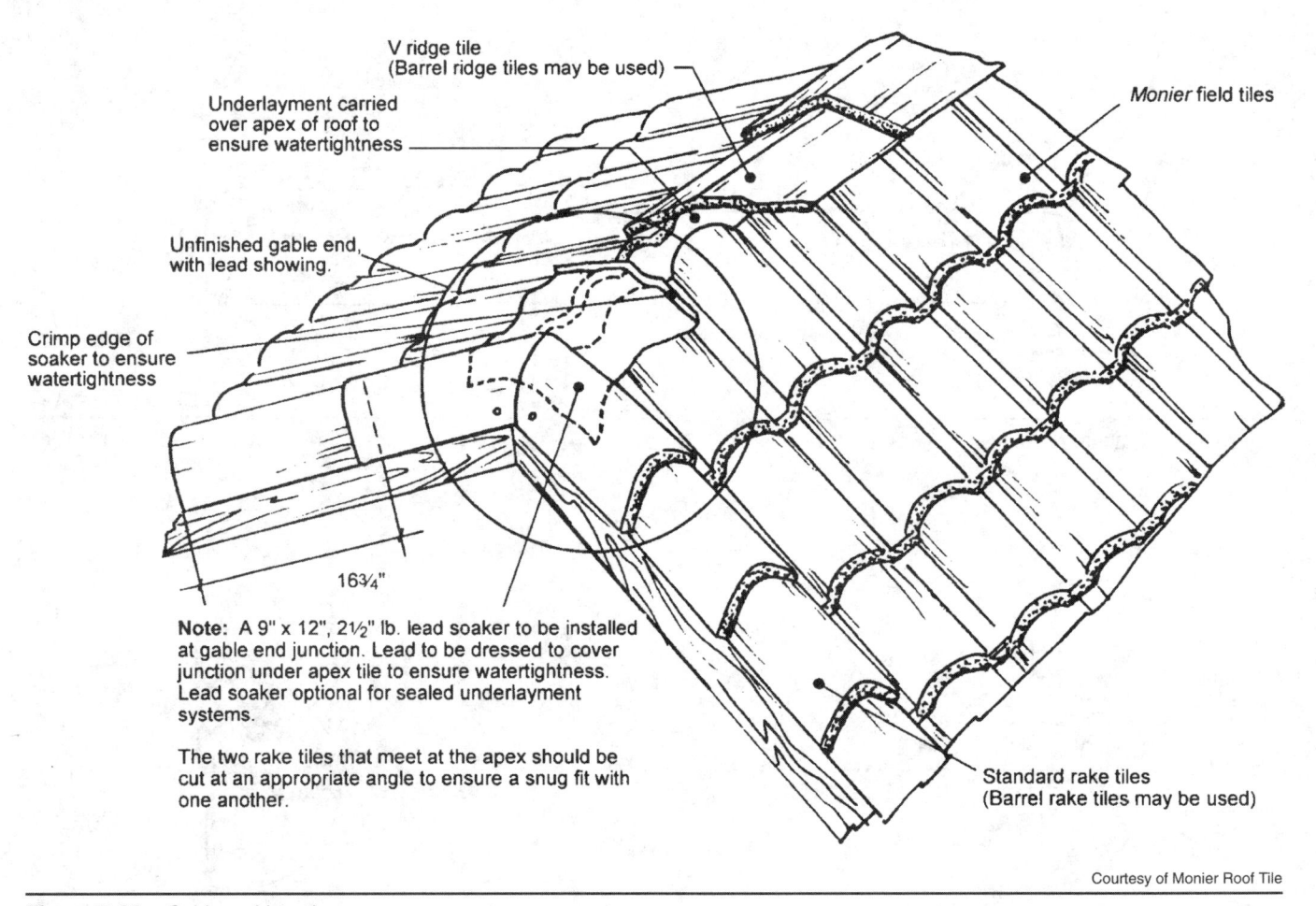

Figure 7-32 Gable end junction

Miter the rake tiles for a snug fit at the peak of the rake. Install the top rake tiles over a 9" x 12" lead soaker (a piece of flashing to help prevent leaks) when you're laying the tiles over an unsealed underlayment system. The lead soaker is optional over a sealed underlayment system. Crimp the inward edge of the soaker to ensure watertightness. These details are shown in Figure 7-32.

Instead of rake tiles, you can install prefabricated metal rake flashing with a 1-inch water guard (water return) as shown in Figure 7-33. Fasten the metal to the sheathing with clips like the ones in Figure 7-34 nailed at 24-inch centers. Never drive nails into the pan area of the flashing (the part that contacts the deck).

You can also install a mortar finish, as shown in Figure 7-35, along the rakes if you're laying flat tiles over a sealed underlayment system.

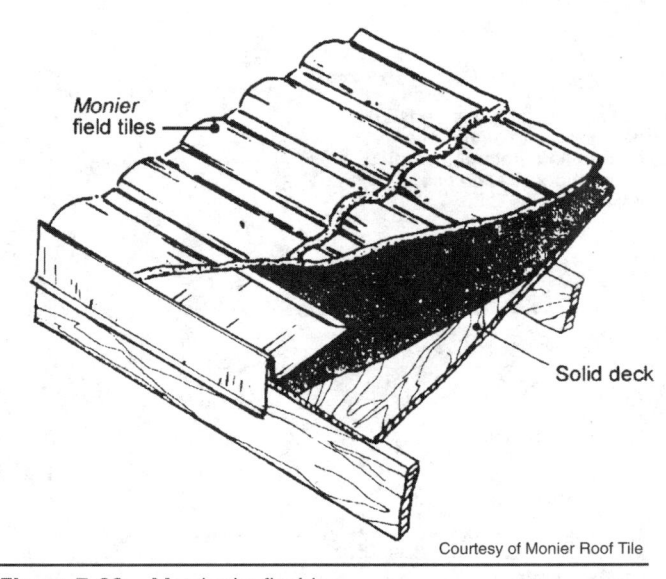

Figure 7-33 Metal rake flashing

Installing Hip and Ridge Tiles

When you come to a hip, miter the tile to form a straight edge over the hip. Cut the tile with a power saw equipped with a Carborundum blade, or trim the tile with a chipper like the one in Figure 7-36. With this tool, you press down on the handle, and the point chips the tile. Space field tiles so there's a 2-inch gap over the ridge. Do the same for field tiles at the hip, as in Figure 7-37 A.

Set hip and ridge units into color-matched Type M mortar, but don't let the mortar cover the center of the hip or ridge. Extend the field tile beyond each bed of

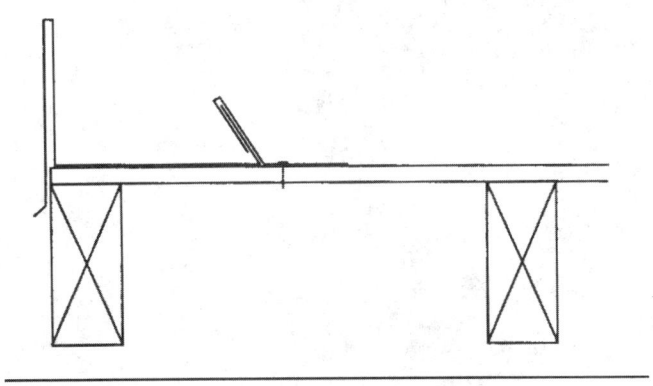

Figure 7-34 Metal fascia clips

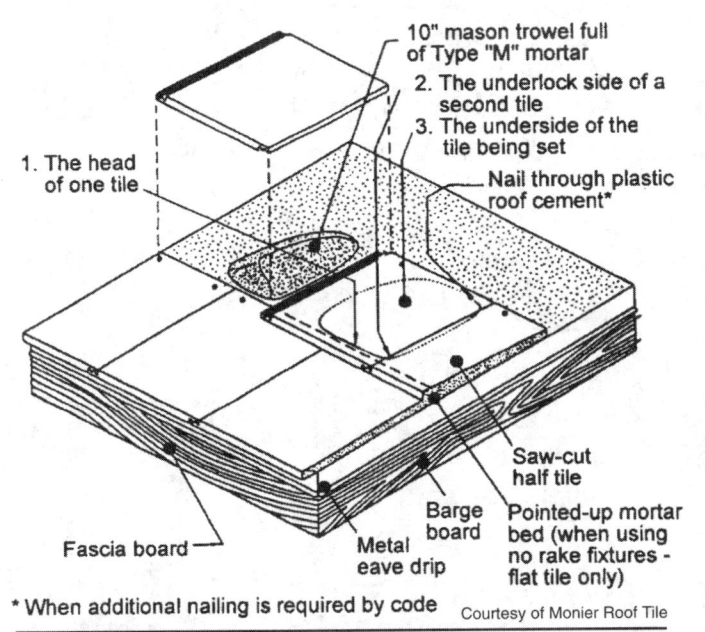

Figure 7-35 Mortar finish at rakes, flat tile

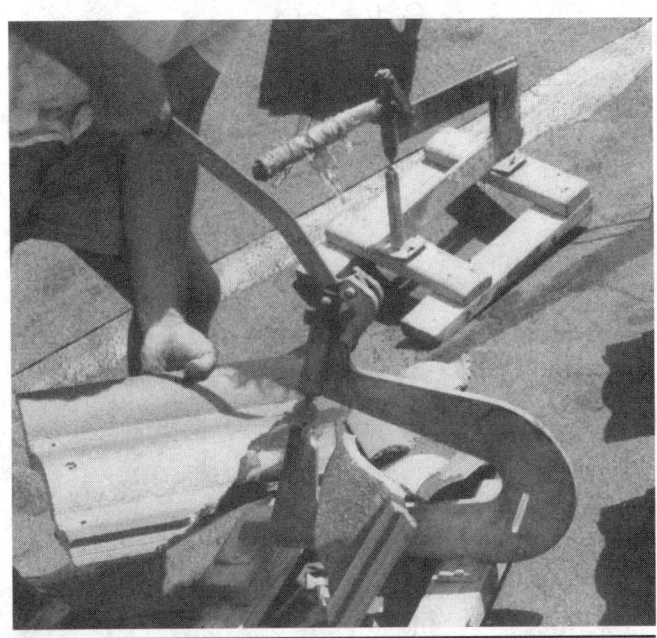

Figure 7-36 Chipper

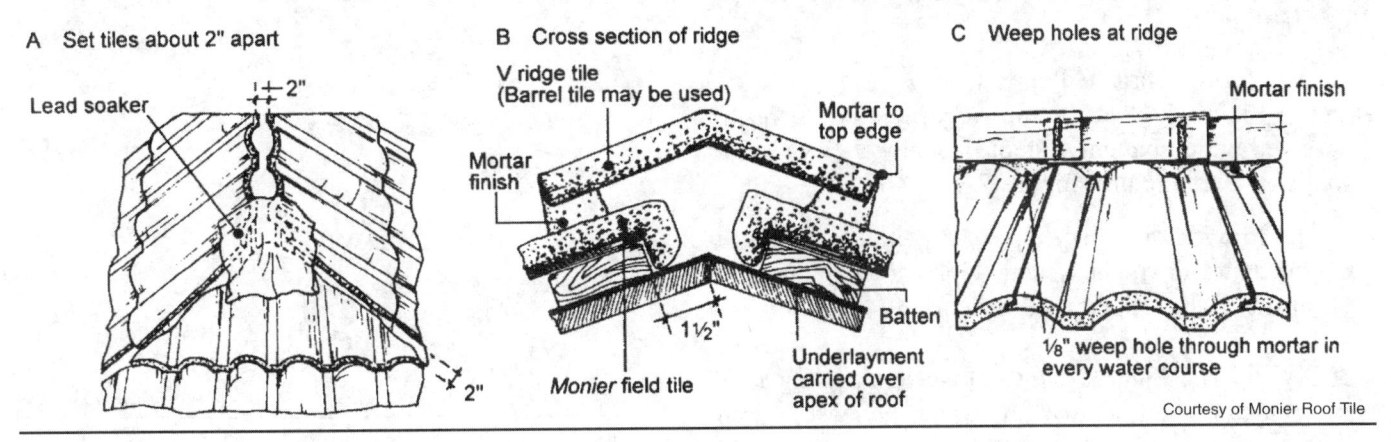

A Set tiles about 2" apart

B Cross section of ridge

C Weep holes at ridge

Figure 7-37 Roof tile at ridge

Figure 7-38 Pointing mortar at a hip

Figure 7-39 Finishing mortar with a dry sponge

mortar by about an inch, as shown in Figure 7-37 B. Before the mortar sets, point the mortar to match the contours of the tiles, as in Figure 7-38. Then rub the mortar with a dry sponge to remove the rough edges. That's shown in Figure 7-39.

Over an unsealed underlayment system, punch $1/8$-inch weep holes in the mortar beneath ridge units at every water course. See Figure 7-37 C.

You can use V-ridge tiles or barrel tiles like those in Figure 7-40 to cover hips and ridges. Install hip tiles before you install ridge tiles.

You can install a prefabricated hip starter like the one in Figure 7-39 at the low end of a hip, or you can use an unaltered-standard hip-ridge tile. You can also install a field-mitered hip-ridge tile and cover the rough edges with mortar.

Install a prefabricated apex trim tile (wye) like the one in Figure 7-41 at the hip-ridge juncture. Over an unsealed underlayment system, install a 9" x 12" lead soaker at the juncture (Figure 7-37 A).

Lap hip and ridge tiles at least 2 inches. Over an unsealed underlayment system, apply a continuous bead of mastic to the lapped part of the ridge tiles.

■ **Dry Hip and Ridge System**   Instead of using mortar, you can install a "dry" hip and ridge system over sealed or unsealed underlayment. To do this, begin by

Courtesy of Monier Roof Tile

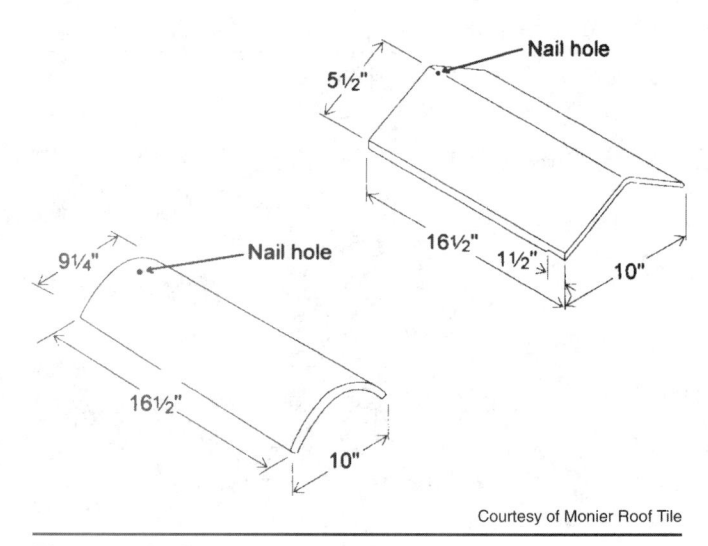

Figure 7-40 Ridge and hip tiles

Figure 7-41 Apex trim tile (wye)

Figure 7-42 A "dry" hip

installing a 2 x 3 (for standard hip-ridge tiles) or a 2 x 4 or 2 x 6 (for barrel hip-ridge tiles) along the centerlines of the hips and ridge, as shown in Figure 7-42. Toenail the boards into the deck or hip rafter.

Then cover the boards with a pressure-sensitive hip-ridge sealer to act as a wind block, as in Figure 7-43. The sealer has an adhesive back that's covered with release paper. It comes in 50-foot rolls, 9 or 12 inches wide. Available colors include terra cotta, dark brown, black and aluminum. Lap the ends of the wind block at least 2 inches.

When you've got the sealer in, nail the hip-ridge tiles to the boards through the wind block material. Use one nail per tile.

Valleys

The type of valley flashing in an open valley on a tile roof depends on the underlayment system. Over an unsealed underlayment system, install preformed 26-gauge galvanized iron flashing or 16-ounce copper with a 2$\frac{1}{2}$-inch water diverter and 1-inch water guards (Figure 7-44 A). Use nails 6 inches on center or clips nailed 24-inches on center. Don't drive nails into the flashing. Treat a valley that ends at a roof plane this way.

If you need more than one piece of flashing, lap the upper piece over the lower by at least 6 inches and seal the lap with roofing cement. Use flashing at least 16 inches wide.

Install a 12" x 24" lead ridge saddle over the flashing at the valley-ridge juncture, as shown in Figure 7-44 B. When the lower end of a valley ends at a roof plane, lap the valley metal over a 24" x 16" lead soaker (skirt) at

[Figure 7-43 illustration]

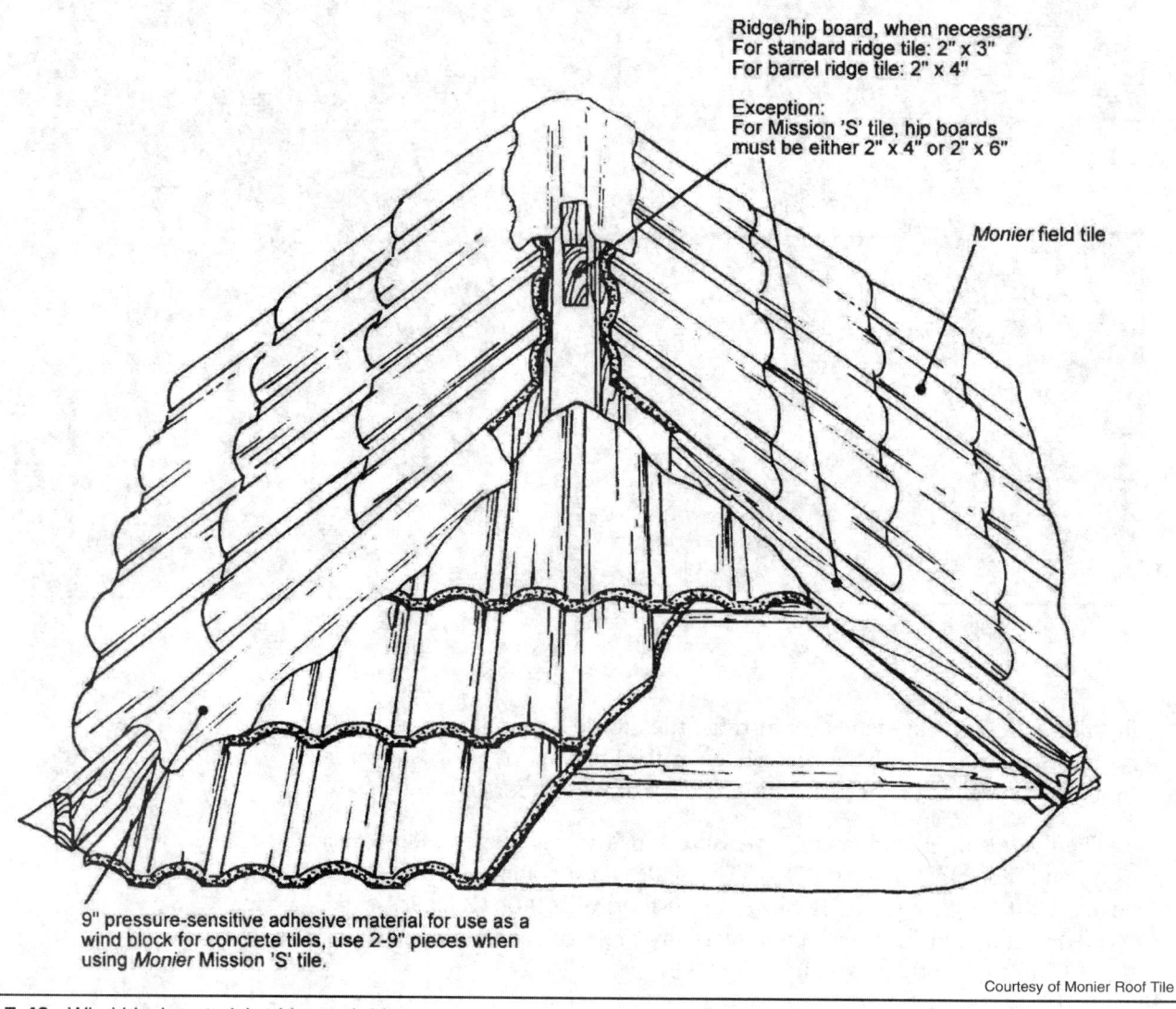

Ridge/hip board, when necessary.
For standard ridge tile: 2" x 3"
For barrel ridge tile: 2" x 4"

Exception:
For Mission 'S' tile, hip boards must be either 2" x 4" or 2" x 6"

Monier field tile

9" pressure-sensitive adhesive material for use as a wind block for concrete tiles, use 2-9" pieces when using *Monier* Mission 'S' tile.

Courtesy of Monier Roof Tile

Figure 7-43 Wind block material at hips and ridge

the lower end of the valley to send water runoff back onto the field tile. That's shown in Figure 7-44 C. Or you can extend the valley metal over the top of the field tile and omit the skirt.

Over a 30/90 hot-mop sealed underlayment system, you can install standard roll-metal valley flashing sandwiched between the 30-pound felt and 90-pound mineral-surfaced roll roofing.

Over a sealed underlayment system, you can use valley flashing without water guards, as shown in Figure 7-45. Nail the flashing to the sheathing, driving nails on 6-inch centers along rows 1 inch from each edge. Seal all nail holes with a strip of 30-pound felt embedded in roofing cement. That's also shown in Figure 7-45. Again, lap pieces of flashing at least 6 inches, and seal the lap with roofing cement.

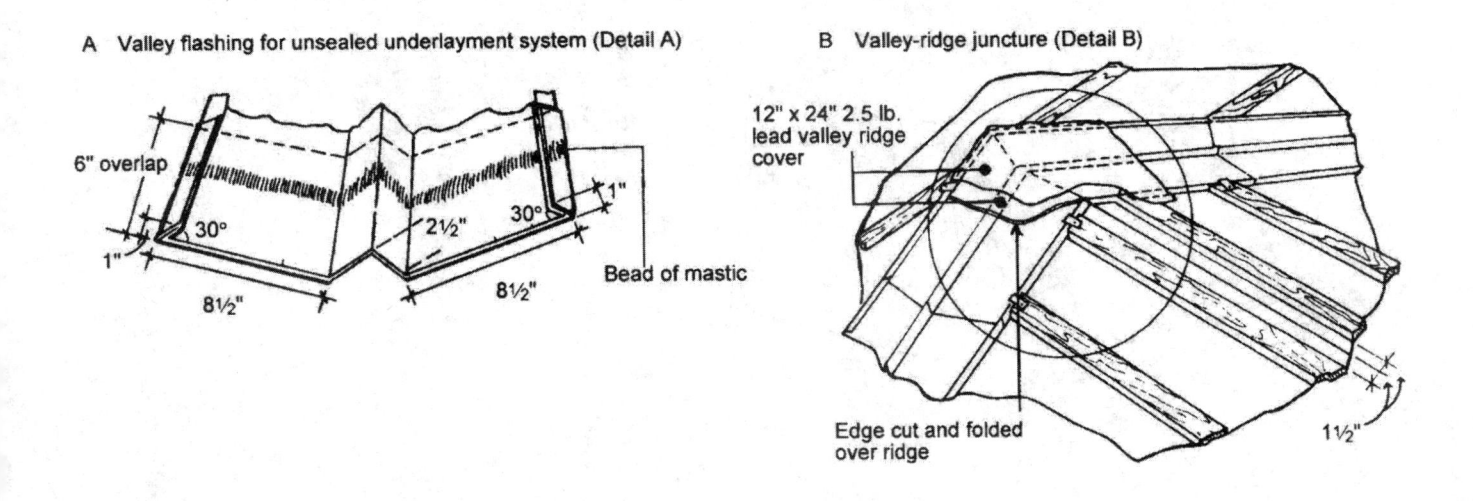

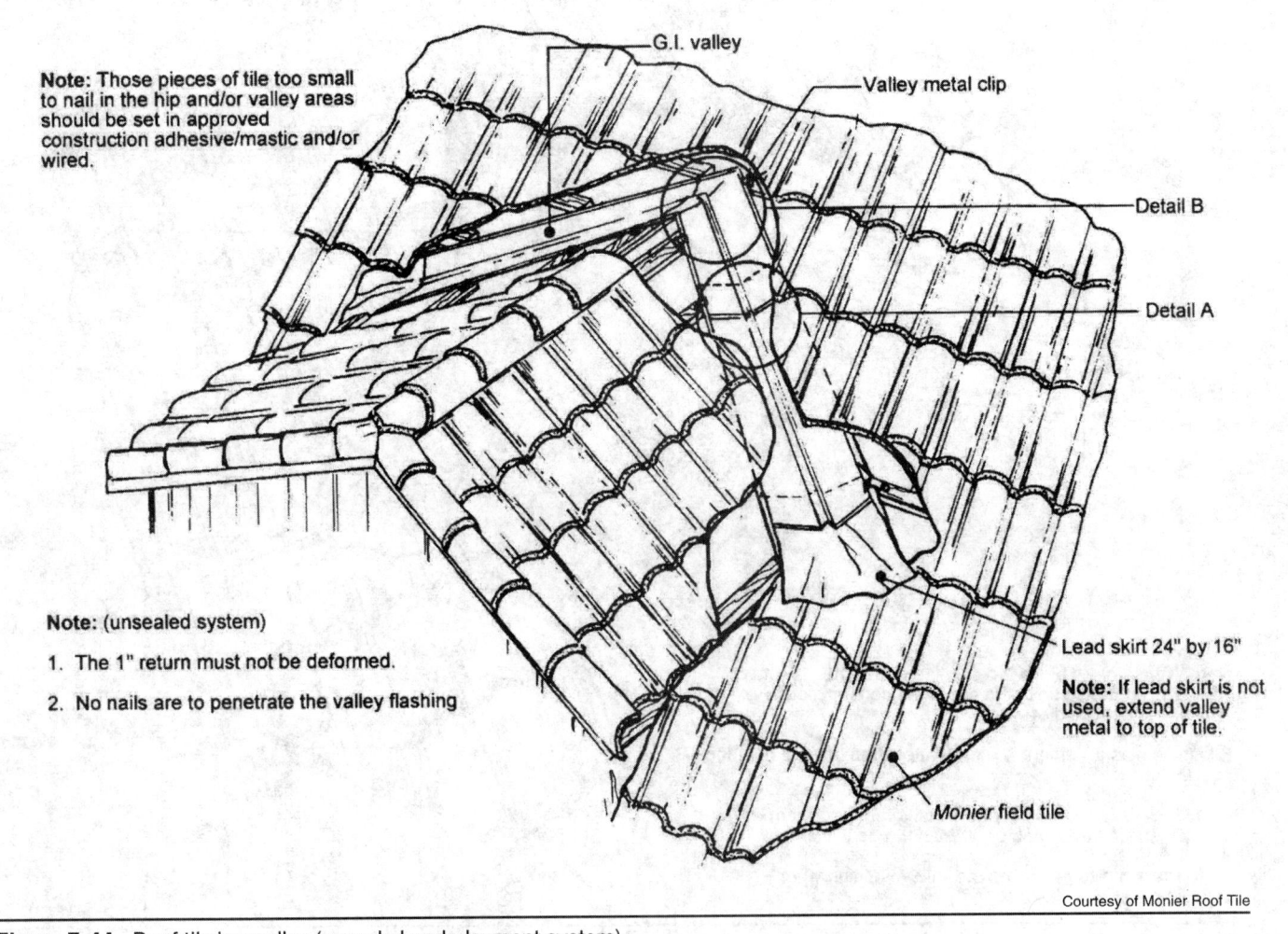

Tile Roofing

A Valley flashing for unsealed underlayment system (Detail A)

B Valley-ridge juncture (Detail B)

12" x 24" 2.5 lb. lead valley ridge cover

6" overlap
30°
2½"
30°
1"
1"
8½"
8½"
Bead of mastic

Edge cut and folded over ridge

1½"

C Lap valley metal over lead skirt

Note: Those pieces of tile too small to nail in the hip and/or valley areas should be set in approved construction adhesive/mastic and/or wired.

G.I. valley
Valley metal clip
Detail B
Detail A
Lead skirt 24" by 16"

Note: If lead skirt is not used, extend valley metal to top of tile.

Monier field tile

Note: (unsealed system)
1. The 1" return must not be deformed.
2. No nails are to penetrate the valley flashing

Courtesy of Monier Roof Tile

Figure 7-44 Roof tile in a valley (unsealed underlayment system)

205

Roofing Construction & Estimating

A Valley flashing for sealed underlayment system (Detail A)

- Overlap of metals
- Sealant
- 2½"
- 9½"
- 9½"

B Tile in valley over sealed underlayment

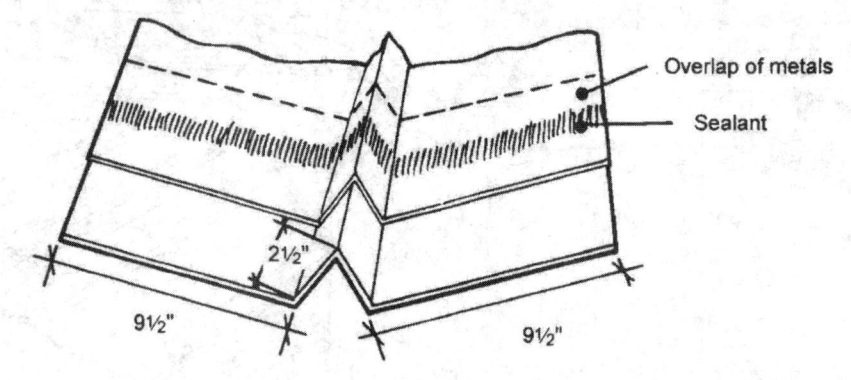

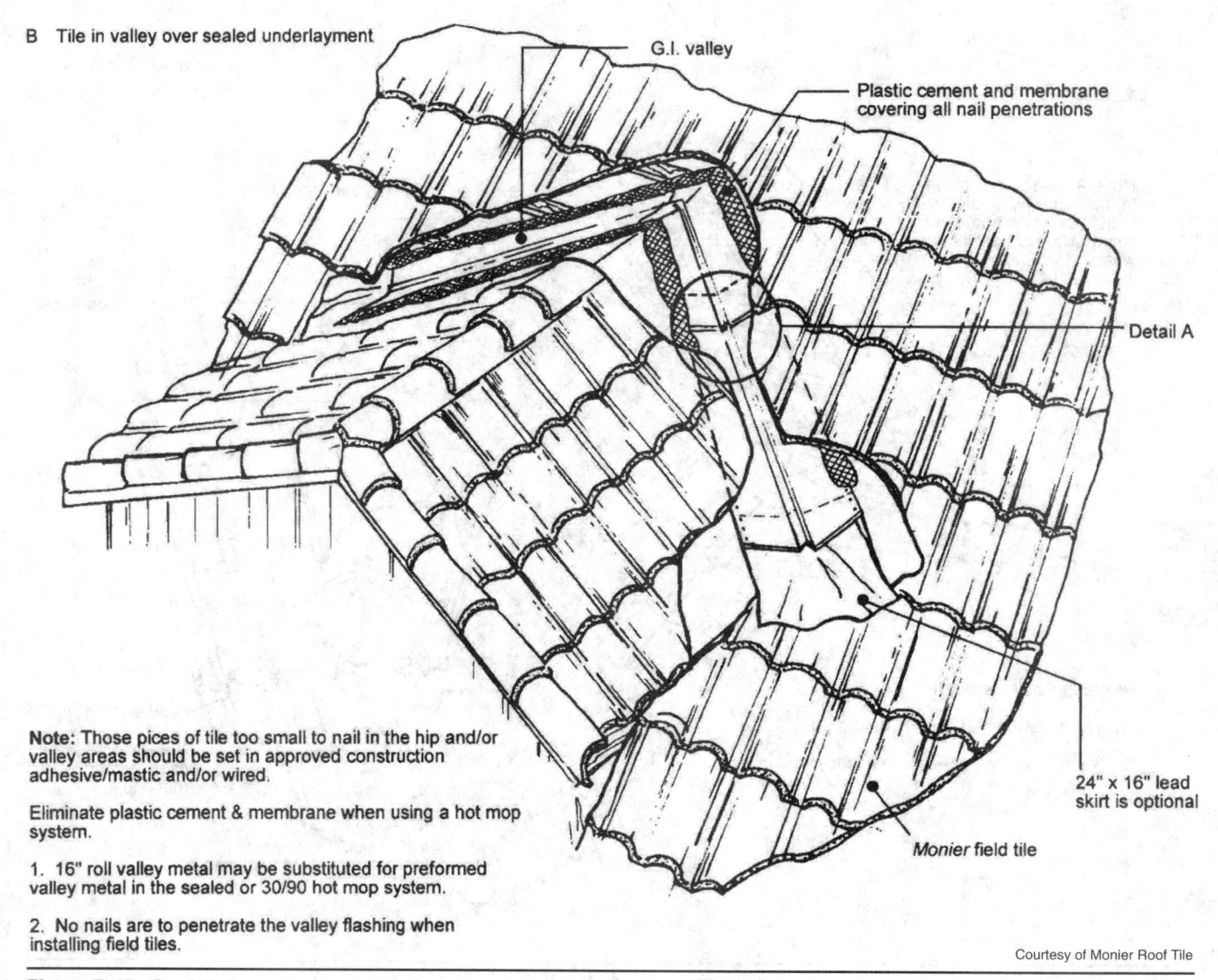

- G.I. valley
- Plastic cement and membrane covering all nail penetrations
- Detail A
- 24" x 16" lead skirt is optional
- *Monier* field tile

Note: Those pieces of tile too small to nail in the hip and/or valley areas should be set in approved construction adhesive/mastic and/or wired.

Eliminate plastic cement & membrane when using a hot mop system.

1. 16" roll valley metal may be substituted for preformed valley metal in the sealed or 30/90 hot mop system.

2. No nails are to penetrate the valley flashing when installing field tiles.

Courtesy of Monier Roof Tile

Figure 7-45 Roof tile in a valley (sealed or two-ply sealed underlayment system)

Tile Roofing

■ **Open Valley Installation** To secure tiles adjacent to an open valley, snap chalk lines at least 2 inches on each side of the centerline of the valley. Apply a bed of mortar along the outside edge of each chalk line and embed trimmed tiles in the mortar to form straight borders, as shown in Figure 7-46.

While the mortar under the valley tiles sets up, place two 2 x 4s on edge down the center of the valley. Apply color-matched mortar between the tiles and the 2 x 4s, pointing the mortar to match the tile contours. After the initial set (when the mortar has gotten "crunchy") slide a trowel along the outside edges of the 2 x 4s to separate them from the mortar. Remove the boards and point the mortar. Then rub the mortar with a dry sponge to remove the rough edges.

Some roofing contractors install hip-ridge tiles over a valley. I don't recommend this because a leak can trap water under the tiles in the valley.

A   Lay trimmed tiles

Flashing at Vertical Walls

The type of flashing you use along the juncture of a sloping roof and a vertical wall depends on the type of underlayment system you've used. Over an unsealed underlayment system, install continuous flashing with a 1-inch water guard on the horizontal leg. That's shown in Figure 7-47 A. Secure the flashing to the sheathing with clips nailed at 24-inch centers. Don't drive nails into the pan of the flashing.

B   Place two 2 x 4s in center of valley

Figure 7-46   Roof tile in an open valley

Begin at the low end of the wall and work up-slope to install the flashing over the underlayment. If you need more than one piece of flashing, lap the upper piece at least 4 inches over the lower one and seal the lap with roofing cement. Extend the flashing at least 6 inches over the underlayment and 5 inches up the wall, as shown in Figure 7-47 B.

When the lower end of the wall ends at a roof plane, you can:

1) Extend the flashing down to the eaves, or

2) Lap the pan flashing over a lead apron to send water runoff back onto the field tile. In this case, the underlayment can handle runoff during the most extreme rainy conditions. Both are shown in Figure 7-48.

Over a sealed underlayment system, you can install continuous flashing without a water guard. Nail the flashing to the sheathing, driving nails on 6-inch centers along a row 1 inch from the edge of the horizontal leg. Seal all nail holes with a strip of 30-pound felt embedded in roofing cement, as

Roofing Construction & Estimating

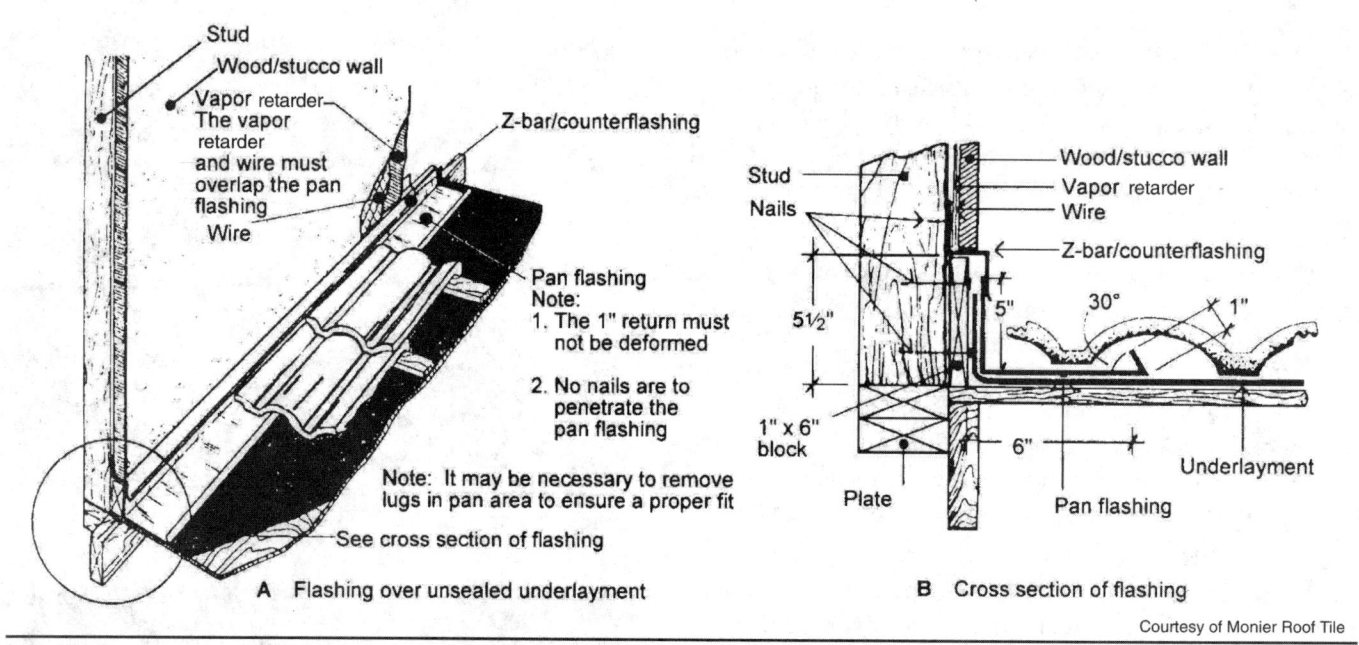

Figure 7-47 Continuous flashing over unsealed underlayment

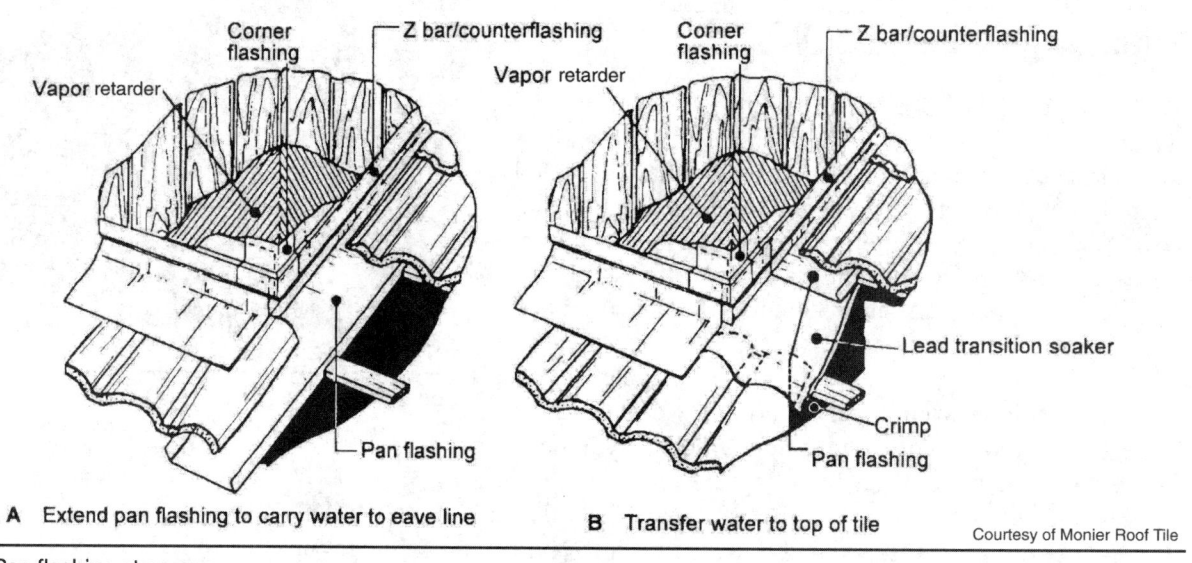

Figure 7-48 Pan flashing at corner

shown in Figure 7-49 A. Begin at the low end of the wall and work up-slope. Lap flashing at least 4 inches, and seal the laps with roofing cement. Extend the flashing at least 6 inches over the underlayment and 5 inches up the wall, as shown in Figure 7-49 B.

There's no need to send runoff water to the eaves or over the field tiles when the lower end of the wall ends at a roof plane. That's because the sealed underlayment system handles the runoff. However, cut the flashing with tin snips and bend it around the corner, as in Figure 7-50, to make sure the flashing at the bottom corners of the wall is watertight.

208

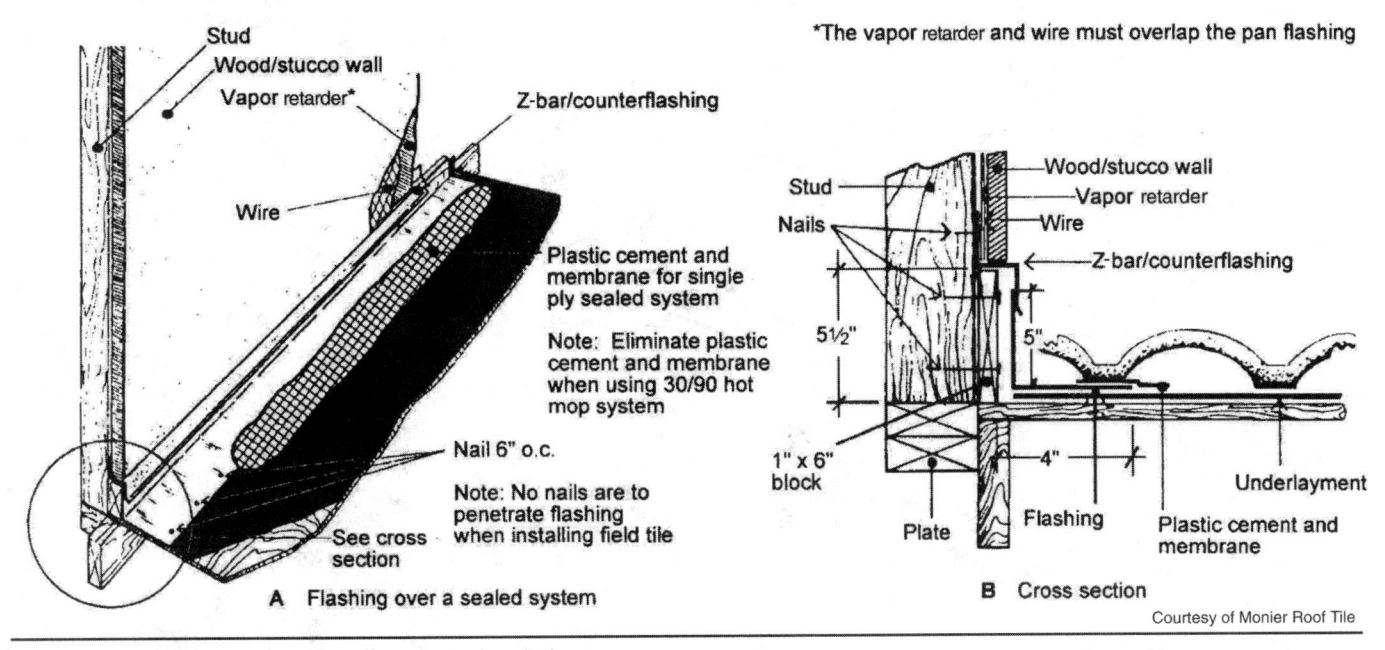

Figure 7-49 Continuous flashing over sealed underlayment

To install counterflashing on vertical walls, you can use either Z-bar flashing, as shown in Figures 7-47 and 7-49, or J-bead flashing, as in Figure 7-51. Lap the counterflashing over the base flashing a minimum of 3 inches.

The flashing method for the front of a dormer also depends on the type of underlayment system. On an unsealed underlayment system, install continuous flashing that overlaps the field tiles, as in Figure 7-52. On a sealed underlayment system, install the flashing as shown in Figure 7-53.

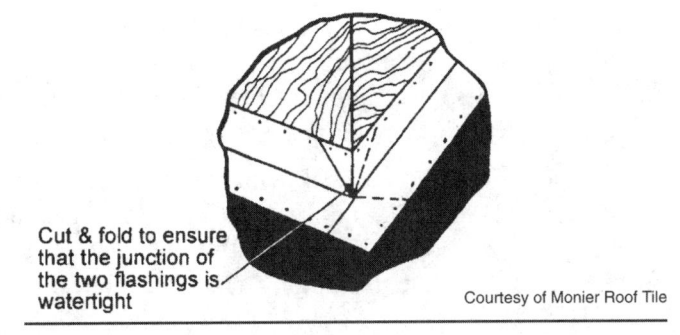

Figure 7-50 Watertight corner

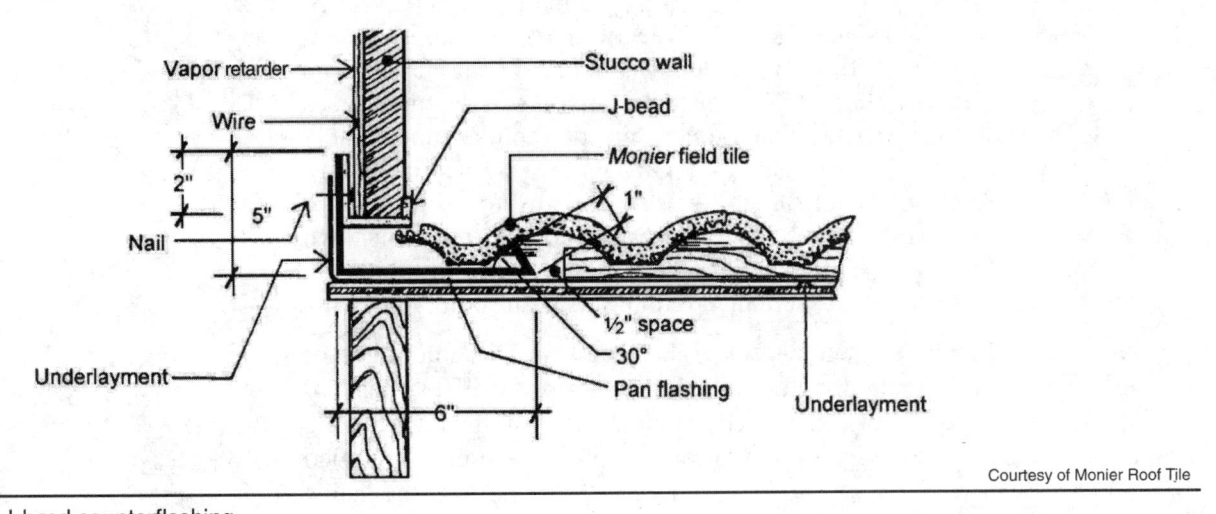

Figure 7-51 J-bead counterflashing

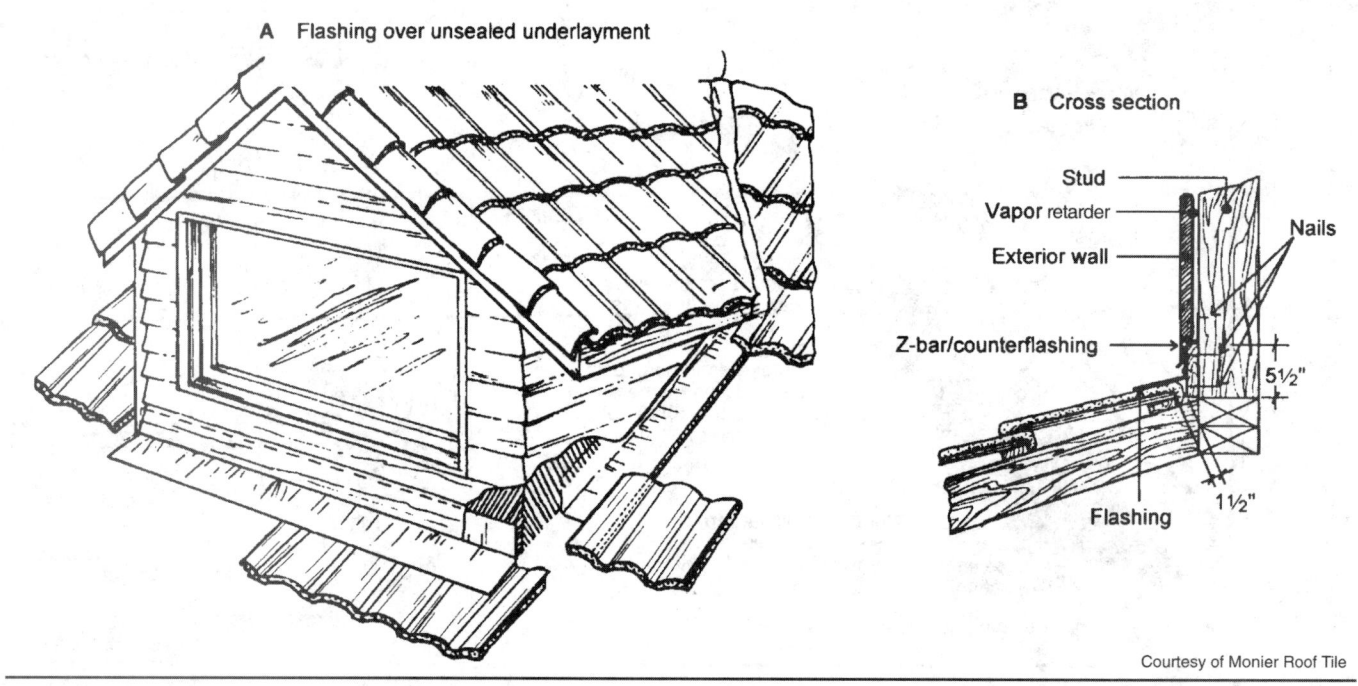

Figure 7-52 Flashing the front of a dormer over unsealed underlayment

Flash around chimneys as shown in Figure 7-54. If the horizontal width of the chimney is greater than 48 inches, install a galvanized iron cricket at the upper side of the chimney. Flashing and counterflashing is enough if the chimney is less than 48 inches wide.

Seal the flashing to the deck with plastic cement and roofing membrane. Refer back to Figure 7-48 for ways to apply the flashing at the sides of the chimney, and Figures 7-52 and 7-53 for the front. Be sure the counterflashing laps over the base flashing by at least 2 inches.

Flashing Soil Stacks and Vents

The method you use to seal a vent pipe depends on the type of underlayment system you've installed. Over an unsealed underlayment system, cut tiles to fit around the pipe, then install a lead vent jack with an 18" x 18" flange (or skirt). Form the flange to the contours of the tile courses covered. Be sure to extend the flange under the tile course immediately above the pipe. This is shown in Figure 7-55. On flat tiles, you can substitute a flat, corrosion-resistant metal flange for lead. Be sure to crimp the up-slope and side edges of the metal flange up at a 30-degree angle as shown in Figure 7-56. You can use pliers to crimp the metal, or have it done at a sheet metal shop. Secure the base flange to the tiles with roofing cement.

Over a sealed underlayment system, embed the vent jack into a bed of roofing cement over the 43-pound base sheet (single-ply sealed system) or the 30-pound felt (30/90 hot-mop sealed system). Then nail the flange along the edges. Cover the nails with felt and roofing cement (single-ply sealed system) or hot-mop 90-pound mineral-surfaced roll roofing over the flange (30/90 hot-mop sealed system). See Figure 7-57.

Tile Roofing

A Single ply sealed system

B Two ply sealed system

C Cross section

Courtesy of Monier Roof Tile

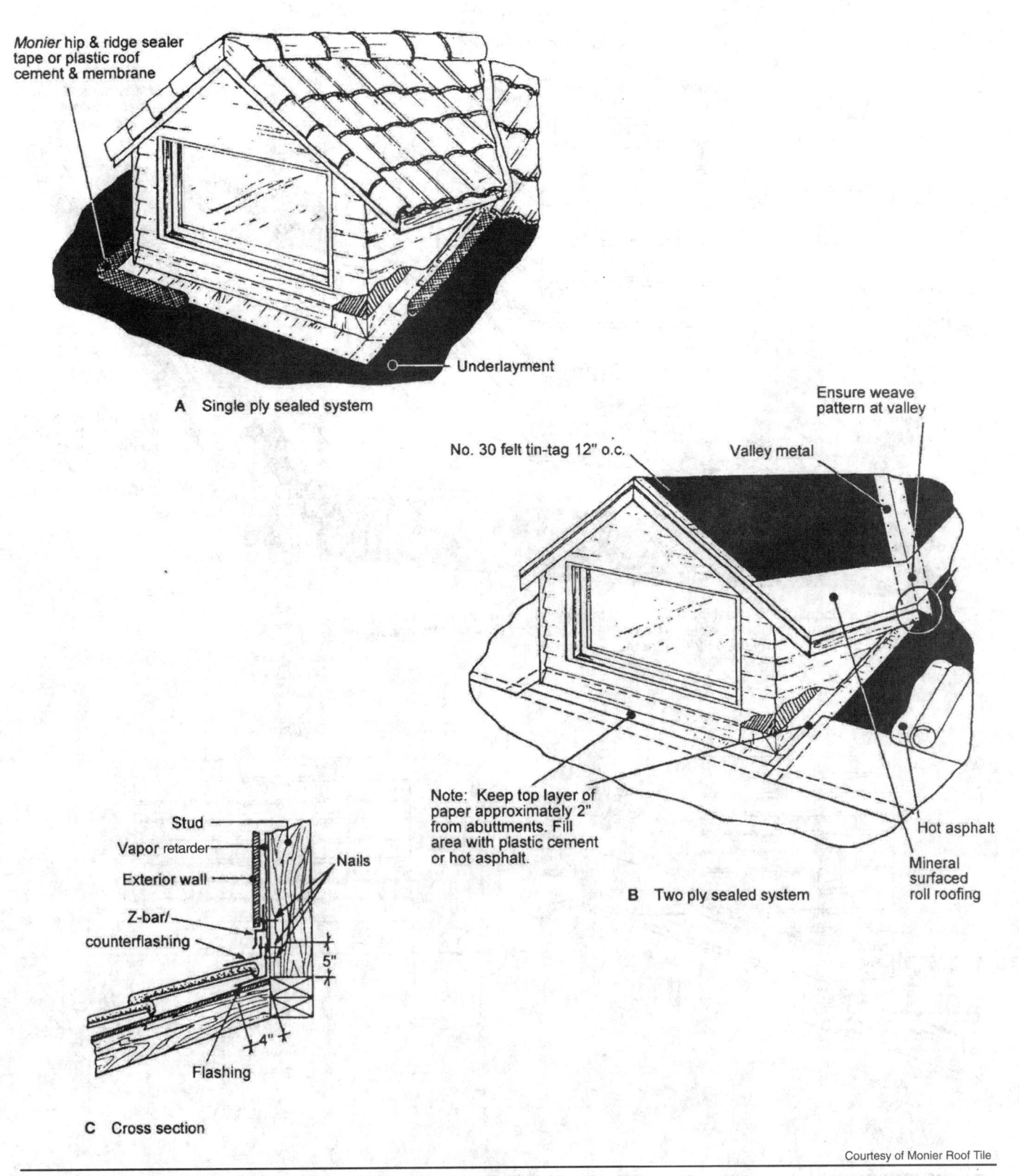

Figure 7-53 Flashing the front of a dormer over sealed underlayment

Roofing Construction & Estimating

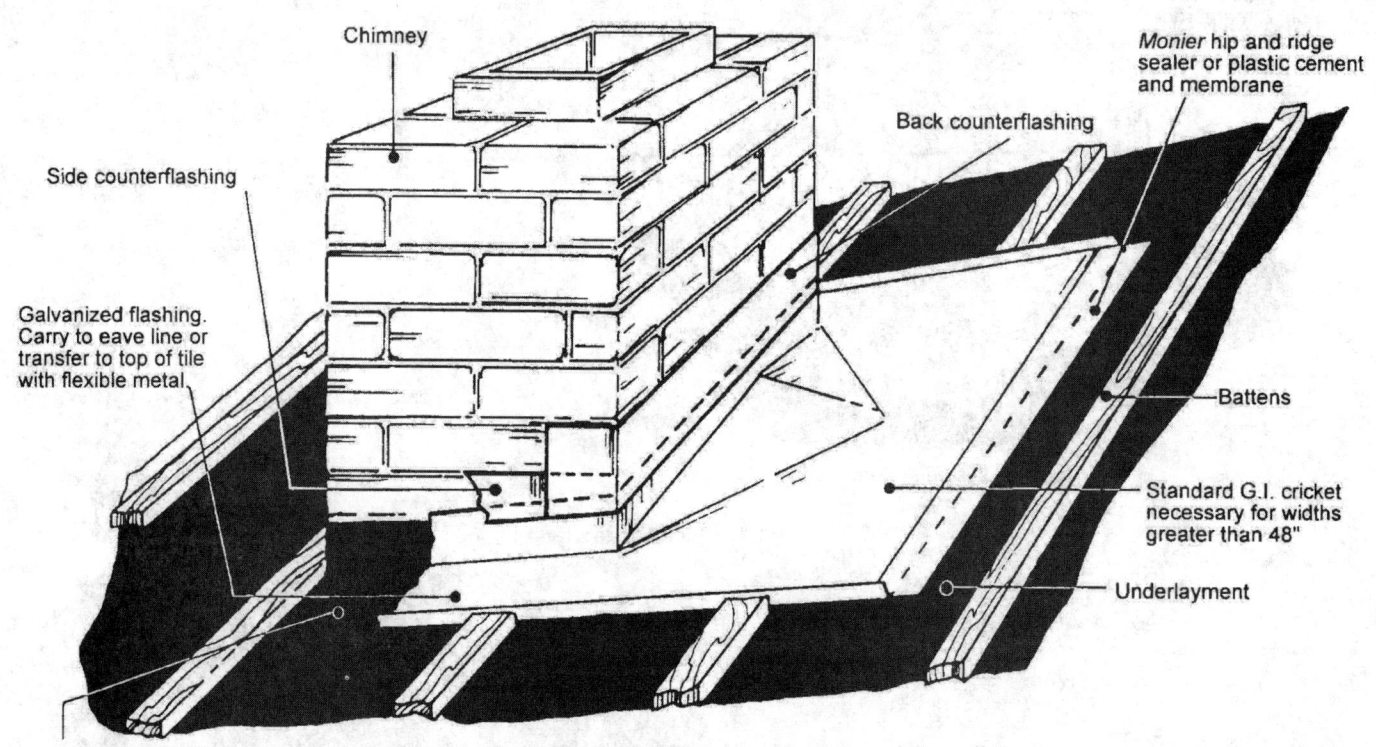

Figure 7-54 Flashing a chimney

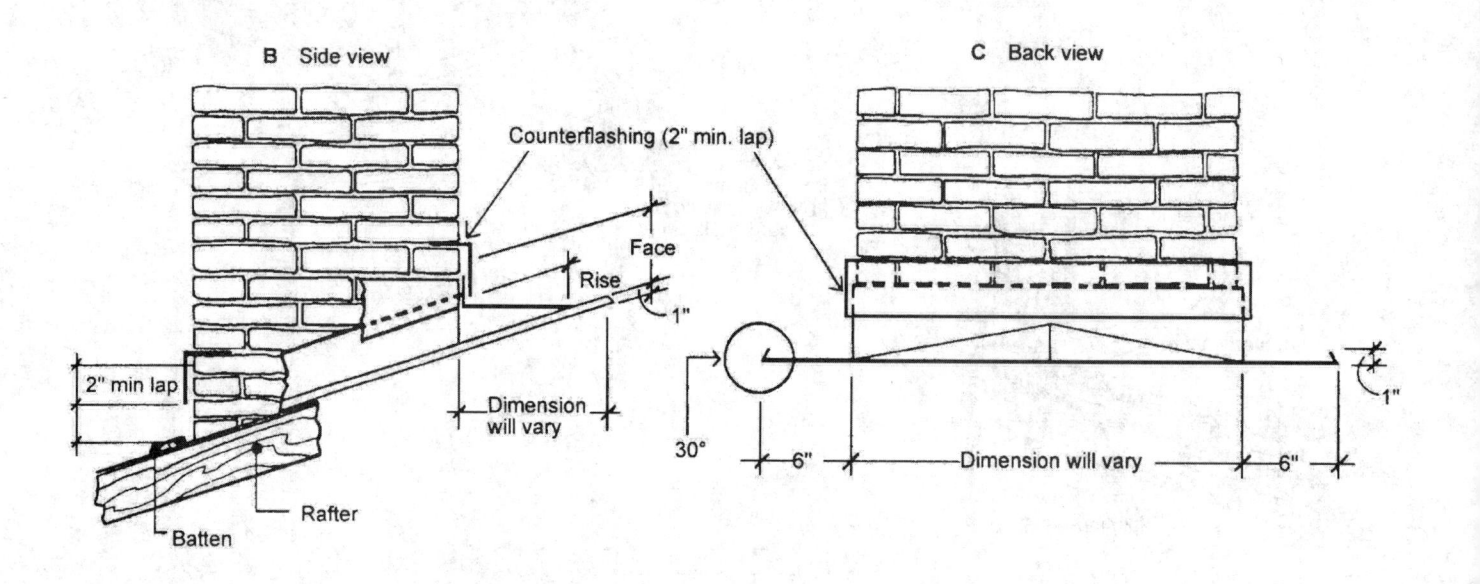

Courtesy of Monier Roof Tile

Tile Roofing

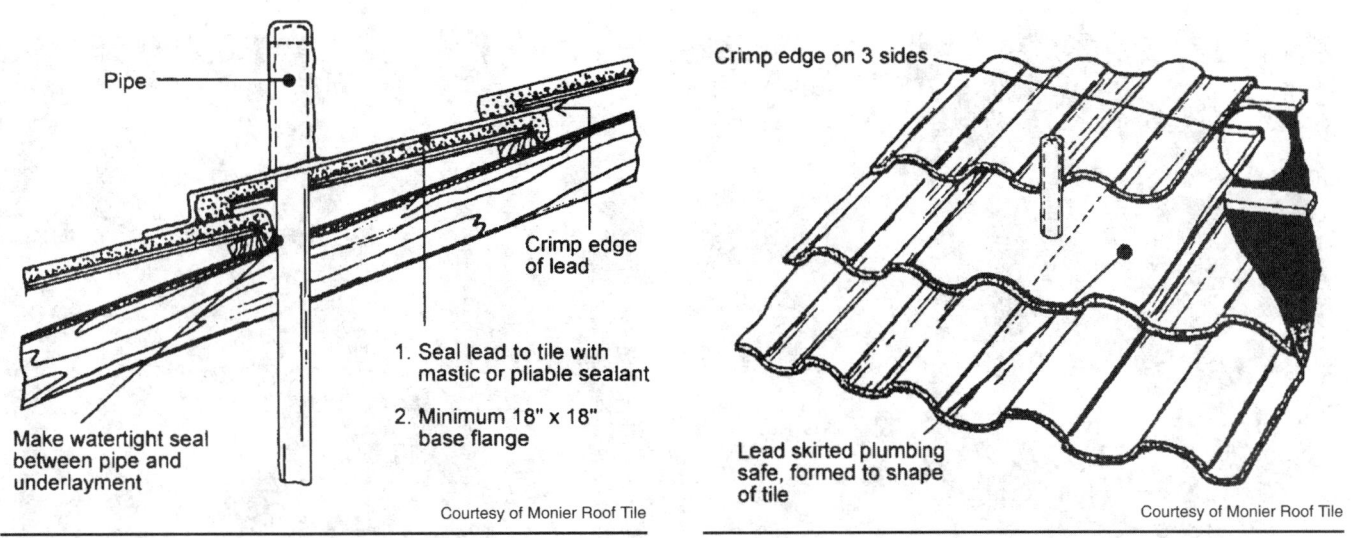

Figure 7-55 Flashing a vent pipe over an unsealed underlayment system

- Pipe
- Crimp edge of lead
- Make watertight seal between pipe and underlayment
- 1. Seal lead to tile with mastic or pliable sealant
- 2. Minimum 18" x 18" base flange

Courtesy of Monier Roof Tile

Figure 7-56 Flange with crimped edges

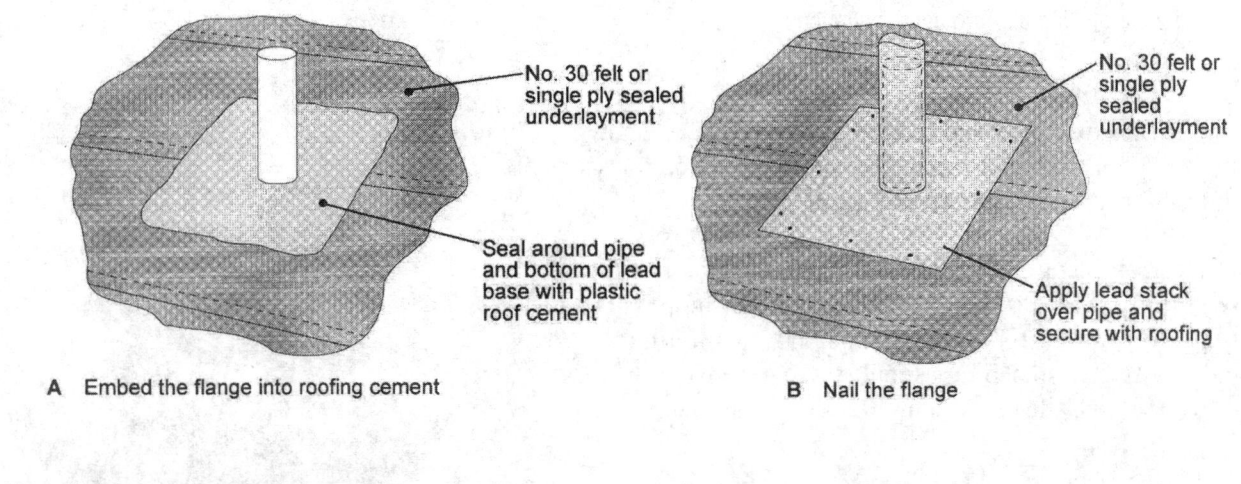

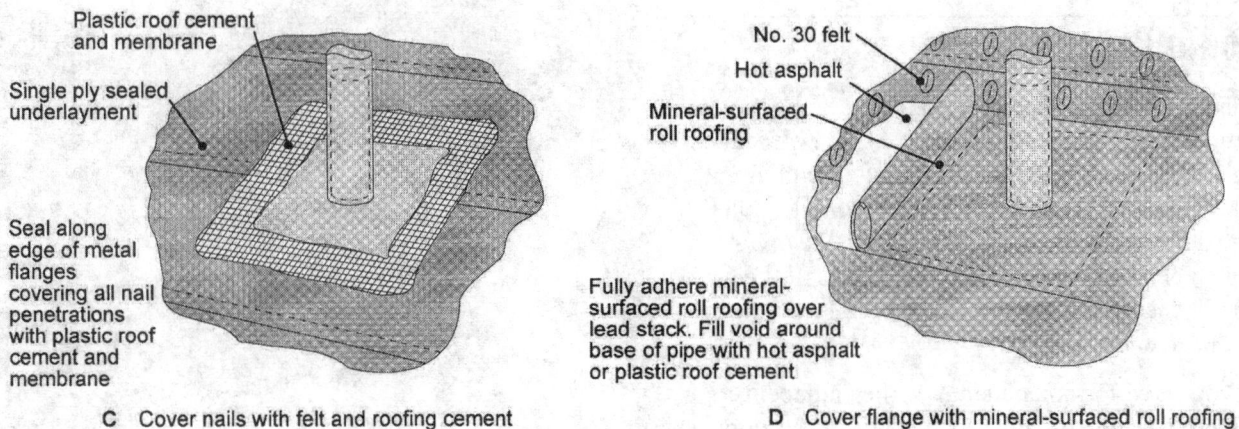

- Crimp edge on 3 sides
- Lead skirted plumbing safe, formed to shape of tile

Courtesy of Monier Roof Tile

A Embed the flange into roofing cement
- No. 30 felt or single ply sealed underlayment
- Seal around pipe and bottom of lead base with plastic roof cement

B Nail the flange
- No. 30 felt or single ply sealed underlayment
- Apply lead stack over pipe and secure with roofing

C Cover nails with felt and roofing cement
- Plastic roof cement and membrane
- Single ply sealed underlayment
- Seal along edge of metal flanges covering all nail penetrations with plastic roof cement and membrane

D Cover flange with mineral-surfaced roll roofing
- No. 30 felt
- Hot asphalt
- Mineral-surfaced roll roofing
- Fully adhere mineral-surfaced roll roofing over lead stack. Fill void around base of pipe with hot asphalt or plastic roof cement

Courtesy of Monier Roof Tile

Figure 7-57 Installing vent jacks

213

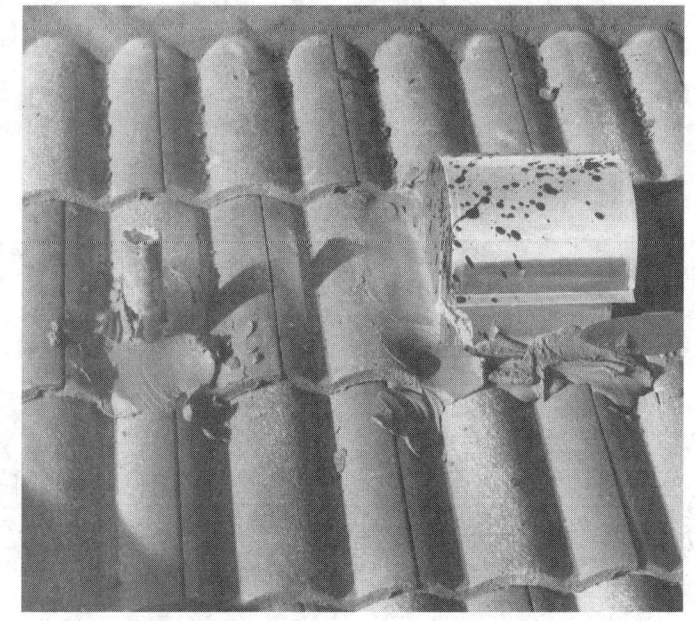

A Cut tiles to fit around penetrations

B Fill voids with mortar

Figure 7-58 Fitting tiles around penetrations

Cut tiles to fit over and around penetrations, as shown in Figure 7-58 A. Fill the voids with color-matched mortar, pointed to match tile contours as the roofer has started to do in Figure 7-58 B.

Paint all metal penetrations to match the color of the tile. Apply paint after you've finished the hot-mop work, but before you install the tile. You don't want to ruin the paint with any splashing asphalt. Also, you won't have to walk on the tiles so much after you install them.

Replacing Broken Tiles

To replace a broken or damaged field tile, break the tile with a hammer so it's easier to get out. If the tile was nailed in, drive the nails flush with the sheathing or batten with a hammer and flat bar. Then apply a minimum of 1 square inch of tile adhesive to the tile in the course below. Also apply $^3/_8$-inch beads of cement along the edges of the new tile and the one it will lay next to. Then install the new tile over the adhesive. See Figure 7-59.

When you have to replace smaller tiles adjacent to a hip or valley, pry up the nose of the tile in the course above the damaged tile with your hand and a mortar trowel and apply a $^3/_8$-inch bead of adhesive along the head of the replacement tile. Then set the new tile.

Figure 7-59 Replacing the tile

Estimating Tile Quantities

When you estimate tile roofs, consult the manufacturer's information for the number of field tiles per square, number of squares per pallet, and the number of tile accessories packaged per box. Keep in mind that field tile is usually sold only in full pallets, and fittings in full boxes. Also, shipping charges for tiles are by the full truckload.

It takes about 90 concrete field tiles to cover a square of roof area when you use a 3-inch top lap. Remember, over a sealed underlayment system, you only need a 2-inch top lap. In hurricane areas where the rafters are more than 20 feet long and the roof slopes from 4 in 12 to 7 in 12, you must use a 4-inch top lap. Take off tile quantities by the square foot, and then convert to the number of tiles required.

You calculate the actual number of tiles required by beginning with the number of tiles per manufacturer's square and using a percentage factor. The formula is the same as for asphalt shingles in Chapter 4, Equation 4-2:

$$\text{Percentage-of-Increase Factor} = \frac{\text{Recommended Exposure}}{\text{Actual Exposure}}$$

▼ **Example 7-1:** Ninety field tiles of the type shown in Figure 7-4 will cover a square of roof area when laid with a 3-inch top lap. To find the number of field tiles required if the tiles are installed with a top lap of 2 inches:

$$\text{Percentage-of-Increase Factor} = \frac{16.5 \text{ in.} - 3 \text{ in.}}{16.5 \text{ in.} - 2 \text{ in.}}$$

$$= \frac{13.5 \text{ in.}}{14.5 \text{ in.}}$$

$$= 0.93$$

To cover a square of roof area, you need 0.93 × 90 tiles, or 84 tiles.

To find the number of field tiles required if the tiles are installed with a top lap of 4 inches:

$$\text{Percentage-of-Increase Factor} = \frac{16.5 \text{ in.} - 3 \text{ in.}}{16.5 \text{ in.} - 4 \text{ in.}}$$

$$= 1.08$$

You need 1.08 × 90 tiles, or 98 tiles.

When you estimate tile roofs, add at least 3 percent to field tile quantities to account for waste and breakage.

Estimating Accessories and Mortar

The types of accessories you need depend on the type of field tile you're installing and the roof design. Tile accessories often include:

- rake tiles, as in Figure 7-30
- hip and ridge tiles, as in Figure 7-40
- ridge start and end, under-eaves tiles, as in Figure 7-15
- apex trim tiles, as in Figure 7-41
- hip starters like those in Figure 7-39

You'll also need portland cement mortar for tile installation and to seal eaves, valleys, hips and ridges. Fill mortar is normally made of 1 part portland cement mortar (Type M) or $1^1/_2$ parts high-strength masonry mortar (Type S) and 4 parts sand. Add color to exposed mortar to match the tile color.

Another suggested mix is $^1/_2$ part portland cement, 1 part masonry cement, 4 parts sand and an oxide for color. The addition of an acrylic polymer bonder is optional. I recommend this mix because of the added strength provided by the binder.

You can use plastic cement or silicone sealant instead of cement mortar to install accessories such as rake tiles. That way you don't have to spend time mixing the mortar. Just snap the end off a tube and apply. The cost is about the same when you consider the time you'll save. You can also add plastic or silicone adhesives to mortar for extra strength when you install ridge tile.

You may also need various types of eaves closures, storm clips, hip-ridge wind block material and flashings.

▼**Example 7-2:** Assume a roof slope of 5 in 12, a 2-inch eaves overhang, a 10-inch-wide flat tile laid at a 15-inch exposure and the use of roofing tiles sold at 110 pieces per square. Find the quantities for field tiles, rake tiles and under-eaves tiles required for the gable roof in Figure 7-60. Also assume a 3 percent waste factor for field tiles.

From the roof-slope factor in Appendix A, you calculate the net roof area (including a 2-inch overhang at the eaves) as:

Net Roof Area = 31' x 26'4" x 1.083 (the roof-slope factor)
 = 884 square feet ÷ 100
 = 8.84 squares

For this job, you need 8.84 squares at 110 tiles/square, times 1.03 (for the 3 percent waste factor), or 1,002 tiles.

From the roof-slope factor table, the rake length is 13 feet times 1.083, or 14 linear feet.

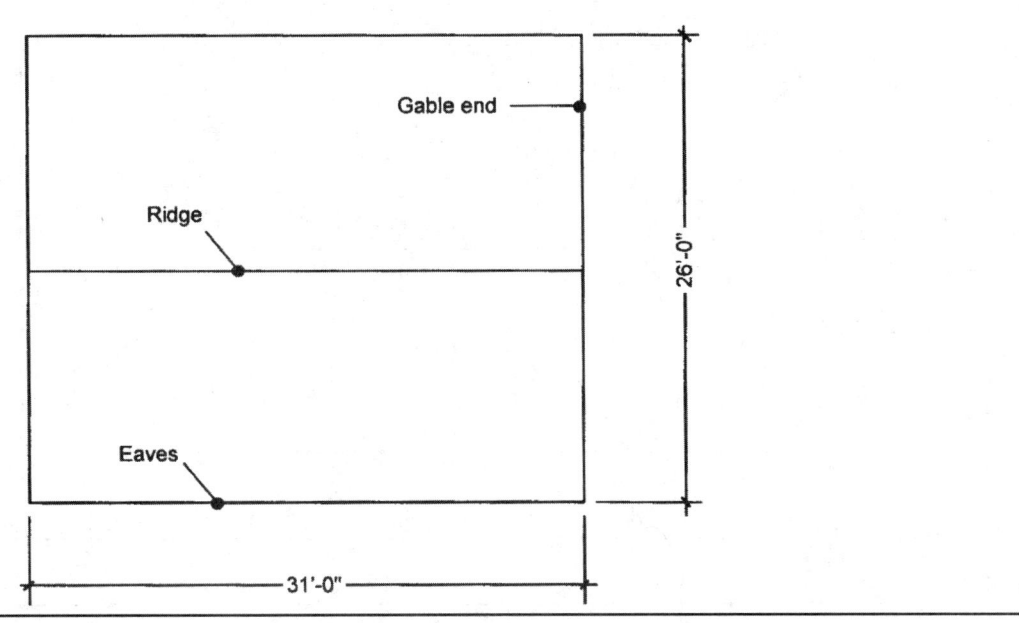

Figure 7-60 Gable roof example

The formula from Chapter 4 for shingle courses also works in this situation. Here's the formula:

Equation 4-3: Number of Courses $= \dfrac{\text{Dimension of Structure}}{\text{Exposure}}$

Just substitute rake length for dimension of structure:

Number of Rake Tiles (Single Rake) $= \dfrac{14 \text{ ft.}}{1.25 \text{ ft.}}$

$= 12$ tiles

There are four rakes, so you need a total of 48 tiles. Notice you round up for each rake, then multiply the result.

The eaves length is 31 linear feet. Use the same formula to find the number of under-eaves tiles:

Number of Under-Eaves Tiles (Each Eave) $= \dfrac{31 \text{ ft.}}{10 \text{ in. (tile width)}}$

$= 38$ tiles

(Note that 10 inches is the same as 0.83 feet.) Double the 38 tiles to get total tiles for both eaves (76 tiles).

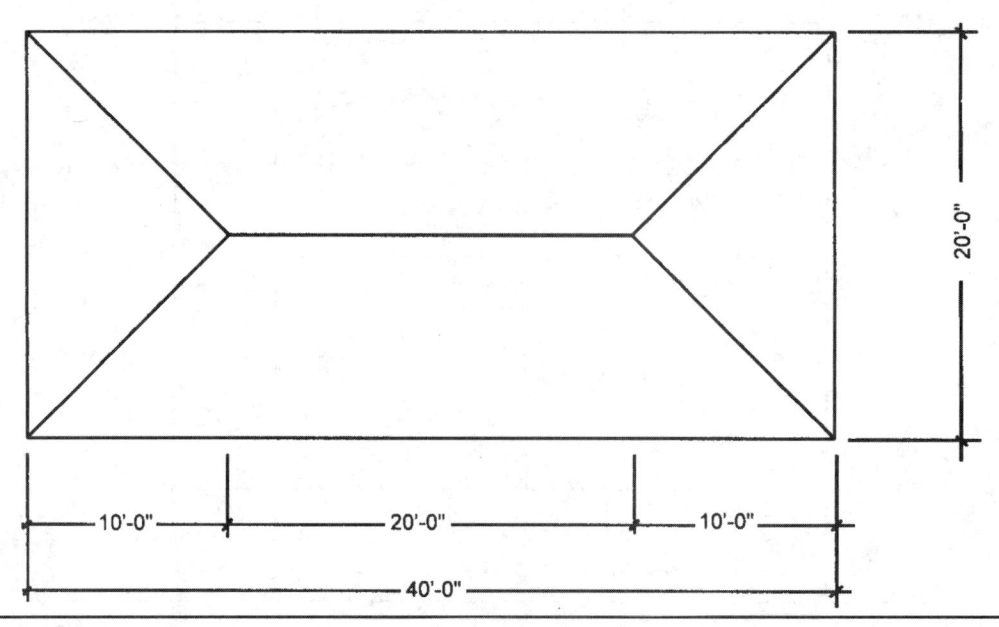

Figure 7-61 Hip roof example

Again, the formula also works for the ridge tiles:

$$\text{Number of Ridge Tiles} = \frac{31 \text{ ft.}}{15 \text{ in.}}$$
$$= 25 \text{ tiles}$$

▼ **Example 7-3:** Assume a roof slope of 6 in 12, and the conditions given in Example 7-2. Find how many field tiles, under-eaves tiles and hip and ridge tiles you need for the roof of the building diagrammed in Figure 7-61. Assume a 3 percent field tile waste.

Using the roof-slope factor from Appendix A, the net roof area, including the 2-inch overhang at the eaves, is:

Net Roof Area = 40'4" x 20'4" x 1.118
 = 917 square feet ÷ 100
 = 9.2 squares

You'll need 1043 tiles: 9.2 squares x 110 tiles per square x 1.03 (3 percent waste).

Also, order 49 under-eaves tiles for each of the long eaves (40'4" ÷ 10" = 49), and 25 under-eaves tiles for each of the short eaves (20'4" ÷ 10" = 25), for a total of 148.

The ridge is 20 feet long, and each of the hips is 15 feet long (10 foot run times the hip factor of 1.5 from Appendix A), for a total of 80 linear feet for ridge and hips. Therefore, you need 64 ridge and hip tiles (80' ÷ 15" = 64). Of these 64 tiles, four are hip starters. And don't forget the two apex trim tiles where the ridge joins the hips.

Estimating Tile Roofing Costs

Let's say you've got to cover the roof shown in Figure 7-60 with "S"-shaped Spanish clay roofing tile. For this example, we'll use the materials costs in the current *National Construction Estimator*, with field tiles, including felt, flashing, fasteners, vents and sales tax at $345.08 per square, ridge units at $1.21 per linear foot, and rake units at $2.31 per linear foot. Allow $1.21 per linear foot for ridge units and $2.31 per linear foot for rake units.

From Example 7-2, you must purchase 8.84 squares of field tile, 31 linear feet of ridge tile and (4 x 14' =) 56 linear feet of rake tile. Therefore we have a material cost of:

Field tiles   = 8.84 square x $345.08/square = $3,050.51
Ridge tiles = 31 LF x $1.21/LF   =   37.51
Rake tiles   = 56 LF x $2.31/LF   =   129.36

The total material cost is: $3,217.38

According to the *National Construction Estimator*, an R1 crew consisting of one roofer and one laborer can install field tile at the rate of 1 square per 3.46 manhours. The crew can install ridge and rake tiles at the rate of 1 linear foot per 0.047 manhours. Assuming the roofer costs $41.66 per hour and the laborer costs $30.31 per hour, including labor burden, the average cost per manhour is: $41.66 + $30.31 = 71.97 ÷ 2 = $35.99 per manhour. Therefore the labor cost is:

Field tiles   = 8.84 square x 3.46 manhours/square x $35.99/manhour
              = $1,100.80

Ridge tiles = 31 LF x 0.047 manhours/LF x $35.99/manhour
              = $52.44

Rake tiles   = 56 LF x 0.047 manhours/LF x $35.99 manhour
              = $94.73

The total labor cost is: $1,100.80 + $52.44 + $94.73 = $1,247.97

So the total cost (excluding overhead and profit) is:

$3,217.38 + $1,247.97 = $4,465.35

Now, let's move on to the next chapter on slate roofing.

8 Slate Roofing

▶ Like tile, slate roofing is among the more expensive types of roofing: The material itself costs more, it takes longer to install, and slate thicker than $3/8$ inch requires a stronger supporting frame. But with the extra cost comes some important advantages.

- If it's well constructed, a slate roof has a normal life expectancy of more than 50 years.
- A slate roof is fireproof, so insurance premiums are lower.
- It's also waterproof, and resists damage from extremes in the weather.
- Slate requires very little maintenance and never needs any kind of preservative.
- The higher cost of a good slate roof is usually reflected in a higher value for the home.
- Slate comes in a wide range of attractive textures and colors.

These advantages make slate a very desirable roofing material, especially with some architectural styles and in some communities.

Slate Size, Color and Texture

Most slates manufactured in the United States are quarried in Virginia, Maine, Maryland, Pennsylvania and Vermont. Slate are manufactured by splitting slate blocks into standard thicknesses of $3/16$, $1/4$, $3/8$, $1/2$, $3/4$, 1, $1 1/4$,

L x W	L x W	L x W
26 x 14	18 x 12	14 x 9
24 x 14	18 x 11	14 x 8
24 x 12	18 x 10	12 x 10
22 x 12	18 x 9	12 x 9
22 x 11	16 x 12	12 x 8
20 x 14	16 x 10	12 x 7
20 x 12	16 x 9	12 x 6
20 x 11	16 x 8	10 x 8
20 x 10	14 x 12	10 x 7
	14 x 10	10 x 6

Figure 8-1   Common slate sizes

$1^1/_2$, $1^3/_4$ and 2 inches. The most common thickness is $3/_{16}$ inch (commercial standard slate). The maximum practical thickness for roofing slate is 2 inches, but slate that thick is rarely used in residential construction.

Natural stone can't be split so that every piece is exactly the same thickness, so slight variations above or below the nominal thickness are acceptable. If your job requires absolute consistency, each piece will have to be measured. That will increase the material cost and may delay delivery.

The slates used on homes are usually about $3/_{16}$-inch thick and one of the standard lengths and widths in Figure 8-1. Slates thicker than $1/_2$ inch are manufactured in lengths up to 30 inches. Random width and length slates are also available. Sizes ranging from 10 x 6 inch through 14 x 10 inch are the most plentiful. If you insist on non-standard sizes, you'll pay more and wait longer for delivery.

Slate trade names are based on color. They're known as black, blue black, gray, blue gray, green, mottled green, purple, mottled purple, purple variegated and red. Other colors are considered "specials."

The trade name usually has the word "unfading" or "weathering" before it. Slates from some quarries weather. The color of weathering slate "mellows" after it's exposed to the elements for a few months. Unfading slate keeps its original natural shade.

Based on grade, the trade names of slates are No. 1 Clear, Medium Clear, No. 1 Ribbon and No. 2 Ribbon. The architectural classifications for slate roofs include standard slate, textural slate, graduated slate and flat slate.

Course	Slate thickness	Length	Exposure
Under-eaves course	3/8" thick	14" long	No exposure
First course	3/4" thick	24" long	10 1/2" exposure
1 course	3/4" thick	24" long	10 1/2" exposure
2 courses	1/2" thick	22" long	9 1/2" exposure
2 courses	1/2" thick	20" long	8 1/2" exposure
2 courses	3/8" thick	20" long	8 1/2" exposure
4 courses	3/8" thick	18" long	7 1/2" exposure
5 courses	1/4" thick	16" long	6 1/2" exposure
3 courses	1/4" thick	14" long	6 1/2" exposure
3 courses	3/16" thick	14" long	5 1/2" exposure
8 courses	3/16" thick	12" long	4 1/2" exposure

Courtesy of Vermont Structural Slate Co.

Figure 8-2 Coursing method for graduated slate roof

Textural slate has a rougher surface texture than standard slate, often with uneven butts and variations in shingle thickness, size, and exposure. Slates may be the same in length but vary in thickness, or the thickness may be the same and the length (and exposure) varies. If the thickness varies, it'll usually be between 1/4 to 3/4 inch, although thicker slates may be installed at random to give the roof a rough textured appearance.

Slates usually come from the quarry with the holes already punched. Some contractors, however, order their slate unpunched. They've invested in slate punching machines so they can keep their crews busy on idle days or during bad weather. Most slates are punched with two holes, but be sure you punch four-nail holes through any slates thicker than 3/4 inch, or longer than 20 inches. While most specifications call for slates to be machine punched, you can hand punch with a punch and maul for slates cut to fit at hips.

When slate blocks are split to make roofing slates, the thicker slates tend to have rougher faces and edges than the thinner slates. The thinner slates split more cleanly so they tend to have more even surfaces. The thinner slates may make the roof look flat when used beside the rougher, heavier slates near the eaves. This becomes especially important close to the ridge because the ridge slates are farther from the eye and will already look smoother than the ones lower down on the roof. You may want to special-order thinner slates with a rougher surface.

Slating a Graduated Roof

A graduated roof is slated with the longest and thickest slates at the eaves. Use slates that are shorter and thinner as you move up to the ridge. Graduated roofs combine textural slate with a wide variety of slate sizes, thicknesses, exposures and color. Figure 8-2 shows one method of coursing for a graduated roof. For example, a graduated roof might use slates from 12 to 24 inches long and $3/16$ to $1^1/_2$ inches thick. The head lap should be a constant 3 inches for all slates, regardless of the length.

Be sure to lay random slates on a graduated roof so the vertical joints are broken and offset from those in the courses above and below them. See Figure 8-3.

Underlayment

A slate roof will actually perform quite well in most areas without any underlayment at all, and if you work in an area with older homes, you may come across some. But today, building codes require the appropriate underlayment, as well as ice and water shield over the eaves in cold areas.

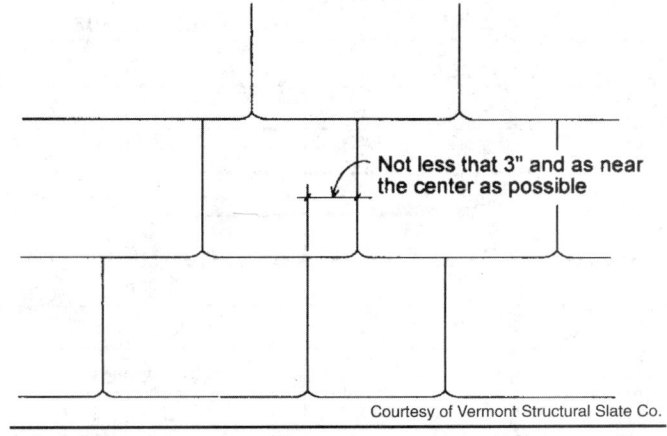

Figure 8-3 Proper jointing

A good underlayment will provide some degree of weatherproofing and protection from leaks should a slate get damaged or come off the roof, but remember that you have to drive nails through it to install each piece of slate. Underlayment requirements depend on the size and thickness of the slate, the slope and architecture of the roof, and also vary from location to location. Cold, wet climates have more-stringent requirements than those in warm, dry areas.

Roofing in Florida, I used 15-pound saturated felt for standard slate on a residence, and 30-pound for a textured roof or slates up to $3/4$ inch thick on a graduated roof. For thicker slates, 45-, 55-, or 65-pound roll roofing was needed. In the northeast and other areas of cold and hard-driven rain, a double layer of 30-pound felt may be necessary. Because there are so many factors involved, it's especially important to check with the local code official beforehand for the requirements for the specific slate and roof you're dealing with.

Synthetic underlayment is more popular nowadays; it has a longer exposure time before being covered, is safer to walk on, and is much lighter and easier to install. The down side of synthetic on a slate roof is that most don't provide the cushioning that felt does. A little cushioning may be helpful on a slate roof. And I always appreciated that felt provided a good surface for chalk lines when laying out the courses.

There are some synthetic hybrids on the market — ice and water shields — that combine benefits of both, but they're not cheap nor quite as easy to install. Again, see what's required in your location.

Installation on a Flat Roof

If the roof slope is less than 4 in 12, slate has to be installed with no lap, in a setting bed of asphalt, pitch, or concrete. You can install $3/16$-inch thick slate on a flat roof.

Slate Roofing

Weights of flat slate roof without concrete bedding slab			
Materials	Weight of materials per sq. (100 sq. ft.)	Total Weight per sq. (100 sq. ft.)	Total weight pounds per sq. ft.
Waterproofing (Weight varies, assumed here to be 150 lbs.)	150		
3/16" slate	250	400	4.0
1/4" slate	335	485	4.85
3/8" slate	500	650	6.5
1/2" slate	675	825	8.25
3/4" slate	1,000	1,150	11.5
1" slate	1,330	1,480	14.8

Weights of flat slate roof with concrete bedding slab			
Materials	Weight of materials per sq. (100 sq. ft.)	Total Weight per sq. (100 sq. ft.)	Total weight pounds per sq. ft.
Waterproofing (Weight varies, assumed here to be 150 lbs.)	150		
3/4" concrete bed	750		
3/16" slate	250	1,150	11.50
1/4" slate	335	1,235	12.35
3/8" slate	500	1,400	14.00
1/2" slate	675	1,575	15.75
3/4" slate	1,000	1,900	19.00
1" slate	1,330	2,230	22.30

Courtesy of Vermont Structural Slate Co.

Figure 8-4 Weights of flat slate roofing

Slate for flat roofs comes in standard sizes of 6 x 6, 6 x 8, 6 x 9, 10 x 6, 10 x 7, 10 x 8, 12 x 6, 12 x 7 and 12 x 8. Set slates that are less than 3/4 inch thick in asphalt or pitch. Set slates 3/4 inch and thicker over a 3/4-inch setting bed of concrete that you've installed over waterproofing felt. Figure 8-4 shows weights of flat slate roofs.

You shouldn't walk on a slate roof. Roof slate is brittle and will crack at the nail holes. If the roofing crew is walking about on it, there's going to be damage that may not show up right away, but will later. A slate roofing job should be properly staged, using roof ladders and scaffolds, so foot traffic on installed slate is unnecessary. But if foot traffic is absolutely unavoidable, use slates that are at least 3/8- to 1/2-inch thick. The 3/16-inch slates have enough wearing surface, but 3/8- and 1/2-inch slates go into the bedding better and stay more securely in place under a roofer's foot. Make sure you order unpunched slate.

Installation on a Sloping Roof

The weight of slate on a sloping roof varies tremendously, depending on the thickness of the slate, the head lap, and the weight of the slate itself, which varies. For instance, the weight of standard slate varies from 700 to 800 pounds per square, but thicker slates vary from 800 to 3600 pounds per square. See Figure 8-5. Rafters that are strong enough to hold wood shingles or shakes are usually strong enough to hold 3/16-inch-thick commercial standard slate when you lay it with a 3-inch head lap.

In residential construction, you usually install slate over solid wood sheathing or plywood. For fireproof commercial work, you can install slates over a variety of substrates, including concrete, gypsum slabs and steel purlins, as shown in Figure 8-6. Over steel purlins, install slates punched with four holes and run wire through each pair of holes and around the steel angle. Then twist the ends of the wire to draw the slate down tightly to the angle.

After you've put in the underlayment, drip edge and valley flashing material, install a starter course (under-eaves slates) at the eaves of the roof. This course is usually applied over a wood cant strip. Allow a 2-inch overhang at the eaves and a 1/2-inch overhang at the rakes.

To figure out how long the under-eaves slates should be, add 3 inches to the exposure you use on slates in the main part of the roof. For example, if you use a 6 1/2-inch exposure at the center of the roof, use 9 1/2-inch slates

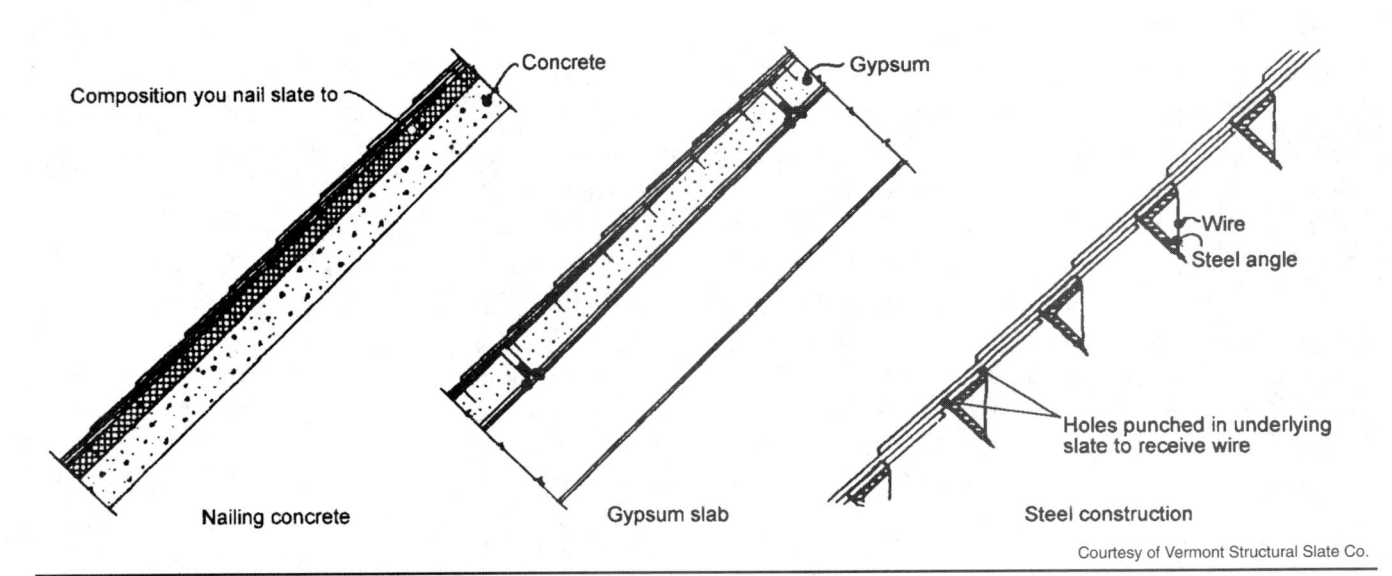

Thickness (inch)	Weight (lbs/square)
3/16	700 - 800
1/4	900 - 1000
3/8	1300 - 1400
1/2	1700 - 1800
3/4	2500 - 2800
1	3400 - 3600

Courtesy of Vermont Structural Slate Co.

Figure 8-5   Weights of slate roofing per square with 3-inch head lap

Figure 8-6   Slates on various types of fireproof construction

at the eaves. You can either lay half slates, or use full slates with the long edge laid parallel to the eaves. You can use under-eaves slates that are the same, or only half as thick, as the field slates.

Install the first slate course over the under-eaves slates with the butts of both courses set flush and the vertical joints offset.

The head lap you use to lay slates will depend on the slope of the roof. Use a 4-inch head lap over slopes between 4 in 12 and 8 in 12. Reduce the head lap to 3 inches on slopes between 8 in 12 and 20 in 12. You can cut the head lap to 2 inches on roofs steeper than 20 in 12, or on vertical walls. Figure 8-7 shows recommended head laps.

The part of each slate exposed depends on the head lap. Here's how you figure the exposure:

$$\text{Exposure} = \frac{L - HL}{2}$$

where: L = Length of shingle
HL = Head Lap

So, the exposure of a 24-inch slate laid up with a 3-inch head lap is:

$$\text{Exposure} = \frac{24 \text{ inch} - 3 \text{ inch}}{2} = 10\frac{1}{2}"$$

Figure 8-8 shows exposures required for various lengths of slate installed with a 3-inch head lap. The exposure used on a graduated slate roof will often vary, as shown in Figure 8-2. It shows a typical variation in exposure.

Install slates so the joints in each course are offset from the joints in courses above and below. If you're installing random-width slates, lay the slates so that a joint of the underlying slate is located as close to the center of the overlying slate as possible. Offset joints in adjacent courses at least 3 inches. Refer back to Figure 8-3.

When you use slates that are all the same width, proper jointing is taken care of automatically by starting every other course with a half slate. For a more interesting and random jointing pattern, install slates as shown in Figure 8-9.

Install slates with staggered joints all the way up the roof. Adjust the exposure of courses below the ridge so the ridge course of slates doesn't have to be trimmed. The

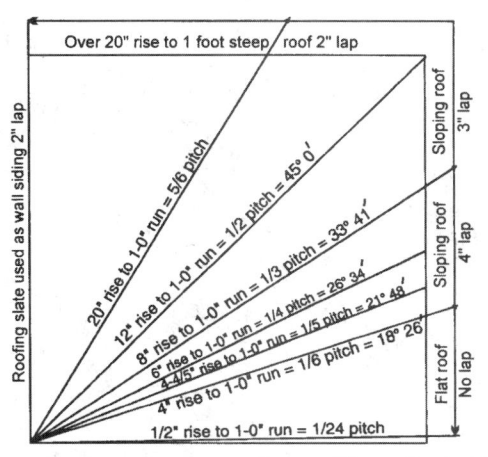

Figure 8-7 Head lap of slate for various roof slopes

Length of Slate (inches)	Exposure at slope 8" to 20" per foot, 3" lap (in inches)
24	10 1/2
22	9 1/2
20	8 1/2
18	7 1/2
16	6 1/2
14	5 1/2
12	4 1/2
10	3 1/2

Figure 8-8 Exposure for sloping roofs

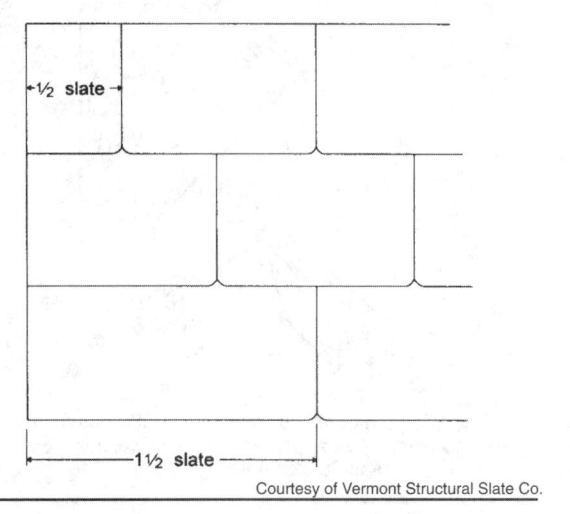

Figure 8-9 Starting slates

Roofing Construction & Estimating

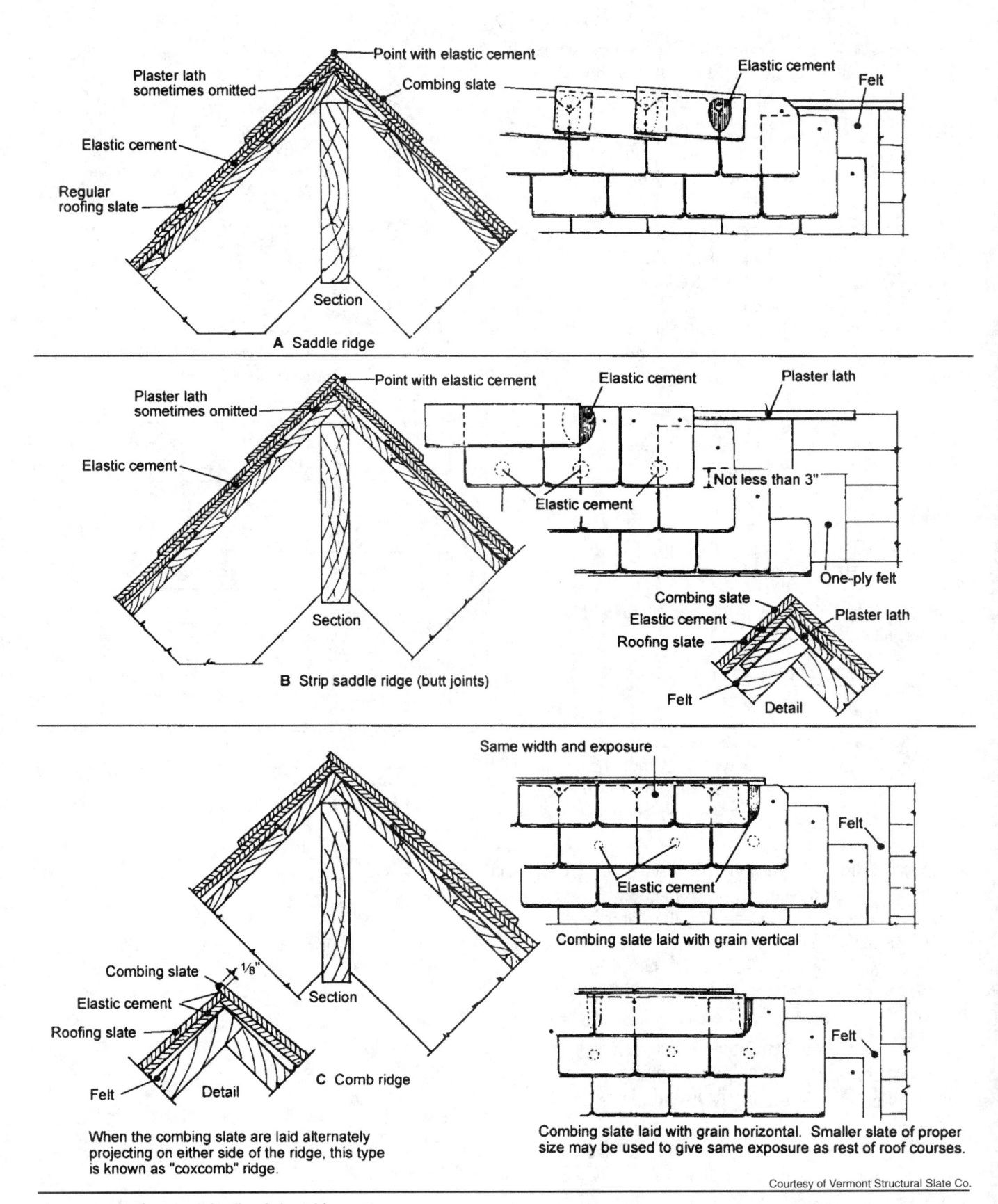

Figure 8-10 Standard details of slate ridges

Length of Slate (inches)	Spacing of lath (inches)
24	10$\frac{1}{2}$
22	9$\frac{1}{2}$
20	8$\frac{1}{2}$
18	7$\frac{1}{2}$
16	6$\frac{1}{2}$
14	5$\frac{1}{2}$
12	4$\frac{1}{2}$

Courtesy of Vermont Structural Slate Co.

Figure 8-11 Spacing of nailing strips

top course should be both nailed and set in spots of roofing cement to keep the wind from lifting them up. Figure 8-10 shows this.

Normally, you nail slates directly to the sheathing. However, in climates without wind-driven rain or snow, you can omit the roofing felt and nail the slates directly on 1 x 2 or 1 x 3 nailing strips installed parallel to the eaves. Space the nailing strips so they're appropriate for the length of slate and head lap you're using. If you use a 3-inch head lap, space the nailing strips as shown in Figure 8-11.

Ridges

There are three basic types of slate ridges: the saddle ridge, the strip saddle ridge and the comb ridge. Look again at Figure 8-10.

To create any of these ridges, you lay slate right up both sides of the roof so slates from both sides butt flush together at the roof top. Then cover the ridge with *combing slate*. A comb is the ridge of a roof, so combing slate is the topmost row of slates, which projects above the ridge line. Usually the grain of this slate should run horizontally. Nail the combing slate through joints in the underlying slates, not through the slates themselves. Use combing slate that's wide enough so you can make the exposure of the uppermost slate course about the same as the exposure on the rest of the roof. Also, be sure the joints of combing slates are offset from the joints of slates underneath them.

Apply elastic roofing cement along the joint at the peak of the roof. Cover all nail heads with cement. Apply more cement under the unnailed ends of combing slates.

■ **Saddle Ridge** Install combing slate from both sides of the ridge with butts flush. Install each combing slate with two nails and overlap adjacent slates so the nails are covered. See Figure 8-10 A.

- **Strip Saddle Ridge** Install each combing slate with four nails and butt the end joints over a bed of elastic cement. See Figure 8-10 B.

- **Comb Ridge** Install about the same way as a strip saddle ridge except you extend the combing slate on one side of the ridge about $1/8$ inch beyond the top of the combing slate from the other side. See Figure 8-10 C.

The *coxcomb ridge* is a variation of the comb ridge. You lay the combing slates so they alternately project on either side of the ridge.

Hips

There are five basic types of slate hips. These are the saddle hip, the strip saddle hip, the mitered hip, the Boston hip and the fantail hip. They're shown in Figure 8-12.

On all of these hips, drive nails through the joints of the underlying slates. Apply elastic cement along the joint at the centerline of the hips, over all nail heads and at the lower ends of the hip slates. In addition to nailing, set the slates underlying the hip slates into spots of roofing cement. This keeps slates from blowing off in high winds.

- **Saddle Hip** Nail each hip slate over a lath strip or $3\frac{1}{2}$-inch cant strip. Install slates that are the same width as the exposure you're using on the main roof. Install each slate with four nails and overlap adjacent slates so the butts of the hip slates line up with the butts of the slate courses on the main roof. See Figure 8-12 A.

- **Strip Saddle Hip** Install the same way as a saddle hip, except you lay the narrower hip slates up with butted joints that don't necessarily line up with butts of the slate courses on the main roof.

- **Mitered Hip** Cut the hip slates into triangular pieces so they fit together, forming a tight joint over the hip. The slate roof courses and hip slates are all on the same plane. I recommend that you install metal flashing under mitered hip slates. Why chance a leak in such an expensive roof? See Figure 8-12 B.

- **Boston Hip** Install the same way as a saddle hip, except cut the hip slates into trapezoidal-shaped pieces. See Figure 8-12 C.

- **Fantail Hip** Install like a mitered hip, except you cut the bottom corner along the hip at an angle to form the fantail. See Figure 8-12 D.

Slate Roofing

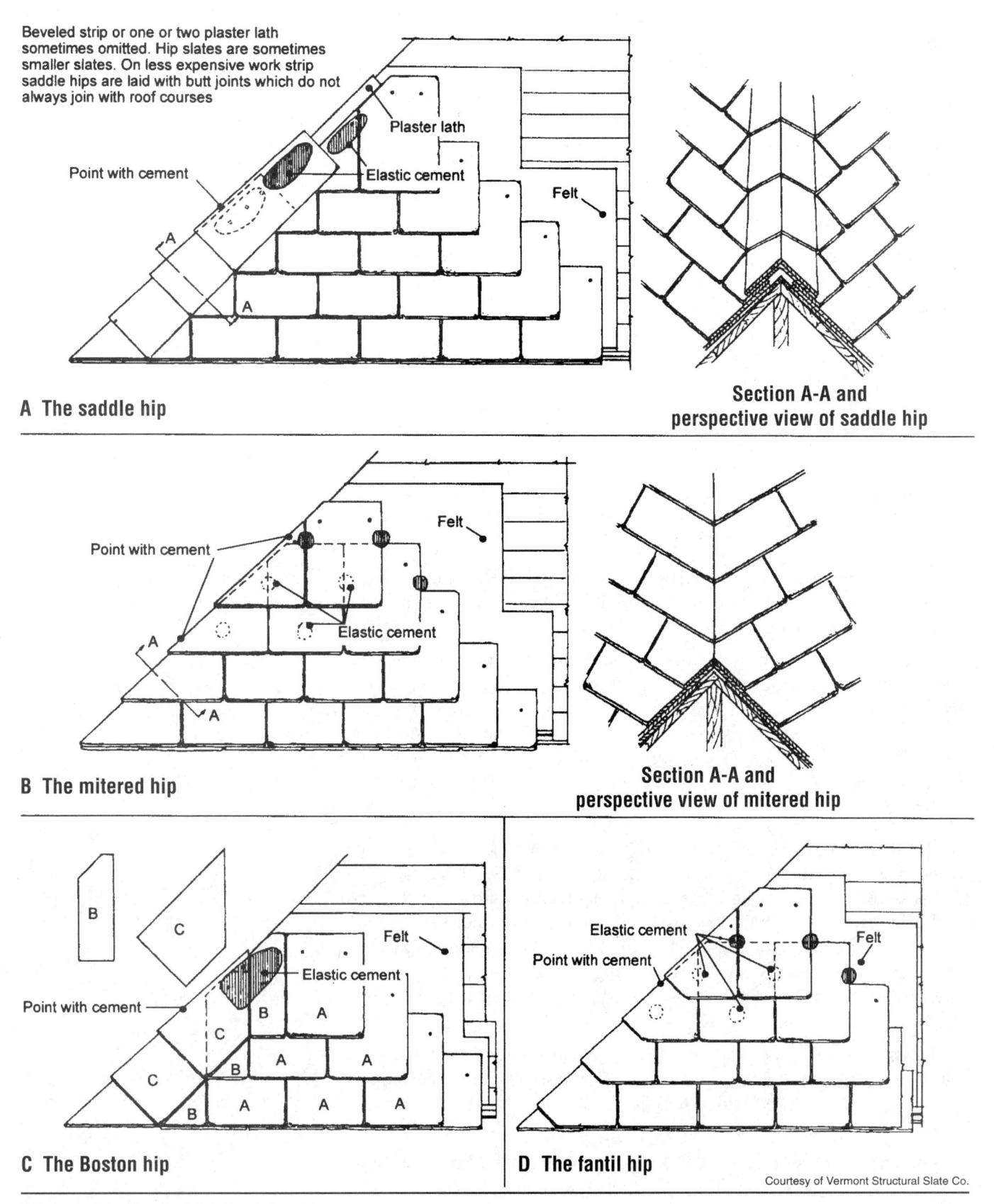

Beveled strip or one or two plaster lath sometimes omitted. Hip slates are sometimes smaller slates. On less expensive work strip saddle hips are laid with butt joints which do not always join with roof courses

Point with cement
Plaster lath
Elastic cement
Felt

A The saddle hip

Section A-A and perspective view of saddle hip

Point with cement
Felt
Elastic cement

B The mitered hip

Section A-A and perspective view of mitered hip

Felt
Elastic cement
Point with cement

C The Boston hip

Elastic cement
Point with cement
Felt

D The fantil hip

Courtesy of Vermont Structural Slate Co.

Figure 8-12 Standard details of slate hips

Valleys

There are four basic types of slate valleys. These are the open valley, the closed valley, the round valley and the canoe valley. See Figure 8-13.

■ **Open Valley** It's the easiest to install. Install slates at the upper end of the valley to within 2 inches of each side of the centerline of the valley. Widen this distance about $1/8$ inch per foot going down the valley. See Figure 8-13 A.

Flash the valley with metal no less than 15 inches wide, enough to extend a minimum of 4 inches under the slate. Six to 8 inches is better. Valleys on slate roofs should be fabricated with 16-ounce copper, or with zinc-coated metal no less than 0.0179 inch thick, with a center crimp and water guards. Use cleats spaced on 8- to 12-inch centers to secure the flashing to the sheathing. See Figure 8-13 B.

■ **Closed Valley** Trim and work the slates tightly into the valley centerline. Usually you flash a closed valley with sheets of copper fabricated with a center crimp but no water guards. See Figure 8-14. Nail the flashing to the sheathing each 18 inches along the edges of the flashing. Be sure that the nails you use to secure the slates don't go through the flashing.

Fabricate copper valley sheets no longer than 8 feet. If the valley is more than 8 feet long, use two sheets and lap the upper sheet over the lower by at least 4 inches.

You can also flash a closed valley by placing copper pieces under each course of slate entering the valley. See Figure 8-13 C. Make each piece long enough so it extends 2 inches above the top of the underlying slate, and overlaps the underlying piece of flashing by 3 inches. Install each piece flush with the butt of the overlying slate. Make each piece wide enough to extend at least 4 inches beyond each side of the centerline of the valley.

■ **Round Valley** This makes an attractive transition between intersecting slopes when installed on a graduated or textural roof. See Figure 8-13 D. To make the valley slates fit, they must be tapered toward the bottom. To make them lie flat, use valley slates 4 inches longer than the slates you use on the corresponding courses on the roof, and set the top edge of the slate in a mortar bed. This is called *shouldering* the slates. Usually you flash the valley with metal or mineral-surfaced roll roofing cut to the proper radius.

The type of foundation you use to support a round valley will depend on the shape of the valley. In a valley with a slight curve, cut tapered edges on 1 x 12s and nail them into the angle formed by the intersecting roofs.

For a round valley of substantial curvature, cut 3-inch blocks to fit the valley angle. Saw them to the proper radius, then nail them horizontally over the sheathing. Space the blocks vertically at about the same distance

Slate Roofing

Section A-A

A Open valley

Section B-B

C Closed valley

B Valley flashings

D Round valley

E Foundation for a round valley

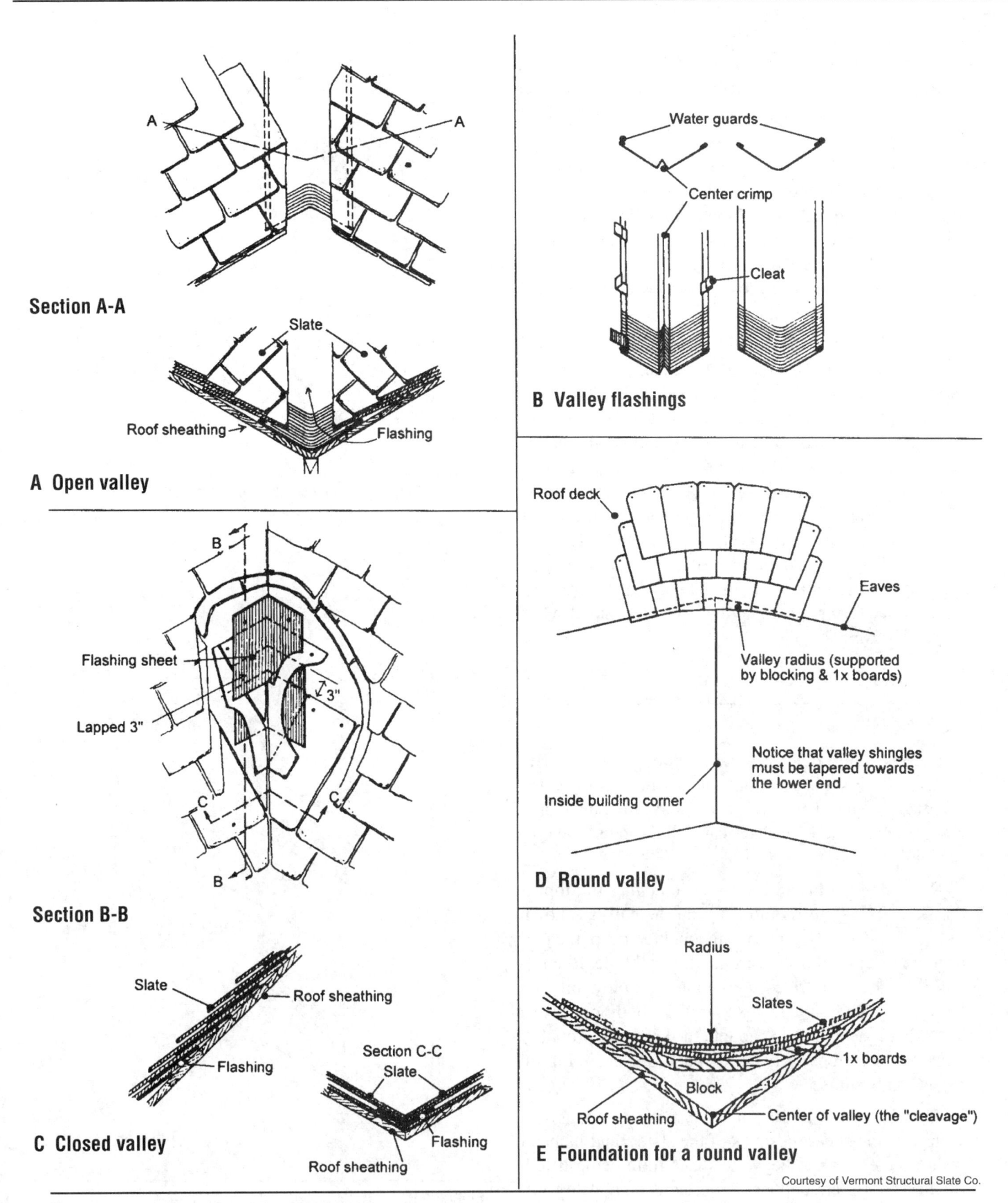

Courtesy of Vermont Structural Slate Co.

Figure 8-13 Standard details of slate valleys

as the exposure of the slates so the blocks form nailing strips under the slate. You'll have to use smaller blocks as the valley gets smaller where it approaches the ridge.

Another method used to form the foundation is a combination of the two methods previously discussed. In this case, you cut 3-inch blocks to fit the curve of the valley. Space them vertically on 20- to 30-inch centers up the valley. Then nail tapered 1 x 2s or 1 x 3s lengthwise up the valley, over the blocks. See Figure 8-13 E.

The radius of a round valley is greatest at the eaves and goes to practically zero at the ridge. The minimum radius allowed at the valley is 26 inches. If roof conditions don't permit this radius, you can use a variation of the round valley called a *canoe valley*. Install a canoe valley the same way as a round valley except that the radius at the eaves and ridge will be practically zero. Then you gradually increase the radius until it reaches its maximum halfway between the eaves and ridge. The result is that the rounded part of the valley is widest halfway up the valley, thinning toward the ridge and eaves. That forms a surface that's canoe-shaped when you see it from above.

Fasteners

Slate is a high-quality, durable roofing material. But it has to be installed correctly to perform as expected for many years. That means you have to punch nail holes properly, nail the slates carefully, and use the correct nails.

Drive nails with the heads *just touching* the top of the slate. The slate should hang on the nail, as shown in Figure 8-15 A. If you drive the nail too deep, the nail head will shatter the slate around the nail hole. Eventually the slate will "ride" up the nail and blow off in a heavy wind. Figure 8-15 B shows this condition. If you don't drive the nail deep enough, the head will probably crack the slate above when someone walks on it. Figure 8-15 C shows this.

The shaft of a roofing nail isn't strong enough to withstand the shearing stress of the slates, and its head isn't large enough to keep the slates from being lifted. It's best to use cut copper slating nails on a slate roof.

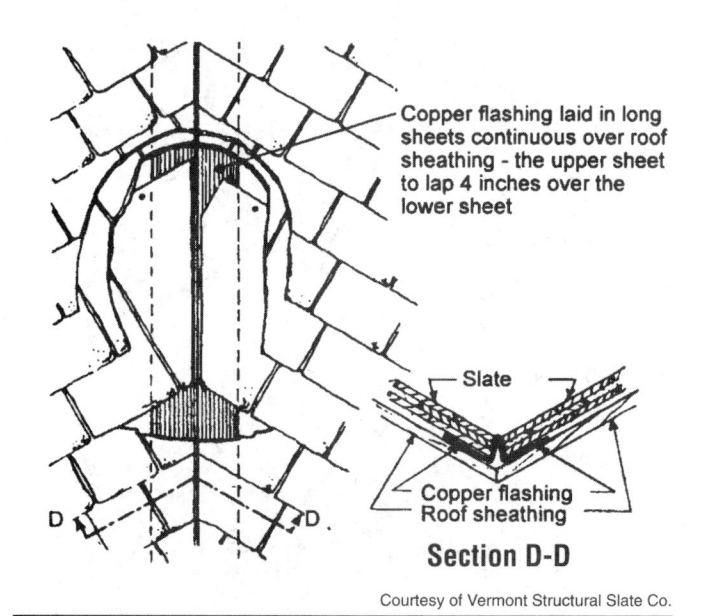

Figure 8-14 Flashing a closed valley using long sheets underneath slate

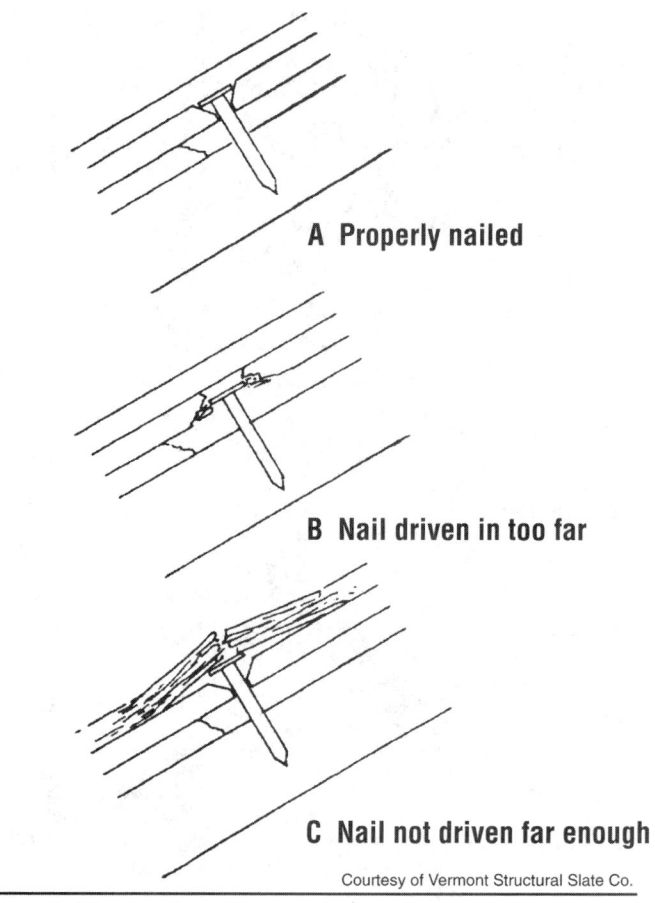

Figure 8-15 Proper nailing for slate roofing

But you could also use cut brass, cut yellow-metal or cut zinc slating nails to fasten slates to the sheathing. In dry climates, hot-dipped galvanized nails are acceptable.

Use 3d nails for commercial standard slates up to 18 inches long, and 4d nails for longer slates. Use 6d nails on hips and ridges. Thicker slates need longer and heavier-gauge nails. As a rule of thumb, use nails 1 inch longer than twice the thickness of the slate. If you're using ³/₄-inch slate, you'd use 2¹/₂-inch nails.

Install each slate with two nails, except for slates thicker than ³/₄ inch or longer than 20 inches, which need four nails.

Flashing

Install flashing on a slate roof as explained in Chapter 4. Figures 8-16 through 8-18 show basic flashing methods for slate roofs. Use top quality flashing that will last as long as the slate roof itself.

When you use copper flashings, install at least 16-ounce soft copper (16 ounces per square foot). Lock and solder the joints of all base flashings (Figure 8-17). You don't need to solder cap flashings. When soldering copper seams, be sure both sides of each sheet to be soldered are tinned 1¹/₂ inches along the edge. Tinning (precoating) means coating a metal with solder or another tin alloy before soldering or brazing to help the solder flow into the joint. Then use a rosin-flux solder, and make sure each joint is thoroughly soldered. Otherwise, the joints may come loose when the metal expands and contracts.

Electrolysis (electrical current passing between dissimilar metals) can dissolve copper. So insulate the joint between copper and iron or steel. Place lead strips between copper and iron or steel, or else heavily tin the iron.

When you use tin flashing, install at least 28-gauge (12-ounce) tin, and lock and solder all base flashing joints. Always paint the underside of tin flashing with metal or bituminous paint.

When you use lead flashing, use 3-pound hard lead for all flashing except for cap flashing, which can be 2¹/₂-pound. Install lead so that it can expand and contract. Never drive nails through lead. Instead, use cleats of 16-ounce soft copper on 3-pound hard lead. Fasten the cleats to the sheathing with hard copper wire flat-head nails at least ³/₄ inch long, or with brass screws equipped with lead shields. Join lead sheets with locked seams beaten with wooden hammers. Don't solder the seams.

Roofing Construction & Estimating

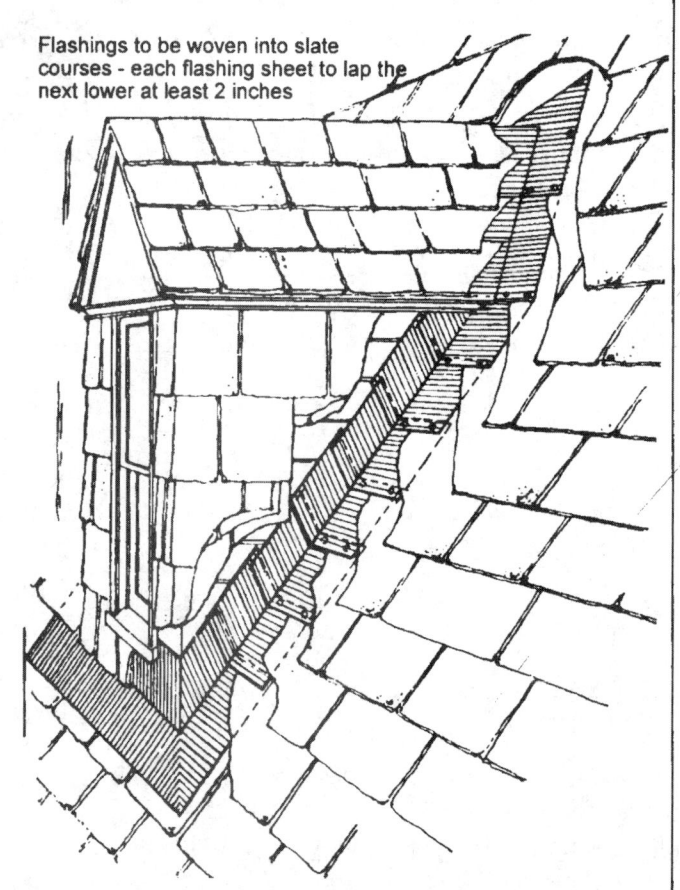

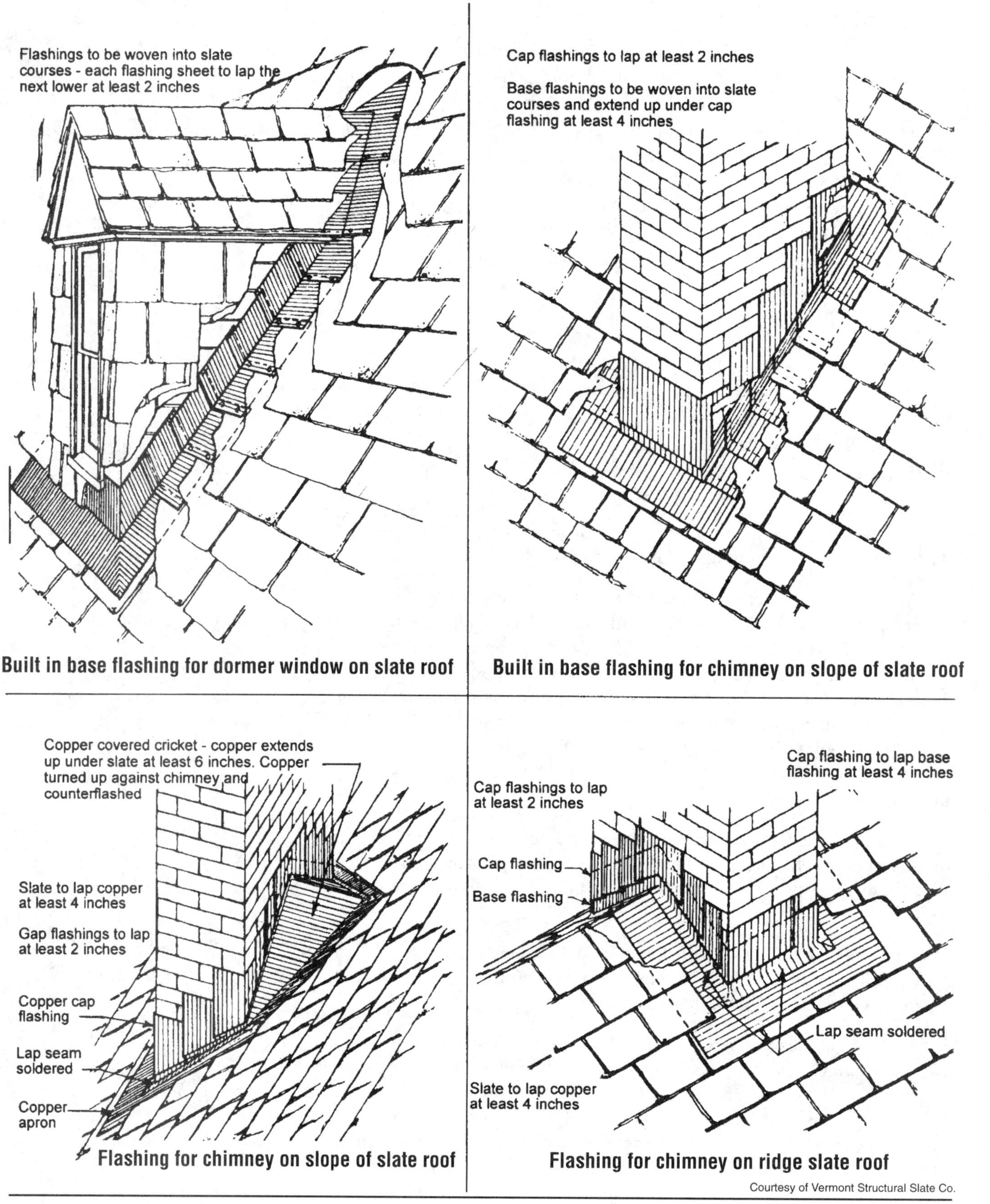

Built in base flashing for dormer window on slate roof
Flashings to be woven into slate courses - each flashing sheet to lap the next lower at least 2 inches

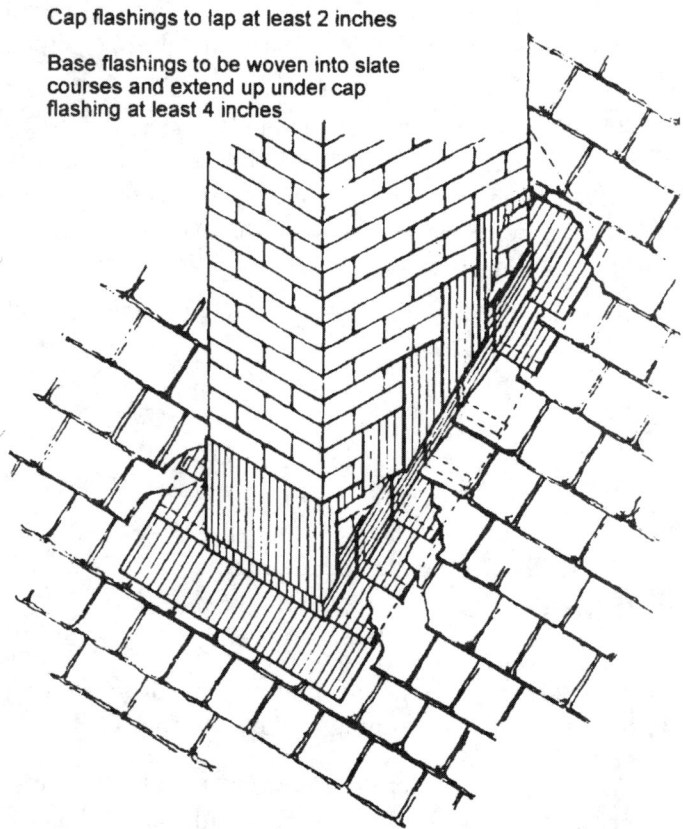

Built in base flashing for chimney on slope of slate roof
Cap flashings to lap at least 2 inches
Base flashings to be woven into slate courses and extend up under cap flashing at least 4 inches

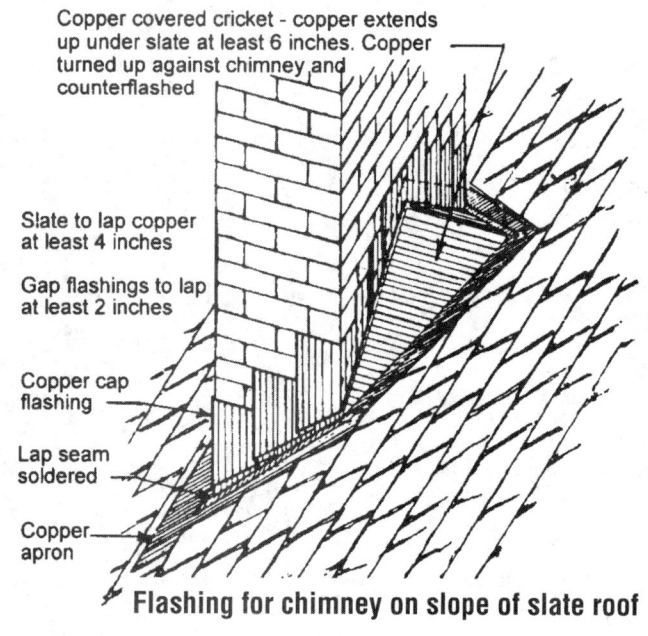

Flashing for chimney on slope of slate roof
Copper covered cricket - copper extends up under slate at least 6 inches. Copper turned up against chimney and counterflashed
Slate to lap copper at least 4 inches
Gap flashings to lap at least 2 inches
Copper cap flashing
Lap seam soldered
Copper apron

Flashing for chimney on ridge slate roof
Cap flashing to lap base flashing at least 4 inches
Cap flashings to lap at least 2 inches
Cap flashing
Base flashing
Lap seam soldered
Slate to lap copper at least 4 inches

Courtesy of Vermont Structural Slate Co.

Figure 8-16 Standard flashing details

Slate Roofing

Section A-A
- Cleat
- Soldered lap seam
- Lock seam secured to roof with cleats
- Roofing hooked over edge strip
- Roof sheathing
- Brass edge strip

Flashing for a recessed dormer window

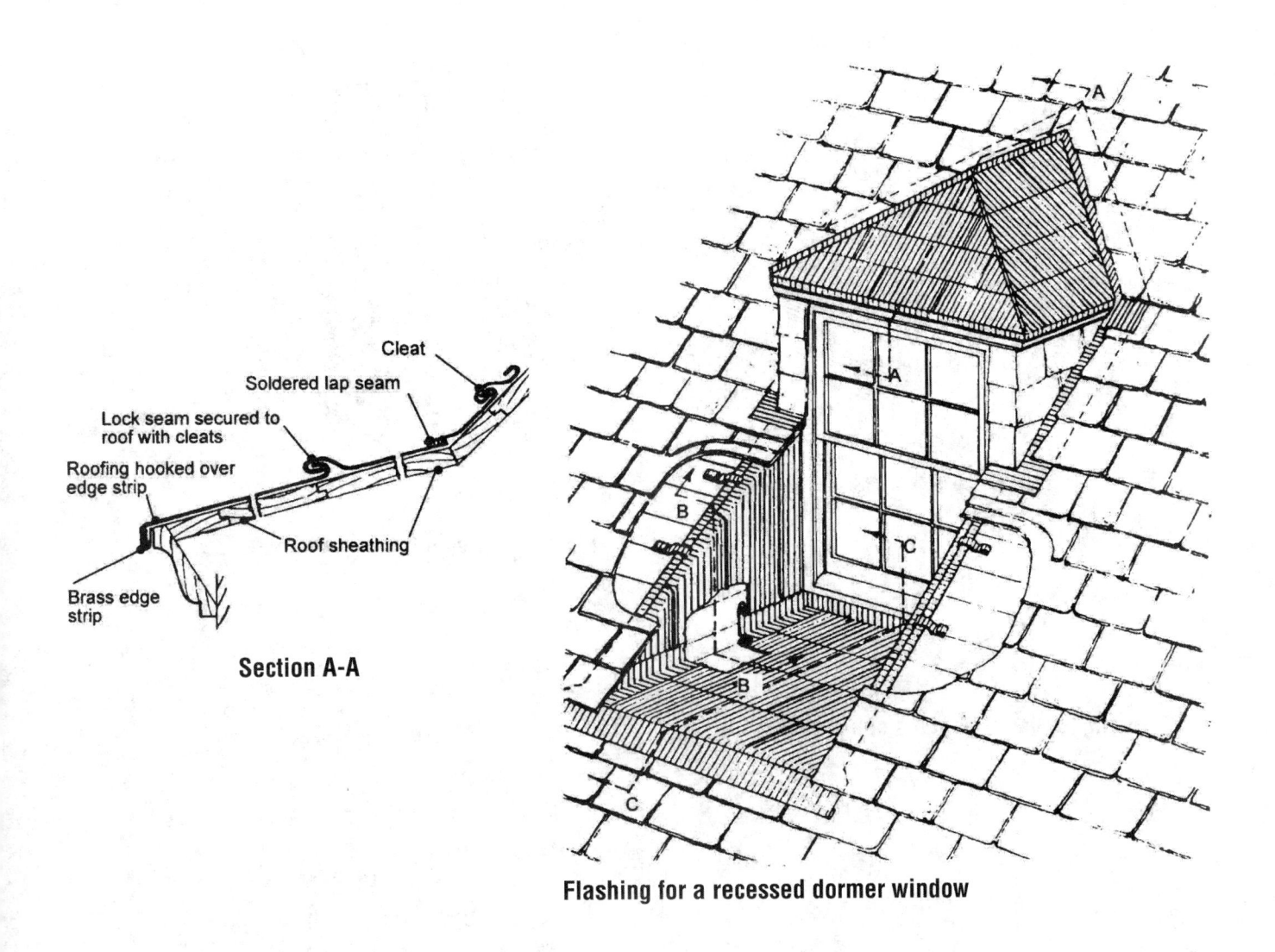

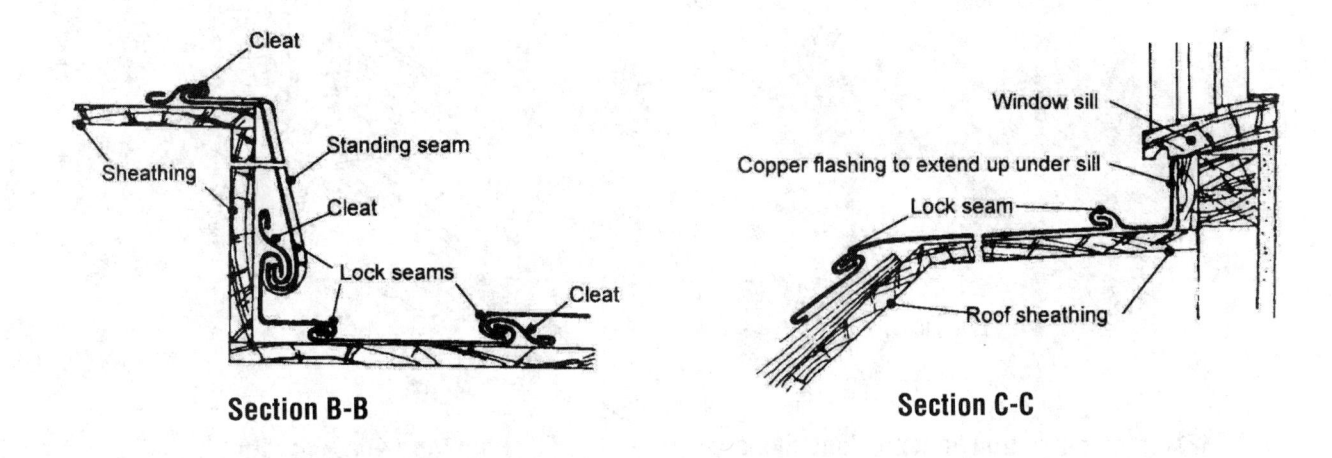

Section B-B
- Cleat
- Sheathing
- Standing seam
- Cleat
- Lock seams
- Cleat

Section C-C
- Window sill
- Copper flashing to extend up under sill
- Lock seam
- Roof sheathing

Courtesy of Vermont Structural Slate Co.

Figure 8-17 Flashing a dormer

Roofing Construction & Estimating

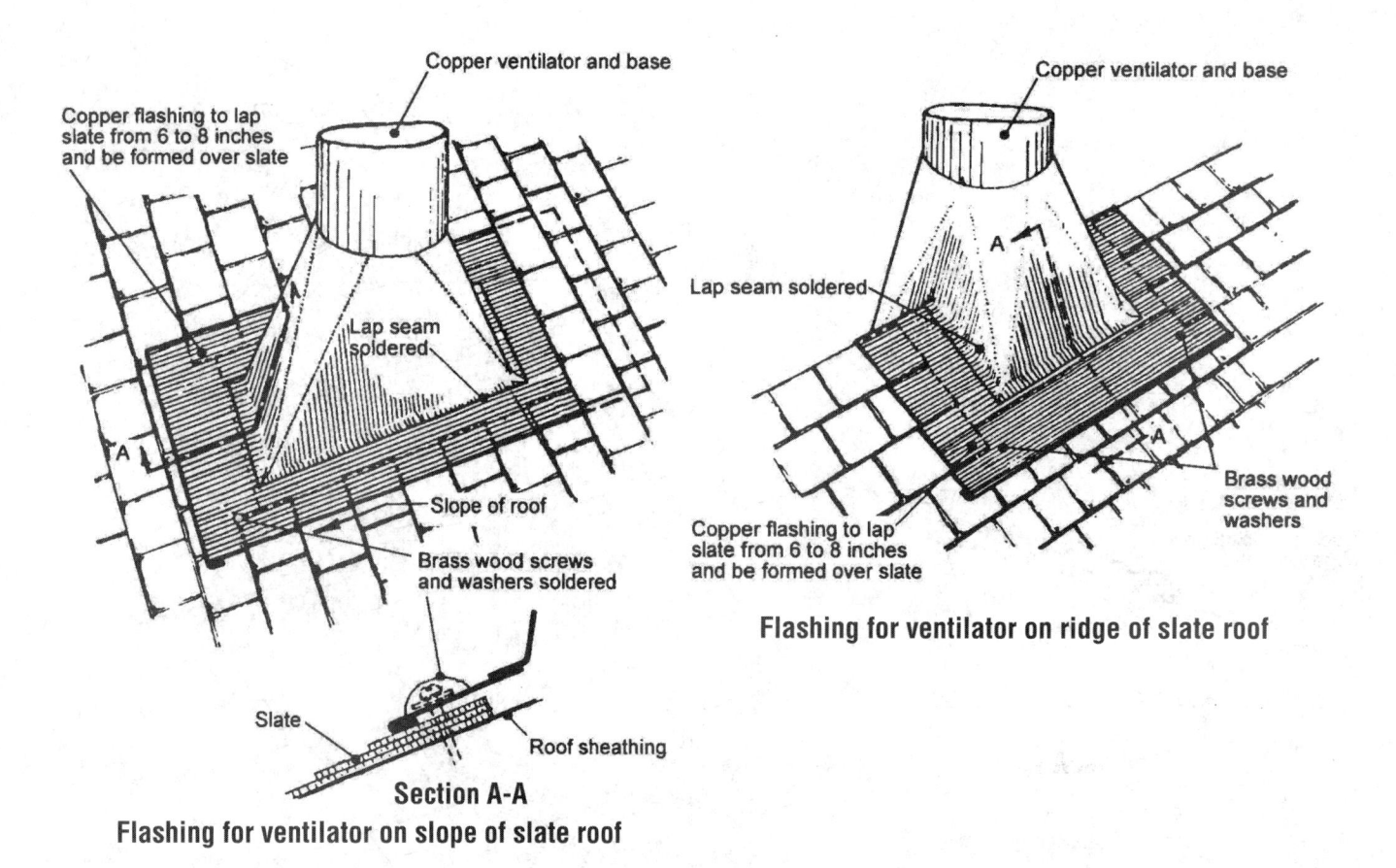

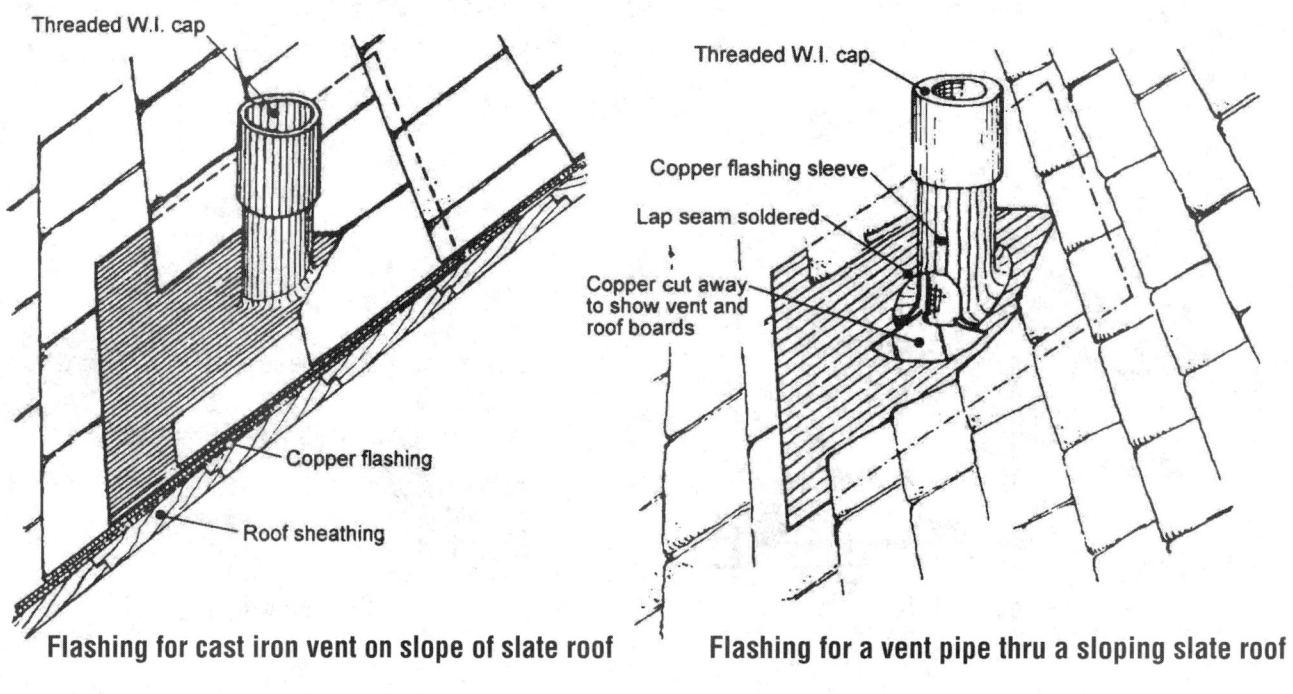

Figure 8-18 Vent flashings

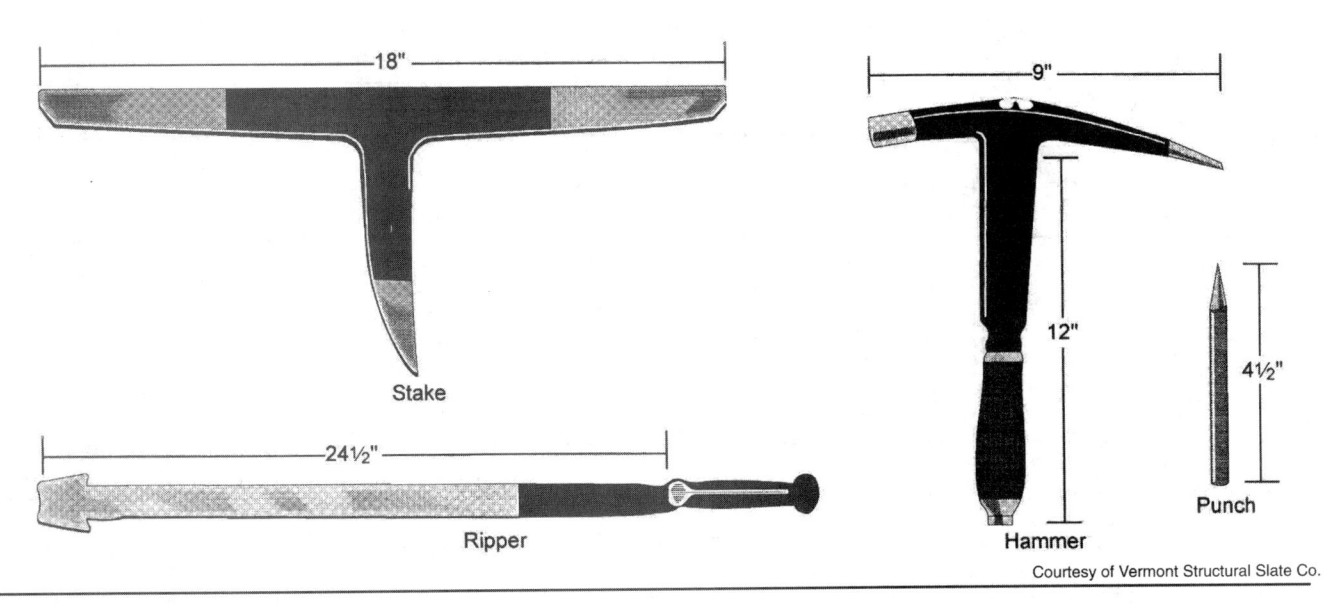

Figure 8-19 Slater's tools

When you use zinc flashing, install at least 11-zinc-gauge (0.024 inch thick) rolled zinc. Don't drive nails through zinc. Use zinc clips or cleats instead (see Figures 8-13 B and 8-17). Use an acid-flux solder to solder all joints.

Slater's Tools

Slate roofers use specialized tools. These include the stake, ripper, hammer and punch shown in Figure 8-19. The punch is $4^{1}/_{2}$ inches long and is used for punching countersunk nail holes in the slates. The hammer also has a sharp point for punching holes in slate. It has a claw at the center of the head for removing nails. One edge of the shank is sharp, for cutting slate. The ripper is about 24 inches long and is used to remove broken slate. The hook at the end of the blade is used to cut and remove slating nails. The stake is 18 inches long and is used as a "work bench" for cutting and punching slates. The short tapered arm is driven into a plank to hold the tool in place. Place slate over the long edge and cut or punch as required.

Estimating Slate Quantities

The first step is to take off slate quantities by the square foot. Then you convert quantities into squares of slates required. Slate prices are based on the number needed to cover a square of roof surface when laid with a 3-inch head lap. Figure 8-20 shows the quantity needed to cover a square of roof surface at each of three different head laps.

Size of slate	Number of slates required per square		
L x W	2" head lap	3" head lap	4" head lap
26 x 14	86	89	94
24 x 14	94	98	103
24 x 12	109	114	120
22 x 12	120	126	133
22 x 11	131	138	146
20 x 14	114	121	129
20 x 12	133	141	150
20 x 11	146	154	164
20 x 10	160	170	180
18 x 12	150	160	171
18 x 11	164	175	187
18 x 10	180	192	206
18 x 9	200	213	229
16 x 12	171	184	200
16 x 10	206	222	240
16 x 9	229	246	267
16 x 8	257	277	300
14 x 12	200	218	240
14 x 10	240	261	288
14 x 9	267	291	320
14 x 8	300	327	360
12 x 10	288	320	360
12 x 9	320	355	400
12 x 8	360	400	450
12 x 7	411	457	514
12 x 6	480	533	600
10 x 8	450	515	600
10 x 7	514	588	686
10 x 6	600	686	800

Figure 8-20 Number of slates required at various head laps

Slate Roofing

Here's how to find the quantity required if you're not using a 3-inch head lap:

Squares Ordered per Square Covered equal:

$$\frac{\text{Slates per 100 SF at your head lap}}{\text{Slates per 100 SF at 3-inch head lap}} \times \text{Roof area (in squares)}$$

Equation 8-1

Here's an example. Suppose you have a roof area of 500 square feet and you're going to use 12 x 10-inch slates with a head lap of 2 inches, 3 inches, or 4 inches:

First, divide the total roof area by 100 to find the area in squares:

Squares of roof area = Roof Area ÷ 100
= 500 ÷ 100
= 5 squares

Using the values from Figure 8-20 for 12 x 10-inch slates with a 2-inch head lap, you need:

$$\frac{288}{320} \times 5 = 4.5 \text{ squares}$$

So, if you're using a 2-inch head lap, you have to order only 4.5 squares of slate to cover the area.

With a 3-inch head lap, you need:

$$\frac{320}{320} \times 5 = 5 \text{ squares}$$

With a 4-inch head lap, you need:

$$\frac{360}{320} \times 5 = 5.63 \text{ squares}$$

Under-eaves Slates

Slate is installed with a double starter course at the eaves. The formula to figure how much under-eaves starter-course material you need is from Chapter 4:

Starter Course (SF) = Eaves (LF) x Exposure Area (SF/LF)

Equation 4-10

For example, material needed for 80 LF of eaves with a 6-inch exposure is:

Starter-course material (in square feet) = 80 LF of eaves x 0.5 SF/LF
= 40 SF

A more practical formula, since you usually order under-eaves slates by the piece, is:

$$\text{Number of Eaves Shingles} = \frac{\text{Eaves Length (ft.)}}{\text{Slate Width (ft.)}}$$

Equation 4-12

For example, the number of 8-inch-wide under-eaves slates needed for 80 linear feet of eaves (regardless of the exposure) is:

$$\text{Number of under-eaves slates} = \frac{80 \text{ LF}}{0.67 \text{ LF}} = 120 \text{ slates}$$

Cutting Waste at Hips and Valleys

You can account for most of the cutting waste and breakage on a slate roof by adding 1 square foot of area to be covered for each linear foot of hip or valley. Waste beyond this will depend on the complexity of the roof and the skill of the roofing crew.

Hip and Ridge Slates

All slate roofs require *ridge* slates. However, the Boston and saddle hips are the only hips that require hip slates. You don't necessarily install hip and ridge slates at the same exposure as field slates, and they aren't necessarily the same size as the field slates. Figure the number of hip and ridge slates you need using the formula adapted from Chapter 4:

$$\text{Courses} = \frac{\text{Dimension of Structure}}{\text{Exposure}}$$

Equation 4-3

Since each side requires a slate, we need:

$$\text{Hip and ridge slates} = 2 \times \frac{\text{LF of ridge and hip}}{\text{Exposure}}$$

Equation 8-2

Here's an example using the roof shown in Figure 8-21. Assume a roof slope of 5 in 12 and a 2-inch overhang at the eaves. Suppose you use 16 x 8-inch field slates and 16 x 9-inch saddle hip and ridge slates. Also assume an exposure of 10 inches for the saddle hip and ridge slates.

First, the total length of the eaves is 2 x (40' + 20'), or 120 linear feet. Since the roof slope is 5 in 12, you'll install the slates with a 4-inch head lap (according to Figure 8-7). Using a 4-inch head lap, the exposure is 6 inches (material length less the head lap, divided by 2).

Since the under-eaves slates must be only 3 inches longer than the exposure, you can use 12 x 9-inch slates laid horizontally at the eaves. The number of 12 x 9-inch under-eaves slates required is:

$$\text{Number of under-eaves slates} = \frac{120 \text{ feet}}{1 \text{ foot}} = 120 \text{ slates}$$

Using the roof-slope factor from Appendix A, the net roof area, including the 2-inch overhang at the eaves, is:

40'4" x 20'4" x 1.083 = 889 square feet

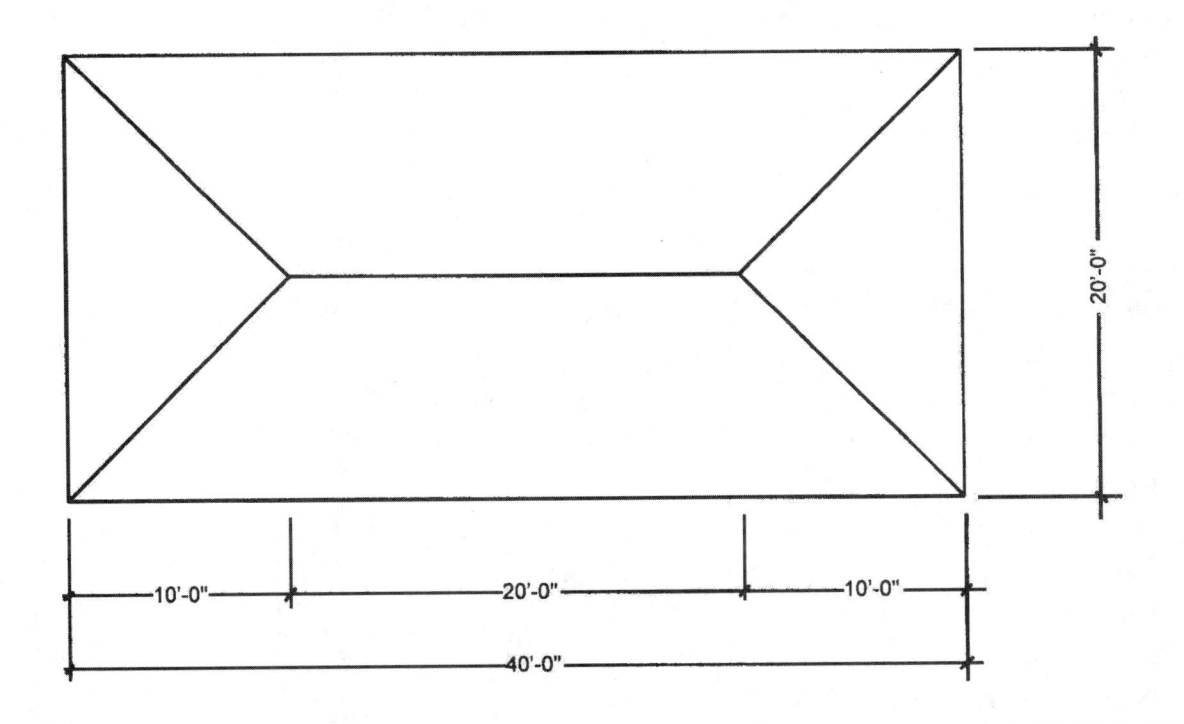

Figure 8-21 Hip roof example

So you need 8.9 squares of field slates. (Use this quantity to determine labor costs.)

The length of any hip is 14.74 linear feet from (Column 3, Appendix A). The total length of all hips is 4 times that, or 59 linear feet. Therefore, include 59 square feet of slate in the estimate for cutting waste based on 1 square foot per linear foot of hip.

Now, the total gross area covered by the field slates, including waste, is:

889 + 59 = 948 SF

Divide that by 100, for 9.5 squares. Now, figure the number of 16 x 8-inch slates for the field (using a 4-inch head lap)

9.5 squares x 300 units per square (from Figure 8-20) = 2,850 slates

But remember, you have to order slate based on coverage using a 3-inch head lap. So to convert, use the formula in Equation 10-1, in Chapter 10:

$$\text{Squares (16 x 8 slates)} = \frac{300}{277} \times 9.5 \text{ squares} = 10.3 \text{ squares}$$

That's the number of *squares* to order to get the number of *slates* you require:

10.3 squares x 277 slates per square = 2,853 slates

The total length of hips and ridge is 79 linear feet (59' + 20').

The total number of 16 x 9-inch hip and ridge slates (using Equation 8-2) required is:

$$2 \times \frac{79 \text{ feet}}{0.83 \text{ feet}} = 190 \text{ slates}$$

Notice we converted the 10-inch exposure to feet for this calculation.

Estimating Slate Roofing Costs

Estimating the material cost for a slate roof is tricky business. Always get a current price quote and information about availability before you bid a job. Your dealer will need the following information:

1) The type, color and texture required
2) The thickness, such as commercial, standard, 1/4 inch, 3/8 inch, etc.
3) The roof classification, such as standard, textural, graduated or flat
4) Number of nail holes required per slate
5) Types of valleys and flashings required
6) Types of hips and ridges required
7) Required delivery date
8) Job location
9) Total quantity of field slate
10) Total linear feet of hips and ridges

Keep in mind that freight from the quarry is a fixed charge based on either carload or less-than-carload lots. The latter carries a freight charge about double that of carload quantities. With this information in hand, your dealer can quote you a realistic price for your materials.

For example, let's assume that you're covering the roof we discussed in the last section. There's a net roof area of 8.9 squares. But to actually cover the roof, you need 10.3 squares of 16 x 8-inch field slates, 120 each 12 x 9-inch under-eaves slates and 190 each 16 x 9-inch hip and ridge slates. Let's assume your dealer quotes a material price of $681.55 per square, including freight and sales tax, for all the slate you'll need. So the material will cost:

10.3 squares x $681.55 per square = $7,019.97

According to Craftsman's *National Construction Estimator*, a crew of one roofer and one laborer can install a square of slate in 11.3 manhours. If the roofer costs $41.66 per hour and the laborer costs $30.31, including labor burden, then the average manhour cost is $35.99 per hour (41.66 + 30.31 ÷ 2 = 35.99). So the labor cost for installing the roof is:

8.9 squares x $35.99 per hour x 11.3 hours per square = $3,619.51

The labor cost is $3,619.51. Assume a 30 percent markup:

(7,109.97 + 3,619.51) x 1.3 = 13,948.32

You'll bid the project at $13,948.32.

9 Metal Roofing and Siding

▶ Any metal sheet heavier than 30 gauge is called *sheet metal*. Many types of sheet metal are used in roofing, the most common being galvanized steel, or *Galvalume®*, a product of Bethlehem Steel, which is a carbon steel base sheet hot-dipped with aluminum and zinc and a tiny bit of silicone to give it more flexibility during manufacturing and storing. Other metals used are stainless steel, copper, zinc, aluminum or a combination of metal alloys. There are still roofs covered with *Monel®* metal, terne metal and lead, but you probably won't be installing any. More on this later, as, depending on what part of the country you're in, you could be called to work on one.

A metal roof can last as long as 80 years if correctly installed and properly maintained. Although some types can rust, they can be restored with a special paint containing a plastic fiber that covers the metal and prevents further deterioration. However, there are environmental issues with some of these coatings, so make sure the type you use isn't going to get you in trouble with the EPA. Metal panels are rigid and highly resistant to wind uplift. But if even a corner works loose, a strong wind can blow the entire panel off.

Metal panels usually have a low fire rating because they are such good conductors of heat. The only way to mitigate this is to surround the panels with fireproofing material. And unless there's insulation between the roof and the building interior, metal panel roofs are very noisy during rain or hailstorms.

Figure 9-1 Modern ribbed metal panel roof

Sheet metal roofing and siding are generally made from cold-rolled metals. Sheet metals are either cold-rolled or hot-rolled. Several properties of the sheet metal depend upon the metal ingot's temperature during rolling. Cold-rolled sheet metal has a lower carbon content. While it's more malleable than hot-rolled sheet metal, cold-rolled sheet metal has the disadvantage of being weaker.

Some materials are factory-coated with other metals or paint to improve the base material's corrosion resistance and appearance. Metal coatings are usually applied with the hot-dip process.

Many building codes require copper roofs to be a minimum of 16-ounce copper, galvanized steel a minimum of 0.0146 inch thick, and aluminum 0.024 inch. Hard lead should be no less than 2 pounds per square foot and soft lead 3 pounds. Use fasteners made either of the same metal as the panels, or stainless steel, which can be used on all types of metal panels.

If you're going to paint a new galvanized metal roof, clean it with muriatic acid. If the roof has already weathered for at least a year, you need clean it only with water before you paint. Be sure the paint is EPA-approved.

Modern Metal Panel Systems

There's an endless variety of modern decorative sheet metal systems available for roofing and siding. Figure 9-1 is an example of a ribbed panel metal roof. There are also systems for mansard roofs, fascia and soffit

Metal Roofing and Siding

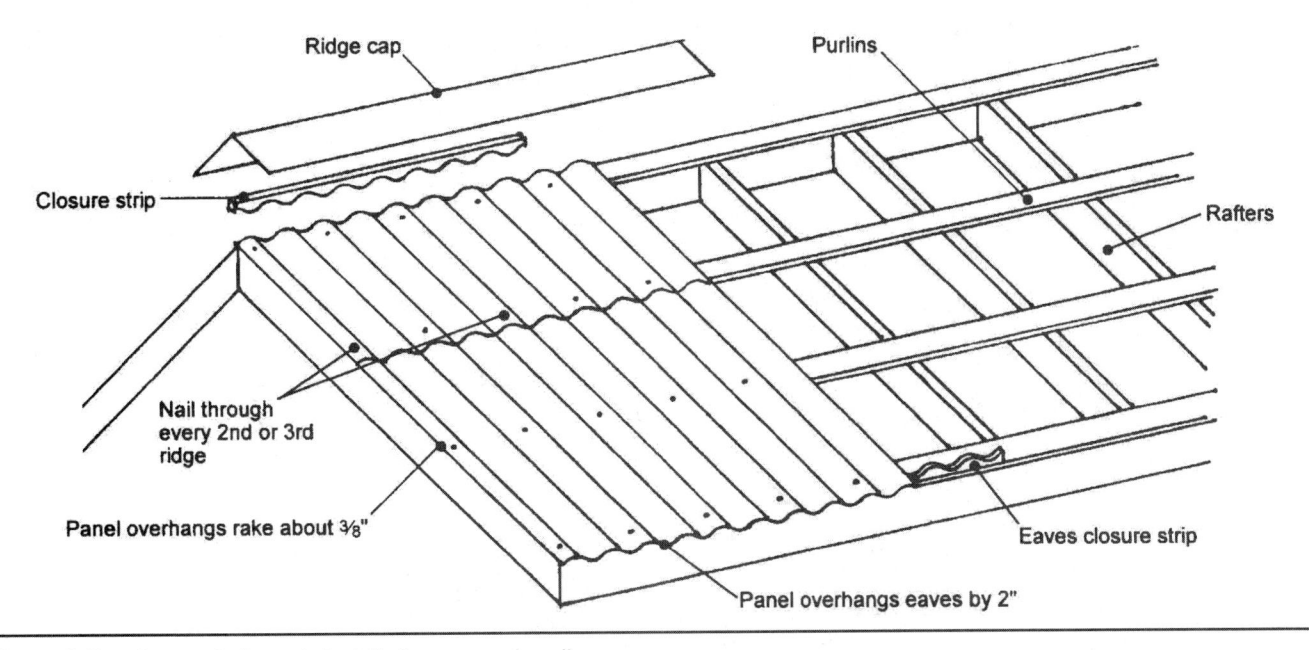

Figure 9-2 Corrugated panels installed over wood purlins

systems. Some systems are designed to be applied over an existing roof. Others are made of a sandwich of metal faces with a foam insulation core. These systems are made of different metals, and they have different surface shapes and finishes. Some systems can be installed on roof slopes as low as 1/4 in 12.

Decking Requirements for Metal Roofing Panels

Any decking required depends on the strength of the panel. Architectural panels usually require solid support, while structural panels can span from support to support. Corrugated steel can be placed over wood purlins, usually 1 x 4s or 2 x 4s spaced at 16- to 48-inch centers, depending on roof slope and any anticipated snow load, as in Figure 9-2. It can also be placed over steel purlins, normally spaced on 4-foot centers and not less than 22 gauge.

Most other roofing panels, such as copper or zinc, have inadequate structural integrity and require solid decking, such as concrete, cement wood fiber, plywood or OSB. Underlayment should be one or two layers of 30-pound felt or appropriate synthetic or hybrid. If you come across a fluted metal deck; that'll need steel panels. See Figure 9-3.

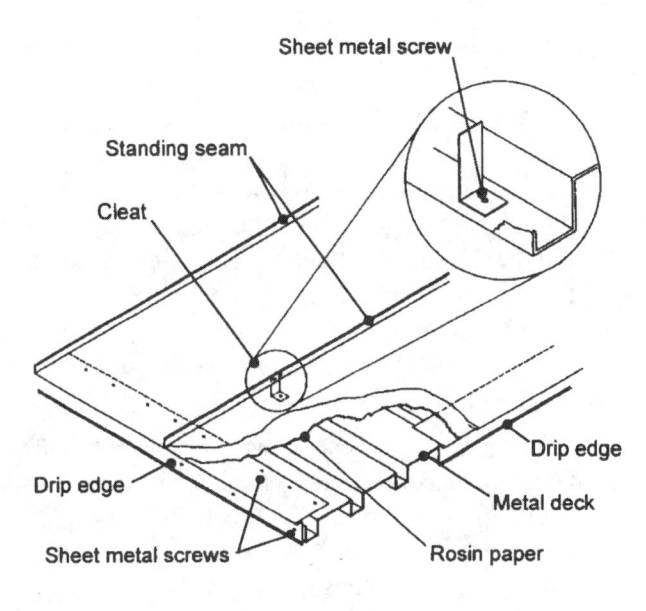

Courtesy of Follansbee

Figure 9-3 Installation over a fluted metal deck

247

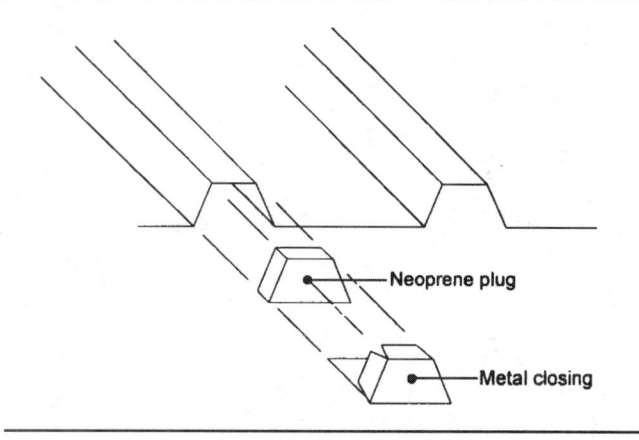

Roofing Construction & Estimating

Figure 9-4 Eaves sealed with closure caps

Figure 9-5 Rakes sealed with an end cover

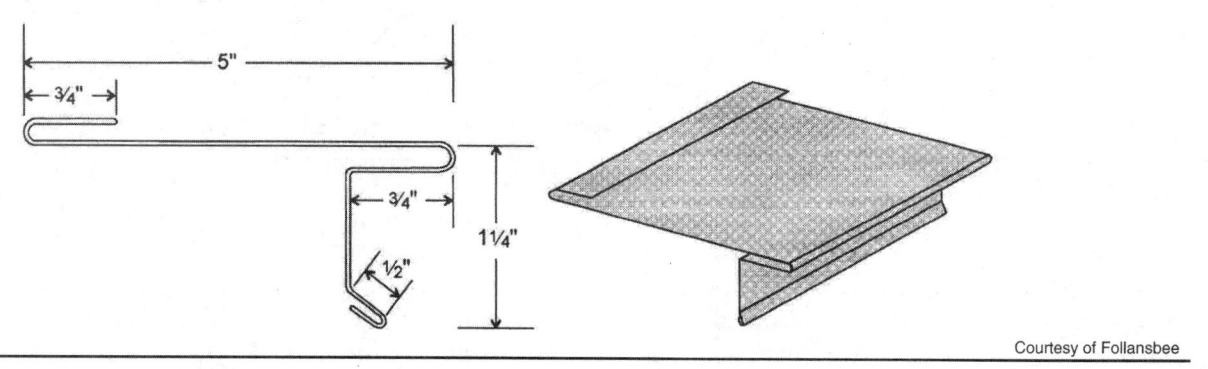

Figure 9-6 Preformed drip edge

Installing Metal Roofing Panels

To apply corrugated metal roofing panels, install a closure strip like the one in Figure 9-2 along the eaves of the roof. Then install the first row of panels along the eaves with the ribs parallel to the slope. Allow a 1-inch overhang at the eaves. At the rakes the size of the overhang doesn't matter. Screw the panels through every other panel ridge into the underlying purlins. I always put the fasteners through the ridges of corrugated panels rather than in the valleys where water flows, but some manufacturers now advise otherwise. Times, materials and methods change, usually for the better. Follow the manufacturer's instructions.

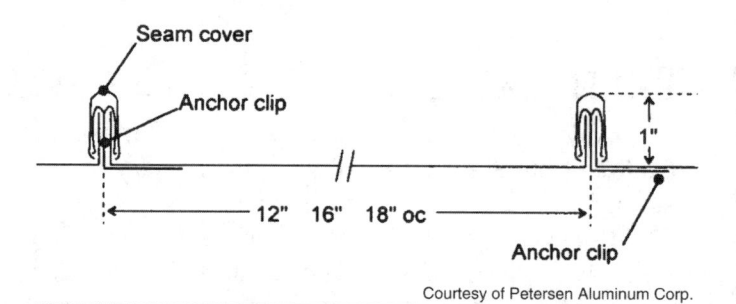

Figure 9-7 Panel ribs with snap-on standing seams

Use self-tapping cap screws or stove bolts to fasten panels to a metal frame. Overlap the panel sides by two corrugations. Install the second row of panels overlapping the ends over the first row by at least 6 inches (on roof slopes 4 in 12 and steeper), or 8 inches (on roof slopes of 3 in 12 to 4 in 12). Seal the end laps with caulking compound on roof slopes less than 4 in 12.

Other metal panel roofs installed without a closure strip may be sealed along the eaves with neoprene plugs covered with metal caps, as in Figure 9-4, and along the rakes with an end cover as in Figure 9-5.

Metal Roofing and Siding

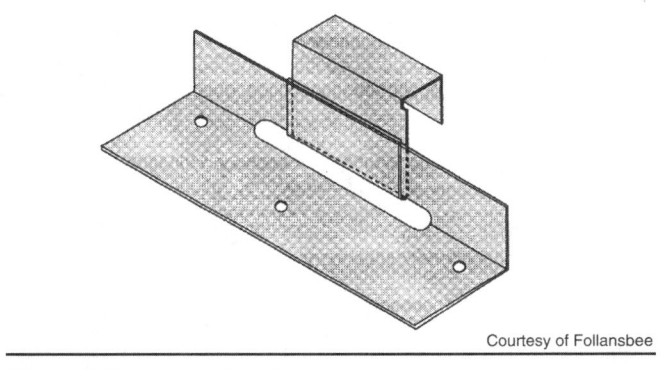

Figure 9-8 Expansion cleat

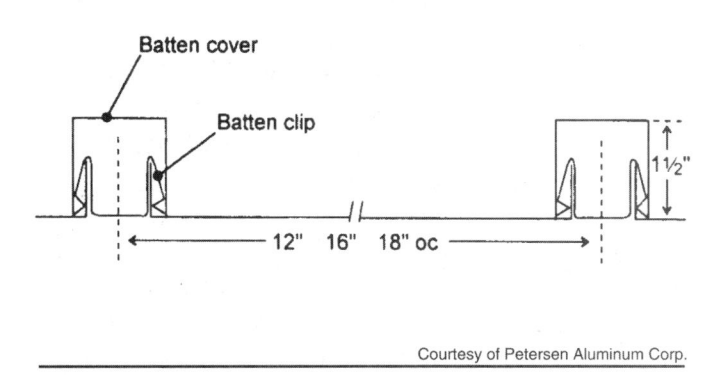

Figure 9-9 Panel ribs with snap-on batten seams

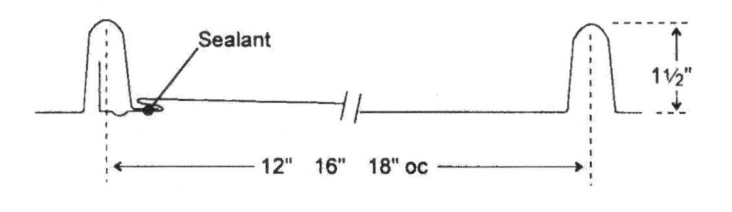

Figure 9-10 Integral standing seams covered by the adjacent panel

Figure 9-11 Integral battens

Some metal roofing systems are anchored over a preformed drip edge along the eaves and rakes. This drip edge is 5 inches wide with a $3/4$-inch hemmed back edge to serve as a water dam. That's shown in Figure 9-6. The drip edge is nailed to wood sheathing or screwed to a metal deck, as in Figure 9-3.

There are many ways to anchor modern metal roofing panels to a deck. Some systems use anchor clips (sometimes called *panel clips*) attached to the deck, or to purlins. The panel rib is attached to the anchor clip and then covered with a snap-on standing seam cover. That's shown in Figure 9-7. Some anchor clips are slotted so the panel can move as it expands and contracts. These clips, called *expansion clips* or *cleats*, are shown in Figure 9-8. Panel laps must be located over purlins since that's where the clips are located.

Some systems use a batten clip nailed or screwed through the rib and into the deck or purlin. Later, you conceal the clip with a snap-on batten seam cover as shown in Figure 9-9. Other systems are nailed or screwed to the deck or purlins and later covered by an adjacent panel (Figures 9-10 and 9-11) which has an integral standing or batten seam. Integral seams are often made with a factory-applied sealant within the seam, as shown in both Figures 9-10 and 9-11. Still other systems use a standing seam that's sealed by crimping the connection with a seaming machine like the one in Figure 9-12.

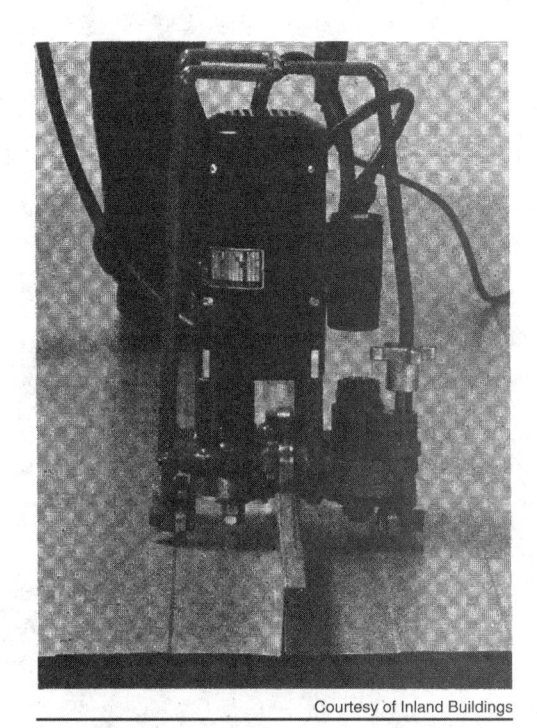

Figure 9-12 Seaming machine used to crimp standing seams

249

Fitting Hips and Valleys

Modern metal panel systems are usually precut at the factory, numbered and delivered with a layout plan showing the locations of the panels. If on-site fabricating is required, flat panels can be cut by a break machine. Ribbed panels can be cut with a portable power saw equipped with a Carborundum blade. Don't worry about making perfect cuts. They'll be concealed by a ridge cap over the ridge and hips as shown in Figures 9-2, 9-13 or 9-14. Caps may be nailed or screwed into the ridge board and hip rafters, or bolted into steel. Then install the lower edge of each side of the ridge cap over neoprene closure strips, as in Figure 9-2, or metal "Z" closure strips as in Figure 9-13. Lap the ridge cap end joints at least 12 inches.

Ridge caps over standing-seam roofs are often installed over wood battens spaced to provide ridge venting. Look at Figure 9-14. For exact spacing requirements, check your local building code. For discussion of alternative venting systems for many types of roofs, look ahead to Chapter 14.

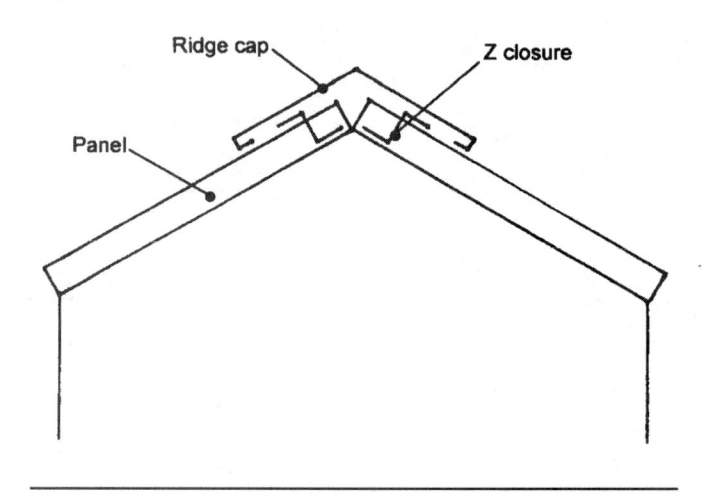

Figure 9-13 Ridge cap with "Z" closure

Valley Flashings

You can flash valleys of metal roofs with preformed galvanized iron flashing, or use the same type of metal as on the main roof. Most building codes require a minimum thickness of 0.019 inch for flashing, Quality manufactured valley flashings come standard with splash diverters and some form of water guard. Otherwise, prefabricate the valley flashing with 1-inch water guards along the edges. For even more protection use a valley similar to the one shown in Figure 3-42 back in Chapter 3. This type of valley has a splash diverter running down its center and water guards on both

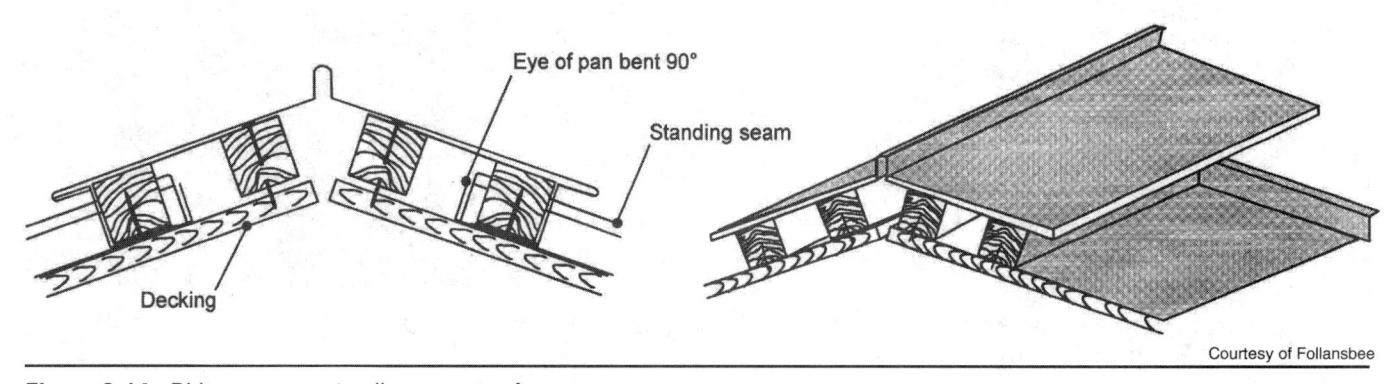

Figure 9-14 Ridge cap over standing-seam roof

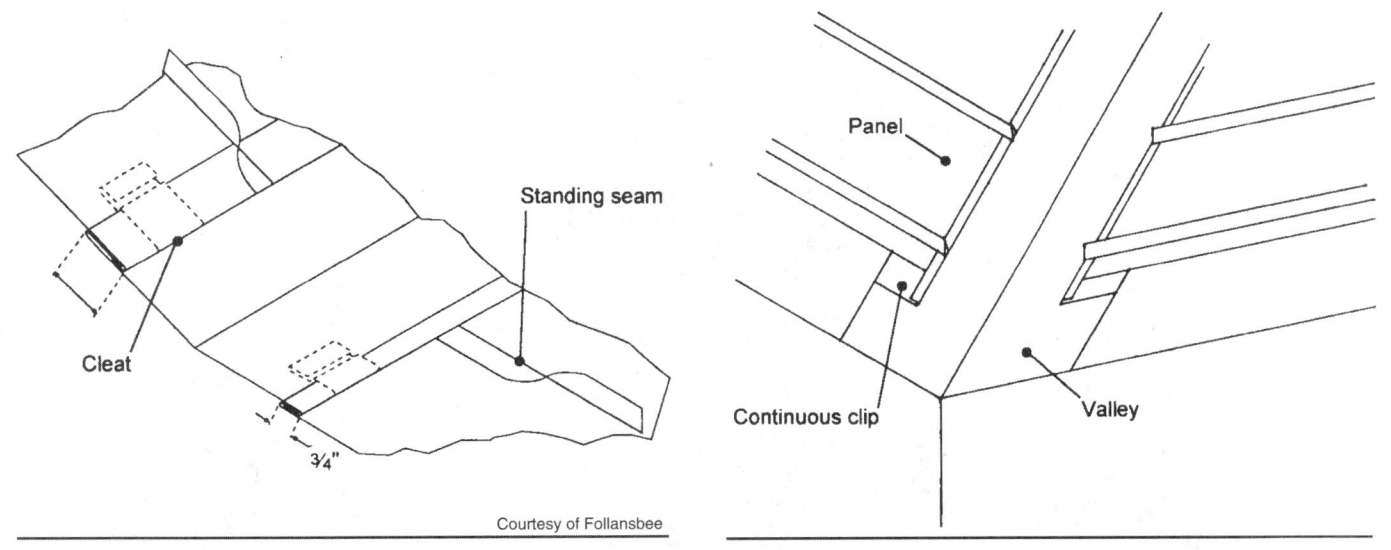

Figure 9-15 Valley flashing installed with cleats

Figure 9-16 Valley flashing installed with a continuous clip

sides as well. The combination provides metal roofs with double protection by preventing water from splashing under the panels adjacent to the valley. Secure a metal valley to the sheathing or the purlins with metal cleats as seen in Figure 9-15, or with a continuous clip. See Figure 9-16. Don't install fasteners into the valley flashing.

If you need more than one piece of flashing over the length of the valley, lap the upper piece at least 6 inches over the lower and seal the lap with cement designed for metal use, or with screws.

Vent Flashing

On flat-panel metal roofs, you can install rubber vent pipe flashing. On a corrugated roof, I used to just pack the opening around the pipe with roofing cement, but that's not acceptable nowadays. The way I've seen it done now is to use rubber flashing that has a square base. You apply a strip of adhesive butyl sealing tape along the bottom edges of the flashing, slightly overlapping at the corners. Then place the flashing over the pipe and diagonally on the roof surface, with the top and bottom corners in the valley and press it down. Then, starting in the top corner and working your way around, use screws with neoprene washers, no more than 1¼ inches apart.

Where the roof intersects a vertical surface such as a wall or a chimney, install either continuous or step flashing, as illustrated in Figures 9-17 and 9-18. Which is better is a matter of preference. I think continuous flashing, if properly done, does a better job. But not all roofers agree with me.

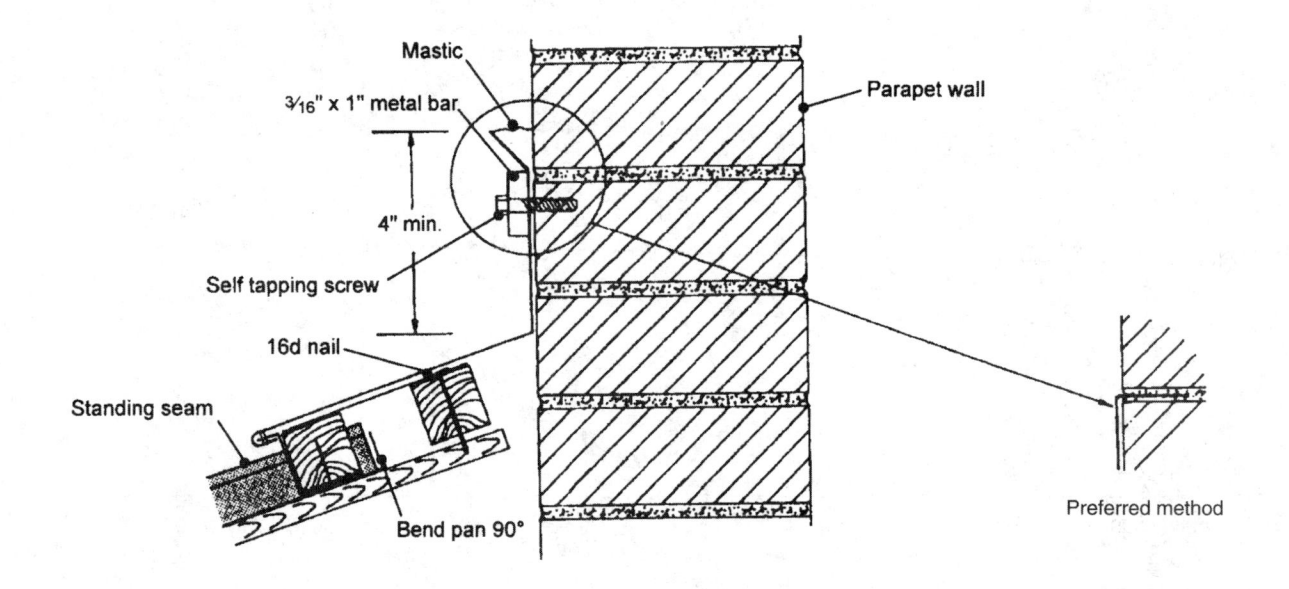

Figure 9-17 Flashing at a wall

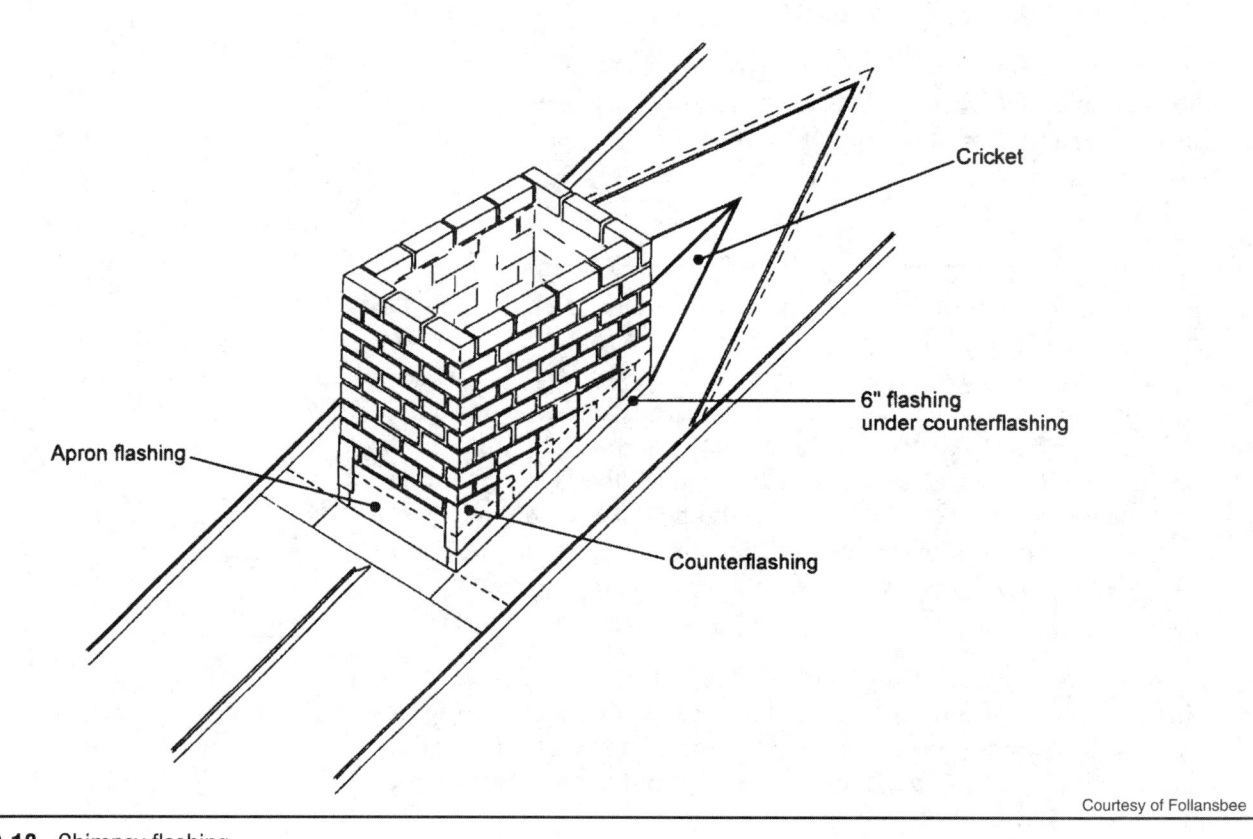

Figure 9-18 Chimney flashing

Metal Roofing and Siding

Fasteners

Use screws and cleats made of the same metal as the roof. The exception is for tin roofs, where you can use steel nails. Use barbed, ringed or screw-shank nails to fasten panels to wood. Use sheet metal screws and rivets to fasten sheet metal to sheet metal. Install neoprene washers under the heads of all exposed screws. In some cases, you can use seams to connect sheets, rather than fasteners.

Closure Strips

Compressible closed-cell neoprene closure strips like the ones in Figure 9-19 seal metal roofing and siding components. Fabricators of metal roofing and siding produce made-to-match closure strips for their product lines. Typically, these closure strips are readily available through suppliers of metal roofing and siding. In exceptional cases the closure strips could be custom-made from wood.

Courtesy of Rubatex Corporation

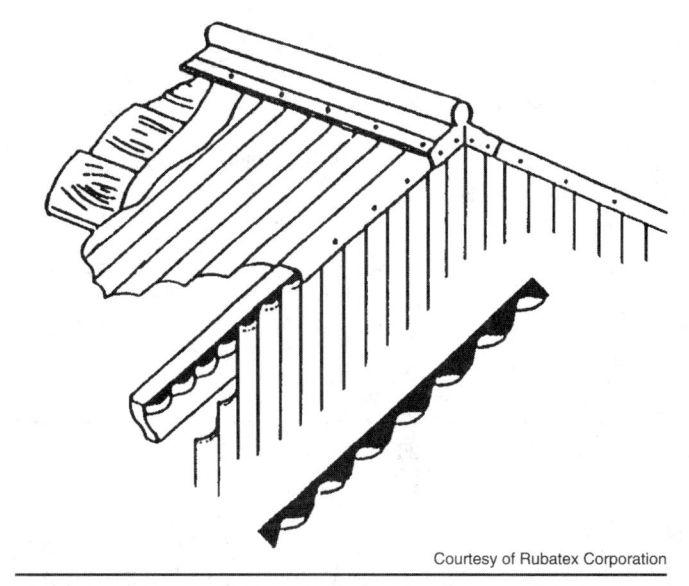

Figure 9-19 Closure strips for metal siding panels

Job-Fabricated Seams

Sheet metal panels such as copper and steel are often joined together by job-fabricated seams. The simplest job-fabricated seam (and the strongest) is the riveted soldered seam shown in Figure 9-20. Use this type of seam for any metal roof that must withstand high thermal stress, heavy

Courtesy of Follansbee

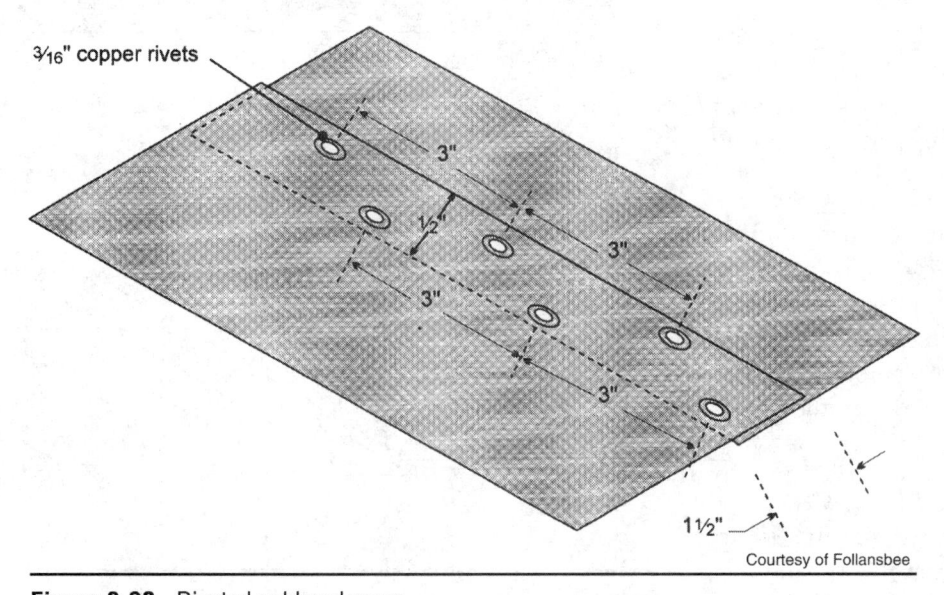

Figure 9-20 Riveted soldered seam

253

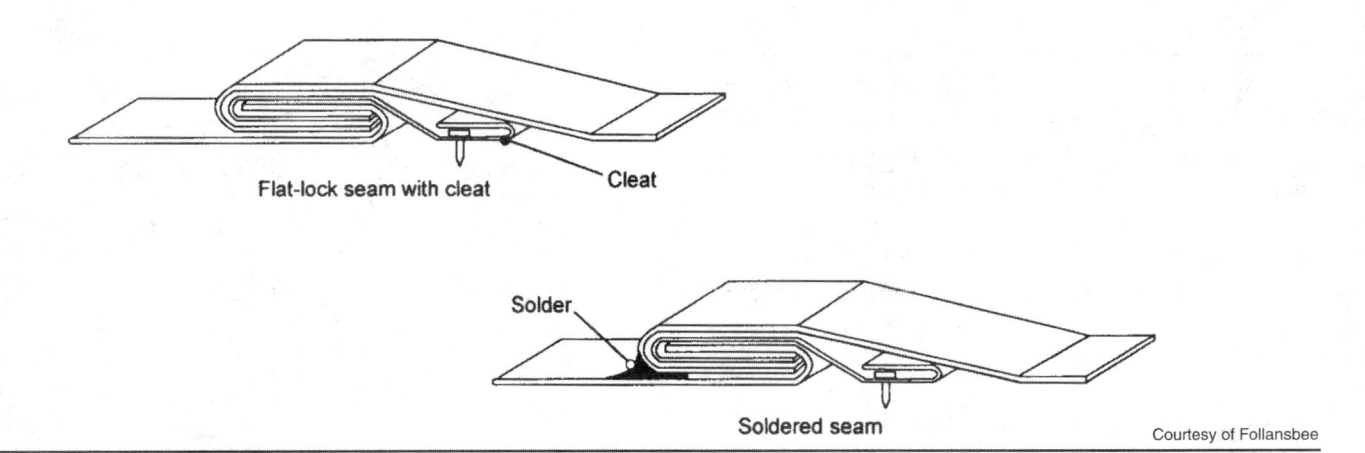

Figure 9-21 Flat-locked seam

snow loading, or a lot of foot traffic. Riveted soldered seams allow a metal roof to expand and contract as a single unit. The result is a very strong roof.

Flat and Flat-Locked Seams

A flat-locked seam is shown in Figure 9-21. Most flat-lock seam metal roofing is factory formed. However, it is possible to site-fabricate flat-lock seams on copper and other metal roofing. You can use a flat-locked seam on roof slopes as low as ¼ in 12. You can also use this type of metal roofing on a curved surface, like the dome shown in Figure 9-22.

To form flat seams on long panels, the long edges are turned back about ¾ inch. While this is usually done at the factory, you can site-fabricate flat-lock seams with special bending tools. Turn one edge up and the opposite edge down, as in Figure 9-23.

Figure 9-23 shows a flat seam and an expansion batten for thermal expansion. An engineer should determine the battens' size, spacing, and what material should be used for the battens. The proper batten depends on the local climate conditions, the kinds and amounts of load expected, and the expansion coefficient of the metal roofing.

To form flat seams in small panels (normally 14" x 20" or 20" x 28"), notch the corners and turn in all four edges of the sheets about ¾ inch. Turn the top and one

Figure 9-22 Copper dome

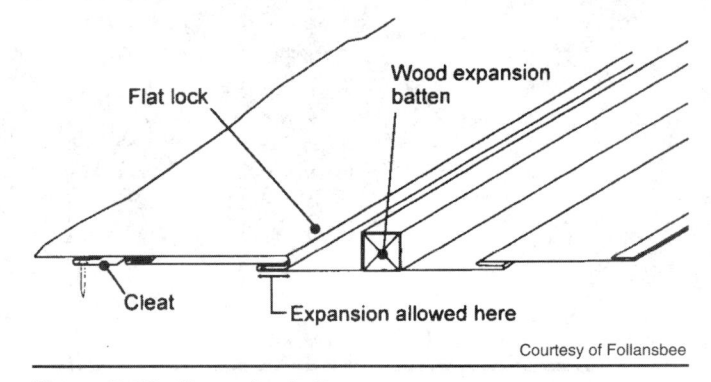

Figure 9-23 Expansion batten

Metal Roofing and Siding

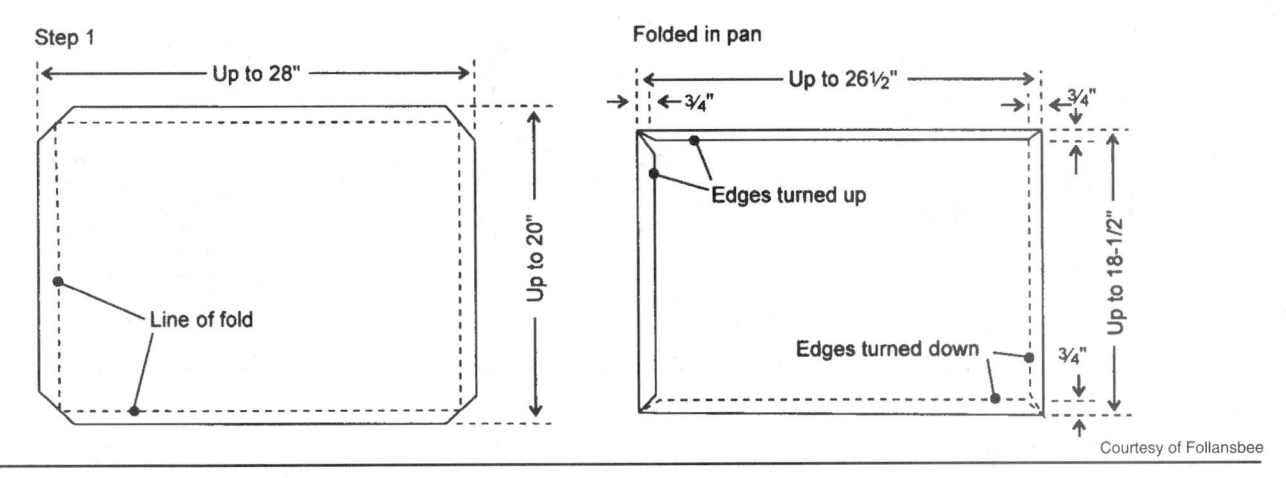

Figure 9-24 Turn back panel edges to form a flat seam

adjacent edge of each sheet up, and the bottom and remaining adjacent edge under to form a flat seam along each panel edge as in Figure 9-24.

On roofs that slope more than ½ in 12, seal the joints of flat seams of most metals with butyl tape or sealant made for use with metal. We used to use caulking compound or white lead, but that's not done any more. On roofs with slopes less than $1/2$ in 12, mallet the joints and sweat them full of solder (Figure 9-21).

Fasten each panel to the roof with cleats locked into the seams on each edge. Fasten the cleats with nails but don't drive the nails through the panels. Place the panels (pans) starting at the low point of the roof. Work up-slope, placing the higher pan over the upper edge of the lower pan. Stagger all joints as shown in Figure 9-25.

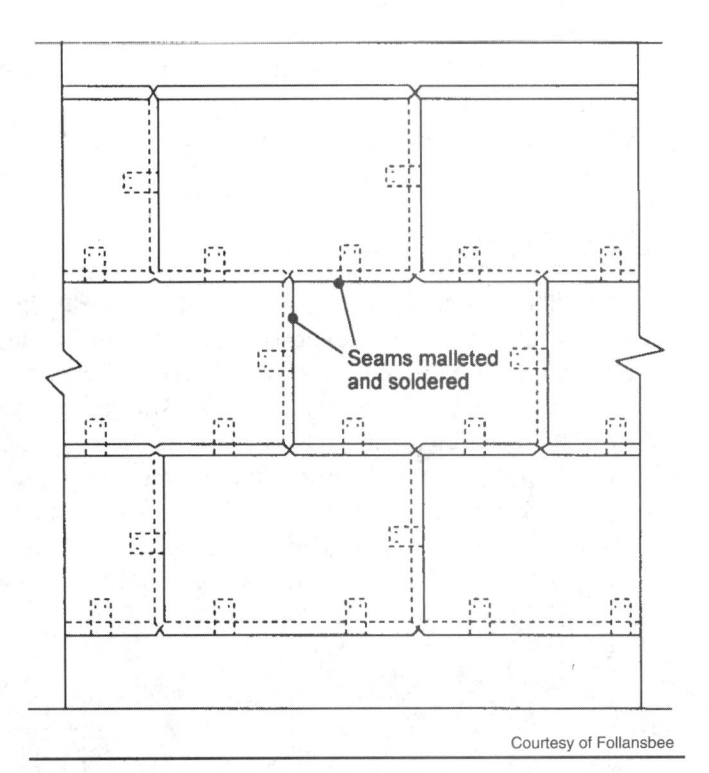

Figure 9-25 Cleats for fastening a flat-seam roof

Standing Seams

The job-fabricated standing seam shown in Figure 9-26 is another way to interlock metal sheets. If you use job-fabricated standing seams, choose panels with preformed edges (preformed pans) as in Figure 9-27. Use a seaming machine like the one in Figure 9-12 to crimp the seams.

Use cleats to fasten job-fabricated standing seams to the deck. Generally, cleats for metal panels are 2 inches wide, spaced on 12-inch centers. Nail each cleat with two $7/8$-inch (minimum) flat-head galvanized roofing

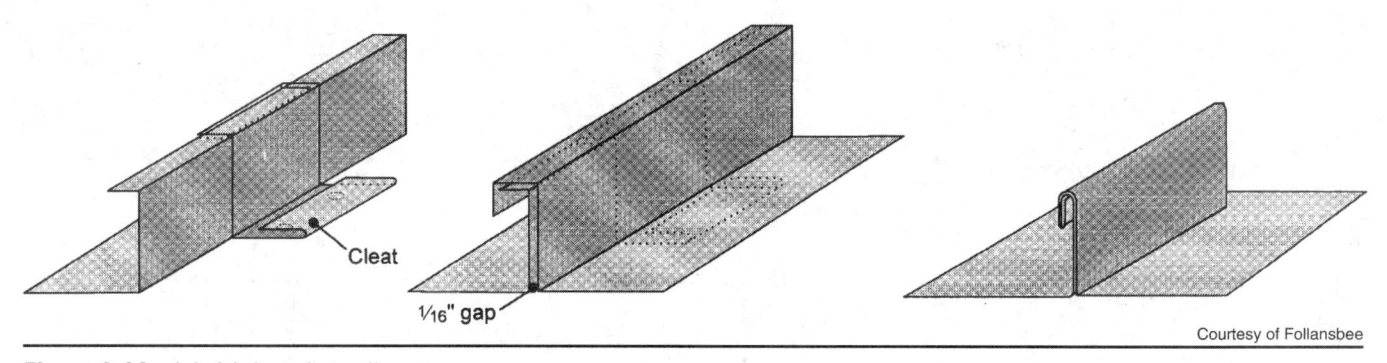

Figure 9-26 Job-fabricated standing seam

nails. The cleat becomes an integral part of the seam as the seam is folded. You can see that in Figure 9-26.

Allow $1/16$ inch between job-fabricated standing-seam panels so the metal can expand and contract. In general, the maximum recommended distance between standing seams is 21 inches. For panels more than 20 feet long, the maximum recommended seam center is 17 inches. For panels more than 30 feet long, install expansion cleats like the ones in Figure 9-8 to allow for thermal movement.

Batten Seams

Another common type of job-fabricated seam is the *batten seam*. Some batten seams are installed over wood batten strips, as shown in Figure 9-28. You can install batten seams on roof slopes of 3 in 12 and steeper. Batten seams are recommended when you're using the heavier sheet metals, or where the coefficient of expansion of the roofing material is significant. (Look ahead to Figure 9-39 for the coefficients of expansion for various metals.) It's a good idea to use batten seams for copper, zinc, or aluminum materials. Install 2-inch-wide cleats on 12-inch centers. Nail the cleats into the wood batten, where they become an integral part of the batten as the seam is folded, as shown in Figure 9-28.

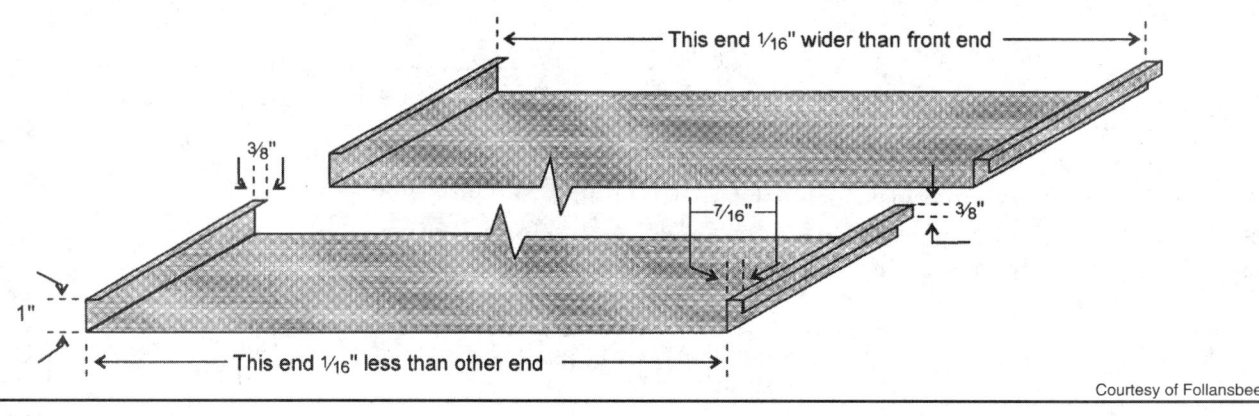

Figure 9-27 Preformed pans

Metal Roofing and Siding

1. Batten seam

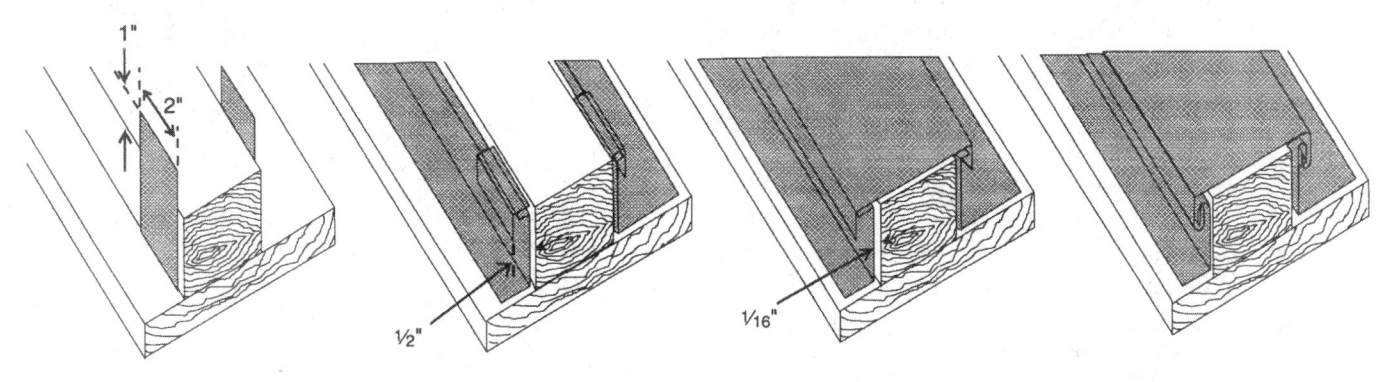

2. Finishing batten end

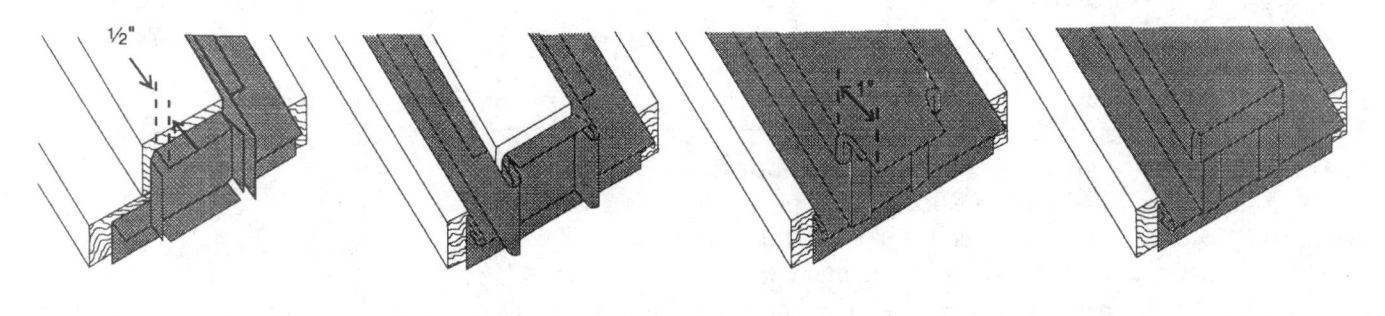

3. Ridge

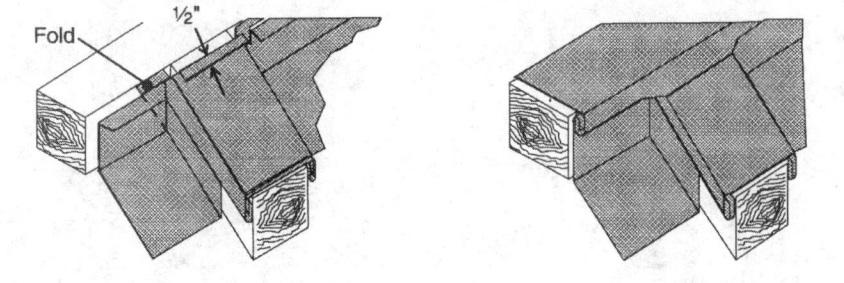

Courtesy of Follansbee

Figure 9-28   Job-fabricated batten seam

Cross Seams

You can get panel systems with the ends of the panels (the cross seams) preformed for a flat-seam lap joint, no matter what kind of side seams are used. Some cross seams are prefabricated (already shaped and with the clip installed) as shown in Figure 9-29. All you have to do is hook the next panel into them.

Other cross seams are job-fabricated, as in Figures 9-30 and 9-31. The type of cross seam used on a metal roof depends on the slope of the roof. On roof slopes between 3 in 12 and 6 in 12, install a solder strip 4 inches below the upper edge of the down-slope panel. A solder strip is a strip of metal that is soldered across a panel that forms an attachment point for an overlying panel, as in Figure 9-30. On slopes of 6 in 12 and steeper, you don't need a solder strip at the seam. Figure 9-31 shows this seam. In either case, fasten the cross seam to the roof structure with two cleats and seal the seam with butyl rubber sealant or tape.

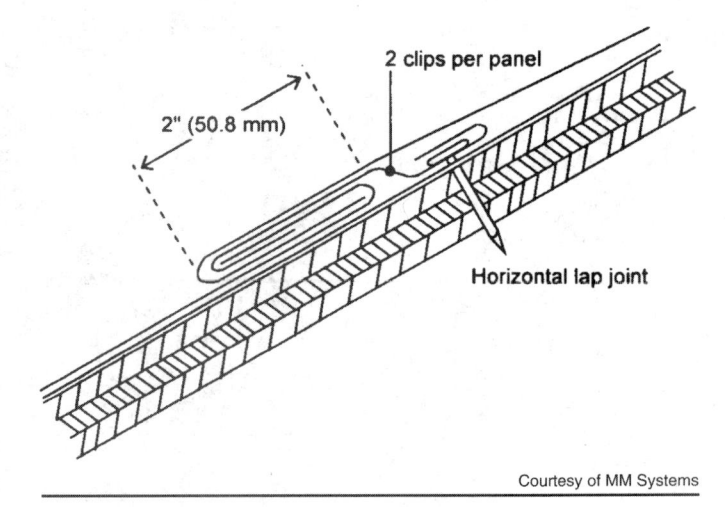

Figure 9-29 Prefabricated cross seam

Sealing Job-Fabricated Joints

Although lead-containing solder is prohibited for use in water systems, 50-50 tin-lead solder is still the most common type of solder used on sheet metal. But since some have concerns about the fumes released when it's melted (though there's not actually any lead in the fumes), it's increasingly being replaced with lead-free solders that use antimony, bismuth, brass, copper, indium or tin, even silver. Some metals tarnish quickly and require surface cleaning with a rosin flux prior to soldering. You can use 50-50

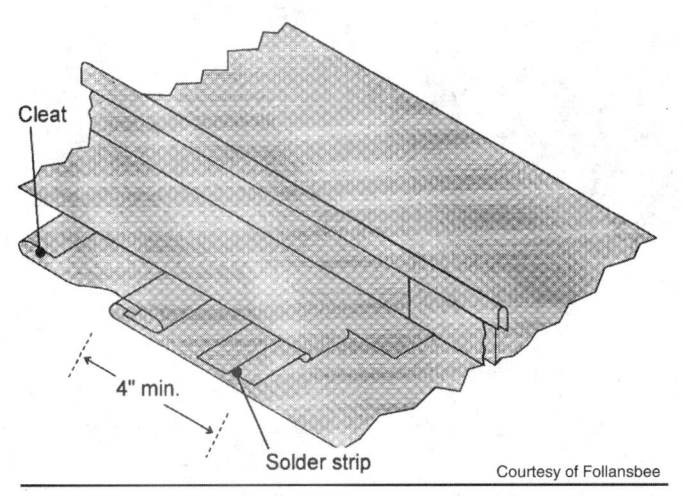

Figure 9-30 Job-fabricated cross seam (low slope)

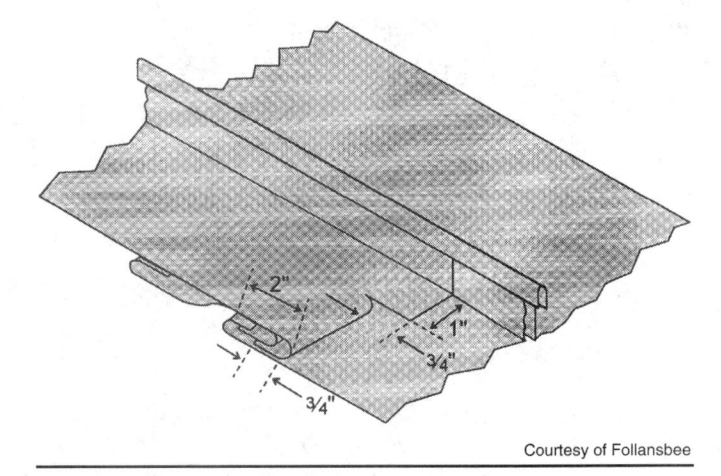

Figure 9-31 Job-fabricated cross seam (steep slope)

solder on most metals, including lead. For stainless steel and nickel alloys, use high-tin (60 to 70 percent) solder. There are special solders for aluminum, but they're seldom used in sheet metal work. Use butyl rubber or similar sealant instead.

Use soft solders only to conceal joints and to make them watertight. They aren't meant to resist stresses.

You can weld joints instead of soldering to make them waterproof. The advantage of welding is that the resulting joints don't require mechanical fasteners or locked seams to resist stresses. The disadvantage is that some metals are hard to weld, and the heat tends to burn off the protective metal coating. I don't recommend that you weld panels to a steel frame when you're going to cover the panels with insulation, or a built-up roof system. You want the roof covering and the panels to expand and contract as a single unit. That's not possible if you've used welded joints.

Brazing isn't practical due to the expense, and contact between dissimilar metals can cause galvanic corrosion.

When you can't solder or weld, you can still produce waterproof joints. You do so by simply applying a sealant between the two metal sheets you're joining. Then you finish the joint by using mechanical fasteners, or by forming locked seams (any seam except the riveted soldered seam in Figure 9-20). Noncuring butyl-based caulk is a good sealant. Give edge-lapped material a bead of caulk along the full length of the joint. You can also buy panels with an integral standing seam that contains a continuous bead of sealant material. (Refer back to Figures 9-10 and 9-11.)

Estimating Metal Roofing and Siding

Take off metal roofing and siding by the square, based on the total area to be covered. As a rule of thumb, you deduct only half of any opening smaller than 100 square feet, but you deduct 100 percent for openings larger than 100 square feet.

You must also account for flashing and accessories, including ridge and hip covers and closures, eaves closures or drip edge, rake end covers, valley flashing, continuous wall flashing and gutters. Other accessories you'll sometimes need include snap-on seams, concealed panel clips and cleats, solder strips, roof curbs and ridge ventilators.

Due to side laps and end laps (cross seams), or because of material lost in making seams, the net area that some metal panels will cover is less than the total area the panel is sold by. However, many modern ribbed metal panels are sold according to the net length and width (exposure) of the panel. If necessary, you can calculate the waste factor with the following formula:

$$\text{Waste Factor} = \frac{\text{Gross Panel Area}}{\text{Net Panel Area}}$$

Equation 9-1

Find the number of panels required per square using:

$$\text{Panels per Square} = \frac{100 \text{ SF}}{\text{Net Panel Area (SF)}}$$

Equation 9-2

If the panel dimensions are given in inches, use:

$$\text{Panels per Square} = \frac{14{,}400 \text{ sq. in.}}{\text{Net Panel Area (sq. in.)}}$$

Equation 9-3

▼ **Example 9-1:** Find the waste factor and the number of panels required per square for 28" x 120" panels installed with flat-seam construction. Assume that you lose $1\frac{1}{2}$ inches in both panel dimensions due to the lap of flat seams.

The waste factor from Equation 9-1 is:

$$\text{Waste Factor} = \frac{28 \text{ in.} \times 120 \text{ in.}}{26.5 \text{ in.} \times 118.5 \text{ in.}}$$

$$= 1.07, \text{ or 7 percent waste due to seams}$$

The number of panels required per square (using Equation 9-3) is:

$$\text{Panels/Square} = \frac{14{,}400 \text{ sq. in.}}{26.5 \text{ in.} \times 118.5 \text{ in.}}$$

$$= 14{,}400 \div 3{,}140.25$$

$$= 4.6 \text{ panels}$$

The charts in this chapter contain waste factors for only the most commonly produced metal panels. They don't include waste due to cutting off the uppermost panel so its end terminates at the ridge. Consult the manufacturer's figures when you estimate a specific metal roof panel system.

You can order many panels factory-cut to the lengths required for your job. Depending on the manufacturer's longest available length, some panels are long enough to run continuously from the ridge to the eaves. Modern panel lengths vary from 10 to 40 feet. Longer panels can be special ordered, but it will require special truck transport permits. Unless the roof is designed in increments of the net panel width, you'll have to cut the last panel on the job, or special order the width of the last panel.

Steel Roofing and Siding Quantities

Steel roofing and siding sheets are coated with zinc, tin, nickel, aluminum, or a combination of materials. Steel coated with zinc is called *galvanized* steel. Steel coated with tin is called *bright plate* and steel coated with a lead-tin alloy was called *terne metal*. Though this particular coating isn't manufactured in this country any more due to concerns about lead, there are alternatives that have similar properties, but with coatings made with zinc instead of lead.

Galvanized Sheet Metal

Galvanized steel is relatively inexpensive and corrosion-resistant. It can be soldered and used in direct contact with concrete, mortar, lead, tin, aluminum and wood (except redwood and red cedar).

Galvanized sheet metal is manufactured in many gauges and in a number of coating classes, depending on the amount of zinc applied to the sheet. The common galvanized sheet metal used for construction purposes has 0.625 ounces of zinc applied per square foot on one side of the sheet. The most common gauge is 24, but it can range from 18 gauge for structural panels to 26 gauge for architectural panels installed over solid decking. Galvanized sheet metal is generally available in 6- to 10-foot lengths (in 1-foot increments) and in 24- to 48-inch widths (in 2-inch increments).

When produced with a flat seam, the edges of the sheet are each turned in about half an inch. This means the sheet covers about $1^{1}/_{2}$ inches less than its size in each dimension. Use Figure 9-32 to determine waste factors and the number of metal sheets required per square for flat-seam galvanized steel roofing.

You'll have more waste when you install a standing seam roof than when you use the flat seam. A standing seams loses $2^{3}/_{4}$ inches of width at each standing seam, and $1^{1}/_{8}$ inches of length at each cross seam. The waste factors and coverage for galvanized steel sheets with a standing seam are given in Figure 9-33. You can install a standing seam roof on roofs with slopes of 2 in 12 and steeper.

Galvanized metal also comes in rolls 14, 20, 24 and 28 inches wide. The minimum recommended gauge for counterflashing is 26 gauge, but I recommend you don't use thinner than 24 gauge. You can use Figure 9-34 to determine waste factors and coverage for galvanized steel rolls with a standing seam.

Ribbed Metal Panel Quantities

There are many types of ribbed metal roofing and siding panels available. A selection of styles offered by one manufacturer appears in Figure 9-35. We'll take a closer look at one of them, corrugated steel panels.

Panel size (inches)	Waste factor	Panels required per square
24 x 96	1.08	6.8
24 x 120	1.08	5.4
26 x 96	1.08	6.2
26 x 120	1.08	5.0
28 x 72	1.08	7.7
28 x 84	1.08	6.6
28 x 96	1.07	5.8
28 x 108	1.07	5.1
28 x 120	1.07	4.6
30 x 96	1.07	5.4
30 x 120	1.07	4.3
36 x 96	1.06	4.4
36 x 120	1.06	3.5
42 x 96	1.05	3.8
42 x 120	1.05	3.0
48 x 96	1.05	3.3
48 x 120	1.05	2.6

These figures allow for a loss of 1½" in panel length and width for making the flat seams.

Figure 9-32 Waste factors and coverage for galvanized steel sheet metal panel (flat-seam method)

Panel size (inches)	Waste factor	Panels required per square
24 x 96	1.14	7.1
24 x 120	1.14	5.7
26 x 96	1.13	6.5
26 x 120	1.13	5.2
28 x 72	1.13	8.1
28 x 84	1.12	6.9
28 x 96	1.12	6.0
28 x 108	1.12	5.3
28 x 120	1.12	4.8
30 x 96	1.11	5.6
30 x 120	1.11	4.5
36 x 96	1.10	4.6
36 x 120	1.09	3.6
42 x 96	1.08	3.9
42 x 120	1.08	3.1
48 x 96	1.07	3.4
48 x 120	1.07	2.7

These figures allow for loss of 2¾" of panel width for making the standing edge seam and 1⅛" in panel length for making the flat cross seam.

Figure 9-33 Waste factors and coverage for galvanized steel sheet metal panels (standing-seam method)

Roll width (inches)	Waste factor	Linear feet required per square
14	1.24	107
20	1.16	70
24	1.13	57
28	1.11	48

This table allows for loss of 2¾" in panel width for making the standing edge seam.

Figure 9-34 Waste factors and coverage for galvanized steel sheet metal rolls (standing-seam method)

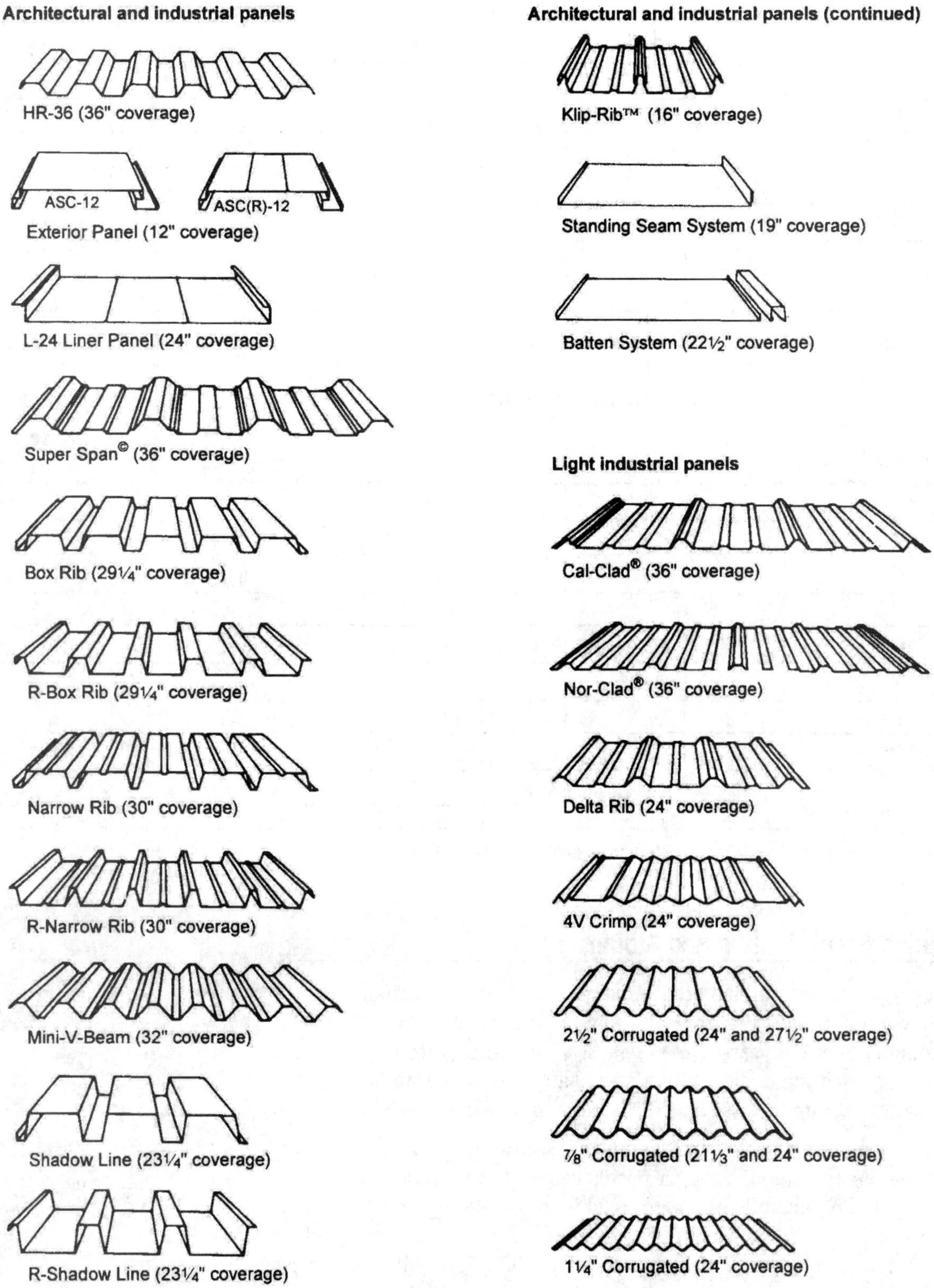

Metal Roofing and Siding

Architectural and industrial panels

HR-36 (36" coverage)

ASC-12 ASC(R)-12
Exterior Panel (12" coverage)

L-24 Liner Panel (24" coverage)

Super Span© (36" coverage)

Box Rib (29¼" coverage)

R-Box Rib (29¼" coverage)

Narrow Rib (30" coverage)

R-Narrow Rib (30" coverage)

Mini-V-Beam (32" coverage)

Shadow Line (23¼" coverage)

R-Shadow Line (23¼" coverage)

Architectural and industrial panels (continued)

Klip-Rib™ (16" coverage)

Standing Seam System (19" coverage)

Batten System (22½" coverage)

Light industrial panels

Cal-Clad® (36" coverage)

Nor-Clad® (36" coverage)

Delta Rib (24" coverage)

4V Crimp (24" coverage)

2½" Corrugated (24" and 27½" coverage)

⅞" Corrugated (21⅓" and 24" coverage)

1¼" Corrugated (24" coverage)

Courtesy of ASC Pacific, Inc.

Figure 9-35 Various ribbed roofing and siding panels

Length of sheets (feet)	26-inch wide (2½-inch pitch) Waste factor	26-inch wide (2½-inch pitch) Sheets required per square	26-inch wide (3-inch pitch) Waste factor	26-inch wide (3-inch pitch) Sheets required per square
5	1.16	10.7	1.14	10.5
6	1.15	8.8	1.12	8.6
7	1.14	7.5	1.11	7.4
8	1.13	6.5	1.11	6.4
9	1.13	5.8	1.10	5.7
10	1.12	5.2	1.10	5.1
11	1.12	4.7	1.09	4.6
12	1.11	4.3	1.09	4.2

Length of sheets (feet)	31¼-inch wide (2⅔-inch pitch) Waste factor	31¼-inch wide (2⅔-inch pitch) Sheets required per square	32-inch wide (3-inch pitch) Waste factor	32-inch wide (3-inch pitch) Sheets required per square
5	1.12	8.6	1.18	8.9
6	1.11	7.1	1.17	7.3
7	1.10	6.0	1.16	6.2
8	1.09	5.2	1.15	5.4
9	1.09	4.6	1.15	4.8
10	1.08	4.2	1.14	4.3
11	1.08	3.8	1.14	3.9
12	1.07	3.4	1.14	3.6

This table allows for a one-corrugation side lap and a 4" end lap

Figure 9-36 Waste factors and coverage for corrugated steel sheets used for siding

Corrugated Steel Roofing and Siding

Corrugated steel is manufactured in many corrugation sizes, in frequently changing configurations. Figure 9-35 shows a number of designs that have been available over the years. *Galvalume®*, which differs from galvanized steel in that the coating is a blend of zinc and aluminum instead of pure aluminum, is now commonly used, as it has better corrosion-resistance.

Standard lengths for corrugated roofing and siding sheets are 6 to 16 feet, in 2-foot increments, though longer panels can be found. Available widths vary widely among manufacturers, from 24 inches to 48, the most common being 26 and 36. Width is measured from edge to edge; it doesn't follow the contours of the ridges. Remember also that the edges will overlap by several inches on both sides.

Corrugated steel sheets come in a variety of finishes, including uncoated (black), painted, galvanized, vinyl and ceramic.

Metal Roofing and Siding

Length of sheets (feet)	27½-inch wide (2²/₃-inch pitch) Waste factor	27½-inch wide (2²/₃-inch pitch) Sheets required per square	27½-inch wide (3-inch pitch) Waste factor	27½-inch wide (3-inch pitch) Sheets required per square
6	1.21	8.8	1.23	8.9
7	1.19	7.4	1.21	7.5
8	1.18	6.4	1.20	6.5
9	1.17	5.7	1.19	5.8
10	1.17	5.1	1.18	5.2
11	1.16	4.6	1.18	4.7
12	1.16	4.2	1.17	4.3

Length of sheets (feet)	32¾-inch wide (2²/₃-inch pitch) Waste factor	32¾-inch wide (2²/₃-inch pitch) Sheets required per square	33½-inch wide (3-inch pitch) Waste factor	33½-inch wide (3-inch pitch) Sheets required per square
6	1.19	7.3	1.20	7.2
7	1.17	6.1	1.18	6.1
8	1.16	5.3	1.17	5.3
9	1.15	4.7	1.16	4.6
10	1.15	4.2	1.16	4.1
11	1.14	3.8	1.15	3.8
12	1.14	3.5	1.15	3.4

This table allows for a two-corrugation side lap and a 6" end lap

Figure 9-37 Waste factors and coverage for corrugated steel sheets used for roofing

Installation Allowance for Corrugated Siding

For siding, fasten side laps on 18-inch centers. For roofing over solid sheathing, fasten side laps on 12-inch centers. When installing panels over purlins, fasten side laps over each purlin. Fasten end laps over each purlin at every second or third corrugation.

When installed as siding, one-corrugation side lap and a minimum of a 4-inch end lap is enough protection. For roofing, general practice is to provide a two-corrugation side lap and a 6-inch end lap (on roof slopes of 4 in 12 and steeper), or an 8-inch end lap (on roof slopes less than 4 in 12). Figure 9-36 shows waste factors and coverage for various-sized corrugated steel sheets used for siding. Figure 9-37 shows the same information for sheets used for roofing.

Uncoated (black)

U.S. Mfr's Gauge	Thickness (inches)	Weight (lbs per square foot) Flat	Weight (lbs per square foot) Corrugated 2²/₃ x ½	Weight (lbs per square foot) Corrugated 3 x ¾
12	0.1046	4.38	4.77	5.05
14	0.0747	3.13	3.41	3.61
16	0.0598	2.50	2.73	2.89
18	0.0478	2.00	2.18	2.31
20	0.0359	1.50	1.64	1.73
22	0.0299	1.25	1.36	1.44
24	0.0239	1.00	1.09	1.15
26	0.0179	0.75	0.82	0.87
28	0.0149	0.63	0.68	0.72
29	0.0135	0.56	0.60	0.65

Galvanized

U.S. Mfr's Gauge	Thickness (inches)	Weight (lbs per square foot) Flat	Weight (lbs per square foot) Corrugated 2²/₃ x ½	Weight (lbs per square foot) Corrugated 3 x ¾
12	0.1084	4.53	4.94	5.23
14	0.0785	3.28	3.58	3.79
16	0.0635	2.66	2.90	3.07
18	0.0516	2.16	2.35	2.49
20	0.0396	1.66	1.81	1.91
22	0.0336	1.41	1.53	1.62
24	0.0276	1.16	1.26	1.33
26	0.0217	0.91	0.99	1.05
28	0.0187	0.78	0.85	0.90
29	0.0172	0.72	0.78	0.83

Figure 9-38 Gauges, thickness and weights of corrugated steel

Figure 9-38 summarizes corrugated steel's physical properties in a compact form.

Miscellaneous Metal Roofing Quantities

You can use corrugated plastic sheets along with corrugated sheet metal panels to let light enter through the roof. You can also use these panels alone as the entire roof covering. Both colored and translucent sheets are available.

Plastic corrugated sheets are made from acrylics. Sheets are commonly 50$\frac{1}{2}$ inches wide and 8, 10, 12, 16, or 20 feet long. The sheets are usually installed with a one-corrugation side lap and an 8-inch end lap.

Reinforced corrugated fiberglass sheets are made from glass fiber-reinforced acrylics. Corrugated sheets come 26, 34, 35, 40 or 42 inches wide and 8, 9, 10, 11 or 12 feet long. Sheets are installed with a one- or two-corrugation side lap and an 8-inch end lap. All plastic sheets (reinforced or not) weigh about 0.4 pounds per square foot.

Corrugated structural glass sheets come 47$\frac{1}{2}$ inches wide and 8, 10 or 12 feet long. Sheets are $\frac{3}{8}$ inch thick and weigh about 6.3 pounds per square foot. Sheets are installed with a two-corrugation side lap and a 6-inch end lap.

Materials (metals and alloys)	Linear expansion per 100 degrees Centigrade	Linear expansion per 100 degrees Fahrenheit
Aluminum, wrought	0.00231	0.00128
Brass	0.00188	0.00104
Bronze	0.00181	0.00101
Copper	0.00168	0.00093
Iron, cast, gray	0.00106	0.00059
Iron, wrought	0.00120	0.00067
Iron, wire	0.00124	0.00069
Lead	0.00286	0.00159
Magnesium, various alloys	0.00290	0.00160
Nickel	0.00126	0.00070
Steel, mild	0.00117	0.00065
Steel, stainless, 18-8	0.00178	0.00099
Zinc, rolled	0.00311	0.00173

Figure 9-39 Coefficients of expansion of various metals

Gauge	Nominal thickness (inches)	Weight (pounds per SF)
26	0.016	0.226
24	0.020	0.285
22	0.025	0.355
20	0.032	0.455
18	0.040	0.568
16	0.051	0.725
14	0.064	0.910
12	0.081	1.152

Figure 9-40 Aluminum sheet data

You can get prefabricated membrane roof and wall panels made of a foam polyisocyanurate insulation core sandwiched between a 22-, 24- or 26-gauge galvanized steel (or aluminum) outer face and a 26-gauge prepainted steel interior face. The outer face comes with a laminated coating of PVC membrane for protection and a more attractive appearance. They may have interlocking edge joints, or be sealed with heat-welded batten strips. Panels are 2 to 6 inches thick and a standard width of 42 inches. You can get them in any desired length.

Aluminum Roofing and Siding

Aluminum panels are alloyed with copper, zinc, manganese, magnesium, chromium or nickel. The core metal is designed for strength and the corrosion-resistant finish (normally an aluminum alloy or commercially pure aluminum) is fused to the core metal under heat and pressure during a hot-rolling manufacturing process.

Aluminum is light, non-corrosive, rigid and durable. But is also has a very high coefficient of thermal expansion, as you can see in Figure 9-39.

Aluminum is produced in sheets and coils, and in a variety of gauges, with properties as shown in Figure 9-40.

You can install corrugated aluminum sheets on roof slopes of 3 in 12 and steeper, and V-beam sheets on slopes as low as 2 in 12. When using concealed clip panels, the minimum slope allowed is 4 in 12 unless a single panel will cover the total span (ridge to eaves). Then the single panel can be applied over roof slopes as low as $1/2$ in 12.

Metal Roofing and Siding

Terne, *Monel*® and Lead Roofing

When the first edition of this manual was written, roofing made of terne metal, *Monel*® metal and lead were still widely available. Not so any more, but they're mentioned here because they last a long time and many still exist. Depending on where in the country you live, you may find yourself having to deal with one.

By and large, they made for long-lasting, attractive roofs. The Faculty Club of Yale University boasts an original terne metal roof that's over 250 years old! Terne was one of the strongest roofing materials known at the time, and because it was lightweight and had a low coefficient of expansion, it was a favored, though expensive, roofing material. You could still get *terne-coated stainless steel* roofing (TCS) until quite recently, though the terne coating was tin/zinc coated, not tin/lead. Alternative roofing materials have come on the market and are still generally available. These include *Roofinox* tin-matte (terne-coated) stainless steel, *Freedom Gray*, which had the same finished look but was made with copper sheets instead of stainless steel, and others.

Monel® metal was a trade name for a white metal, primarily a nickel and copper alloy, that looked like stainless steel. It was expensive and hard to work with, requiring shop fabrication. Not surprisingly, it's off the market, at least as far as roofing is concerned.

You can still buy sheet lead, but it's rarely used nowadays for whole roof covering. Where it's in use, it can be installed batten-seam, standing seam and flat-seam on roof slopes as low as ¼ in 12. However, because of its pliability and because it can be drawn and stretched to cover warped surfaces, sheet lead is still widely used as a flashing material for tile and slate roofs. It's generally sold by weight in pounds per square foot, and is usually hard lead, weighing 2½ to 3 pounds per square foot and containing 6 to 7½ percent antimony.

Some steel or copper sheets are coated with lead, producing a material with a lightweight core whose surface has the color and provides the protection of lead.

▼ **Example 9-2:** Assume a roof slope of 5 in 12, then find the quantity of corrugated steel panels required for a gable roof with a plan area of 80 feet by 40 feet. Assume that you'll use 27½-inch-wide panels (3" pitch), 12 feet long, with a two-corrugation side lap and 6-inch end lap.

Solution: The net roof area to be covered is:

Net Roof Area = 80' x 40' x 1.083 (from Column 2 of Appendix A)
               = 3466 SF ÷ 100
               = 34.66 squares (Use this quantity to figure labor costs.)

The total length from the eaves to the ridge is:

Total Required Panel Length = 20' x 1.083
                              = 21.66'

The gross roof area (excluding cutting excess panel length at the ridge) is:

3466 SF x 1.17 (from Figure 9-37) = 4055 SF

Using two 12-foot panels with a 6-inch end lap (23.5 feet), we must increase the end lap, or cut off the top of the second panel by:

Panel loss = 23.5' - 21.66
= 1.84'

The total square feet of excess panel length is:

2 x 80' x 1.84' = 294 SF

The gross roof area is:

Gross Roof Area = 4055 SF + 294 SF = 4349 SF
= 4349 SF ÷ 100
= 43.49 squares

The waste factor is:

$$\text{Waste Factor} = \frac{4349 \text{ SF}}{3466 \text{ SF}}$$
$$= 1.25$$

Other items that will probably be required are:

Eaves closure strip = 2 x 80' = 160 feet
Ridge cap = 80 feet
Rake trim = 4 x 20' x 1.083 = 87 feet

Chrome-Nickel Stainless Steel Roofing

Stainless steel, which is highly resistant to corrosion, has chromium and nickel added during the steel manufacturing process. There are six types of stainless steel commonly used as building materials, all with 14 to 20 percent chromium and 6 to 14 percent nickel. The type of stainless steel most widely used for exterior work is Type 302 (Grade 18-8). It has 17 to 19 percent chromium and 8 to 10 percent nickel. Type 430 contains no nickel and is used mostly indoors. Type 316 (Grade 18-12) is used in corrosive marine environments. Some stainless steel is manufactured with a coating of terne metal.

Because of its high strength, stainless steel is produced in relatively thin sheets. But it's so hard that you have to do all fabricating in the shop. You can install stainless steel with a standing seam and a continuous weld over roof slopes as low as $1/4$ in 12.

Zinc Roofing

Zinc is lighter and stiffer than lead, but it's more susceptible to damage from acids and has a high coefficient of expansion. Rolled sheet zinc is used for roofing and flashing, and zinc is often used as a surface coating for sheet steel. Zinc roofing comes in thicknesses of 9 through 16 gauge, as shown in Figure 9-41.

Avoid using zinc flat-seam construction except over surface areas less than 200 square feet (10-gauge is recommended). Install standing seams on slopes of 2 in 12 and steeper. I recommend 10-gauge zinc for flat-seam and standing-seam installations. Use batten seams on slopes between 3 in 12 and 6 in 12. The gauge used in batten-seam construction depends on the on-center batten spacing:

- For spacing of 18" oc, use 11 or 12 gauge.
- For spacing of 30" oc, use 12 or 13 gauge.
- For spacing of 40" oc, use 13 or 14 gauge.

Solder cross seams on roof slopes less than 4 in 12. Stock sheet sizes are 20, 30, 36 and 40 inches wide, and 7 or 8 feet long.

Gauge	Thickness (inches)	Weight (pounds per SF)
9	0.018	0.67
10	0.020	0.75
11	0.024	0.90
12	0.028	1.05
13	0.032	1.20
14	0.036	1.36
15	0.040	1.50
16	0.045	1.68

Figure 9-41 Physical properties of zinc roofing

Copper Roofing

Copper used in sheet metal work is called *tough pitch copper*. It comes in hot-rolled (soft or dead soft) or cold-rolled (hard or cornice temper) forms. Soft-rolled copper is easier to work with because it's more pliable.

Cold-rolled copper, however, is stronger, harder, and less pliable, but also less expensive. Most flashing work is done using cold-rolled copper. Sheet metal copper comes in weights of 16, 20, 24 and 32 ounces per square foot.

Roofing sheets of copper are 20, 24, 30 and 36 inches wide and 8 and 10 feet wide. Strip copper is 10 and 20 inches wide and 8 and 10 feet long. Rolled copper comes in rolls 6 and 20 inches wide, 50 to 100 feet long, and weighing between 80 and 100 pounds per square.

Install 20-ounce cold-rolled copper plates no larger than 16" x 18" with ³⁄₄-inch flat, locked and soldered joints. Use 2-inch-wide copper cleats to install the panels on the deck. You can use flat seams on roof slopes as low as ¹⁄₄ in 12.

After folding the seams, a 16" x 18" sheet will cover only 14¹⁄₂" x 16¹⁄₂". That produces a waste factor of 1.20, so you need about 60 sheets per square. You can install standing seams on roof slopes of 2¹⁄₂ in 12 and steeper. Waste factors and coverage for standing-seam construction are given in Figure 9-42. You can install batten seams on roof slopes as low as 3 in 12.

Sheet size (inches)	Waste factor	Sheets required per square
20 x 96	1.25	9.4
24 x 96	1.20	7.5
30 x 96	1.16	5.8
36 x 96	1.13	4.7
20 x 120	1.25	7.5
24 x 120	1.20	6.0
30 x 120	1.16	4.6
36 x 120	1.13	3.8

This table allows for a loss of 3¾" of panel width for making the standing edge seams and 1½" in panel length for making the flat cross seams.

Figure 9-42 Waste factors and coverage for copper roofing using a standing seam

Install 16-ounce copper when seam widths are on 20-inch centers, or less. Use 20-ounce copper when seam centers are farther apart. Don't rivet or solder standing seams. You don't need to solder cross seams when the roof slope is greater than 3 in 12.

For counterflashing, use a minimum of 16-ounce copper.

Copper-Bearing Steel Roofing

Copper-bearing steel roofing is still available, but not common. It comes in two sizes of 24-gauge sheets:

- $26\frac{1}{2}$ inches by 50 feet when installed by the standing-seam method

- 25 inches by 12 feet when installed by the pressed standing-seam method

You can install standing-seam rolled panels on slopes as low as 2 in 12 and pressed standing-seam panels on slopes as low as $\frac{1}{4}$ in 12. Use cleats and nails to install this type of steel roofing.

Titanium-copper-zinc alloy (T-C-Z) roofing usually comes in 20- or 24-inch widths, and in 10-foot lengths. It's usually installed using the batten- or standing-seam methods. The weight of T-C-Z roofing varies from 125 to 150 pounds per square.

Metal Shingles

Metal shingles have gained popularity in recent years, more-commonly on commercial and public buildings, as homeowners tend to think of metal roofs as belonging on carports or barns. But today, they come in a wide variety of shapes and styles, with enduring coatings in colors and textures that make them very attractive. New designs are coming on the market all the time in what has become a quite competitive field. You'll now find them on high-end architect-designed homes. They're far longer-lasting than asphalt shingles; manufacturers claim they last a lifetime. But they're expensive.

Before you install a metal roof on a building in a housing development, make sure there isn't a homeowners association ban on them. Not all HOAs update their rules as building materials change and improve, and a metal roof will look different.

Metal shingles come as single units or multiple strips (Figure 9-43) and can be used safely for all single-coverage applications. The most-common metals and alloys used in their manufacture are galvanized steel, *Galvalume(R)*, and aluminum, though copper and zinc shingles are also available. For the politically and environmentally conscious customer, they're made in America and composed largely of recycled materials.

Metal Roofing and Siding

The most-common metal shingles are made from 0.019-gauge aluminum, finished in a wood shake pattern. The strips are 36 or 48 inches long and 10 inches (nominal) wide. The strips are designed to interlock at the edges and ends. Weights of aluminum shingles vary from about 40 to 60 pounds per square. One problem with all-aluminum shingles is they can be dented by hail, so building codes in some areas require a heavier-gauge aluminum. Check the local code beforehand.

You used to be able to get shingles that were porcelain enamel on iron that came in giant 10-inch by 10-inch individual shingles. They were glass, heat-fused to metal, and a square weighed almost a quarter of a ton. I doubt if they're made any more, but you may still see them on some old commercial buildings. They were a recognizable feature on some Howard Johnsons restaurants.

Figure 9-43 Metal strip-shingle roof

A variety of shingles are made with a core of 0.019-gauge aluminum or 22- through 29-gauge galvanized steel, coated with a siliconized polyester finish. They come in a wide variety of colors. Styles include mission tile, Spanish tile, and shakes. These shingles are relatively lightweight. Aluminum-core shingles weigh from 40 to 65 pounds per square; steel-core shingles significantly more, from 90 to 250 pounds per square.

Another variety of metal shingle is the interlocking copper shingle, commonly made from rolled 16-ounce, or 24-gauge, copper. There's a lighter-weight version, made from 12-ounce rolled copper, but some building codes don't allow them, requiring the heavier material. You may not be able to get the lightweight ones any more. Copper shingles can also be installed on a re-roofing job, over asphalt or built-up or other types of roofing. But don't install them on roof slopes less than 3 in 12.

Metal shingles and panels are usually installed with special accessories such as sidewalls and end-wall flashing, rake and eaves trim, and hip and ridge caps. Figure 9-44 shows some of the accessories you need to install metal shingles.

Metal Shingle Fasteners

To avoid a chemical reaction between two metals that could cause major problems with the roof, use fasteners of either the same metal as the shingles, or else galvanized or stainless steel, which can go with any metal, except don't use galvanized with copper roofing. I usually went for stainless in case there was some other metal included in the shingle that I may not

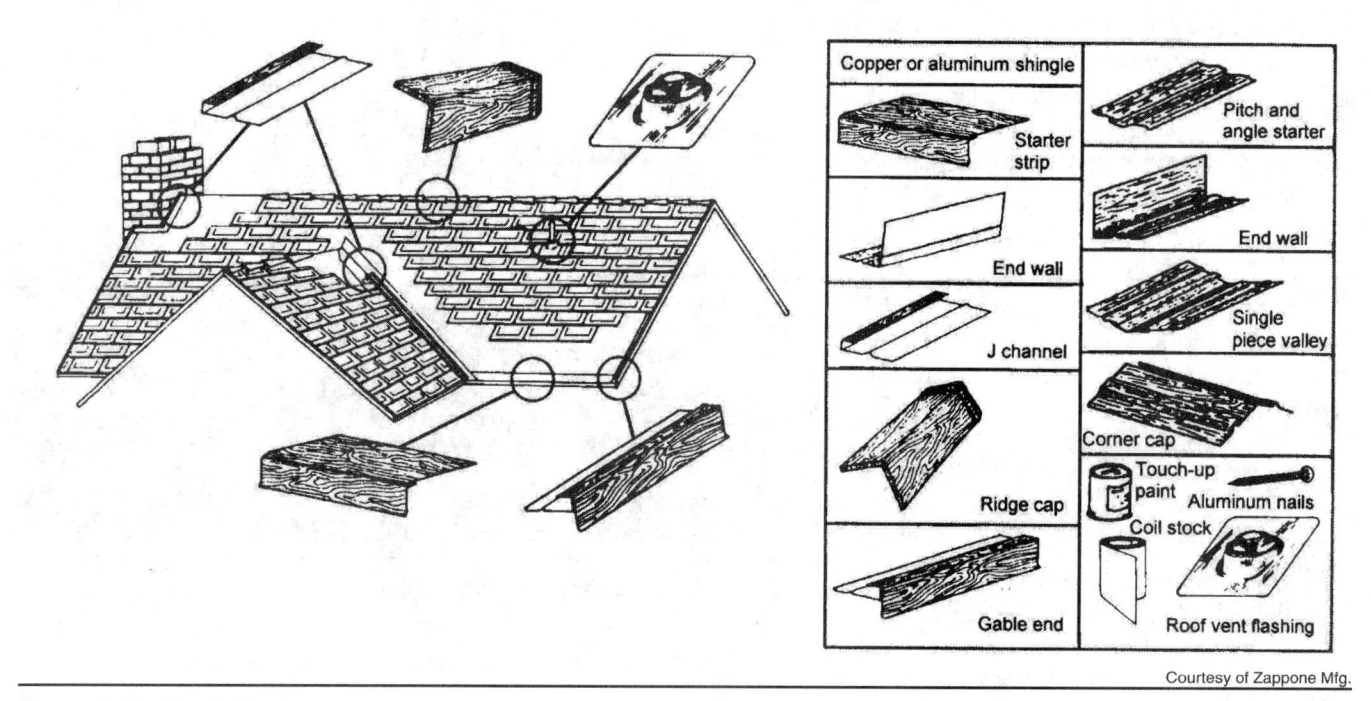

Figure 9-44 Metal shingle accessories

know about. Some metal shingle panels are fastened with self-drilling nylon-headed steel screws with bonded aluminum or neoprene washers. The supplier of the shingles should be able to advise you on the right type to use for the particular shingle you're buying.

Be aware that metal, especially aluminum and copper, have a high coefficient of expansion. Make sure you take into account significant expansion and contraction of any metal roofing you install.

Underlayment for Metal Shingles

Some roofers claim you don't need underlayment on a metal roof with a slope greater than 3 in 12. I don't agree with that. While a vapor retarder or similar layer may not be essential in maintaining the metal's structural integrity, it provides important benefits for a roof system as a whole. Any roof, even a sturdy metal one, is subject to deterioration from entrapped moisture. The right underlayment will help prevent moisture from infiltrating and keep condensation from collecting inside the attic.

There are three common types of underlayment for metal roofs: asphalt-saturated felt, self-adhering membrane, and synthetic. Felt is still the most common, largely because it's what roofers are used to and comfortable with. It's also quite cheap. Under a metal roof you'll need one layer of 30-pound felt or two layers of 15-pound. But here's something to consider; an asphalt shingle roof will last 20 to 30 years, and so will felt underlayment. When the shingles start to deteriorate, the underlayment is also at about the same stage. A complete tear-off and a new roof are needed. But a metal roof will last a lifetime. You want to install an underlayment with a lifespan that will match that of the roofing.

Self-adhering membrane underlayment is a rubberized asphalt or butyl-based adhesive on a polyethylene sheet. You peel off the protective paper and stick the underlayment directly onto the roof decking. Longer lasting than felt, it gives better protection at eaves, valleys, and dormers, and holds up better than felt in most circumstances, such as in high-temperature locations. But it doesn't do as well in extreme cold. It also takes some skill and experience to install. If the job isn't done properly, it will eventually start to peel back, making the roof vulnerable to infiltration.

Synthetic underlayment for metal roofing is gradually becoming more prevalent. It's made from polyethylene or polypropylene fibers woven in a polymer to form a sheet. Synthetic has definite advantages over the other options. It has the longest lifespan, is more resistant to tearing and other damage, is easier, safer and faster to install, and can take high temperatures. While synthetic is more expensive than felt, the labor cost to install it will be less, in most cases balancing it out.

Because synthetic is so effective at keeping moisture out, making sure the attic has good ventilation is especially important when you use this type of underlayment.

Labor for Installing Metal Roofing

Let's assume you're going to cover the roof described in Example 9-2 with 5-V crimp corrugated roofing that costs $2.11 per square foot, including sales tax. From Example 9-2, you must purchase 4,349 square feet of panels. So the total material cost is:

4,349 SF x $2.11/SF = $9,176.39

According to the *National Construction Estimator*, one sheet metal worker can install a square foot in 0.027 manhours. If the employer's cost for one hour is $42.16, including labor burden, then the labor for installing the roof will be:

3,466 SF x 0.027 hours/SF x $42.16/hour = $3,945.42

The total cost is:

$9,176.39 + $3,945.42 = $13,121.81

10 Built-up Roofing

▶ A built-up roof (BUR) is membrane roofing, made up of asphalt- or tar-saturated felts, coated felts, fabrics or mats with alternate layers of asphalt or pitch bitumen. The broad membrane roofing classification also includes elastomeric roofing, which we'll cover in detail in the next chapter.

There are three types of built-up roofing:

- Aggregate-surfaced roofing, which uses a layer of aggregate embedded in a flood coat of bitumen

- Smooth-surface roofing, which uses a glaze coat of bituminous or fibrated aluminum materials

- Mineral-surfaced roofing, which uses a granule-surfaced sheet

Bitumen makes a BUR waterproof. Felts make it strong and flexible so the roof can expand and contract. If properly installed and regularly maintained, the life expectancy of a BUR is 20 years or more. Besides long life, they have another advantage: BUR systems are extremely resistant to wind uplift.

The way you install a BUR system depends on several things. Before we discuss the mechanics of applying a built-up roof, let's see how to figure out the best installation method.

Roof Slopes

The maximum slope of a BUR depends on what type interply and surface bitumen and roofing surface you use. Don't install pitch bitumen over roof slopes more than 1/4 in 12. Don't apply surface aggregate over slopes greater than 3 in 12. You can install smooth-surfaced or mineral-surfaced roofing (using asphalt bitumen) on slopes up to 6 in 12. On non-nailable decks with roof slopes that are more than 1 in 12 (using asphalt bitumen) or 1/2 in 12 (using pitch bitumen), you must install wood nailers to hold the roofing felts or insulation in place. We'll discuss that in detail later.

Substrate Design

The substrate (insulation, or deck) must be compatible with the BUR system that covers it. Decking for BUR systems must be strong enough to resist deflecting under normal live loads such as snow, standing water and foot traffic. Decking that does deflect excessively under a load causes cracks in the overlying BUR.

Report any defects you find in the deck (to the general contractor in writing) and make sure they are corrected before you install a BUR. Look for improper or inadequate fastening of the deck, weak spots or holes, warped boards or improper drainage. If the deck is plywood, look for unsupported edges or inadequate thickness.

Remember, if the roof you installed fails due to an inadequate or defective deck, there's no way you're going to convince the customer that it's not your fault. You should also check that the roof deck is designed to adequately shed and drain water. For good drainage, the roof must have a minimum slope of 1/8 inch per foot; however, a slope of 1/4 inch per foot is better. Make sure there are plenty of roof drains, that leaders and gutters are large enough to handle the flow of water, and that the roof deck is sloped to carry water to drainage areas.

Don't install metered roof drains (or any drain designed to restrict water flow) unless the roof is also designed to retain water. Check that any ponded water areas can evaporate within 24 hours after a rain. Roofs designed to pond water may not be warranted, depending on the manufacturer. Water is heavy: water just 1 inch deep weighs about 5 pounds per square foot. That weight is a primary cause of deck deflection. Sometimes the deflection is permanent. If this happens too often, the roof structure could eventually collapse.

A properly-designed roof slopes toward one or more roof edges to carry drainage water off. You can contour sprayed-on foam insulation to slope toward roof drains, or install tapered rigid insulation to provide good drainage. A roof deck with a proper slope will show no signs of standing water 48 hours after a rain.

Metal Decks

If you use a metal deck for a BUR substrate, be sure it's 22 gauge (minimum) with 1½-inch-deep ribs. The ribs are designed with a maximum deflection of $1/240$ of the span between supporting members under a 300-pound load placed on a 1-square-foot area at mid-span.

Install the deck so there's no more than a $1/16$-inch horizontal deviation across any three adjacent top flanges. Use a straightedge to check the deviation. If the deck fails this flatness tolerance, you have to use mechanical fasteners to secure insulation to the deck. Over steel decking, use fasteners long enough to penetrate the deck. Figure 10-1 shows several types of fasteners, and the substrate materials you use them with.

The metal deck beneath built-up roofing must be galvanized or painted. Stagger the deck end laps and weld, crimp or button-punch the side laps.

The type of metal deck and rib determine both the type and the thickness of the insulation you install. Use a minimum ¾-inch-thick insulation over a metal deck. Due to potential deflection and/or irregular surfaces, some manufacturers recommend that you install insulation over a metal deck using one mechanical fastener for each 2 to 4 square feet, or 12 to 15 pounds of hot steep asphalt per square (hot bitumens are described in detail later in the chapter). Manufacturers recommend using mechanical fasteners.

If you use a vapor retarder base sheet beneath the insulation, install it in hot steep asphalt applied in continuous ½-inch-wide ribbons placed on 6-inch centers parallel with the ribs. You can also apply bitumen over the entire deck in contact with the vapor barrier. You can then install the overlying insulation with mechanical fasteners, or in a solid mopping of hot steep asphalt.

If you're using more than one layer of insulation, you can install both layers in one operation with mechanical fasteners. Or you can install the second layer in 12 to 25 pounds of hot steep asphalt per square. Always install the upper layer of insulation with end and side joints staggered over the joints of the bottom layer.

Lightweight Insulating Decks

Lightweight insulating decks (wet-fill decks) include lightweight aggregate concrete (20 to 40 pounds per cubic foot dry density), foamed concrete, gypsum concrete and thermosetting asphaltic decks.

Over lightweight insulating concrete fill or poured gypsum concrete decks, you install a coated or vented base sheet using mechanical fasteners. Don't mop hot bitumen over lightweight concrete or gypsum concrete. If you don't use insulation board, install a ¾-inch protection board into a solid mopping of hot bitumen between the base sheet and the subsequent roofing membranes. For insulation over lightweight concrete or gypsum

Deck type	Special instructions	Fastener recommendations
Wood, tongue & groove & plywood		1, 3, 4, 7, 8, 11
Lightweight insulating concrete	See note "1"	9, 10
Existing gypsum decks (reroofing)	See note "2"	6, 9, 10, 12
Structural wood fiber	See note "3"	2, 5, 6
Fiberboard roof insulation		1, 3, 4, 7, 8, 12

Note "1" Hydro-Stop™ vapor barrier applied with Celotex® Anchor Bond® LWC™ fasteners must be used as the first ply over new lightweight insulating concrete decks.

Note "2" Due to variation in hardness of existing gypsum decks, fastener selection should be field tested to determine penetration and holding power.

Note "3" A Vaporbar™ base sheet must be used as the first ply or embedment course over structural wood fiber decks if roof insulation is to be hot mopped to the deck.

Use Celotex® Anchor Bond® roof insulation fasteners for attachment of roof insulation to decks.

Detailed description of fasteners & sources of supply

1	Roofing nail 11 or 12 ga. 3/8"-1/16" diam. head — National Nail Co.
2	Insuldeck Loc-Nail — E.G. Building Fasteners Corp.
3	Roofing nail annular thread 11 ga. 3/8" diam. head — Independent Nail Co.
4	Roofing nail spiral thread 11 ga. 3/8" diam. head — W.H. Maze Co.
5	Capped Es-Nail 1" cap — ES-Products
6	Tube-Loc nail 1" diam. cap — Simplex Nail & Mfg. Co.
7	Squarehead cap nail annular thread 1" diam. cap — Simplex Nail & Mfg. Co.
8	Squarehead cap nail spiral thread 1" diam. cap — Independent Nail Co.
9	Nail-Tite type A 1¼" diam. cap — ES-Products
10	Anchor Bond LWC fasteners Zonolite or Nail-Tite Mark III ES-Products; Olympic lightweight concrete fasteners
11	Roofing staple for power driven application only — Bostitch Spotnails
12	Do-All nail hardened — E.G. Building Fasteners Corp.

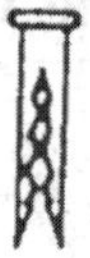

Figure 10-1 Mechanical fasteners

Courtesy of Celotex

concrete, mechanically fasten a vapor retarder sheet between the fill and the insulation board. Also, vent the fill with roof relief vents like those in Figure 10-2.

When top venting is required, install pressure relief vents. As a rule of thumb, install one vent per every 900 to 1000 SF of roof area. Some manufacturers recommend that you install 4-inch-diameter (minimum) hooded pressure relief vents 20 feet from the roof edges and at 40-foot centers thereafter. Make sure the vent openings penetrate the roof fill a minimum of 2 inches. I recommend filling the penetration with loose fiberglass insulation, as shown in Figure 10-2, to prevent condensation from forming on the roof deck.

Freshly poured lightweight insulating concrete decks must cure before you install a BUR system. As a rule of thumb, allow at least ten days of dry weather to pass between the pour and installing a BUR system. This rule of thumb makes one assumption: the deck that the concrete was poured on must have underside venting, like a perforated steel deck. Unfortunately, this isn't always the case. If the insulating deck is poured over a nonporous structure, concrete takes longer to cure. As a rule of thumb, allow a full 30 precipitation-free days curing time to pass before you install a BUR system.

Install gypsum concrete decks a minimum of 2 inches thick and pour the deck over cement-fiber or glass-fiber form boards that vent the underside. Be sure also, if you paint the underside, that you use a breathable paint. You can use slotted corrugated panels instead of form boards if their slots provide at least 1.5 percent open area. Never install lightweight insulating concrete to receive a BUR system over a vapor retarder, or over unvented metal decks, poured concrete decks, existing built-up roofs or any substrate that doesn't let the fill dry from below. Never install lightweight concrete decks on roofs with slopes greater than 1 in 12. Also be aware that some manufacturers of BUR products categorically refuse to warranty systems installed over lightweight insulating concrete decks or lightweight aggregate-asphaltic compacted fills.

Seal joints between precast gypsum panels with an 8-inch-wide felt strip embedded in roofing cement. Then you can install a vapor retarder sheet with mechanical fasteners. Over this, install insulation board or felts in a solid mopping of hot bitumen.

Over thermosetting insulating fill, install the base sheet in a hot mopping of bitumen. If you use additional insulation, install a vapor retarder between the fill and the insulation board, and vent the fill.

Mechanical fasteners like the ones in Figure 10-1 (used over gypsum and insulating concrete decks, or over the insulation board or other nailable decks) are made by various manufacturers for use with specific materials. Make sure you choose the right fastener for the material you're using.

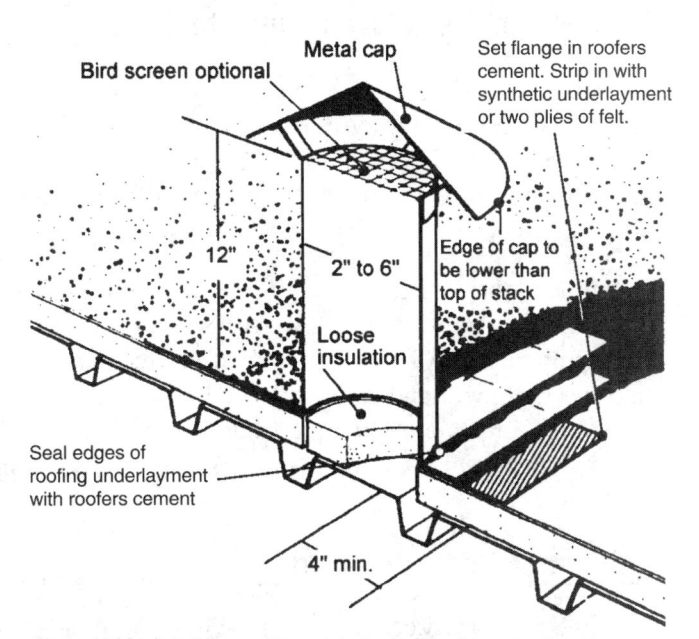

Note: This detail is used to minimize vapor pressure from insulation. The moisture may have entered due to leaks, faulty vapor retarders or during construction. The spacing is determined by the type of insulation used and the amount of moisture to be relieved. It is sometimes used for new roofs when vapor retarders are used and a venting system is desired.

Figure 10-2 Roof relief vent

Precast and Prestressed Concrete Decks

Over a precast or prestressed concrete deck, prime the deck and install insulation board into a hot mopping of bitumen at 30 pounds per square. If joints between the concrete members aren't sealed, don't mop within 4 inches of the joints to prevent the bitumen from dripping into the joints.

Never install a membrane directly over a deck. Many manufacturers recommend that you install a minimum of 2-inch-thick, 2500 PSI concrete fill or a ¾-inch protection board over prestressed concrete roof decks before you apply roofing. That's to compensate for surface irregularities due to slab misalignment and camber. If you pour a concrete deck, don't fill the joints between concrete members because it keeps the fill from drying from below. If you seal the joints, you must vent the fill from above.

Reinforced Concrete Decks

Prime reinforced concrete decks with 7½ pounds (¾ gallon) of asphalt primer per square. Use a special thin liquid asphalt primer that you can use over non-nailable decks. This comes in both 5-gallon pails and 55-gallon drums. After the primer has fully dried, apply hot bitumen at 20 to 30 pounds per square. Install the first roofing ply into the hot bitumen.

If you use a base sheet, spot-mop it in place over the deck. Spot-mopping allows venting of any remaining moisture in the concrete. However, be aware that some manufacturers' warranties only apply when the base ply is solid-mopped to the deck.

Wood and Plywood Decks

Use mechanical fasteners to install the base sheet over wood board or plywood decks. Choose 11-gauge galvanized nails with ⁷⁄₁₆- to ⅝-inch heads, or drive nails through caps with an equivalent diameter. Use large-head annular ring-shank nails over plywood decks. The large heads help prevent the wind uplifting the roofing underlayment. Use nails that are long enough to go at least ⅝ inch into the wood board, and at least ½-inch into the plywood. One advantage of nailing is that you can install the base sheet to the deck and still allow horizontal passage of water vapor trapped between the deck and the roof system.

Structural Wood Fiber Decks

Use structural wood fiber decks manufactured with a factory-applied felt surface. Strip all joints in the deck with 6-inch-wide felt embedded in roofing cement. Nail a vapor retarder base sheet over structural wood fiber decks. Then install at least 1 inch of rigid insulation into a solid mopping of hot

Built-up Roofing

steep asphalt. If you use decking without the felt surface, install rosin paper between the deck and the base ply. Install only as much decking as you can cover with roofing that same day.

Back Nailing

If a built-up roof slopes more than 1 in 12 (using asphalt bitumen) or more than $1/2$ in 12 (using pitch bitumen), you'll have to install roofing underlayment, either felt or synthetic specifically designed for this purpose, with mechanical fasteners in addition to the interply bitumen mopping. That's to stop the membranes from slipping. This is called *back nailing*. Some manufacturers recommend back nailing only on slopes exceeding 2 in 12 with asphalt bitumen. Don't apply coal-tar pitch over slopes exceeding 1 in 12.

Nailable decks include wood, plywood, gypsum (poured or precast plank), lightweight insulating concrete and structural wood fiber. Over those decks, back nail each felt with the appropriate fasteners as specified in Figure 10-1. Apply fasteners on 12-inch centers along a row 1 inch below the top edge of each underlayment strip.

Over non-nailable decks, back nail each ply into high-density fiberboard insulation at 12-inch centers along a row located 1 inch below the top edge of each strip of underlayment, or back nail into treated wood nailers 1 inch (or about 10 and 12 inches) below the top edge of each strip. Some manufacturers recommend that you install nailers parallel with the slope and space the nailers face-to-face at a maximum of 48 inches, as shown in Figure 10-3.

Install the membranes perpendicular to the wood nailers and be sure that at least two roofing plies cover each nail head. Also install nailers at ridges, eaves, gable ends and around openings. Where there's no insulation, install the tops of the nailers flush with the top of the deck. Over insulated decks, install the tops of the nailers (now called insulation stops) flush with the top of the insulation boards. Use insulation stops that are at least 2 inches wide (some recommend 4 inches) and as thick as the insulation. Install 6-inch-wide nailers at the bottom of the slope. That's also shown in Figure 10-3.

Other manufacturers recommend that you install the nailers perpendicular to the slope and the membranes parallel to the slope. This is called the "strapping" method. The spacing you use for the nailers depends on the roof slope and the type of roof surface installed. Figure 10-4 lists nailer locations for various applications.

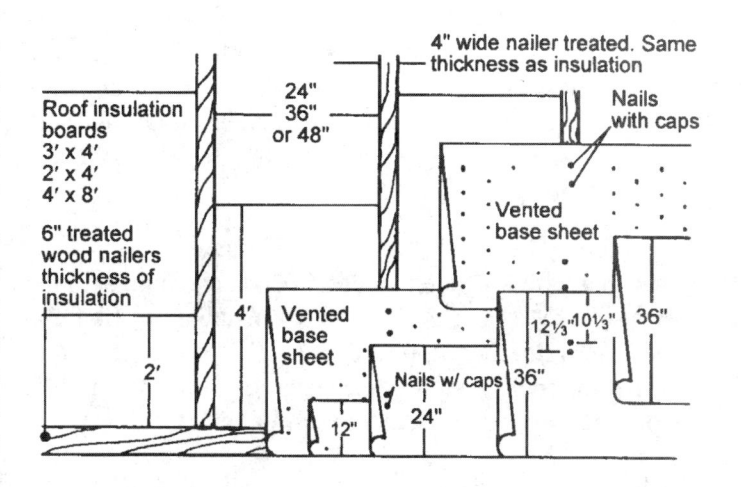

Figure 10-3 Wood nailers

Roof slope	Type of surface	Nailer locations (face-to-face)
1 to 2 in 12	Smooth	None required
	Aggregate or cap sheet	20'
2 to 3 in 12	Smooth	20'
	Aggregate or cap sheet	10'
3 to 4 in 12	Smooth	10'
4 to 6 in 12	Smooth	4'

Figure 10-4 Nailer placement recommendations

Still other manufacturers recommend that you use the strapping method and install the nailers 20 feet face-to-face over slopes through 3 in 12, and 4 feet face-to-face over slopes greater than 3 in 12.

Base Sheets (Vapor Retarders)

The first (and normally the heaviest) membrane you install in a BUR system is the base sheet. In most cases, the base sheet acts as a vapor retarder which protects the insulation or subsequent membranes from damage due to moisture from within the building.

A vapor retarder is normally a coated sheet or a vented membrane. A coated sheet (also called smooth roll-roofing) is a roofing felt (organic or inorganic) coated on both sides with bitumen and finished with a heavy coat of non-sticking mineral powder. You can use a coated sheet as a vapor-retarding base sheet, or as the top ply on a smooth-surfaced built-up roof. Coated sheets are 36 inches wide and 36 to 144 feet long. They weigh 45 to 80 pounds per square.

One type of vented base sheet is made of an asphalt-impregnated glass fiber mat with $1/4$-inch diameter holes located on 5-foot centers. These holes let moisture escape at the time of installation. Some systems require that you install fiberboard, perlite board or fibrous glass insulation between the insulation board and the base sheet to cut down on blisters made by air and moisture trapped between the two layers.

Another type of vented base sheet is made of an asphalt-saturated glass fiber mat whose bottom surface is covered with mineral granules embossed to provide channels for moisture vapor to escape. The top surface is smooth to receive the bitumen and overlying roof membranes.

A vented base sheet weighs about 72 pounds per square. This type of sheet is highly recommended over vented fills such as gypsum decks. Manufacturers also recommend it when you're re-roofing over existing roof membranes.

Another kind of base sheet has $5/8$-inch perforations spaced on 3-inch centers in both directions. This sheet weighs 60 pounds per square. It's secured by the asphalt that penetrates the vent holes as you install the subsequent roofing felt or insulation board with a solid hot bitumen mopping.

Turn down the edges of vented base sheets over the edge of the roof to allow venting under the metal edge strip, as shown in Figures 10-5, 10-6 and 10-7. Turn the edges up along vertical surfaces and spot-mop the sheets at 2-foot centers to allow venting between the base flashing and metal counterflashing, as shown in Figure 10-5. Provide additional venting with relief vents when any dimension of a roof is more than 60 feet.

Built-up Roofing

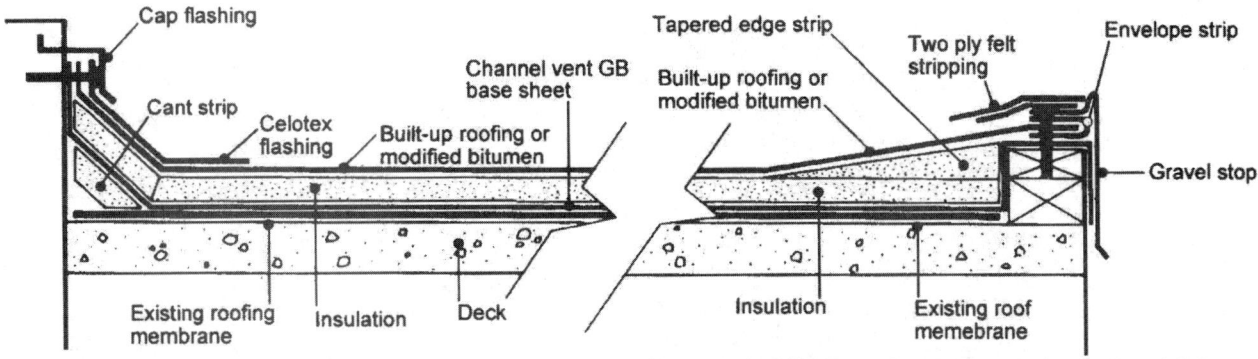

Figure 10-5 Cross section of a BUR system

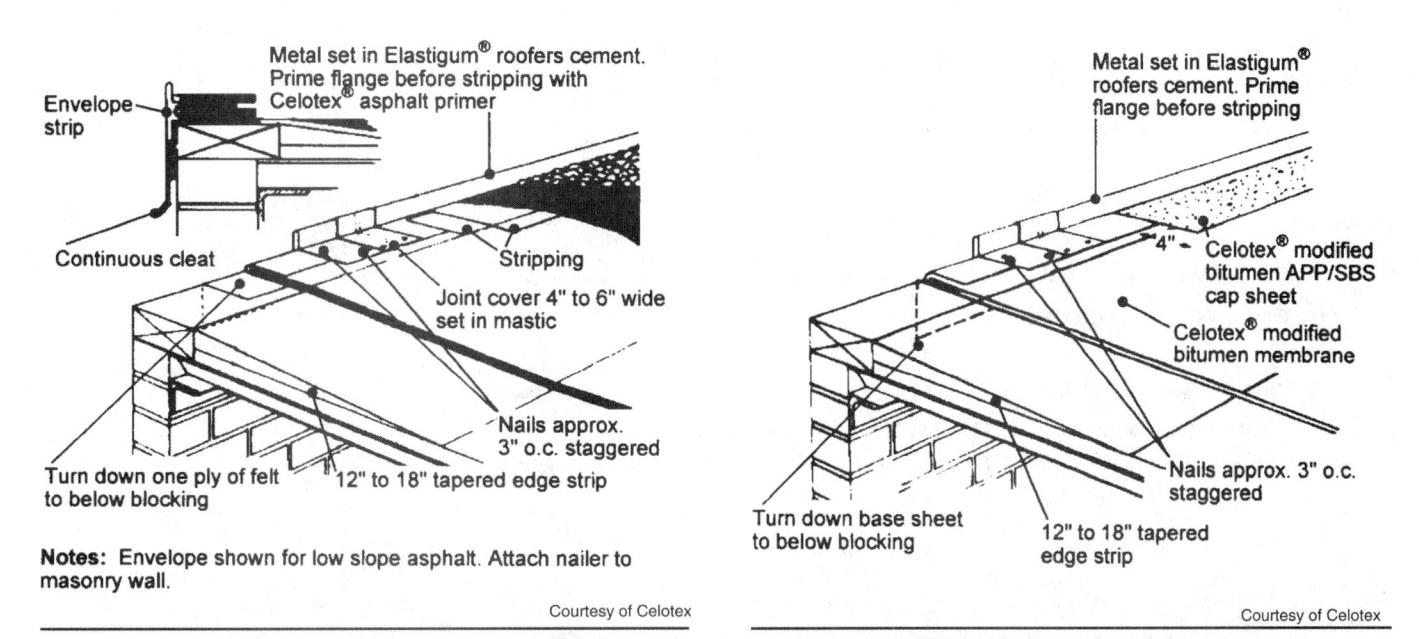

Figure 10-6 Light metal roof edge detail stripped with felts

Figure 10-6 Light metal roof edge detail stripped with cap sheet material

Install a vapor retardant base sheet whenever these conditions exist:

- The average outside winter temperature is 40°F or less
- You expect unusually high relative humidity generated from within the building (45 percent or greater during winter)
- The attic air space is used for a return-air plenum for heating or air conditioning

Since water vapor tends to go from warm to cooler areas, always install the vapor retarder on the side of the roof assembly that's warmer most of the time.

Install the base sheet with 4-inch side laps and 6-inch end laps. Over nailable decks such as wood, plywood, structural cement-fiber or poured gypsum, nail all laps on 9-inch centers and stagger-nail the center of the sheet in two rows on 18-inch centers. Locate each row 12 inches from the edges of the sheets. Drive about 100 nails per square. Drive nails through flat metal discs at least 1 inch in diameter.

On wood decks, install the base sheet over rosin paper. Prime non-nailable decks using 1 gallon of asphalt primer per square. Embed the base sheet in hot steep asphalt using 23 pounds per square. Don't install vapor retarders with cold-applied adhesives.

Over nailable or non-nailable decks, turn the sheets up parapet walls, curbs and other vertical surfaces at least 4 inches above the top of the insulation. Use roofing cement or steep asphalt to adhere the sheet to a vertical surface. At pipe penetrations, flash the base sheet with two plies of asphalt-saturated felt and roofing cement.

When you expect extreme humidity inside the building (50 percent at 70°F or greater), install two base sheets. Embed the second base sheet in a solid mopping using 23 to 25 pounds of asphalt per square.

Roofing Membranes

Membranes used in built-up roofing include roofing felts, fabrics, coated sheets, rosin-sized paper and mineral-surfaced sheets. Most manufacturers recommend that you apply roll material with its long dimension perpendicular to the slope of the roof.

Rosin-sized Paper

In places where hot bitumen might drip through cracks or spaces in the deck, nail rosin-sized sheathing paper or unsaturated felts over decks (especially wood) before you apply the bitumen. Install rosin paper between a wood roof deck and the roofing membrane to prevent damaging chemical reactions between wood resins and the membrane.

You can also install rosin paper over a deck as a slip sheet, to prevent adhesion of the base sheet to the deck. This helps relieve expansion-contraction problems. Nail the base sheet over the rosin paper. Lap rosin paper sheets 2 inches and secure them with enough nails to prevent slippage and wind uplift. Glass fiber felts don't require rosin paper, so you can nail them directly to the deck. Rosin paper, or some other separating layer, is required over plywood decks, and you must install blocking beneath the end joints of the plywood. Rosin paper weighs about 5 pounds per square.

Roofing Felts

Roofing felts make a BUR system strong. They also help protect the lower layers of bitumen from debris, dirt, water, air, and the drying and weathering effects of the sun. Roofing felts are absorbent, so they help prevent bituminous materials from melting and flowing away during hot weather. Felts are perforated so trapped air can escape from under the sheets when you apply them. Glass fiber felts don't have perforations because they're already porous.

Some felts come with a coating of fire-resistant materials. Felts are often produced with a coating of fine mineral powder on one or both sides to keep them from sticking to each other when they're packaged in a roll. If you're using uncoated felts, keep them cool so they don't stick together.

Apply 25 pounds of asphalt or pitch per square between felts with an application tolerance of plus or minus 15 percent. Too thin a layer of bitumen weakens the lap, and too thick reduces friction and allows the felts to slip on sloped roofs.

Porous substrates such as low-density decks and insulation tend to absorb bitumen. In this case, you may need more than 25 pounds of bitumen per square to make sure the felt adheres. You've used enough if there's fluid bitumen remaining on the surface after absorption.

Control the number of plies by varying the widths of the side laps. Apply felts in a shingle-like fashion, starting at the low end of the roof. Felt exposures vary, depending on the number of plies you install (Figure 10-8). You can determine the felt exposure required for any BUR system with the following formula (the 34 inches at the 36-inch felt width minus a 2-inch edge lap):

$$\text{Exposure} = \frac{34 \text{ inches}}{\text{No. of Plies}}$$

Equation 10-1

▼ **Example 10-1:** To find the exposure of each ply of a BUR system when two plies of felts are applied in a shingle-like pattern:

$$\text{Exposure} = \frac{34 \text{ inches}}{2}$$
$$= 17 \text{ inches}$$

Using the same formula, three plies require an exposure of $11^{1}/_{3}$ inches, and four plies require an exposure of $8^{1}/_{2}$ inches.

Roofing felts are either organic or inorganic. The most commonly used organic felt weighs 15 pounds per square. An equivalent glass fiber (inorganic) felt weighs $7^{1}/_{2}$ pounds per square. Roofing felts are usually typed by their weight. For example, Type 8 means a $7^{1}/_{2}$-pound glass fiber felt and Type 30 means a 30-pound felt. Full information about roofing felts is in Chapter 3.

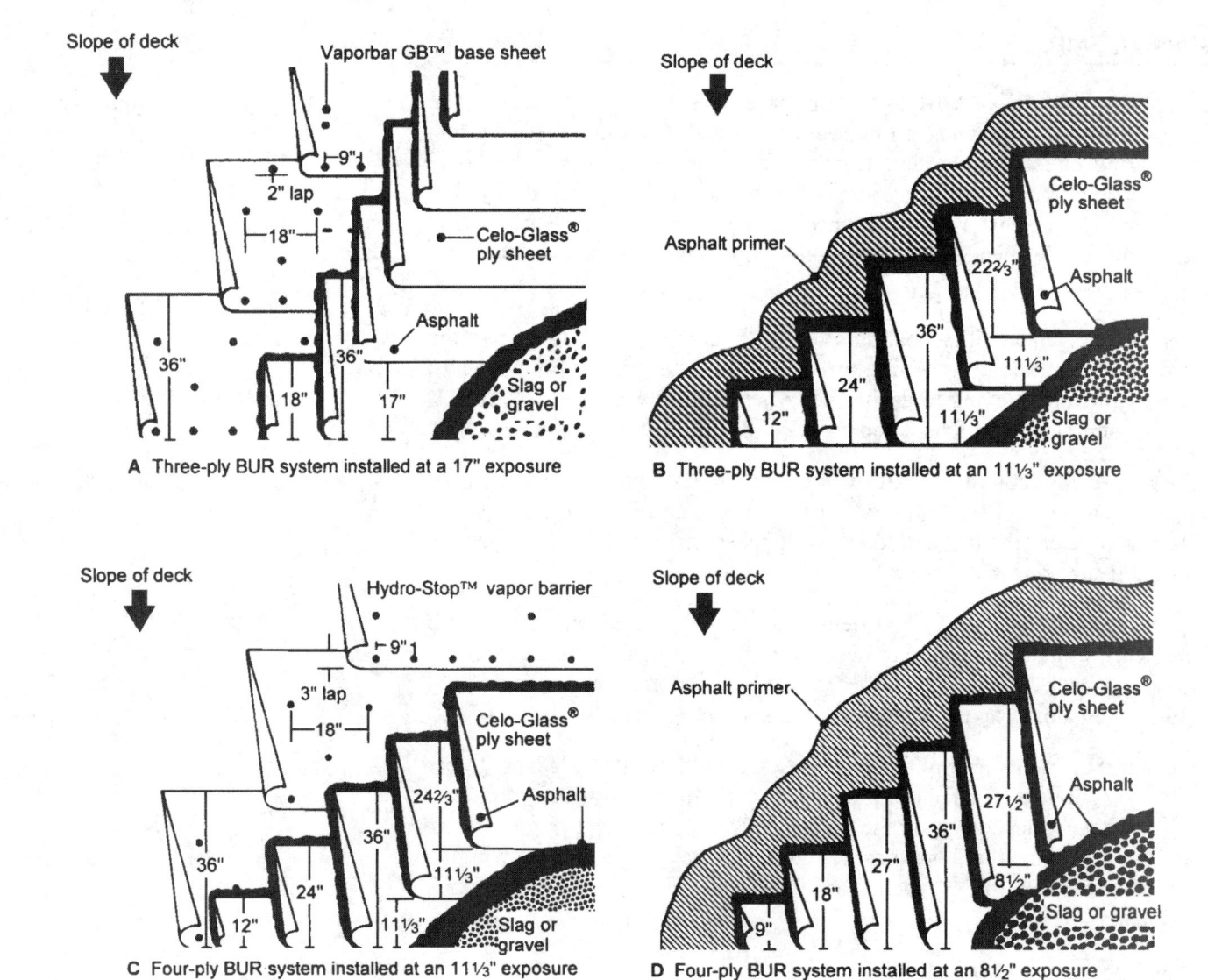

Figure 10-8 Felt exposures

Some roofing felts are coal-tar saturated and you install them with coal-tar bitumen (pitch). Others are asphalt-saturated and you install them with asphalt bitumen.

Three-ply BUR System

A three-ply BUR system normally consists of:

- Base sheet (nailed or mopped) followed by two layers of roofing felts laid like shingles with a 17-inch exposure, as shown in Figure 10-8 A; or,

- Three plies of felts laid like shingles with an 11^1/$_3$-inch exposure, as shown in Figure 10-8 B.

Four-ply BUR System

A four-ply BUR system normally consists of:

- A base sheet (nailed or mopped) followed by three layers of roofing felts laid with a 11$\frac{1}{3}$-inch exposure, as shown in Figure 10-8 C; or

- Four plies of roofing felts laid with an 8$\frac{1}{2}$-inch exposure, as shown in Figure 10-8 D.

Fabrics

Fabric membranes made of bitumen-saturated cotton, glass fiber or jute are sometimes used for roofing membranes. For more detailed information about these products see Chapter 12. Uncoated polyester filaments bonded under heat and pressure produce a roofing fabric that's lightweight (0.9 pounds per square foot) and elastic. Polyester fabric is usually glued and surfaced with cutback asphalt.

Hot Bitumens

Bitumen is defined as an amorphous, semi-solid, organic hydrocarbon mixture. Asphalt and coal-tar pitch are the most widely-used bitumens in the roofing industry. Asphalt occurs naturally, and it's also made during petroleum processing. It weighs an average of 8.7 pounds per gallon.

Coal-tar bitumen (coal-tar pitch) is a product of the distillation of coal tar in the manufacture of coke or gas. It's softer than asphalt and has excellent self-healing properties after it's cooled. It's virtually unaffected by water. It weighs about 10.6 pounds per gallon.

Coal-tar pitch is more expensive than asphalt. But despite its higher cost, it's recommended on dead-level roofs and low-slope roofs with slow drainage and occasional ponding. Coal-tar bitumen is required on roofs designed to retain water. It's also recommended in highly corrosive environments.

There are three types of coal-tar materials:

- Type I (coal-tar pitch) softens at 126° to 140°F. This type of coal-tar is normally used on built-up roofs.

- Type II (waterproofing pitch) softens at 106° to 126°F. Usually used for waterproofing basement walls and promenade decks.

- Type III (coal-tar bitumen) softens at 133° to 147°F; used on roofs with slopes up to ¼ in 12.

Asphalt and coal-tar pitch are generally incompatible, with a couple of exceptions:

- You can install asphalt flashing material with coal-tar pitch
- You can use hot Type II asphalt over coal-tar saturated felts.

When you apply glaze coats of bitumen, it produces a fusion between plies because of the partial melting of asphalt within the plies. Bitumens behave like fluids when they're heated: during warm weather, they're elastic; during cold weather, they harden.

Bitumens are normally applied with a mop and the membranes are pushed into the hot bitumen with a push broom. This is called *brooming-in*. To avoid displacement of the membrane, you drag the broom across the membrane from the unmopped side rather than pushing from behind the roll. It's important that you broom-in the membrane while the bitumen is hot to avoid blisters caused by vapors trapped beneath a membrane. If fishmouths (wrinkles at the edge of the felt) or buckles develop during the broom-in, immediately cut and repair them.

Four types of asphalt are available, with varying softening-point temperatures:

- Type I (dead-level) asphalt softens at 135° to 151°F. This type of bitumen is self-healing, and you can install it over roof slopes up to 1/4 in 12.

- Type II (flat) asphalt softens at 158° to 176°F and you can use it on slopes up to 1/2 in 12.

- Type III (steep) asphalt is the most commonly used. It softens at 180° to 205°F and you can apply it on roof slopes up to 3 in 12. Due to the intensity and length of the summer season, parts of some southern states such as Texas, Florida, New Mexico, Arizona and California require Type III asphalt on roof slopes ranging from dead level to 1/2 in 12.

- Type IV (special steep) asphalt softens at 205° to 225°F. Use this type of asphalt on roof slopes up to 6 in 12 where there is a potential during hot weather for the asphalt to melt and run, or for the roofing felts to slide away. You can use Type IV asphalt on smooth roof surfaces, but it lacks the self-healing properties of the low-melt asphalts.

Using the correct type of asphalt will result in a roof that softens and self-heals, yet doesn't sweat melted bitumen during the heat of the day. The temperature of the lower roof plies can exceed 175°F on a 100° day. In winter, the temperature of a built-up roof over insulation will be close to the outdoor temperature.

The flash point of asphalt is the lowest temperature at which it gives off enough vapor to form an ignitable concentration. Most asphalts have a flash point ranging from 437° to 500°F. You can safely heat dead level asphalt or pitch to 400°F and flat and steep asphalts to 450° and 475°F. Don't heat asphalt to within 25°F of the flash point.

Bitumens don't waterproof as well if you overheat them or heat them too long. If you won't use a bitumen for four hours or more, shut off the kettle or reduce the temperature to about 325° to 350°F. Don't apply Type I and II asphalts or coal-tar bitumen at a temperature lower than 350°F, and Types III and IV at lower than 400°F.

Bitumen congeals quickly in cold weather, which can result in poor adhesion as well as felt slippage. But don't try to compensate for cold conditions by overheating the bitumen. This would likely result in inadequate coverage, poor waterproofing and voids between the plies of roofing felt. Instead, keep the roofing roll close behind the mop (approximately 5 feet) and broom-in the membrane as quickly as possible.

The equiviscous temperature (EVT) of a bitumen is the optimum temperature and viscosity to apply the material so it will stick properly and be waterproof. This is the temperature at which 125 centipoise is attained. Apply roofing bitumen at a temperature within 25°F of the EVT.

Store and protect bituminous materials from rain and snow. If moisture gets into them, they foam up in the kettle and you'll have problems applying them.

Cold-applied Bitumens

There are many cold-applied bituminous materials available. Asphalt mastic consists of asphalt and graded mineral aggregate. Bituminous grout is made of bitumen and fine sand. You can pour either product when heated, but they must be troweled when cool.

Asphalt emulsions are made of fine droplets of water dispersed in asphalt with the aid of an emulsifier such as bentonite clay. Some emulsions also contain glass fibers or other materials. Asphalt emulsions are so thin that you usually spray them on. Use emulsions for surfacing bitumen on slopes up to 6 in 12. Apply surface emulsions at the rate of 3 gallons per square. Since emulsions contain water, don't let them freeze.

Cutback bitumens are thinned with organic solvents and light oils to make them flow freely. Cutbacks are used mainly for interply applications and less frequently as a surfacing material. When used as a flood coat for embedding aggregate, apply cutback asphalt at the rate of 6 to 7 gallons per square. Cutbacks contain flammable solvents; don't expose them to an open flame. Liquid emulsions and cutbacks solidify when the water or solvent in them evaporates, leaving only the bitumen behind.

The most common system used in cold-process roofing uses three layers of 53-pound cold-process felts saturated with a cold asphalt emulsion. Secure the felts with nails and asphalt adhesive applied at the rate of $2\frac{1}{2}$ gallons per square. Then cover the surface with a layer of cold-applied asphalt-fibrated emulsion at the rate of 3 to 4 gallons per square. This roof system is sometimes referred to as a Type 5 roof.

Flashing Sealers

Asphalt cements are thick pure asphalts with no fillers and only enough solvent to permit application. Asphalt cements are used mainly for sealing the laps of flashing materials.

Plastic cements, flashing cements, all-purpose roofing cements and cutback cements are trowelable mixtures of asphalt or coal tar with fillers such as glass fibers, powdered aluminum, rubbers and solvent. These products are used for securing flashing material.

Store asphalt cement and other related products in a warm place until you use them. If you need to warm the cement, place the unopened container in hot water until it's pliable. Never heat the container directly over a flame.

Cold-applied Roof Coatings

Many cold-applied asphalt roof coatings are available. While their composition varies, the main ingredients are always emulsified asphalt and mineral colloids. Acrylic- and neoprene-based adhesives are other popular cold-applied products. These materials are very durable and elastic, and can be applied by brush or spray at a rate of approximately 2 to 6 gallons per square, depending on the specific product.

Roof coatings consisting of asphalt or coal tar, fiber and nonvolatile penetrating oils are sometimes applied over an existing built-up roof to rejuvenate the old asphalt (or tar) to its original condition. This type of coating is called a *resaturant*.

Asbestos

There are still roofing products, including some roof coatings, that contain asbestos fibers. Evidence to date indicates that the asbestos in these materials is contained, so it probably doesn't present the health risk of free asbestos. Nevertheless, many building owners won't allow asbestos in any form in their building due to potential liability, and most designers won't specify products containing asbestos. As a result, asbestos-free roofing materials have been developed that perform just as well and are widely available.

Surface Aggregate

Aggregate is often applied over a hot flood coat of bitumen on the surface of a BUR system. The flood coat is normally applied at the rate of 60 to 70 pounds per square (asphalt), or 70 to 75 pounds per square (coal-tar pitch). In high-wind areas (70 mph and more), an additional 80 pounds of asphalt or 90 pounds of coal-tar pitch will sometimes be specified per square.

On water-retaining roofs, apply aggregate at the rate of 400 pounds per square (300 pounds for slag) into a flood coat of pitch applied at 70 pounds per square. Then sweep off all loose gravel or slag. If the roof doesn't have a controlled flow drainage system, apply a second coat of pitch at 85 pounds per square with aggregate at 300 pounds per square (200 pounds for slag). Again, sweep away any loose aggregate or slag and roll lightly to ensure the remaining aggregate is embedded in the pitch.

Some manufacturers maintain that felt plies can be left uncoated for up to six months before aggregate is installed. But when you apply aggregate, it's very important that you do so while the flood coat of bitumen is still hot. If you can't apply the aggregate soon after installing the membranes, or choose not to because other workers must walk on the roof, apply a glaze coat of hot bitumen to the surface at the rate of 8 to 10 pounds per square. When you can finally place the aggregate, sweep the glaze coat clean before you install the flood coat for the aggregate.

Aggregate does several important things:

- It protects the underlying roof system from the heating, drying and weathering effects of the elements.

- It prevents damage from foot traffic.

- It serves as a ballast to resist wind uplift and improves the roof's fire resistance.

- It reduces the roof surface temperature. Because of its high reflectivity, white aggregate such as marble chips (dolomite) was used to save on energy costs. Now, we have ceramic-coated granules, which boast an even-better reflectivity, including for dark colors, some manufacturers claiming as much as 80 percent reflectivity.

A properly installed aggregate-surfaced BUR system has a life expectancy of 20 years or more. The most common aggregates are ceramic-coated granules, rounded gravel, slag, or crushed rock. Less common are marble chips, scoria, pumice and crushed tile, brick and limestone. Grade gravel aggregate from $1/4$ to $1/2$ inches, and slag from $1/4$ to $5/8$ inches.

Make sure the aggregate is clean and dry so that it will stick well to the bitumen. The maximum acceptable moisture content (by weight) is 0.5 percent for crushed stone or gravel, and 5.0 percent for crushed slag.

Apply gravel or crushed stone embedded in a flood coat of bitumen at a rate of 400 pounds per square, slag at 300 pounds per square and marble chips at 400 to 500 pounds per square. Adhere approximately 50 percent of any aggregate to the bitumen. In high-wind areas (70 mph and more), apply an additional 300 pounds of gravel or 200 pounds of slag per square.

When embedded in cutback bitumen, apply gravel, crushed stone and slag at the rates of 400, 450 and 350 pounds per square, respectively. Don't install aggregate over asphalt-emulsion coatings. Avoid using low-density aggregates like bauxite, kaolin, gypsum, caliche and shale. They're too easily blown and washed off of the roof.

The quantity of ceramic-coated granules per square is much less; usually only a thin layer is required. But since it's a manufactured product, the amount per square varies according to the composition of the material and the usage. For this, follow the manufacturer's instructions.

Choose slag for roof slopes greater than 2 in 12 because it stays embedded better than gravel. I wouldn't install aggregate on roof slopes steeper than 3 in 12.

Mineral granules of the type used to make composition shingles can also be used for roof surfacing over cutback asphalt or mastic. If mastic is used, apply it at the rate of $2^1/_2$ to 3 gallons per square and deposit the granules at the rate of 50 to 60 pounds per square. You can install a granule surface on roof slopes up to 6 in 12.

Smooth-surfaced Roofing

You can install smooth-surfaced (black) roofing on roof slopes as steep as 9 in 12. On nailable decks, nail the top edge of each membrane on 12-inch centers for slopes greater than $^1/_2$ in 12. Install a 43-pound asphalt-coated base sheet, then three plies of 15-pound asphalt-saturated felt solidly mopped with bitumen applied at the rate of 25 pounds per square.

Over non-nailable decks install four plies of 15-pound felts into solid moppings of asphalt applied at the rate of 25 pounds per square. On inclines greater than $^1/_2$ in 12, install nailers for fastening the top edge of each membrane, driving nails at 12-inch centers. Surface the roof with a glaze coating of 20 pounds of hot asphalt per square, followed by 3 gallons of cold-applied asphalt emulsion per square. Instead of an asphalt coating, you can cover the roof surface with 1.2 gallons of fibrated aluminum coating per square. Use a roller to apply this coating.

Smooth-surfaced roofs are lightweight and easy to inspect and repair. Re-roofing can often be done without removing the old smooth-surface membrane. However, all smooth-surfaced roofs are more susceptible to damage and leaking due to hail impact than aggregate-surfaced roofs. The life expectancy of a smooth-surfaced roof is only six years.

Cap Sheets

You can install heavy-duty cap sheets instead of an aggregate surface. Cap sheets are made from a variety of materials, but mineral-surfaced sheets and reinforced polyester sheets are the most common.

Refer back to Chapter 5 for information about the physical features of mineral-surfaced asphalt sheets. Mineral-surfaced sheets are inexpensive, but not very durable. The life expectancy of a mineral-surface roof is only six years. Mineral-surfaced sheets are usually installed on roof slopes between $1/2$ in 12 and 6 in 12.

When you apply a standard roll-roofing BUR system to a nailable deck, nail a sheet of rosin paper to the deck. Cover the rosin sheet with two plies of 15-pound felt nailed and mopped with hot asphalt bitumen applied at the rate of 25 pounds per square.

Unless you're concerned about bitumen dripping into the building interior, no rosin sheet is required over a non-nailable deck or over insulation. Cover the deck with two plies of 15-pound felt solidly mopped with hot asphalt bitumen applied at the rate of 25 pounds per square.

Over either type of deck, install an 80-pound cap sheet over roof slopes of $1/2$ in 12 to 6 in 12. Mop a 55-pound cap sheet over roof slopes of 3 in 12 to 9 in 12. Precut cap sheets into 12-foot lengths before you install them. Lap the sides 2 inches and the ends 6 inches. Offset adjacent end laps at least 3 feet. Over nailable decks, nail end laps over slopes greater than 2 in 12.

A heavier type of roll-roofing BUR system is composed of one layer of rosin paper (over wood decks), followed by a nailed 15-pound asphalt-saturated felt. Following this, mop two layers of 15-pound felt followed by two layers of 120-pound slate-surfaced felt.

Another type, the three-ply glass fiber cap-sheet membrane system, consists of a base sheet, followed by one ply of felt and a cap sheet, as shown in Figure 10-9 A.

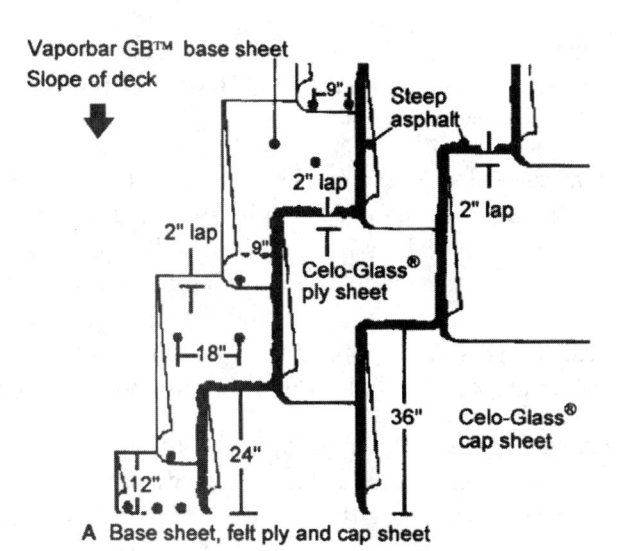

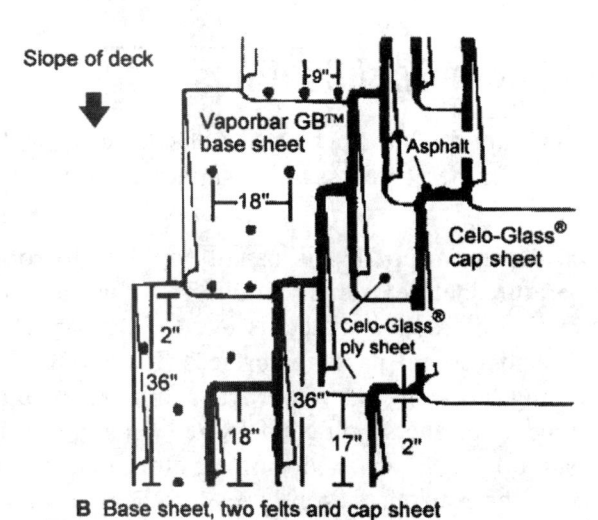

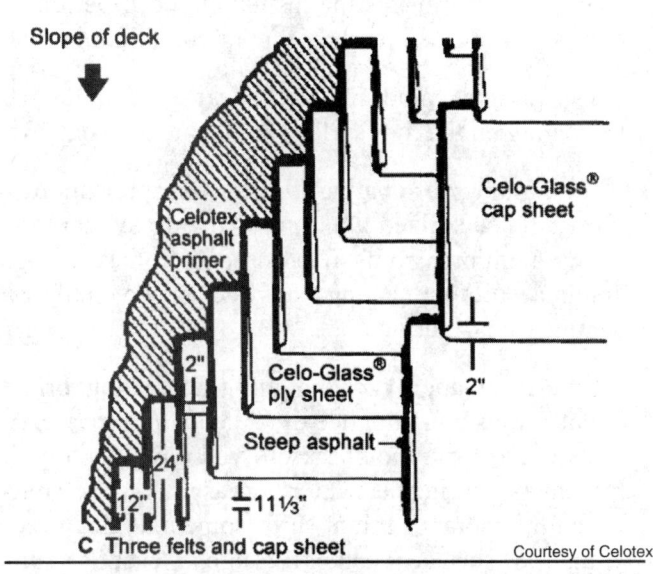

Figure 10-9 Three- or four-ply glass fiber cap sheet membrane systems

A four-ply system normally consists of a base sheet, two plies of felt and a cap sheet, as shown in Figure 10-9 B, or three plies of felt overlaid with a cap sheet, as shown in Figure 10-9 C.

You can also use cap sheets made from a blend of synthetic and glass fibers impregnated with modified (rubberized) asphalt, surfaced with mineral granules. It's installed over a fiberglass base sheet mechanically fastened (over nailable decks) or spot-mopped (over primed non-nailable decks or an existing aggregate surface), followed by the cap sheet solidly mopped onto the base sheet. This sort of cap sheet is approximately 0.1 inch thick and weighs about 73 pounds per square.

Cap sheets can also be used for flashing at parapet walls, roof edges and roof penetrations. Look ahead to the illustrations in the flashing section later in this chapter (starting on page 302).

Aluminum Roof Coatings

Aluminum roof coatings consist of selected asphalts, asphalt emulsions, reinforced mineral fibers, fillers, mineral spirits and pure aluminum pigment flakes.

An aluminum coating is usually applied to roofs where the outside temperature gets extremely high. The coating reduces expansion and contraction of roofing materials due to heat, which stabilizes roof membranes and vertical flashing surfaces. It prolongs the life of the roof by reflecting heat and ultraviolet rays and slows oxidation. It makes the building more energy-efficient by reducing the heat load by 45 percent compared to dark-surfaced roofing, other than ceramic-coated granules. Because the roof insulation is less efficient at high temperatures, this reduced heat also makes the insulation work better, so you need less. And there's a decreased load on the air conditioning system.

This product comes in 1-gallon cans or 5-gallon pails. You spray or brush the coating on the roof surface at the rate of about 1 to 2 gallons per square.

Aluminum roof coating is ideal for application over metal roof decks. There's a specialized aluminum coating available for insulating and protecting aluminum roofs. This product comes in 5-gallon pails. Instead of aluminum roofing coatings, a white acrylic emulsion is sometimes applied over an asphalt surface.

Before you apply an aluminum roof coating, prime the roof surface with asphalt-emulsion at the rate of $3/4$ gallon per square. Allow the newly-coated roof to cure for a month before you apply an aluminum coating. Allow a three-month curing period for new asphalt- or solvent-coated roofs. Because roofing membranes might shift, some manufacturers recommend that you let the roof surface weather through at least one summer before you apply an aluminum roof coating. Don't apply an aluminum surface coating over areas subject to water ponding — the water will eventually degrade it.

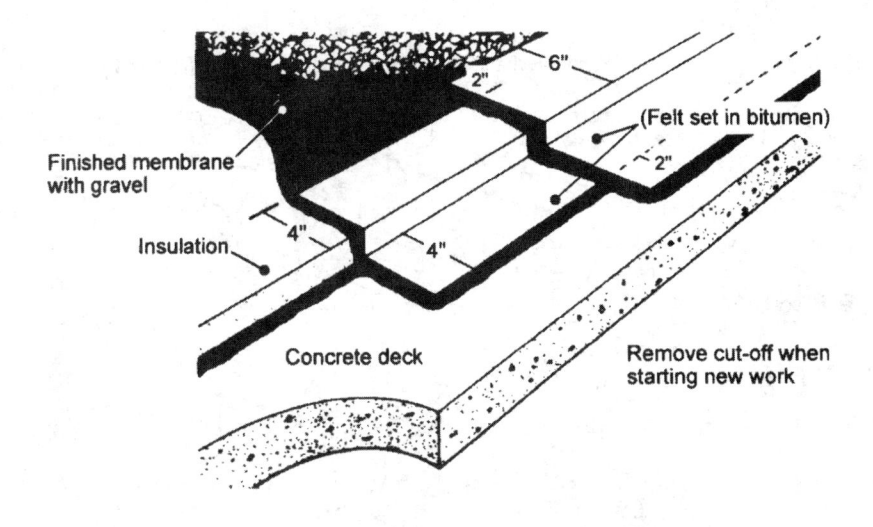

Figure 10-10 Water cut-off

Fibrated aluminum roof coatings are often applied over roof flashing for extra protection. You'll see it in the figures accompanying the section on roof flashing a little later in this chapter.

Phasing

The practice of installing only a portion of a BUR and allowing an unfinished area to remain exposed to weather for a period of time is called *phasing*. But I consider this to be poor roofing practice. Plan the BUR installation so that you don't install more insulation than you can completely cover with membranes and aggregate by the day's end.

If you can't finish applying the gravel, cover all installed insulation with the membranes and apply an asphalt glaze coat with a squeegee at the rate of 10 pounds per square (for organic felts) or 20 pounds per square (for glass fiber felts). You can omit the glaze coat when the bitumen is coal-tar pitch or dead-level asphalt. Then place the surface aggregate no later than one week after you install organic felts, or 30 days after you install glass fiber membranes.

Water Cut-offs

When you must phase roof construction, install temporary water cut-offs as shown in Figure 10-10 to protect the insulation at the end of the day. A cut-off is just a felt strip mopped over the exposed edge of the insulation to provide temporary protection. When you resume work, completely remove the water cut-offs so that insulation joints can be firmly butted together.

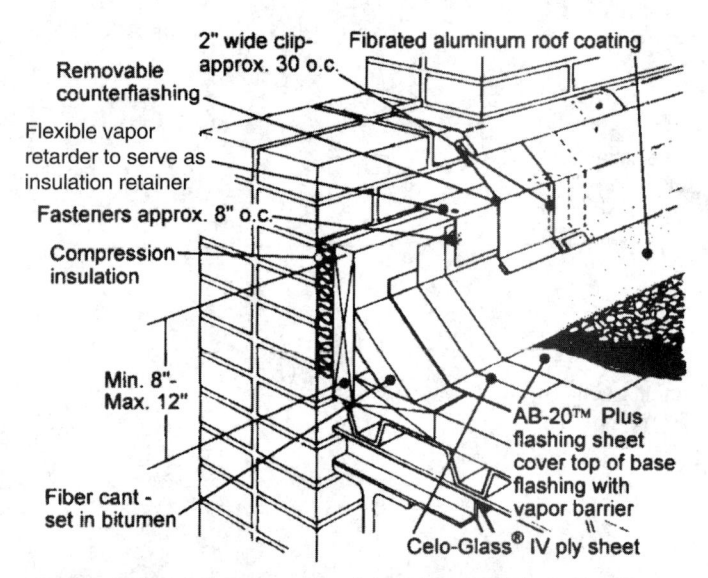

Figure 10-11 Parapet flashing where the deck and wall move independently

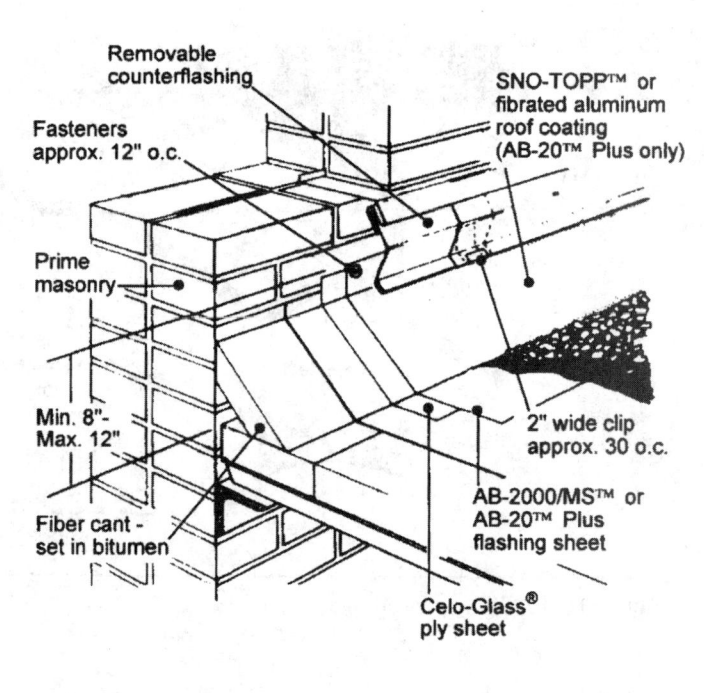

Figure 10-12 Parapet flashing where the deck is supported by the wall

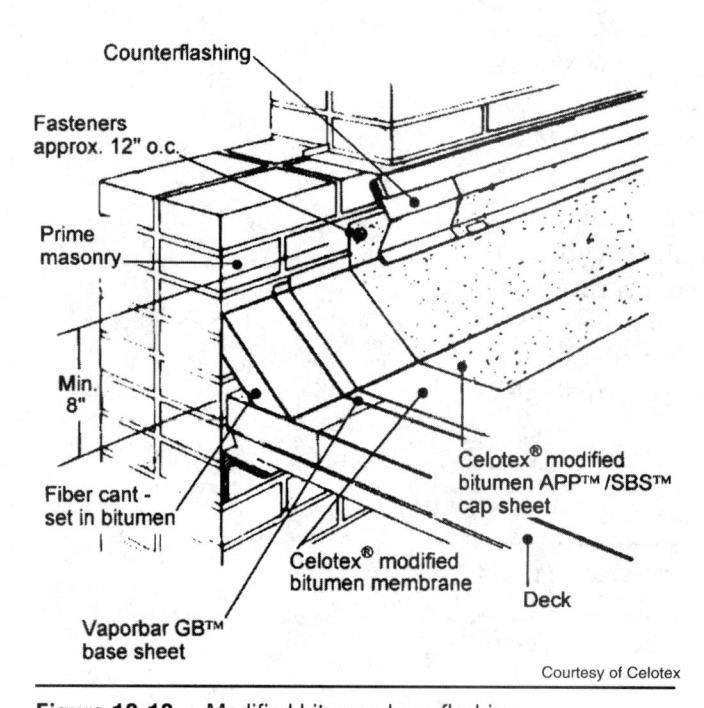

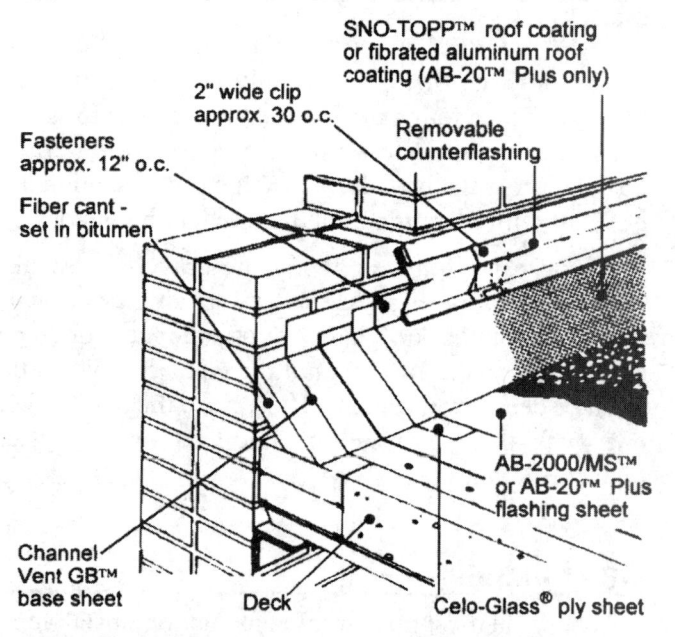

Figure 10-13 Modified bitumen base flashing

Figure 10-14 Base flashing for vented base sheet

Cant Strips

Install cant strips where the built-up roof intersects vertical surfaces in order to break the sharp angle between the wall and roof deck. You can see the cant strips in Figures 10-11 through 10-14. Install cant strips over roof insulation (or deck, in the absence of insulation) with one edge flush against the vertical surface. Run the cant strip out over the roof at least 3 inches and up the vertical surface at least 5 inches.

Prime masonry surfaces with an asphalt primer before you install a cant strip. Use cant strips that are compatible with the roof membrane and bond well with mopped asphalt to the horizontal and vertical substrates. The vertical substrate you install depends upon whether the roof and wall are designed to move independently as shown in Figure 10-11, or the wall helps support the roof deck as shown in Figures 10-12 through 10-14. Nail cants in place over nailable decks or wood nailers, or embed them into a mopping of hot steep asphalt or roofing cement over insulation.

Cant strips are made from a variety of materials including preservative-treated wood, rigid fiberboard, perlite board or concrete. Concrete cant strips are cast along with the wall and deck. Don't use metal cant strips because there's too much expansion and contraction.

Tapered Edge Strips

Tapered edge strips are sometimes installed at the inside edge of wood nailing strips to route water away from the roof's edge as shown in Figure 10-5. As with cant strips, nail or mop edge strips into place. Edge and cant strips are made from the same materials. Install edge strips at least 18 inches wide that provide an incline of approximately 1 inch per foot.

Envelope Strips (Bitumen Traps)

Install envelope strips at the edge of a roof and around openings to prevent bitumen from dripping onto surfaces beneath the roof. Make the strips using 12-inch-wide felt or 6-inch-wide soft metal. Allow half of the strip to overhang the edge or opening. Install the envelope strips before you install the roofing. After you install the membranes, fold the strip back flush with the top of the roof and secure it with roofing cement or bitumen as in Figures 10-5 through 10-7.

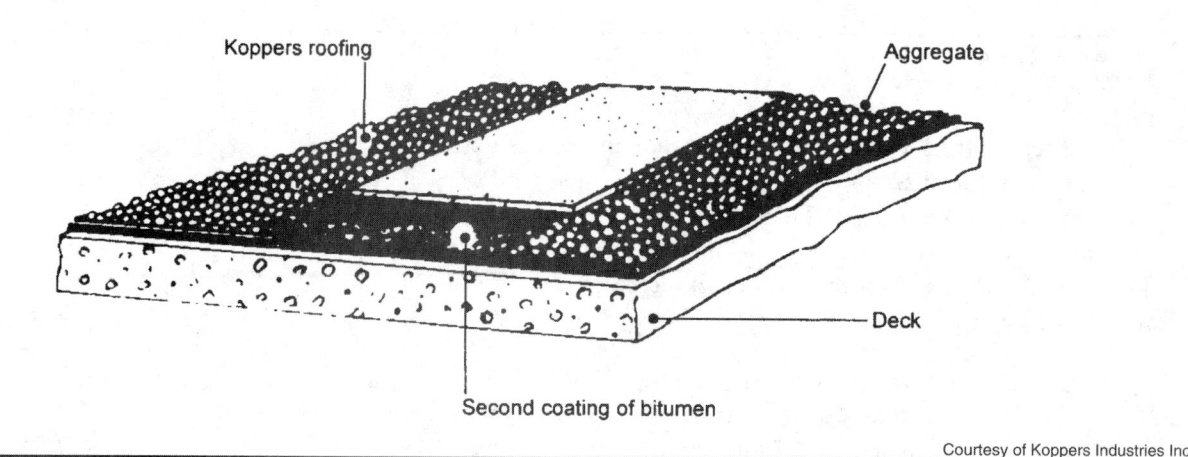

Figure 10-15 Traffic pad

Temporary Roofs

Install a temporary roof when the building interior has to be kept "in the dry" during construction before the permanent roof is installed.

To install a temporary roof over a non-nailable deck, prime the deck and install a coated base sheet in hot steep asphalt. Over a nailable deck, nail a coated base sheet over rosin paper. In either case, the base sheet remains in place to serve as a vapor retarder. Never install a temporary roof over insulation which will remain as part of the permanent roofing system. The temporary roof goes on before the insulation.

Sometimes you may not realize that you need a temporary roof to protect the contents of a building you're working on. For example, I know of a 12,000-square-foot building that was built with a roof deck of prestressed double tees. Because cranes were used to erect the structure, the slab-on-grade was going to be poured after the tees were erected. Shortly after the last roof tee was installed and before the first slab pour, 4 inches of rain fell within 48 hours. With no temporary roof, all of the rainwater poured through the joints between the tees. The compacted fill (now in the shade) inside the building became a muddy quagmire. Among other nightmarish problems created by the flooding, the project was delayed for weeks because of the time it took the fill to dry.

Roof Traffic Pads

Roof traffic pads (roof walkways) as shown in Figure 10-15 provide a protective walkway over built-up or elastomeric roofs. Install traffic pads for paths to areas where security or service personnel need access. As a rule of thumb, install a walkway when more than five people a month will walk over the same path on a roof. Also install pads in areas subject to damage from dropped tools and parts.

Roof walkways for BUR systems are generally made of a uniform core of asphalt, plasticizers and inert fillers bonded by heat and pressure between two saturated and coated sheets of organic felt. The bottom is smooth while the top surface is faced with non-slip mineral or ceramic granules.

Asphalt traffic pads come in 1 x 2-, 3 x 3- or 3 x 6-foot panels, and $1/2$, $3/4$ and 1 inch thick. Weights of $1/2$-, $3/4$- and 1-inch asphaltic traffic pads are 3.5, 5.2 and 7.0 pounds per square foot, respectively. Don't install this type of pad in high-traffic areas, especially where there may be walkers wearing high heels.

Rubber walkway pads are also available. These pads resemble a heavy-duty door mat and are manufactured in 3' x 3'8" x $3/8$" panels. Colors are normally black or white.

Install traffic pads after you've completed the BUR system. Sweep away all loose aggregate and embed the panels in industrial roofing cement, or into the same material (asphalt or pitch) you used to embed the aggregate. Some manufacturers recommend that you install roof walks before you apply the surface aggregate. Whichever way you do it, make sure it's well secured.

Don't install roof walks on slopes steeper than 2 in 12, or over membranes installed over fiberglass roof insulation. Space traffic pads at least 3 inches apart to allow drainage between them.

You can also use precast rod- or mesh-reinforced paver blocks for roof walkways, with non-skid and exposed-aggregate surfaces if you choose. Sizes vary from 12 x 24 inches to 24 x 36 inches. Install asphalt roof traffic pads, laid dry, under the pavers as a protective layer between the roof and the paver.

Water-retaining Roofs

Some roofs are designed to be cooled during hot weather by water on the surface, either held in ponds or sprayed on. Pond depth is governed by controlled-flow drainage systems, installed by the mechanical contractor. Before cold weather begins, the system must be shut down and the water drained off.

Because of the abnormally high dead loads, you must install a water-retaining roof over poured or precast concrete, or other heavy load-bearing decks. The structure must be designed for a minimum 50 PSF dead load with a deflection not exceeding $1/360$ of the span between supporting members.

Use coal-tar bitumen and tar-saturated felts in a water-retaining roof system. Install insulation over two plies of vapor retarder material that's solidly mopped to the deck. Over a non-nailable deck, embed four plies of 15-pound tar-saturated felts in solid moppings of coal-tar bitumen. Install a flood coat of coal-tar bitumen at the rate of 75 pounds per square, followed by either 400 pounds of gravel or 300 pounds of slag per square. Sweep away all loose aggregate. Then apply a second surface flood coat of 85 pounds per square of coal-tar bitumen. Into the second flood coat, embed a second layer of gravel or slag at 300 or 400 pounds per square, respectively. Then sweep away all loose aggregate.

Flashing on Flat Roofs

You need flashing on flat roofs at:

- parapet walls
- the roof edge
- roof penetrations such as skylights, structural members, piping and air conditioning units
- equipment stands
- roof drains
- expansion joints

On any roof, complete all openings and roof penetrations before you install the roof system. Install flashing as roofing progresses to prevent damage to the insulation at roof openings of the deck beneath.

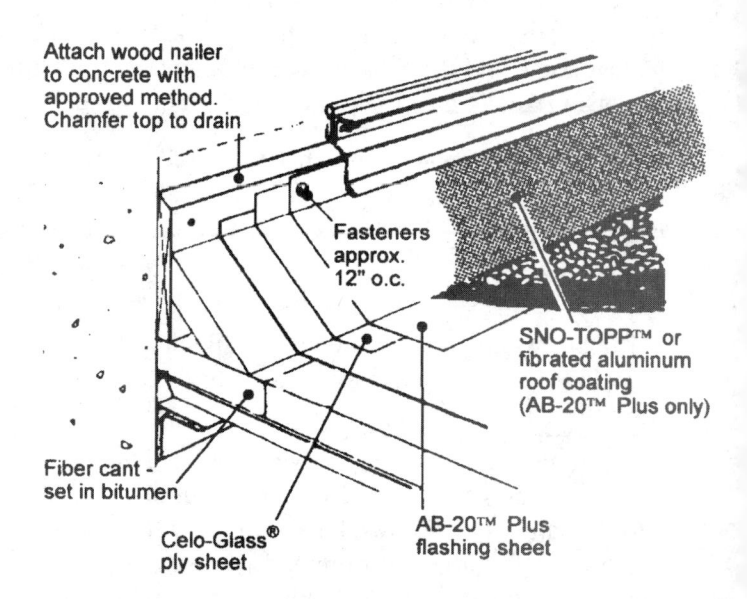

Figure 10-16 Counterflashing for concrete walls or parapets

Flashing at Parapet Walls

Water is most likely to get into a flat roof where the roof deck and a vertical surface meet. Install flashing at the intersection of a BUR and a vertical wall to make the connection waterproof.

Install base flashing over a cant strip set at a 45-degree angle. Extend the flashing 8 to 12 inches up the parapet wall and at least 5 inches out over the roofing surface, overlapping the roof membrane. You can use a heavy fiberglass flashing material (Figures 10-11 and 10-12) or a modified bitumen cap sheet (Figure 10-13).

Fiberglass flashing comes in rolls 12, 18 and 36 inches wide, and 72 feet long. It weighs about 38 pounds per 100 square feet. Be sure the flashing material is compatible with the roof membrane and that it has similar expansion and contraction characteristics. Don't use metal base flashings. Install 35 pounds of base flashing per 100 square feet in two layers with roofing cement (or hot steep asphalt). Lap the ends at least 3 inches and cement the laps. When you install felts next to metal, nail or spot-cement them (don't solidly mop) to allow for movement. I recommend that you use roofing cement to attach felt base flashing to metal because it's permanently soft and waterproof.

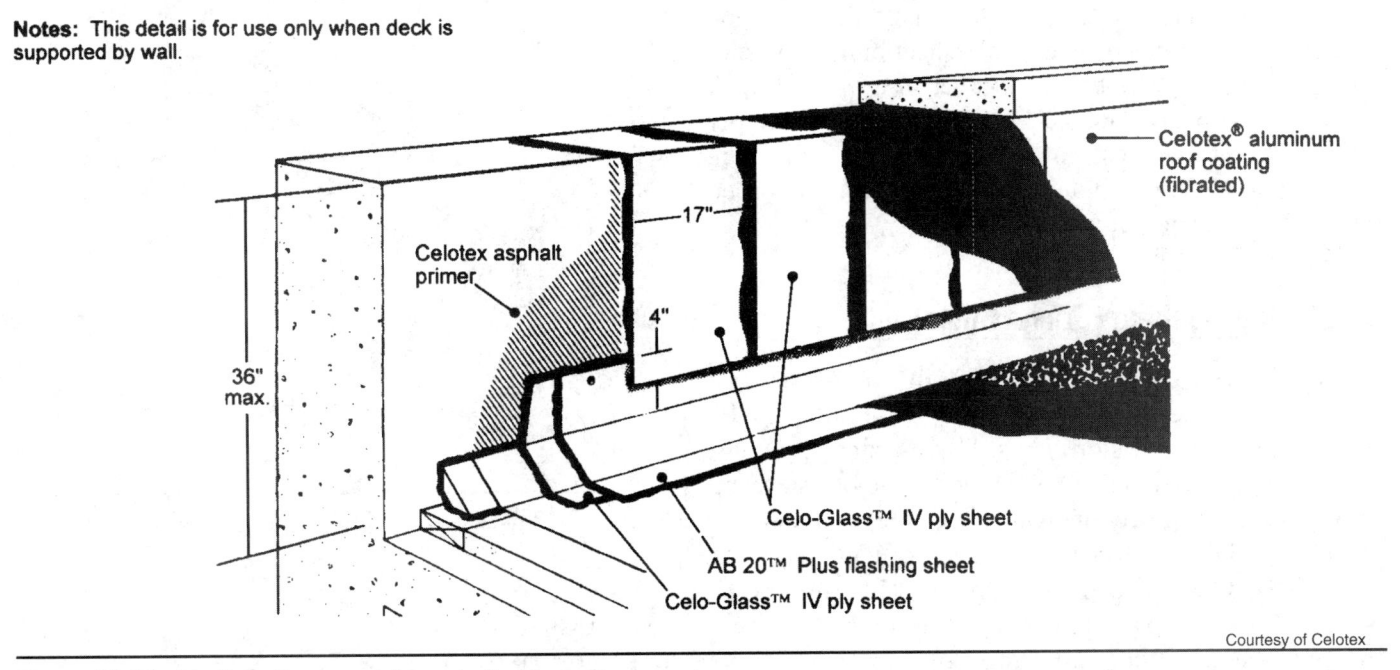

Notes: This detail is for use only when deck is supported by wall.

Figure 10-17 An alternative to metal counterflashing

Nail the top of the second base flashing strip, on 12-inch centers, to the vertical substrate 1 inch below the top edge of the flashing, driving the nails through tin discs. If you're nailing into wood, use 11-gauge, galvanized, barbed roofing nails with $5/8$-inch-diameter heads, long enough to penetrate the full depth of the nailing strip. For masonry walls, you use case-hardened nails. Drive them through 28- or 30-gauge tin caps at least $1^{3}/_{8}$ inches in diameter. Or, you can use large-head Simplex nails without tin caps. After you install the base flashing, cover it with 1 to 2 gallons of an aluminum roof coating per 100 square feet. One method for installing base flashing over a wet-fill roof is shown in Figure 10-14.

Install metal counterflashing (cap flashing) over the base flashing. Bend the upper edge and insert it into a reglet (concrete wall) or a $1^{1}/_{2}$-inch deep raggle or slot in the mortar joint (masonry wall). In new construction, install the cap flashing in mortar joints as shown in Figures 10-11 through 10-14. Seal the joint with mortar or roofing cement. The minimum recommended thicknesses for counterflashing are 24 gauge for galvanized steel, 26 gauge for stainless steel, or 16-ounce copper.

Some counterflashing is designed to be surface-mounted to the parapet wall as shown in Figure 10-16. Install counterflashing about 8 to 12 inches above the surface of the finished roof and extend it down 4 inches over the base flashing. One method of installing counterflashing over a concrete parapet wall is shown in Figure 10-16. Lap counterflashing end joints at least 3 inches and secure the lap with a clip. Don't solder the joint. Install two-piece counterflashing as shown in Figures 10-11 through 10-14 so the roof can be redone without destroying the original counterflashing.

As an alternative to using metal counterflashing, you can install two plies of glass fiber sheet either horizontally or vertically. Embed each layer in a 1/8-inch-thick troweled-on coat of roofing cement. Coat the second ply with roofing cement as shown in Figure 10-17. Flashing made of plasticized polymeric vinyl or PVC is also used for flashing at parapet walls.

Roof Edge Flashing (Gravel Stops)

Install metal flashing at the edge of a flat roof. The flashing serves as a fascia cover and gravel stop. Make the bead of a gravel stop rise at least 3/4 inch on an aggregate roof, and at least 3/8 inch over a smooth-surfaced roof. As added protection against weather, you can install metal flashing over flexible vinyl or PVC flashing material. On roof slopes less than 1 in 12, prime both sides of the flashing. Then install it into a bed of roofing cement you apply over the top of the completed roof membrane as shown in Figures 10-6 and 10-7. Make the nailing flange of the edge flashing no wider than the nailer (usually 3 1/2 inches). Nail the nailing flange to wood nailers or insulation stops in two staggered rows on 3- to 4-inch centers. Never nail the

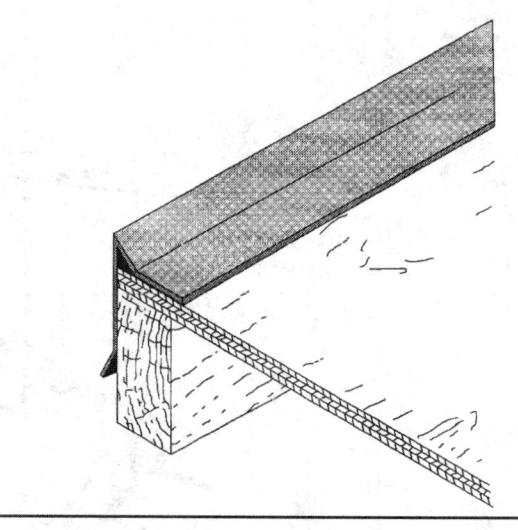

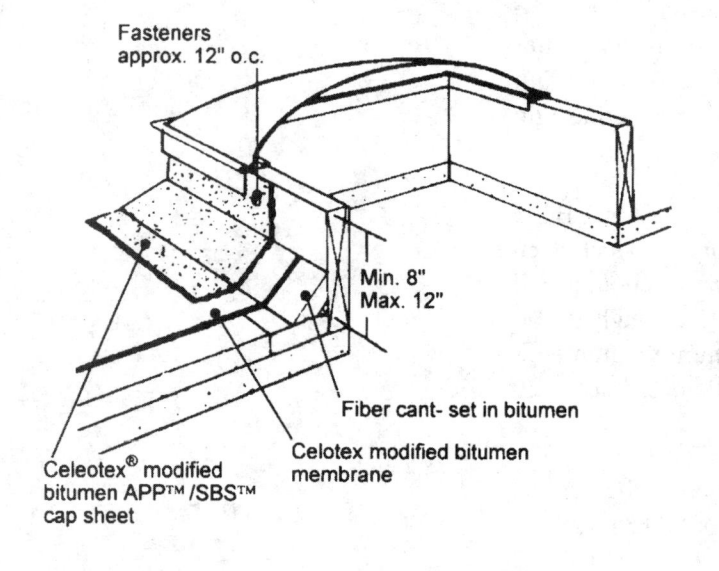

Figure 10-18 Installed roof flashing with gravel stop

Figure 10-19 Flashing around a skylight

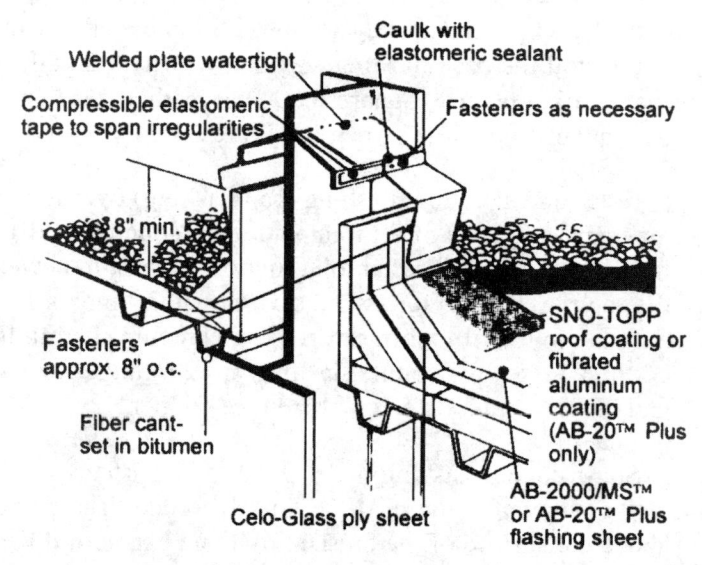

Figure 10-20 Flashing a structural member through roof deck

flange into insulation. Then cover the flange with two strips of 15-pound felt. Make the lower and upper strips 8 and 12 inches wide respectively, as shown in Figure 10-6. You can install cap sheet material instead of felt strips at the roof's edge as shown in Figure 10-7.

On roof slopes of 1 in 12 and steeper, install the roof edge flashing before you apply the roofing material as shown in Figure 10-18. Nail the nailing flange in two staggered rows on 3- to 4-inch centers. Mop all membranes covering the nailing flange up to the lips of the flange.

Flashing at Roof Penetrations

You'll usually find counterflashing built into a skylight unit as shown in Figure 10-19. If it's not, you install flashing here using the same method as you would for a parapet wall.

Framed roof penetrations such as openings for strucural members (Figure 10-20), piping (Figure 10-21) and curbs for rooftop air-handling units (Figure 10-22) require you to use shop-fabricated flashing.

You flash small-diameter roof penetrations by setting a primed metal flange over the top ply of membrane into a bed of roofing cement. Follow that with two plies of felt stripping. Set flanges of small vent pipe roof penetrations in mastic over the base ply. On nailable decks, you nail the

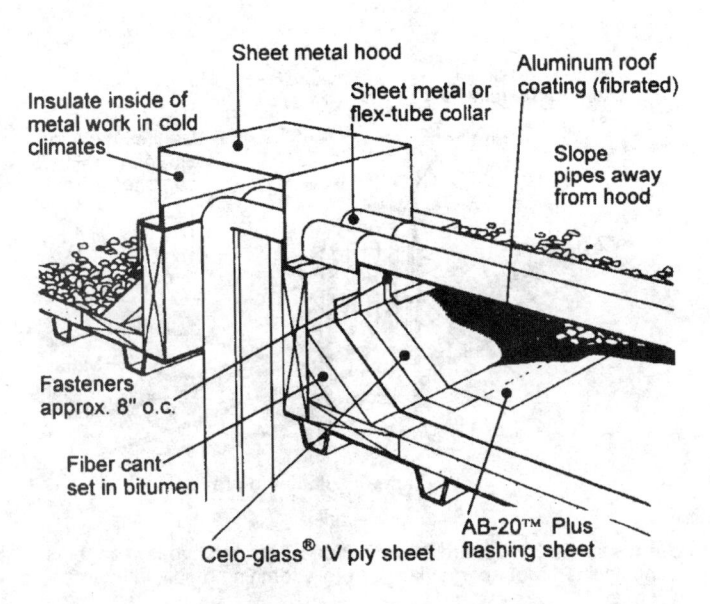

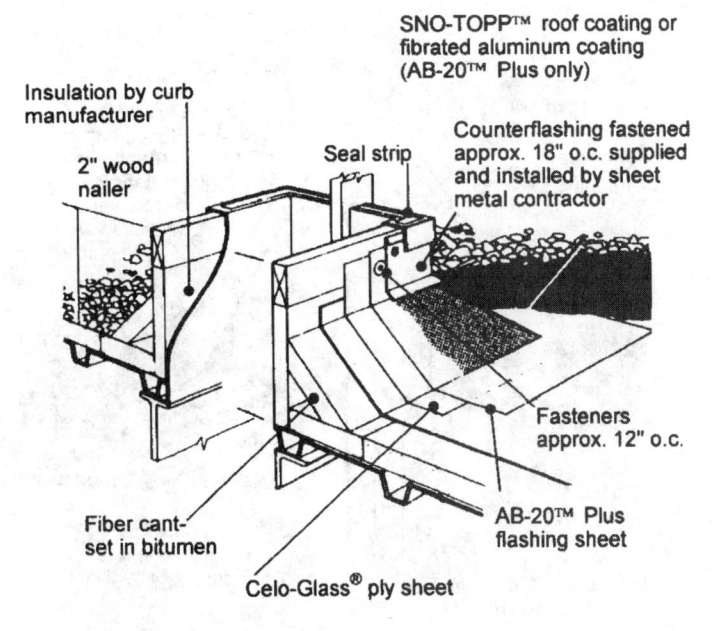

Figure 10-21 Piping through roof deck

Figure 10-22 Curb detail for rooftop air-handling units

Roofing Construction & Estimating

Width of equipment	Height of legs
Up to 24"	14"
25" to 36"	18"
37" to 48"	24"
49" to 60"	30"
61" and wider	48"

Note: This detail is preferred when the concentrated load can be located directly over columns or heavy girders in the structure of the building. This detail can be adapted for other uses such as sign supports

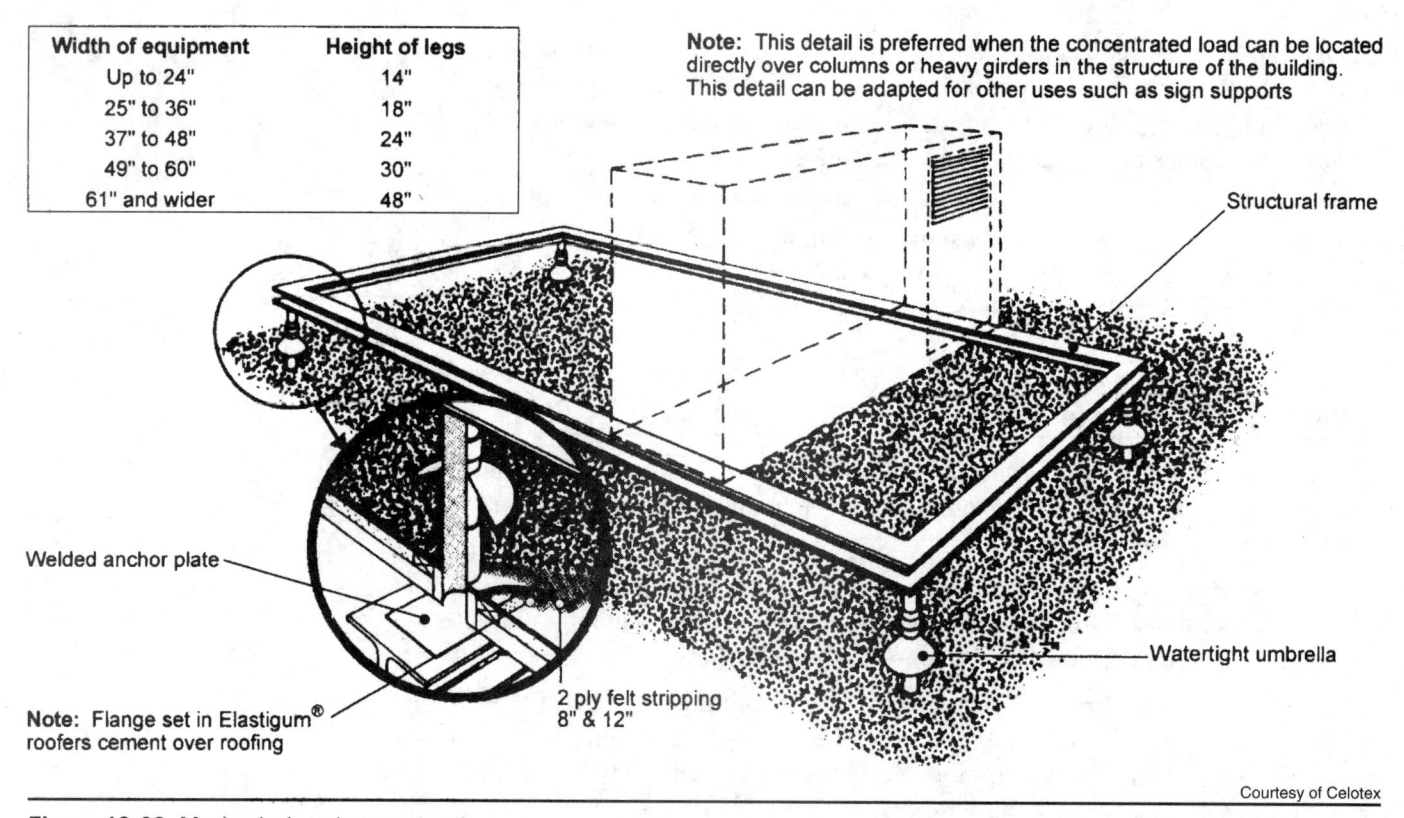

- Structural frame
- Welded anchor plate
- Watertight umbrella
- 2 ply felt stripping 8" & 12"

Note: Flange set in Elastigum® roofers cement over roofing

Courtesy of Celotex

Figure 10-23 Mechanical equipment stand

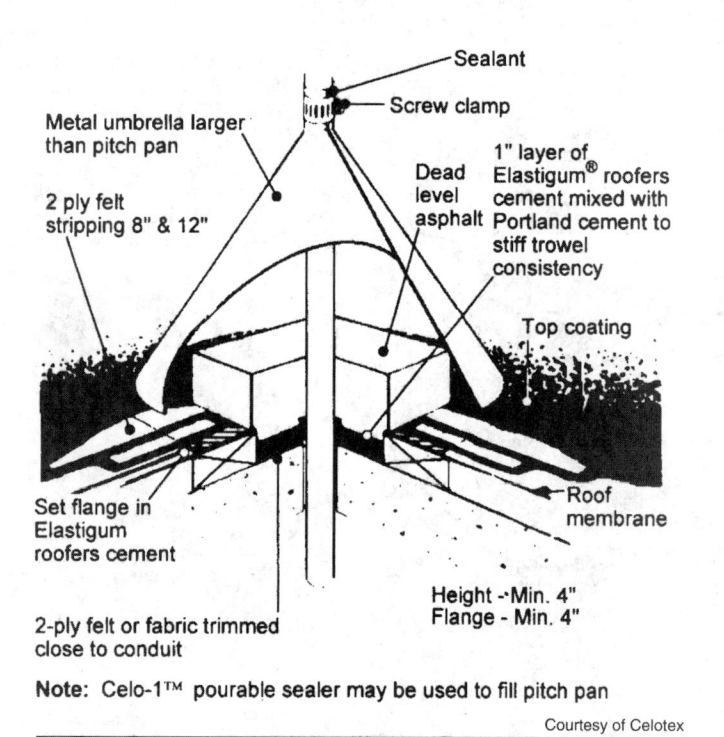

Figure 10-24 Hooded pitch pan

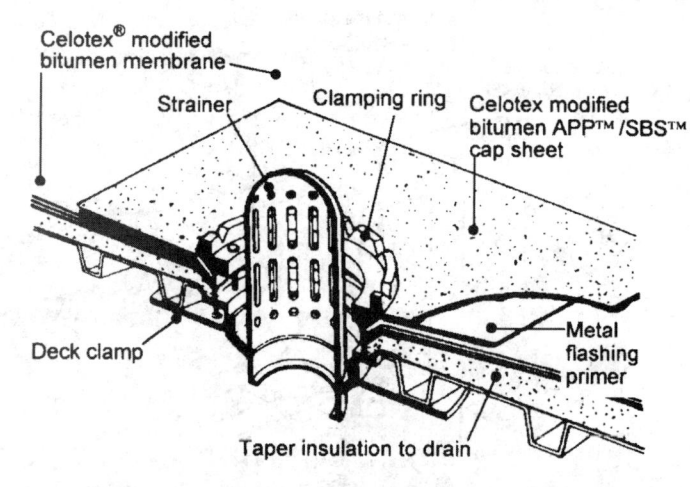

- Sealant
- Screw clamp
- Metal umbrella larger than pitch pan
- Dead level asphalt
- 1" layer of Elastigum® roofers cement mixed with Portland cement to stiff trowel consistency
- 2 ply felt stripping 8" & 12"
- Top coating
- Set flange in Elastigum roofers cement
- Roof membrane
- 2-ply felt or fabric trimmed close to conduit
- Height - Min. 4"
- Flange - Min. 4"

Note: Celo-1™ pourable sealer may be used to fill pitch pan

Courtesy of Celotex

Figure 10-25 Roof drain

- Celotex® modified bitumen membrane
- Strainer
- Clamping ring
- Celotex modified bitumen APP™/SBS™ cap sheet
- Deck clamp
- Metal flashing primer
- Taper insulation to drain

Notes: Min 30" square 2½# lead or 16 oz. soft copper flanges set on finished roof - prime flange before torching modified bitumen sheet.
Membrane plies, metal flashing and cap sheet stripping extend under clamping ring.
Extend Celotex® modified bitumen stripping 4" min. beyond edge of metal flashing

Courtesy of Celotex

flange edges on 3-inch centers. Strip the flange with 6-inch glass fiber felt or cap sheet material set in bitumen. Then install subsequent membrane plies over the roof. Don't install penetrations within 15 inches of the lower edge of a cant strip. Typical small-diameter roof penetrations include:

- roof relief vents (Figure 10-2)
- mechanical equipment stands (Figure 10-23)
- hooded pitch pans (Figure 10-24)
- roof drains (Figure 10-25)

Be sure to allow at least 12 inches workspace between parapet walls and roof penetrations and mechanical equipment. Hopefully, the building designer also considered working space beneath any rooftop equipment when specifying the height of its supporting legs.

Roof Expansion Joints

You install expansion joints in roofs to let the roof expand and contract with changes in temperature. Here's what happens to a BUR roof that doesn't have any expansion joints. High temperatures cause the roof to expand and since there aren't any expansion joints, the felts split. Low temperatures, meanwhile, cause the roof to contract, pulling the felts into ridges in the process. The useful life span of this roof is obviously very short.

Align roof expansion joints with the structural expansion joints located throughout the building. Extend each expansion joint to the full length or width of the roof right to the edge of the deck. Never bridge expansion joints with either insulation or roofing membrane. You always elevate expansion joints above the finished roof by at least 4 inches. Never install expansion joints flush with a roof surface. When you design a roof, make sure that water won't flow over or around expansion joints. Finally, never install an expansion joint in a valley.

Here are some obvious places for expansion joints:

- the structural framing or decking changes direction
- the composition of decking material changes
- new construction intersects existing construction
- there's a difference in the elevations of two adjoining decks
- the roof deck intersects a wall
- interior heating or cooling conditions change
- wings where L-, U- or T-shaped buildings intersect
- there may be seismic movement between dissimilar structures
- canopies or exposed overhangs are attached to an air-conditioned building

Roofing Construction & Estimating

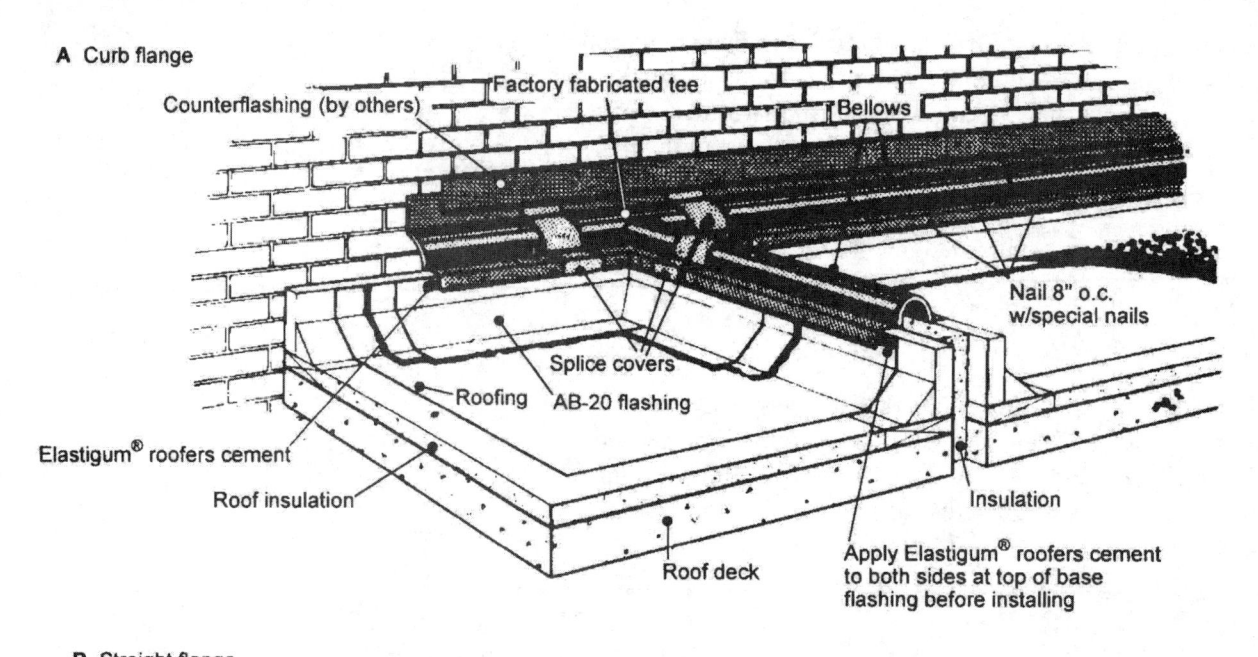

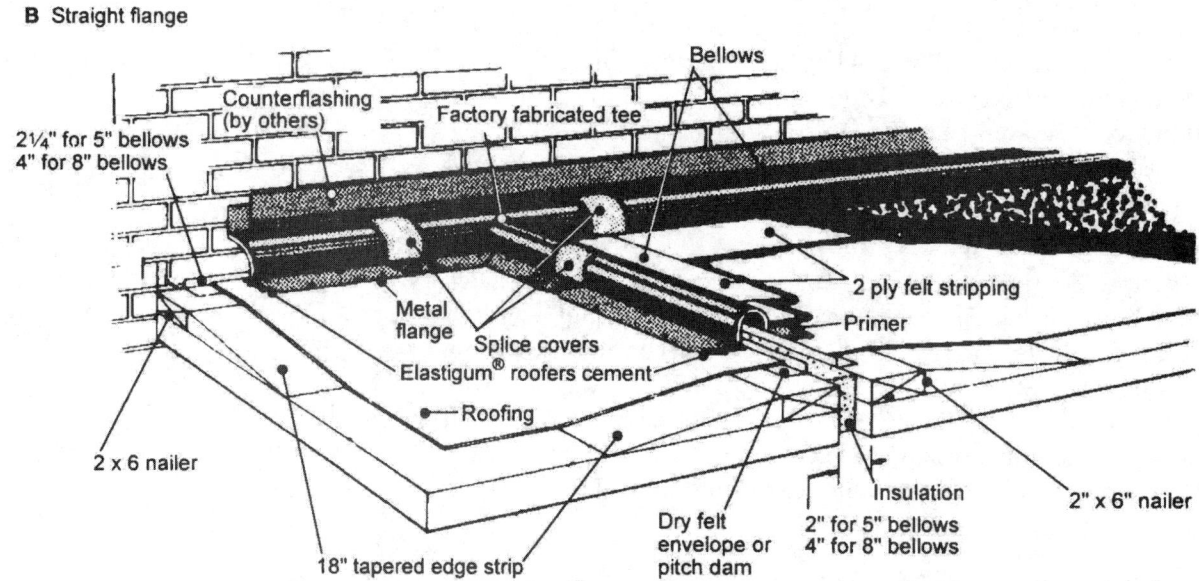

A Curb flange

B Straight flange

Figure 10-26 Expansion joints

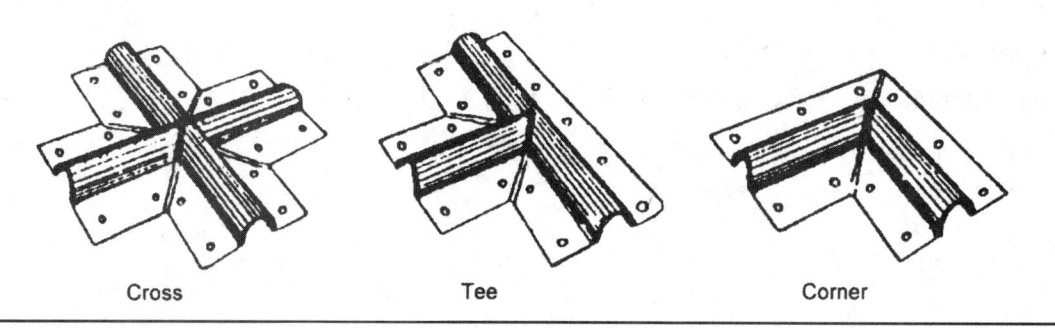

Cross   Tee   Corner

Figure 10-27 Expansion joint accessories

308

Expansion joint shields (or covers) insure that the joints are waterproof while still allowing movement throughout the expansion joint. Expansion joint shields are manufactured in three basic configurations as shown in Figure 10-26:

1) the curb flange
2) the straight (low-profile) flange
3) the curb-to-wall flange

Expansion joint covers, regardless of their shape, are essentially a bellows. This bellows is made out of Dacron-reinforced, chlorinated polyethylene laminated to a core layer of closed cell foam. In order to install expansion joint covers, you connect the bellows to a pair of metal (galvanized iron, copper, stainless steel or aluminum) nailing flanges. Expansion joint shields come in 50-foot rolls. Expansion joint accessories include splice covers and connections for crossovers, tees and corners as shown in Figures 10-26 and 10-27.

Coefficient of Linear Expansion

As you've seen over and over throughout the last ten chapters, all roofing materials (metal especially) expand and contract after you install them. You've also seen that if the roof isn't able to do so as a single unit, it won't last for very long. Now we'll look at how you go about determining how much expansion room you need to allow for the particular roofing material. To answer that question, you must know the material's coefficient of linear expansion. Simply put, the coefficient of linear expansion is the measured change in length, per unit of length, when the temperature changes by one degree. Assuming that the ends of the material are free to move, the total change in length due to an increase or decrease in temperature is expressed as follows:

Change of Length = E x T x L | Equation 10-2

Where: E = coefficient of linear expansion
            (from Figure 9-39, Chapter 9)
      T = change in temperature
      L = length of member

▼ **Example 10-2:** A continuous roll of zinc flashing is 200 feet long at 60°F. Let's assume the metal is free to move, and find the change in the length of the material when the temperature increases from 60° to 95°F (an increase of 35°):

From Figure 9-41, the coefficient of expansion for rolled zinc is 0.00173 per 100°F.

Change in Length = 0.00173 x 35/100 x 200
                      = 0.12 linear feet

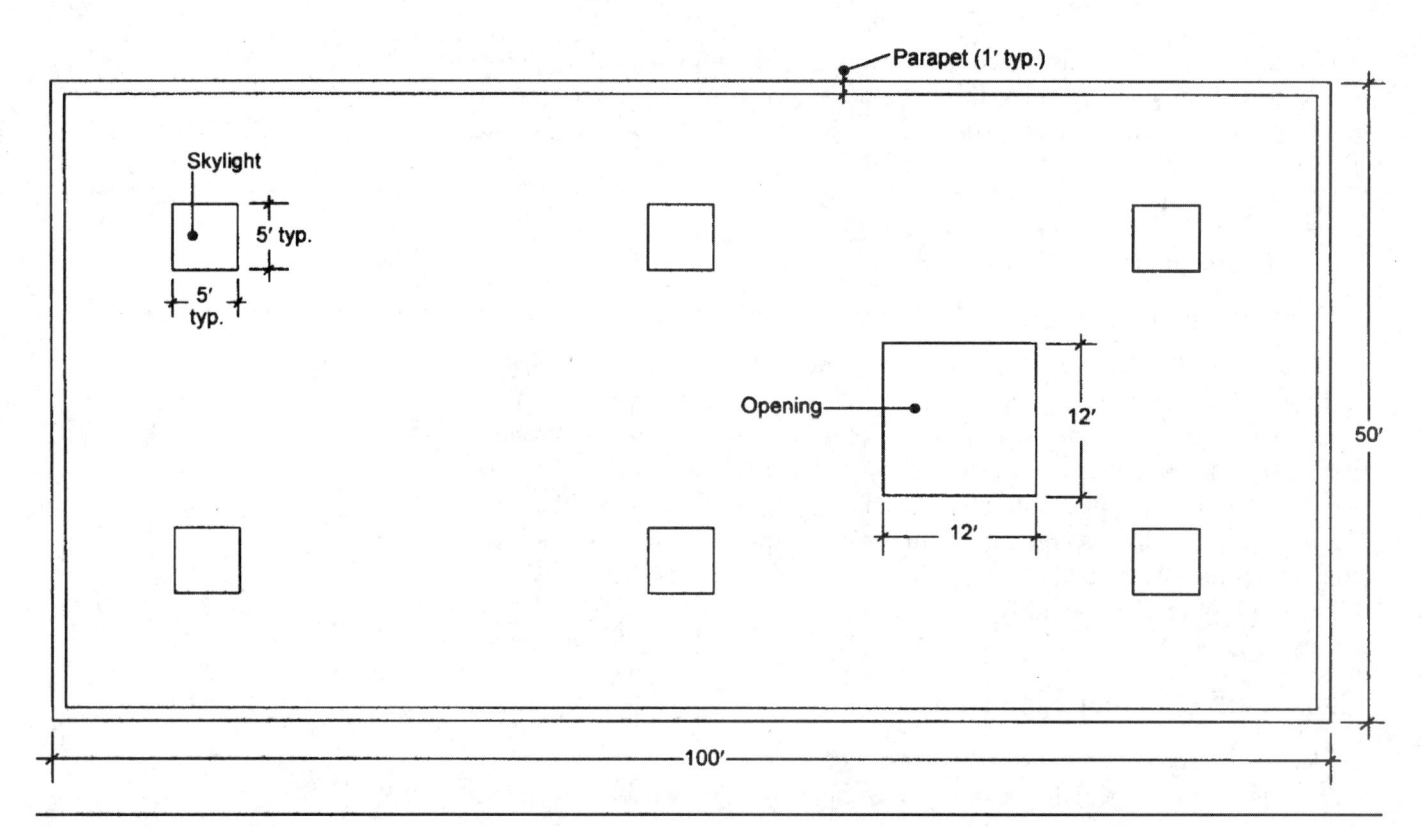

Figure 10-28 Built-up roof example

Converting to inches, we have:

Change in Length = 0.12' x 12 in./LF
                        = 1.44 inches
                        = $1^{7}/_{16}$ inches

Estimating BUR Systems

When you estimate the cost of BUR systems, note the type of roof, including components, roof deck, and any unusual requirements. Your description, for example, might say:

2" insulation, 4-ply (glass base sheet plus three No. 15) plus gravel

Take off the area of the roof in square feet. Then convert this area into squares. Find the roof area from roof edge to roof edge on roofs without parapet walls. On roofs bordered with parapets, measure the area from outside the parapet walls to allow for material you turn up the wall.

Don't deduct anything for openings less than 100 square feet and deduct only half of the area for openings larger than 100 square feet and less than 500 square feet. Deduct the entire opening area for openings 500 square feet and larger.

You must also take off accessories such as fasteners, cant strips, edge strips, envelope strips, flashing, counterflashing, expansion joint cover assemblies, roof relief vents, temporary roofing and glaze coats, water cut-offs, pitch pans, roof drains, skylights and other roof penetrations, primer coats, aluminum coatings, roofing cement, roof traffic pads, additional aggregate, and so forth.

Roof slope, roof size, height of the roof above the ground, and the number and types of roof penetrations will also affect your estimate. For example, the roof slope affects the type of bitumen used. The size of the roof also has an effect. You add a higher percentage of profit for a small job than for a large one. If the roof is high enough to require a crane to raise materials instead of a mobile placer, it's more expensive. And every penetration requires flashing, which adds to the cost.

▼ **Example 10-3:** Find the net area of four-ply built-up roofing required for the roof of the building in Figure 10-28. Assume a wood deck.

Gross Area	= 100' x 50'	= 5,000 SF
Less	½ x 12' x 12'	= (72 SF) (half the openings)
Net Area		= 4,928 SF ÷ 100
		= 49.3 squares

I'd round it up to 50 squares for figuring the total amount of material required for the BUR system.

Specific quantities for such a roof would be:

Item	Weight (pounds/square)
Rosin paper (nailed to wood deck)	5
Coated base sheet (1 ply, nailed)	43
No. 15 felts (3 plies, mopped)	45
Asphalt (3 applications @ 25 pounds each)	75
Surface flood coat	75
Gravel	400
Total weight	**643**

Depending on the job, your take-off items could also include:

Insulation	= 98' x 48'	= 4704 SF
Less: 6 x 5' x 5'		= (150 SF)
Less: 12' x 12'		= (144 SF)
Net installed		= 4410 SF ÷ 100 = 44.1 SQ
(10% waste)		x 1.10
Purchased		= 4851 SF ÷ 100 = 48.5 SQ

Base flashing:
    At Parapets   = 2 x (98' + 48')   = 292 LF
    At Small Openings   = 6 x 20'   = 120 LF
    At Large Openings   = 4 x 12'   =   48 LF

Net Installed   460 LF
(5% waste)   x   1.05

Purchased   = 483 LF

Counterflashing (as above)   = 460 LF Installed
                                          = 483 LF Purchased

Cant strip (as above)   = 460 LF Installed
                                    = 483 LF Purchased

To price out this job, let's assume the following material prices:

2", R-5.30 perlite insulation board   = $233.28/SQ
4-ply fiberglass BUR (base sheet, 3 felt plies + gravel) = $132.57/SQ
Fiber cant strip   = $0.29/LF
Base flashing   = $0.78/LF
Counterflashing   = $1.61/LF

    The total material costs would be figured as follows:

    2" insulation   = 48.5 SQ x $233.28/SQ = $11,314.08
    4-ply BUR   = 50 SQ x $132.57/SQ = $6,628.50
    Cant strip   = 483 LF x $0.29/LF   = $140.07
    Base flashing   = 483 LF x $0.78/LF   = $376.74
    Counterflashing   = 483 LF x $1.61/LF   = $777.63

Total material cost $19,237.02

    According to the *National Construction Estimator*, an R-3 crew consisting of two roofers and one laborer can install BUR roofing materials at the following rates:

    2" insulation: 1 SQ in 0.82 hours
    4 ply BUR: 1 SQ in 2.55 hours
    Gravel surface: 1 SQ in 0.718 hours
    Cant strip: 1 LF in 0.019 hours

    Assuming a roofer costs $41.66 per hour and a laborer costs $30.31 per hour, including labor burden, the average cost per manhour is: [(2 x $41.66) + $30.31] = $113.63 ÷ 3 = $37.88 per manhour. Then the labor costs for the items above would be:

    2" insulation   = 44.1 SQ x 0.82 manhour/SQ x $37.88/manhours
                                = $1,369.82
    4 ply BUR   = 49.3 SQ x 2.55 manhours/SQ x $37.88/manhour
                                = $4,762.08

Gravel surface = 49.3 SQ x 0.718 manhours/SQ x $37.88/manhour
= $1,340.85
Cant strip = 460 LF x 0.019 manhours/LF x $37.88/manhour
= $331.07

An SM crew consisting of one sheet metal worker can install flashing material at the following rates:

Base flashing = 1 LF in 0.033 hours
Counterflashing = 1 LF in 0.047 hours

Assuming the sheet metal worker costs $42.16 per hour, including labor burden, the labor for installing the flashing would be:

Base flashing = 460 LF x 0.033 manhours/LF x $42.16/manhour
= $639.99
Counterflashing = 460 LF x 0.047 manhours/LF x $42.16/manhour
= $911.50

The total labor cost is: $1,369.82 + $4,762.08 + $1,340.85 + $331.07 + $639.99 + $911.50
= $9,355.31

The total material and labor cost is: $19,237.02 + $9,355.31 = $28,592.33.

Testing BUR Systems

Sometimes the designer will ask you to make a test cut in the roof to make sure you applied the amount of interply bitumen he or she specified. Do these cuts before you apply the final surfacing so you can completely repair them.

Cut out an area at least 10 by 42 inches at right angles to the length of the felts to show that the felts have been properly lapped and laid to the correct exposure. To pass the test, the weight of the test-cut components must be within 15 percent, plus or minus, of the specified requirements.

I don't recommend test cuts. Avoid doing them if you can. They're not very accurate and they don't reflect the true condition of the overall roof. Knowing that a test cut is required, a roofing contractor may use too much interply bitumen so that the roof is prone to sliding, even on low slopes. Also, a test cut leaves a patched area that's weaker than the original roof and prone to leaks. Furthermore, making a test cut often voids any warranty. You can avoid having to make a test cut by having an inspector watch the application.

Built-up Roofing Warranties

When a designer specifies a warranted (bonded) roof, an approved roofing contractor must be selected to perform the work. An approved roofing contractor is one whose reputation for good workmanship and financial stability qualifies him for a bond by the material manufacturer furnishing the bond.

When the roof is finished, the roofing contractor sends a "Statement of Compliance" to the manufacturer. The manufacturer then issues a "Limited Service Warranty" to the roofing contractor. The manufacturer keeps a job file during the life of the warranty and sends a representative to make an inspection before the roof is two years old. A limited service warranty usually runs for 10 years and provides a specified dollar amount of coverage for five years and prorated coverage for the remaining five. Some manufacturers offer warranties for 5, 10, and 20 years. The longer the warranty, the more costly. The length depends on how much the owner is willing to spend, since the warranty costs must be included in your bid.

Roofing systems on as many as three separate buildings can sometimes be covered by one warranty if:

- they're for a single owner
- they're all located on the same site
- their completion dates are all within a 12-month period
- all of the roofing systems qualify for the same warranty

A warranty won't cover damages due to contractor errors, omissions or poor workmanship. Damage due to ponded water, natural disasters or building settlement or other structural movement isn't covered. The roof must be capable of shedding all ponded water within 24 hours after a rain. This depends on roof slope, roof deflection and proper roof drain and gutter design and maintenance. Ponded water allows seepage through bare spots, blisters and minute cracks in the roof's surface. Ponded water provides an environment for the growth of bacteria and fungi which can weaken the roof membranes.

The warranty can be voided if the type of building usage is changed from its original intended purpose. Test cuts will often void a roof warranty. It can be difficult, if not impossible, to get a warranty for a building of variable occupancies, residences, apartment buildings, condominiums, nursing homes, or any building where the total area of the roof is less than 5,000 square feet. Buildings this size and smaller often have a high turnover of occupants, each making alterations in the building to suit unique needs. Buildings such as storage silos, soybean or milk-processing plants as well as buildings located outside of the U.S. are also regularly excluded in the terms of the warranty. Roofs over domed structures, heated tanks, dry kilns, car-wash buildings, swimming pools and other structures with abnormally high interior humidity conditions also can't be warranted.

Roof decks which are not properly anchored to the structure, roofs whose insulation is not properly installed, roofs applied over lightweight insulating concrete decks or roof applied directly over prestressed concrete decks usually can't be warranted.

An existing warranty can be voided if improper fasteners or other unapproved construction methods or materials were used in any phase of the roof or substrate construction. These unacceptable conditions include phased roofing or materials not produced by the manufacturer issuing the warranty. The subsequent installation of conduit or pipes over the roof can also void the warranty. Other actions that can void a warranty include failure due to repairs or alterations performed by a non-sanctioned contractor, the installation of roof penetrations, or alterations such as aerials, signs, water towers, antennae, and building expansion or additions.

Some limited service warranties cover only the cost of replacing the membrane. Others cover the cost of replacing the membrane and insulation. The liability covered in the warranty doesn't include damage to the building interior and contents, roof deck or lost profits and/or rents.

The warranty automatically expires at the end of the specified time period or when the dollar amount of the warranty is used up.

Built-up Roofing Repairs and Re-roofing

As soon as you install a built-up roof, it begins to degrade. Loose aggregate gradually and continually washes away, bit by bit, Over time, the roof develops bare spots and the exposed bitumen oxidizes, exposing the underlying cap felt layer to ultraviolet radiation. As a result, the entire roof surface gradually hardens. It's easier for moisture to penetrate the roof, and the alligatoring of felts soon follows. As preventive maintenance, apply resaturant to the roof surface to lower the surface bitumen softening point and to restore flexibility to felt membranes. The resaturant also provides a base for new aggregate to be embedded.

Before you apply resaturant, remove all loose aggregate and sweep the roof surface clean. If you're going to reuse the aggregate, clean it with water and let it dry. After you've made minor roof surface repairs, apply about 7 gallons of the resaturant per square. Replace the aggregate immediately.

You can generally resurface and rejuvenate a roof that delaminates when it's flexed and shows blistering, and has insulation that's not damaged or wet.

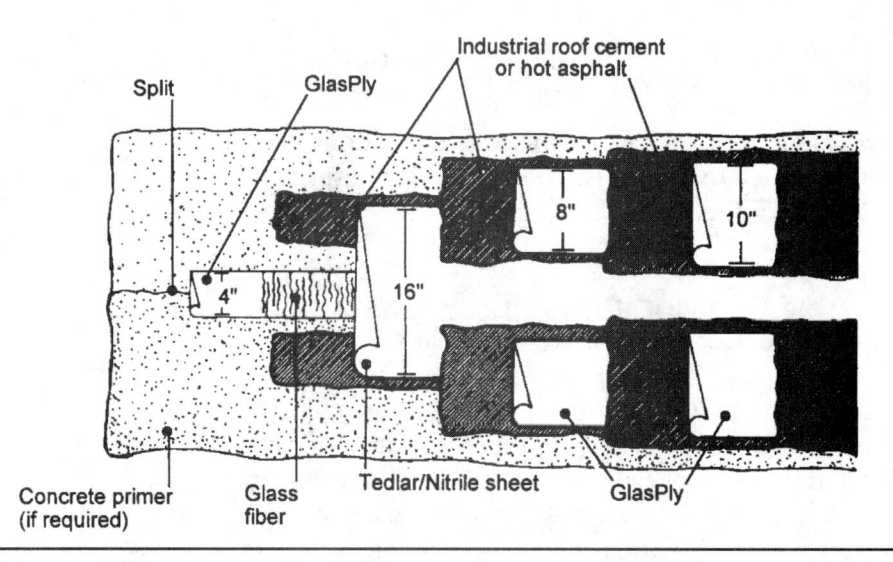

Figure 10-29 Repairing a split

Moisture trapped within voids between felt plies can expand and make blisters. The voids happen if:

- You didn't broom the felts tightly.
- The bitumen was too hot when you applied it.
- You didn't use enough interply bitumen.
- You didn't store the materials properly and they absorbed moisture.

Large bubbles caused by these problems are often the source of roof leaks, especially if they're located in a low spot. To repair a large bubble, remove aggregate at the problem area and cut an "X" in the bubble. Embed the flaps of the cut material into roofing cement and nail a patch of heavy membrane material over the hole. Drive nails on 1-inch centers along the edges of the patch, then cover the patch with roofing cement.

A roof will usually split if it moves excessively because it doesn't have enough roof expansion joints or they aren't in the correct places. They also occur if the roof membrane or insulation isn't properly attached. Use a spud to remove aggregate adjacent to the split and cut off the curled edges of the membrane. To allow for expansion and contraction, install a loose 4-inch glass fiber strip covered with a 16-inch Tedlar sheet cemented at the edges. Cover the edges of the Tedlar sheet with fully cemented 8- and 10-inch glass fiber sheets. Then cover the sheets with roofing cement or hot asphalt as shown in Figure 10-29.

Alligatoring is the result of a drying and deteriorating roof surface. Apply asphalt emulsions to help eliminate this problem.

Fishmouths are caused by inadequate brooming of felt edges or improper alignment of the felts. To repair fishmouths, cut the loose material and embed it into roofing cement or hot asphalt.

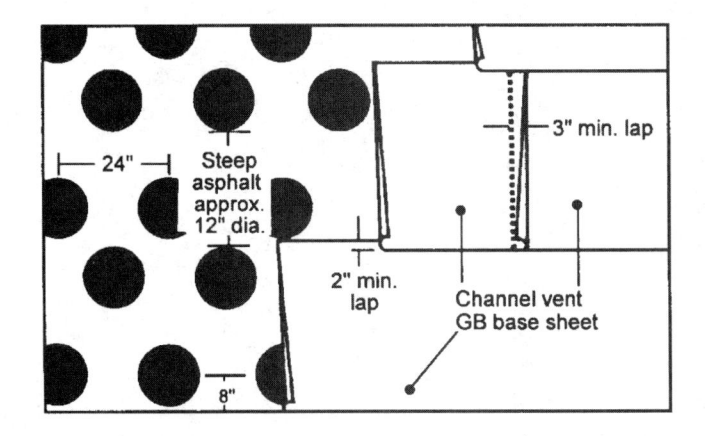

Figure 10-30 Vented base sheet installed over an old roof

Figure 10-31 Venting an old roof

To repair splits, blisters, alligatoring and fishmouths, embed a patch of heavy membrane material into $1/8$-inch-thick cold-applied mastic. Then put on a top coat of $1/8$-inch mastic. Make the patch 6 inches wider than the area you're repairing.

Flashing failure can be caused by equipment vibration and/or thermal expansion and contraction of dissimilar flashing materials such as membrane and metal. To repair flashing, embed and cover the loose material with roofing cement.

A primary location for leaks in a built-up roof is at the roof edge. As with other flashing repairs, cover the problem area with roofing cement.

Insulation can get wet if the roof covering or flashings aren't waterproof. Also, the insulation won't dry if it's not properly vented.

If you catch these things early on, you can avoid replacing the whole roof. A roof that has a lot of splitting, brittle surface material and saturated insulation will have to be replaced.

If there's no insulation under the existing roof covering, remove all of the roof covering before you install any new roof covering. You'll also have to completely remove the existing roof system if the extra weight of the additional roof system exceeds the safe design load.

If there is insulation under the roof, remove all aggregate, blisters and loose felts and sweep the roof clean. Replace all wet or damaged insulation.

Some manufacturers recommend that you remove all insulation to avoid problems with trapped moisture. If you damage good insulation when you remove the original roof, sandwich a recovery board between the insulation and the new BUR system.

Over non-nailable decks, prime the old roof membrane and install a vented base sheet embedded in spot moppings of hot steep asphalt. Install the mopped spots on 18-inch centers as shown in Figure 10-30. Over nailable decks, nail the base sheets at laps on 9-inch centers and stagger-nail the sheets in two rows along the center on 18-inch centers.

When you're re-roofing adjacent to a roof requiring no work, protect the existing roof from damage due to traffic and impact. Never solidly mop a new roof over an old roof. Also, when you're in doubt about trapped moisture in the old insulation, cut holes in the old roof at 3-foot centers in both directions so trapped moisture in the old roof can escape before you install a vent sheet. That's shown in Figure 10-31.

You can test questionable substrates for dryness and adhesion by pouring one pint of bitumen, heated to application temperature, over a representative area of the deck. If the bitumen foams or bubbles, there's too much moisture. If you can strip the cooled sample cleanly from the deck, it didn't stick properly. Apply a primer when a deck can't be cleaned to permit proper adhesion.

Now that we've covered built-up membrane roofing, we're ready to move on to the other membrane system, elastomeric roofing. That's the subject of the next chapter.

11 Elastomeric Roofing

▶ An elastomer is any material that has the elasticity to return to its original shape after being repeatedly stretched to twice its size at room temperature. Some elastomeric roofing materials will stretch up to 450 percent, compared to a felt built-up roof system which stretches only $1/2$ to $1 1/2$ percent.

Elastomeric roofing is compatible with clean concrete and exterior-grade plywood decks, smooth metal, glass, flagstone, wood, and in some cases, asphalt or tar. Some types of elastomeric roofing aren't compatible with lightweight concrete or gypsum decks due to blistering that results from the residual moisture within the deck. However, some varieties are vapor permeable and allow moisture to escape.

Many elastomeric materials require a primer brushed, rolled, squeegeed or sprayed over the substrate. The rate of application varies from $1/2$ to $3/4$ gallon per square. I strongly recommend a primer when the roof deck is especially dry or if felts show excessive alligatoring.

Always follow the manufacturer's instructions with regard to substrate preparation. For example, petroleum-based materials such as bitumen or coal-tar pitch adversely affect PVC membranes. You have to install a slip sheet (also called a separation sheet) between these materials. Also, lumber in contact with PVC material must be wolmanized (pressure-treated).

If you're installing roof walkways, elastomeric roofing manufacturers recommend that you install an additional layer of membrane beneath the walkways. Install walkway panels over an elastomeric roof with the adhesive recommended for the particular elastomeric roofing system.

The Advantages of Elastomeric Systems

Elastomeric roofing material is more expensive than a BUR, but elastomeric roofing also has many advantages. It:

- expands and contracts without tearing
- remains flexible at low temperatures
- bonds well to the substrate
- weighs less (some systems)
- conforms to any roof shape or slope
- withstands water ponding (many systems but not all)
- is self-healing and self-flashing (many systems but not all)

Elastomeric roofing also keeps your labor costs down because it's easier to install than a built-up roof. Since there are fewer operations, there are fewer mistakes and your workers turn out a better job. It's also safer. Hot roofing is responsible for at least 23 percent of all roofing injuries, but elastomeric roofing doesn't require hot kettles. There's no risk of asphalt stains on the building face, and you can install polystyrene foam insulation — the most efficient insulation available.

Another advantage is that you can prefabricate elastomeric sheets off-site. Just cut the rolls to the proper lengths, re-roll them and take them to the site. Then unroll and install them in the proper order. That reduces time on the job, which can help you meet a tight project deadline.

On re-roofs over BUR systems, you don't have to completely remove the old roof before you install many elastomeric roof systems. You only have to remove loose aggregate and install a layer of rigid insulation before you apply the membrane.

Most elastomeric roofs are virtually maintenance-free and easy to repair. You can install many loosely laid systems in any type of weather. You can put them over an icy, snow-covered, wet or humid surface, so bad weather won't delay the job.

Elastomeric roofing comes in a wide variety of colors, including some with a reflective surface which deflects solar heat. It's highly resistant to ultraviolet light and industrial environments containing dirt and mild acid fumes. Builders also use elastomeric materials to waterproof basements, and for flashing.

Other than its higher initial cost, the only disadvantage of elastomeric systems is the limited number of roofers who know how to install them. If you're one of the ones who has experience, it can open a new world of opportunities for you.

Liquid-applied Elastomers

Although this chapter focuses on the application of sheet roofing materials, an overview of some liquid-applied elastomers is included here as part of the subject of elastomeric roofing. *Liquid-applied* elastomers are a highly effective and practical roof coating, and in this edition are included in a separate chapter. For detailed coverage of liquid-applied elastomers, including its uses, benefits, application instructions, and estimating methods and example estimates, see Chapter 13, *Roof Coatings*.

You can get liquid-applied elastomers in four consistencies: caulk, trowel, brush and spray. You can apply many liquid elastomers to virtually any substrate, including aged tar or asphalt. However, alligatored areas sometimes require a primer coat. Most liquid elastomers cure within 24 hours. Then they're hard and dry enough to walk on without it feeling tacky.

Apply the troweled variety over reinforced mats at a rate of about 4 gallons per square. Next install a reinforced polyester mat, then cover that with an additional coat of troweled material, also at a rate of 4 gallons per square. Reinforced mats come in 5-, 10- and 30-square rolls 36 inches wide.

Use the brushed variety to rejuvenate dried-out felts. Apply both the troweled and brushed materials $1/8$ to $3/16$ inch thick.

Silicone, Urethane and Vinyl

Some of the most commonly used liquid elastomers include silicone (polysiloxane), silicone rubber, urethane or vinyl liquid applied directly to the substrate. Done correctly and in accordance with the manufacturer's instructions, these can provide a UL Class A fire rating. The number of coats recommended for each type of roof varies, though two is normal. The required total thickness also varies, but is unlikely to be less than 15 dry mils. Use a micrometer to make sure. Ultimately, you need to check the manufacturer's instructions for mil thickness, as they won't warranty a coating that doesn't meet their recommended thickness.

You can embed a layer of ceramic-coated granules in the upper coating to make the roof more attractive, minimize discoloration, and provide a non-skid surface that's resistant to foot traffic, UV radiation and hail. Silicone rubber repels water from the outside, but allows vapor to escape from the substrate below. You can also repair damaged surfaces with silicone building sealant.

Urethane is also used as a protective coating over polyurethane foam insulation. Apply urethane in two coats at the rate of $1 1/2$ gallons per coat per square. Apply a total minimum coating thickness of 25 dry mils. Urethane is a "breather" type of coating which allows water vapor to escape. Urethane coatings stretch up to 400 percent and can span structural cracks up to $1/8$ inch wide. Standard colors are black, white or aluminum.

Figure 11-1 Loosely-laid and ballasted system

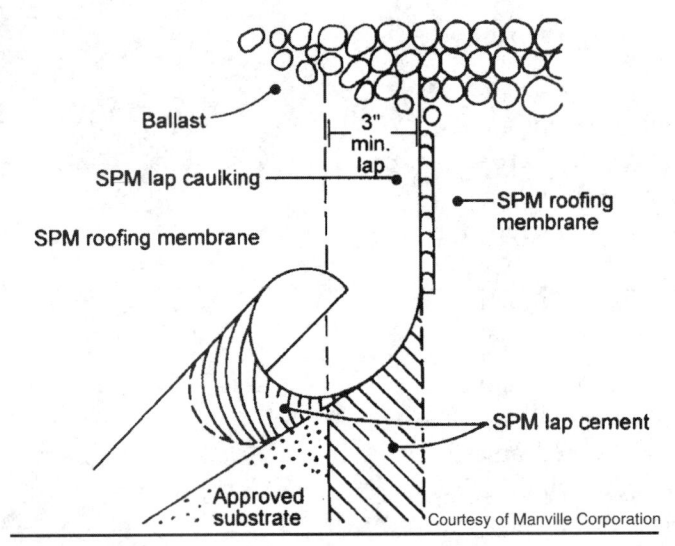

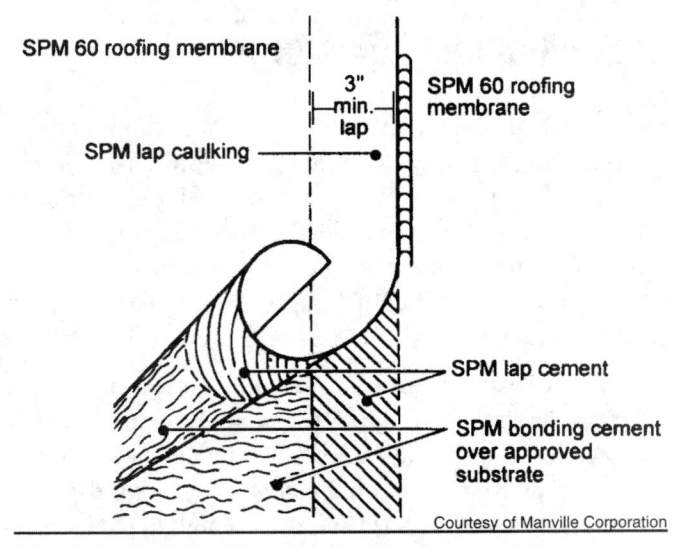

Figure 11-2 Fully-adhered system

Neoprene

Another popular liquid-applied elastomeric system includes neoprene (chloroprene). It's applied in much the same way as other liquid elastomers. See Chapter 13, *Roof Coatings* for specifics on most types of liquid-applied roof coatings and their application. Neoprene roofing is also available in sheets, which are fastened to the deck with adhesives. Polyester-reinforced fabrics are also popular. Cover the liquid or sheet with liquid coats of hypalon (chlorosulfonated polyethylene) or vinyl acrylic to get the minimum required dry film thickness. Fiberglass embedded in the elastomeric material is an excellent reinforcement for roof joints under liquid roofing.

Single-Ply Roofing Systems

The most common elastomeric sheet roofing materials are butyl rubber, neoprene, neoprene-hypalon, chlorinated polyethylene (CPE), polyvinyl chloride (PVC), ethylene propylene diene monomer (EPDM), polyisobutylene (PIB) and chlorosulfonated polyethylene (CSPE). A coal-tar elastomeric membrane reinforced with polyester fiber is also manufactured.

You can install preformed elastomeric sheets in any of three ways:

1) Loosely laid and ballasted with aggregate or pavers (Figure 11-1)

2) Fully adhered (Figure 11-2)

3) Mechanically fastened (Figure 11-3)

In any system, you cement or tape the joints between sheets or "weld" them with hot air or solvent. You must also caulk the edge of the joint.

Elastomeric Roofing

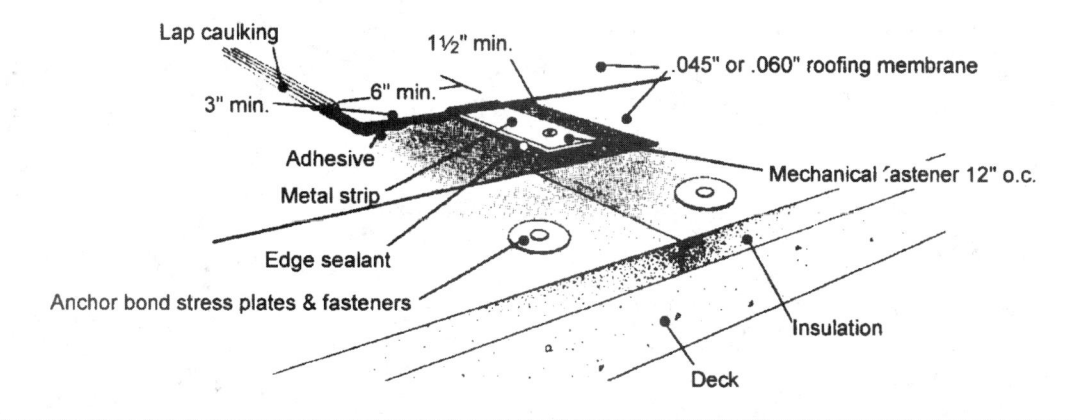

Figure 11-3 Mechanically-fastened system

Ballasted Roof Systems

If the roofing system is loosely-laid and ballasted with aggregate or pavers, the ballast adds about 10 pounds per square foot of dead load over the structure. This is a very important design consideration, especially when you're re-roofing over a structure originally designed for a specific roof dead load. When in doubt, consult a structural engineer. It's OK to install an unballasted elastomeric roofing system for retrofit on roof decks that can't support the weight of the ballast.

One disadvantage of an aggregate-ballasted roof system is that the membrane is vulnerable to physical damage from the ballast. Also, there are height limitations (80 feet), slope limitations (2 in 12), and the system can't withstand high winds. Furthermore, the aggregate ballast collects dirt, oil, chemicals and pollutants which eventually damage the membrane.

Ballast systems have advantages, too. First, you can remove the system and reuse it when more stories are added to an existing building. Second, the membrane moves independently of the desk and allows moisture venting. Since many elastomeric sheets are virtually impervious, venting is sometimes necessary.

Fully-Adhered Roof Systems

You can install a cemented system two ways. Either bond the entire membrane to the deck, or spot-bond only 40 percent to allow for expansion and contraction in all directions. Check the specifications of the bonding adhesive you plan to use. Some can't be applied when temperatures are extremely high or low, or in high humidity conditions. Many manufacturers recommend that you install a slip sheet (or separation sheet), typically made of unsaturated felt, between the substrate and the elastomeric sheet.

When you install elastomeric roofing directly to a deck, you must provide adequate air circulation between the deck and roofing material to prevent moisture from getting trapped. When you install elastomeric roofing over insulation, use a cold-process adhesive and put a vapor retarder between the deck and the insulation. You must install rigid insulation between elastomeric membranes and steel roof decks or aggregate-surfaced roofs.

EPDM Elastomeric Roofing

EPDM (ethylene propylene diene monomer) consists of vulcanized rubber-based membranes that are compatible with almost any roofing surface, including foam or asphalt-faced insulation products. But EPDM membranes are not compatible with and should never be installed with either plastic cement or dead-level asphalt. Never apply EPDM roofing to any roof that's subject to chemical discharge or covering a cold storage or freezer facility. The chemicals may react with the EPDM, and cold spaces vent too much moisture.

EPDM is very durable, with a service temperature range of -60° F to 200° F. Sheets come in a variety of sizes: Available widths are 7½, 10, 20, 40 or 50 feet; standard lengths are 50, 100, 125, 150 or 200 feet. You can also special-order membranes that measure as much as 50 x 300 feet. Standard thicknesses are 0.045 inch (45 mils), 0.06 inch (60 mils), and 0.09 inch (90 mils), all stretchable by 400 percent or more. Reinforced EPDM sheets are also available.

Your color choice for EPDM is limited; it's only available in black (containing carbon black) or white (containing titanium dioxide) added according to whether UV rays need to be converted to heat or deflected from the surface. If the job requires a light-reflective surface, you can apply a hypalon coating over the face of EPDM membranes on fully-adhered or mechanically-fastened systems. However, before you apply the coating, allow the membrane to weather in place for at least two weeks. Then you apply the color coating to the EPDM membrane in two coats. You can put on the second coat as soon as the first one is dry.

Some systems come with a 15-year limited service warranty, although some membranes last up to 20 years or more.

You can install EPDM systems loosely laid and ballasted, fully adhered, or mechanically fastened. But no matter which method you use, install a butted insulation protection course over any substrate that has cracks or joints ¼ inch wide or wider. On re-roofs, remove all gravel before you install insulation. Overlap adjacent EPDM sheets at least 3 inches (6 inches if mechanically fastened), and offset joints at least 12 inches. Position all sheets with a 2- to 3-inch overhang at the roof's edge and a 2- to 3-inch run-up at all vertical surfaces.

Loosely-Laid EPDM Systems

Install a loosely-laid system in new construction or as re-roofing over plywood, wood plank, approved insulation (confirm with manufacturer) and smooth-surfaced roofs. The system is also compatible with lightweight concrete decks, provided:

- The deck is at least 3 inches thick.
- It has a dry density of at least 22 pounds per cubic foot. A testing lap can determine this.
- It has a minimum compressive strength of 125 pounds per square inch.
- It's installed over a substrate that allows the deck to dry from underneath, for example, form boards or vented metal panels.

Insulation materials such as urethane, fiberglass, polyisocyanurate, phenol formaldehyde, perlite and polystyrene are all compatible with EPDM roofing membranes. Install a slip sheet between the membrane and incompatible substrates, like insulation covered with dead-level asphalt.

Install a ballasted EPDM system over loosely-laid insulation boards or mechanically fasten the insulation board to the deck or install it in hot asphalt. The method depends on the deck type and roof slope.

Membrane thickness for a ballasted system is normally 0.045 inch. Some owners may feel the extra protection the 0.06-inch thickness provides is worth the cost. Seal and caulk all membrane laps as shown in Figure 11-1. Use $3/4$-inch river-washed gravel or paver blocks weighing up to 60 pounds each for ballast. Using crushed stone for aggregate ballast will damage the membrane. Use smooth aggregate designated as $1/2$ to $1 1/2$ inch, where no more than 10 percent is retained in a sieve with 2-inch openings, and no more than 10 percent passes through a sieve with $1/2$-inch openings.

On roofs 80 feet or higher above the ground, you must ballast the roof with 2' x 2' x 2" precast concrete pavers installed to provide a minimum of 10 pounds per square foot. Install aggregate or pavers at the end of each day, or at times necessary to prevent wind damage.

The ballasted system is given a UL Class A fire rating.

Fully-Adhered EPDM Systems

Here are some installation tips for a fully-adhered EPDM system:

- Install the system in new construction or as re-roofing on inclines through 6 in 12. You can install this system over wood, plywood, high-density fiberboard, concrete or approved insulation (confirm with manufacturer).
- If you install the system over insulation, mechanically attach the insulation to the deck with one fastener for each 2 square feet to make sure the insulation won't move.
- Over non-nailable decks, install the insulation in a solid mopping of hot steep asphalt.

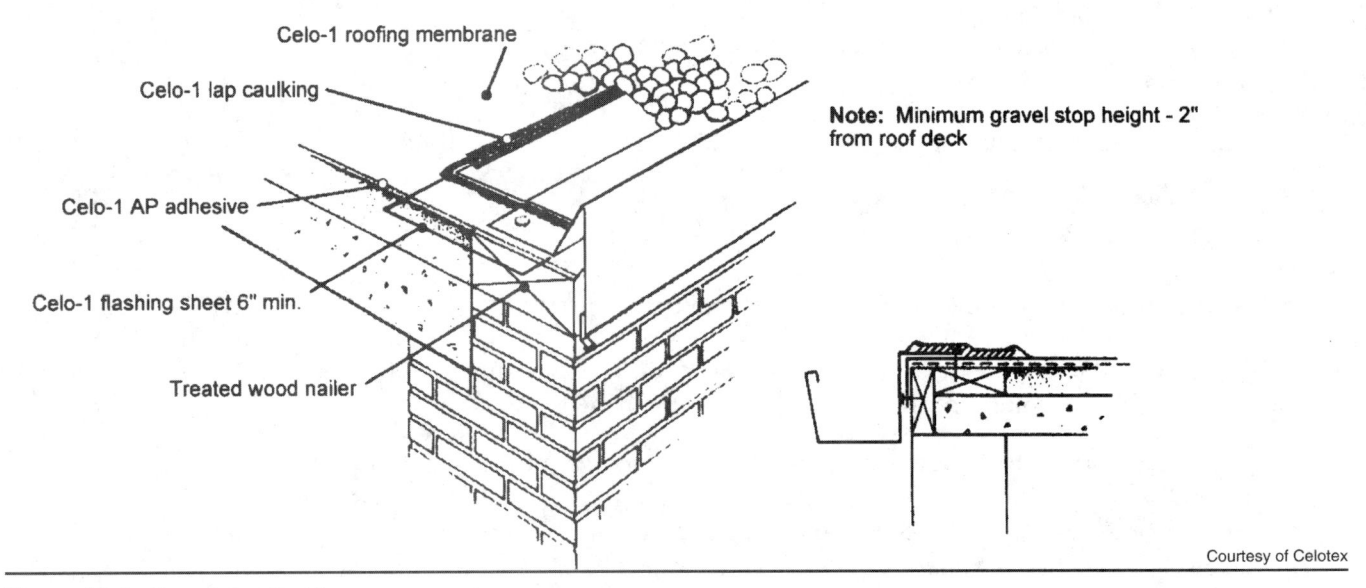

Figure 11-4 Gravel stop flashing

- Mechanically fasten a fully-adhered system at the roof perimeter and around openings of 16 square feet or more.

- Apply membrane 0.06 inch thick. Apply the adhesive used for bonding the membrane to the deck or insulation (bonding adhesive) to both the bottom of the membrane and the top of the substrate at the rate of $1^2/_3$ gallons per square. That's about 1 gallon for every 60 square feet of membrane. Note: A fully-adhered system weighs about $1/_3$ pound per square foot.

- Apply the adhesive using a 9- to 12-inch-wide paint roller with a solvent-resistant core.

Mechanically-Fastened EPDM Systems

Install a mechanically-fastened (bar-anchored) EPDM system in new construction or as re-roofing on inclines up through 6 in 12 as follows:

- Install this system over wood plank, plywood, approved insulation (confirm with manufacturer) or smooth-surfaced roofs over any deck that will accept and hold mechanical fasteners.

- Install membrane 0.045 inch thick (0.06 inch in high-wind areas). Note: This system weighs approximately $1/_3$ pound per square foot.

- Secure the underlying lapped portion of each sheet under an aluminum or galvanized cap strip (termination bar or anchor bar) set in sealant and mechanically fastened to the deck on 12-inch centers.

- Lay the lapped-over portion of the successive sheet over the metal strip, adhere it with an adhesive and seal the lap with caulk.

- Anchor the membranes under metal strips at the roof perimeter and around openings.

Elastomeric Roofing

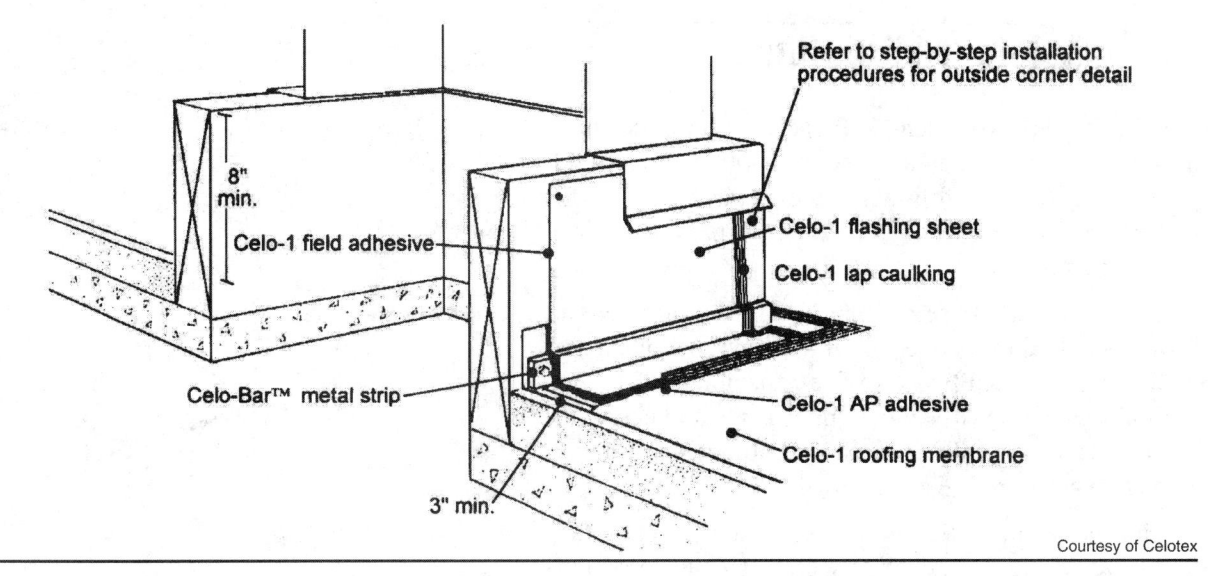

Figure 11-5 Curb flashing (vertical nailer)

- Use mechanical fasteners long enough to penetrate at least 1 inch into concrete decks and 3/4 inch into wood or plywood decks.
- Use narrow EPDM sheets (5'6" wide) in a mechanically-fastened system.

Mechanically fasten the membrane on 8-inch centers over treated wood nailers located at the roof perimeter (Figure 11-4), at roof penetrations (Figure 11-5), and vertical abutments (Figure 11-6). Use nailers (No. 2 or better lumber) or plywood. Either must be pressure-treated with salt preservatives (wolmanized) and at least 1/2 inch thick. Creosote- or asphalt-treated lumber is incompatible with the membrane. Use 1-inch galvanized nails at wood nailers with 1-inch-diameter (minimum) heads.

You can also install metal strips instead of wood nailers as shown in Figures 11-7 and 11-8.

Adhere all laps with an approved lap cement or joint tape and caulk the joint with a 1/4-inch-diameter bead of seam caulk. Brush-apply lap cement at the rate of 1 gallon per 150 to 200 linear feet for a 3-inch lap, or 1 gallon per 75 to 100 linear feet for a 6-inch lap. Apply seam caulk at the rate of 1/10-gallon tube per 20 to 25 linear feet. Be sure that lap cement (at seams) and bonding adhesives (in the field) are compatible with the EPDM material. Don't use coal tar, asphalt- or oil-based roofing cements. Use lap cement (seaming cement) only for field seaming membranes and flashing; never use it to adhere membranes and flashing to any other surface.

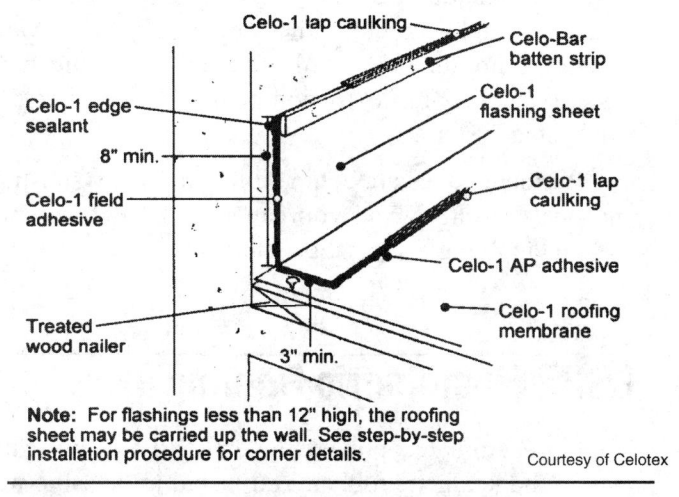

Figure 11-6 Roof-to-wall flashing (horizontal nailer)

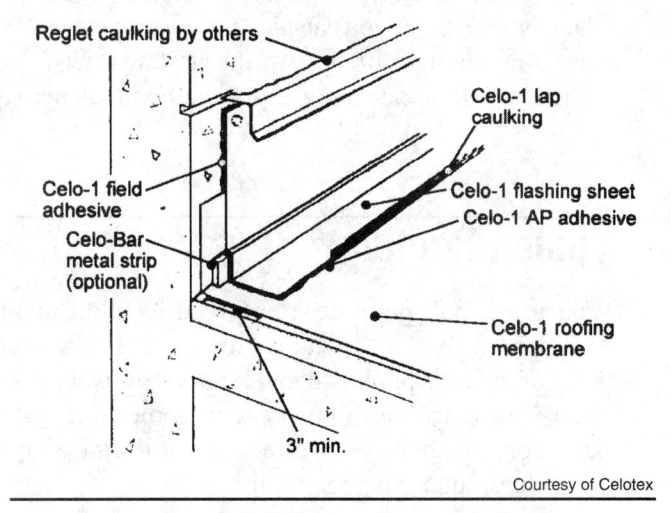

Figure 11-7 Roof-to-wall flashing with counterflashing

CPE Elastomeric Roofing

Chlorinate polyethylene (CPE) elastomeric roofing is usually mechanically attached to the roofing system with plates and screws, following the same rules as for EPDM. Seams are heat-welded to form a one-piece roofing membrane.

CPE membranes are resistant to both ultraviolet and infrared radiation, ozone, oil and many chemicals and microorganisms. CPE is also fire-resistant. CPE retains flexibility at below-freezing temperatures and can withstand heat up to 150° F. Some manufacturers offer a 10-year warranty.

The inclusion of titanium dioxide during manufacture gives a highly reflective white color to the membrane which saves air conditioning costs. CPE roofing is normally manufactured in rolls 5'2" wide and 103'6" long and weighs about 5 ounces per square foot. Manufacturers produce a 40-mil-thick roll for a mechanically-fastened roof system and a 32-mil extrusion-coated polyester reinforced membrane for a ballasted system.

CPE membranes are compatible with many substrates including asphalt and coal-tar pitch. And you can install CPE membranes over insulation without applying a separation sheet.

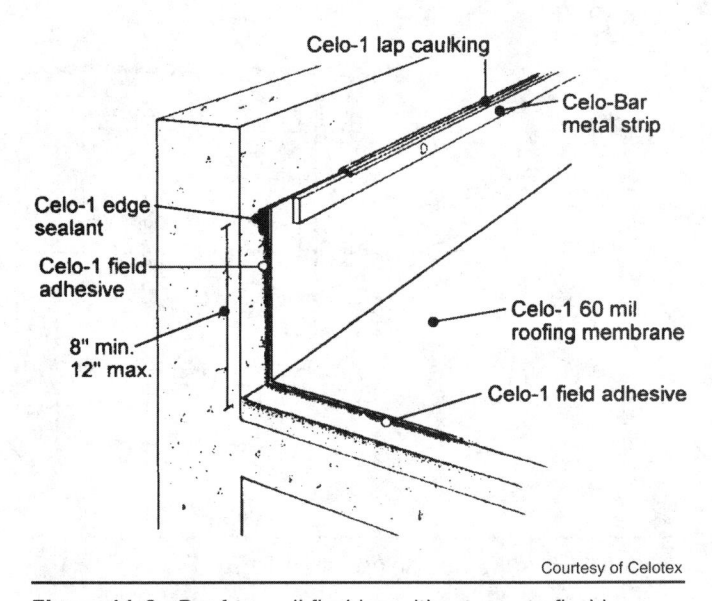

Figure 11-8 Roof-to-wall flashing without counterflashing

CSPE Elastomeric Roofing

Chlorosulfonated polyethylene (CSPE) is basically a combination of plastic and synthetic rubber. You install it the same way as EPDM roofing. CSPE is compatible with bituminous materials and is resistant to water, ozone and ultraviolet light. CSPE membranes are most commonly available in black or white, but you can also get custom colors. White is recommended in the summer heat of the southern states. CSPE comes in a nominal thickness of 0.045 inch and weighs 0.29 pounds per square foot.

Hypalon Roofing

You can install polyester-reinforced hypalon membranes mechanically attached or fully adhered to a substrate. This material is unaffected by asphalt or coal-tar pitch and can be bonded using these materials. Heat- or solvent-weld lap joints. Some systems come with a 10-year warranty. Sheets usually come with a white surface, but you can spray coat the surface with hypalon to produce another color.

PVC Elastomeric Roofing

Polyvinyl chloride (PVC) elastomeric roofing is normally manufactured in rolls 5'10" wide and 72 feet long. Rolls are 0.045 or 0.06 inch thick and weigh 0.3 pounds per square foot. Stabilized PVC is resistant to ultraviolet light and ozone and is virtually impervious to water vapor. But PVC frequently loses its plasticizer with heat aging and becomes more brittle over time. Before heat aging, some PVC membranes stretch up to three times their original size in any direction. You can also use reinforced PVC material impregnated with woven glass or polyester fabric.

PVC elastomeric roofing is appropriate for ballasted, mechanically-fastened or fully-adhered systems. You can fully adhere PVC roofing to concrete, plywood or approved insulations. You can install it loosely laid or mechanically attached over concrete, wood, metal or approved insulation. Compatible insulations include perlite and polyurethane boards.

PVC is incompatible with bituminous material (including bituminous-coated insulation), foamed glass or polystyrene insulation, unless you separate them with a slip sheet before you install the PVC membrane. But never fully adhere PVC to incompatible substrates even when a slip sheet is installed.

Lap and seal membrane seams with a seam solvent, or heat-weld and caulk the seams. Use 3-inch laps for the ballasted system and 6-inch laps for the bar-anchored system.

Composite Roofing Systems

Elastomeric roofing composites are plastic films laminated to fabrics, paper, metal, felts or rubberized asphalt of various types. Metals incorporated in composite elastomeric membranes include aluminum, copper and stainless steel. Some composite membranes are coated with reflective white vinyl acrylic after installation. It reflects heat better — and looks better. Many manufacturers recommended that you reapply the coating over flat roofs at 5-year intervals.

Composite membranes, like many other elastic membranes, are installed ballasted, mechanically-fastened or fully-adhered. Some materials come with release paper covering the bottom surface. You just remove the paper and adhere the membrane to the substrate without additional adhesives. Bond the other types of composite materials to the substrate with cold-applied adhesives.

Rubber-vinyl composite membranes come in 50-mil-thick rolls 74'6" long and 3 feet wide. Each roll weighs 83 pounds which equates to a system weighing about 37 pounds per square. Rubberized composite membranes are self-healing and remain watertight even under ponded water.

Plastic-bitumen composites come in rolls 3'7" wide and 33 feet long. The material is 160 mils thick and each roll weighs 99 pounds. Plastic-bitumen-aluminum composites come in rolls 3'7" wide and 33 feet long. These are 120 mils thick and each roll weighs 76 pounds. You can install a plastic-bitumen composite loosely-laid and ballasted. Use asphalt cement or steep asphalt to adhere plastic-bitumen-aluminum membrane. Never apply the aluminum sheet directly to a wood deck because wood resins react with it.

TPO Roofing

TPO (thermoplastic polyolefin) is a single-ply roofing membrane developed in the early 1990s as an economical alternative to PVC roofing. The name is a bit misleading. TPO isn't actually a plastic. It's a blend of different types of rubber, usually polypropylene and ethylene-propylene. TPO has three layers: a TPO polymer base, a polyester-reinforced fabric center (scrim) and a thermoplastic polyolefin compounded top ply.

While TPO may not technically fall in the category of elastomeric roofing, its usefulness, properties and installation methods are similar. Some TPO fabrics include EPDM, an elastomer.

The key to a good TPO roof is properly welded seams. Each sheet of TPO laps a few inches over the lower layer. The top of the lower layer and the bottom of the upper layer are softened with a specially designed heat source and then pressed firmly together. Done correctly, the two layers bond together, forming a single waterproof membrane that extends across the entire roof.

TPO is touted to have the benefits of PVC roofing but at a lower cost. If you work in an area where TPO is readily available and in common use, consider recommending TPO for flat or low-slope roofs. The cost may be less than alternatives, especially on smaller jobs. Plus, TPO has energy-saving advantages over other roofing materials.

Most TPO manufacturers provide a warranty when the material is applied by an approved installer following the manufacturer's recommendations. A warranty is often required on commercial buildings. TPO is also used on flat or low-slope residential jobs, largely because of the low cost.

A good-quality TPO roof can be:
- As UV-resistant as EPDM.
- As heat-resistant as EPDM.
- As heat-weldable as PVC.

There are good reasons to recommend TPO roofing:
- Less damaging to the environment than EPDM.
- Exceeds EPA Energy Star requirements.
- Has earned a listing with the Cool Roof Rating Council.

- Has a Class 4 impact rating (resists damage from hail to 2 inches in diameter).
- Can carry a Class A fire-rating under specific conditions.

As with any roofing material, TPO has advantages and disadvantages. TPO is a comparatively new roofing material. Manufacturers are still improving the chemical composition of TPO to increase durability while maintaining a competitive price. The chief advantage of TPO is low cost, usually less than EPDM or other types of rolled rubber roofing. The other big positive for TPO is its light-reflectivity, largely due to the white surface color. This may make TPO less attractive for homeowners. White isn't a popular roof color for homes and won't be allowed by the owners' association in some communities. Some manufacturers offer light grey or black reflective TPO with good UV-resistance and a "cool" rating. It's no longer true that a low-cost, reflective roof has to be white. This makes TPO a better option for homeowners who want an attractive, inexpensive roof that lowers cooling costs.

TPO has another appeal for environmentally conscious customers. It's completely recyclable. The old material can be ground up and recycled.

Compared to other thermoplastic membrane roofing, TPO is as good or better at resisting mold and algae, accumulated dirt, corrosion and breakdown on contact with other materials. TPO is also more flexible. As the roof settles, expands or contracts with changes in temperature, TPO membranes stretch and shrink more effectively than other single-ply roofing products. The result is fewer rips and punctures.

Ease of installation

TPO membranes are manufactured in sheets as wide as 12 feet or more for commercial jobs. Even with wide sheets, the rolls are relatively lightweight. Wider rolls make the material easier to install and reduce the number of seams, lowering the installation cost. With fewer seams, there aren't as many areas where strain from expansion and contraction can loosen the bond between sheets.

TPO Installation Step-by-Step

Most TPO is installed with fastener plates and screws. Fastener installation costs less, reduces the labor cost and can be completed in nearly any weather condition. In high-wind areas, TPO should be installed with adhesive. If you plan to use adhesive, you'll need about one gallon for each square of roofing. Don't try to apply TPO with adhesive if there's a chance of rain or if the air temperature is below 40 degrees. Peel-and-stick TPO is also available at a higher cost. In any case, follow the manufacturer's recommendations.

1. Clean and prepare the roof deck. Usually this means tearing off the old roof and any existing insulation. Scrape and blow away the debris. You want to start with a clean, dry, smooth deck that slopes evenly to the roof edge or roof drains.

2. Since TPO roofing has almost no insulating value, you'll usually need to install a layer of rigid foam insulation over the existing roof surface. Attach foam board using either mechanical fasteners or adhesive.

3. Attach metal flashing and roof edging as needed.

4. Roll out the sheets of TPO. Sheets higher on the roof should overlap sheets lower on the roof. Work usually starts by boxing the perimeter with a 4-foot or 6-foot-wide roll. That prevents uplift from wind at the roof perimeter. Crease and fold sheets up any perimeter wall. Then roll out full sheets to fill the area between perimeter sheets.

5. When each sheet is cut to length, you're ready to apply adhesive or drive fastener plates. The edge of most TPO sheets is marked each 12 inches for fasteners.

6. If you're using adhesive, fold back one-half of the width of the TPO sheet. Roll adhesive on both the exposed bottom sheet and the exposed roof deck. Keep adhesive off the overlap area. While the adhesive is still tacky, pull the folded half sheet back on the roof. Then fold back the other half of the same sheet. Apply adhesive as before. Then pull that half-sheet back into the adhesive you applied on the roof deck. Finally, run a weighted roller over the full sheet to flatten out any bubbles and ensure good adhesion.

7. Required overlaps vary from 3 inches for adhered applications to 6 inches for sheets secured with fasteners. Control lines on the TPO usually show a recommended overlap. In any case, the sheet above has to lap over fasteners on the sheet below. Side lap on a perimeter sheet is usually 4 to 5 inches. End lap should be at least 3 inches.

8. Whether you use fasteners, adhesive or self-stick TPO, good seam and edge sealing is critical. Poor sealing is the most common cause of a leaking TPO roof. Seal all seams, roof edges and flashing, including around chimneys and other penetrations.

9. Seal seams with heat and pressure. Be sure the seam lap is clean and dry before starting work. A powered heat welder is best for seam sealing. Powered welders can be set to advance from about 10 feet per minute to about 30 feet per minute. Heat can be set to temperatures from about 400 degrees to nearly 900 degrees. The best roller speed and heat setting depends on job conditions. A hand-held seam welder and small hand roller will be suitable for sealing seams on smaller jobs. Either way, be sure to seal at least 1½ inches of each TPO seam.

10. Seal vents and other roof fixtures with boots and collars made to be compatible with TPO roofing.

TPO Roofing Cons

The main drawback with TPO is that it's a laminated product. Most laminates have weak areas that can shrink, crack, craze, or deteriorate over time. The quality of TPO laminates can vary. Select products from a manufac-

Elastomeric Roofing

turer with a reputation for good-quality, long-lasting materials. But paying more for TPO doesn't necessarily guarantee a better product.

TPO comes in thicknesses from about 45 mils to 90 mils. Generally, the thicker the wear layer of a material, the more durable it is. But that rule of thumb doesn't necessarily apply to TPO. Thicker TPO may provide extra protection against damage from roof traffic, but doubling the wear layer of TPO doesn't necessarily double the life expectancy. A 40 mil TPO roof may last nearly as long as a 60 or 90 mil TPO roof.

When first developed, TPO roofing wasn't a good choice for hot, dry climates. TPO couldn't take extreme thermal and solar loads. This problem has been largely eliminated. But if you work in the southeast or southwest, check with the manufacturer before selecting any TPO roofing product.

Under most conditions, a TPO roof can be expected to last about 20 years. Data sheets offered by the manufacturer will be your best guide to durability of any TPO product. Buy from a dealer or manufacturer responsive to your needs and with a reputation for delivering quality products.

Flashings for Elastomeric Roofs

Flashings installed on elastomeric roofs vary with the particular roof material. In most cases, you'll install elastomeric sheet base flashing at vertical surfaces by continuing the membrane up the wall as in Figure 11-7. Extend the base flashing up the vertical surface at least 8 inches and 3 inches onto the field of the roof (Figure 11-6). I only recommend separate base flashing for areas that are hard to flash, like corners and tight areas.

Install flashing material in adhesive applied at the rate of 1 gallon per 120 to 140 square feet. Lap seams at least 4 inches into seaming cement and caulk the joint and all edges of flashing material. Use flashing cement compatible with the membrane material. Don't use coal-tar, asphalt- or oil-based roofing cements.

Use a prefabricated vent flashing boot to waterproof vent stacks as in Figure 11-9 and flash roof drains as shown in Figure 11-10.

Flashing material used on EPDM roofs consists of uncured (or unvulcanized) neoprene or EPDM rubber. EPDM flashings usually have a nominal thickness of 0.06 inch. They come 6, 12, 18, 24, 36 and 48 inches wide, in 50-foot lengths. Install these uncured materials so they'll adapt to irregular shapes and surfaces. With aging, they vulcanize and become tough rubber membranes.

Courtesy of Celotex

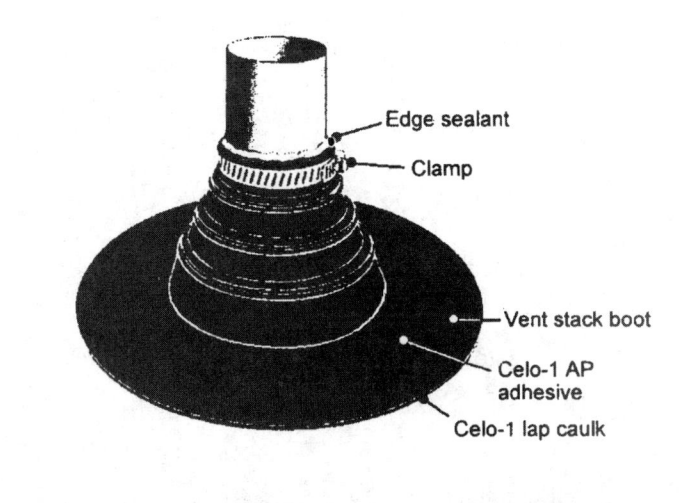

Figure 11-9 Vent stack boot

Courtesy of Celotex

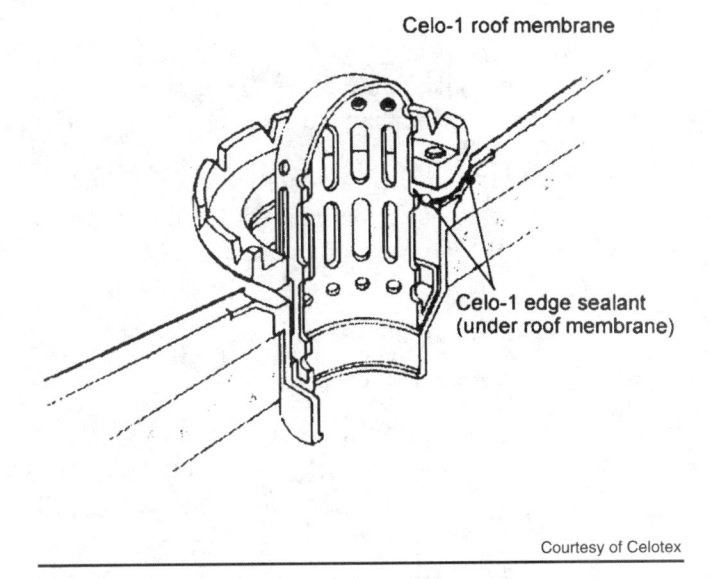

Figure 11-10 Roof drain flashing

Estimating Elastomeric Roofing

Basically, you can use the same rules to estimate elastomeric roofing that you use for BUR systems. But since elastomeric roofs aren't as common as BUR systems, you've got to consider whether or not you'll be able to use left-over membrane for future projects. If not, figure the most cost-efficient roll width to use on the current project.

For example, let's assume that you're going to install 100-foot-long rolls of fully-adhered EDPM elastomeric roofing (using a 3-inch lap) over the roof shown in Chapter 10, Figure 10-28. Using the standard roll widths available, you can calculate the waste.

For 4'4" rolls:

$$\text{No. Rolls} = \frac{50 \text{ feet}}{4.08 \text{ feet}} = 12.25 \text{ rolls}$$

The waste factor is: $\frac{13 \text{ rolls}}{12.25 \text{ rolls}} = 1.06$ or 6% waste

For 5'6" rolls:

$$\text{No. Rolls} = \frac{50 \text{ feet}}{5.25 \text{ feet}} = 9.52 \text{ rolls}$$

The waste factor is 5%

The waste for other roll widths would be:

Roll Width	No. Rolls	Waste
7'	8	8%
10'	6	17%
20'	3	19%

Using 5'6" rolls, you'd have 10 seams to splice, but the high cost of the membrane more than makes up for the added labor.

▼ **Example 11-1:** Figure the cost for installing 5'6" rolls of fully-adhered EPDM elastomeric roofing over the roof shown in Figure 10-28 (in Chapter 10). Use the surface area and flashing quantities from Example 10-3.

The insulation material will cost $11,314.08.

Other materials will cost:

```
EPDM membrane  = 49.3 SQ x $165.55/SQ           = $8,161.62
Lap cement     = 10 seams x 100 LF x $0.33/LF   =   $330.00
Sheet flashing = 483 LF x 1 SF/LF x $4.64 SF    = $2,241.12

Total material cost                               $10,732.74
```

According to the *National Construction Estimator*, an R3 crew consisting of two roofers and one laborer can install roofing materials at the following rate:

EPDM membrane: 1 SQ in 1.05 hours
Lap cementing: 1 LF in 0.00223 hours
Sheet flashing: 1 SF in 0.021 hours

From Example 10-3, the average cost per manhour is $46.76. Then the labor costs would be:

2" insulation (from Example 10-3)   = $1,369.82

EPDM membrane = 49.3 SQ x 1.05 manhours/SQ x $37.88/manhour
               = $1,960.86
Lap cementing = 1000 LF x 0.00223 manhours/LF x $37.88/manhour
               = $84.47
Sheet flashing = 460 LF x 0.021 manhours/LF x $37.88/manhour
               = $365.92

The total labor cost is: $1,369.82 + $1,960.86 + $84.47 + $365.92 = $3,781.07.

The total cost is: $10,732.74 + $3,781.07 = $14,513.81.

In the next chapter, we'll take an in-depth look at other materials important to the roofing contractors — insulation and vapor retarders.

12 Insulation, Vapor Retarders and Waterproofing

▶ As a roofing contractor, you're usually responsible only for installing roofing insulation when it's required. But in this chapter I'm including a general discussion of building insulation and vapor retarders. In view of rising energy costs and the need for universal conservation, I think all technical construction authors should use every chance to stress the importance of a properly insulated and moisture-resistant building.

I'll also discuss waterproofing and dampproofing. While those operations most often apply to slabs, foundations, basements and exterior walls, the materials and techniques are familiar to the roofing contractor. These jobs can increase your business in slack times, or open the door to work in your specialty. Let's begin with insulation.

Insulation Materials

You can buy building insulation as batts or blankets, loose fill, foam and rigid board. Both the batts and rigid insulation may have reflective surfaces. Insulation materials prevent heat loss and resist the transmission of water vapor, dust and sound. The different types of insulation are shown in Figure 12-1.

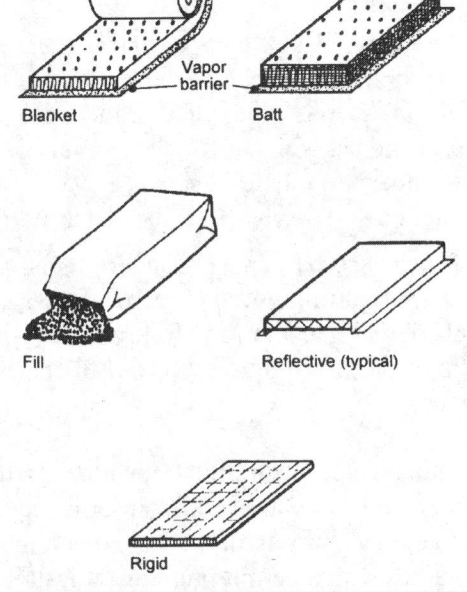

Figure 12-1 Types of insulation

337

Thickness	Kraft paper faced	Foil faced
2	R-7	R-7
3	R-11	R-11
3 5/8	R-13	--
5 1/4	R-19	R-19
6	R-22	R-22

Figure 12-3 R-values for rock wool insulation

A Unfaced or Kraft-faced fiberglass batt insulation	
Insulation thickness (inches)	R-values
3 1/2 or 3 5/8	R-11 and R-13
6 1/4 or 6 1/2	R-19
7 1/2	R-22
9 1/4 or 10 1/2	R-30
12 or 13	R-38
B Mobile home fiberglass batt insulation	
Insulation thickness (inches)	R-values
1 5/8	R-5
2 1/2	R-7
3 1/2 or 4	R-11
4 1/2	R-14
C Foil-faced fiberglass batt insulation	
Insulation thickness (inches)	R-values
3 1/2 or 4	R-11
6 1/4 or 6 1/2	R-19
D High-density fiberglass batt insulation	
Insulation thickness (inches)	R-values
3 1/2	R-15
6	R-21
8	R-30

Figure 12-4 R-values of fiberglass batt insulation

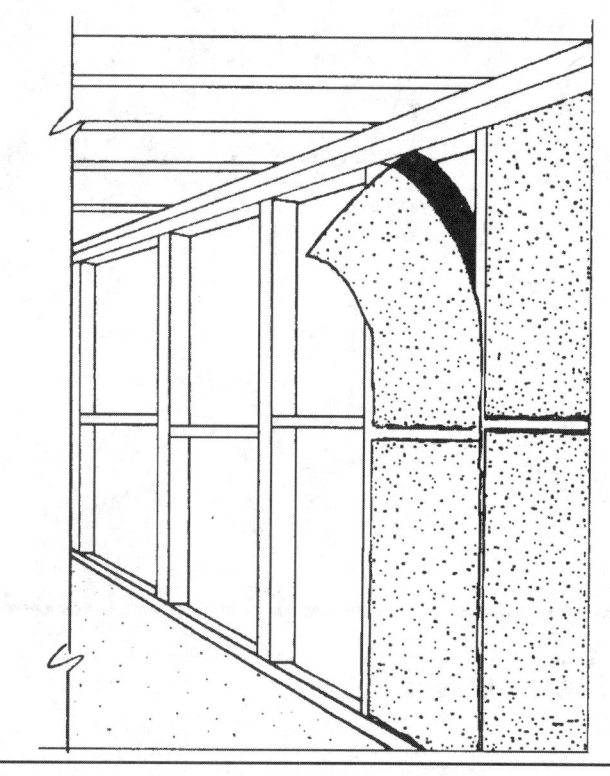

Figure 12-2 Fiberglass batts between studs

Batts and Blankets

Builders install batt and blanket insulation between the structural members of the floors, walls and ceilings in new construction. Figure 12-2 shows batt insulation installed between studs.

The two types of batts or blankets are glass fiber and rock wool. Rock wool (mineral wool) is made from the fibers of various types of volcanic rock. Rock wool is denser and has a higher R-value per unit of thickness than glass fiber. Figure 12-3 shows R-values for rock wool insulation. I'll discuss R-values later in this chapter.

Fiberglass batt insulation products come in many forms, including unfaced, paper-faced, foil-faced and high density insulation. Figure 12-4 shows R-value ratings for the common types of fiberglass batt insulation.

■ **Unfaced Batt Insulation** Use unfaced (rigid-fit) insulation alone where no vapor retarder is required, or install a separate vapor retarder. Unfaced batt insulation comes in widths which permit pressure-fit installation in cavity walls: 15, 15 1/4 and 16 inches for 16-inch stud spacing, or 23, 23¼ and 24 inches for 24-inch stud spacing. It comes 47, 48 and 93 inches long or in rolls approximately 39 feet long.

Unfaced fiberglass batts, 2½ to 2¾ inches thick (R-7 to R-8), 3½ to 4 inches thick (R-11), or 6¼ inches thick (R-19) are often used to reduce sound transmission between adjacent rooms when thermal insulation isn't a factor. These batts are also called *sound attenuation blankets* or *sound control batts*. Blankets come 16 and 24 inches wide, and 96 inches long. Sound batts also come with a Kraft paper vapor-retarder facing.

You can also get batts for use between furring strips over masonry walls as shown in Figure 12-5. They're 15 or 23 inches wide, and ¾ to 1⅛ inches thick with an R-value of 3.4. You can install sill sealers between a foundation wall and sill plate in pier-and-beam construction. The standard size is 1 inch thick, 3½ or 6 inches wide, in 100 foot rolls.

Another unfaced variety of fiberglass insulation is available for mobile homes.

■ Paper- or Foil-faced Batt Insulation Kraft-faced batt insulation has the paper facing coated with asphalt to provide a vapor retarder. The facing paper has edge flanges to staple the batts to the studs. These batts come in widths of 15, 15¼ and 16 inches, or 23 and 24 inches. Lengths are 47, 48 and 93 inches, or rolls about 39 or 70 feet long.

Special Kraft-faced fiberglass batts are available for installation over suspended acoustical ceilings. Standard thickness for these batts is 3½ (R-11) or 6¼ (R-16) inches. They come 24 inches wide and 48 inches long.

Foil-faced fiberglass batt insulation has an aluminum foil facing that provides a vapor retarder and a reflective face to inhibit infrared heat passing across an air space. The aluminum facing has edge flanges for stapling to studs. Foil-faced batts come 15 and 23 inches wide, in rolls about 39 or 70 feet long. Flame-resistant reinforced foil-faced fiberglass batt insulation is also available. It comes in 4-foot lengths.

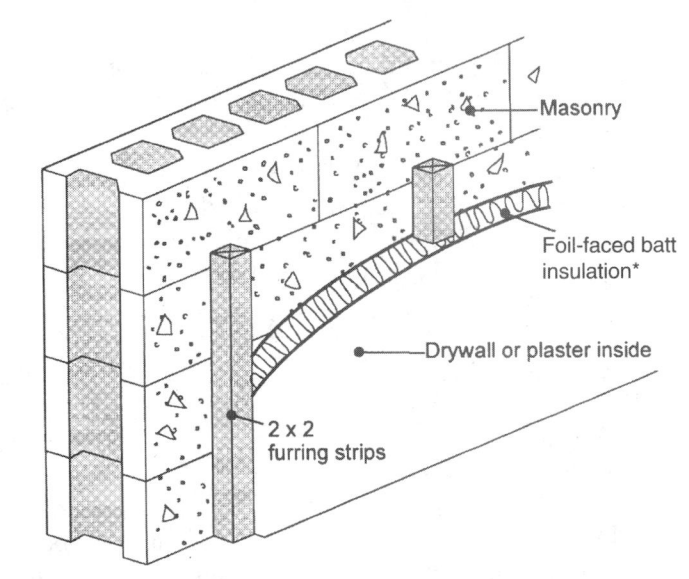

* **Note:** Use only insulating products whose thermal efficiency is not affected by moisture. Check with your local building official or state energy code for insulation requirements if a basement is to be used as a habitable living space.

Figure 12-5 Masonry wall insulation

	R-value per inch	
Material type	**Poured**	**Blown**
Glass fiber	2.2	2.2
Rock wool	2.2 - 2.8	2.8
Cellulose fiber	3.7	3.7
Vermiculite	3.0	--

Figure 12-6 R-values of loose fill insulation

Loose Fill Insulation

Loose fill insulation is made of glass fibers, rock wool, cellulose fiber, vermiculite or perlite. The R-values per inch thickness of loose fill insulation are shown in Figure 12-6.

Roofing Construction & Estimating

■ **Insulating Frame Walls** You can either pour or blow loose fill insulation. It's usually poured into accessible horizontal spaces in floors and ceilings as shown in Figure 12-7. All the insulating materials I've mentioned, except for vermiculite, can be blown. Blow insulation through holes into inaccessible spaces such as between studs in finished walls.

Figure 12-8 shows material and manhours required for poured insulation. You'll find the material and installation requirements to achieve required R-values with blown fiberglass or rock wool insulation in an attic in Figure 12-9.

■ **Insulating Masonry Walls** You can get concrete block with polystyrene core inserts, or sometimes polystyrene scrap is poured into cores. But inserts don't completely fill a core, and polystyrene beads don't completely fill the wall because of electrically-repelling static within the polystyrene material.

Water-resistant inorganic vermiculite or silicone-treated perlite insulation is a better choice to insulate masonry walls. You can pour it into the cores of concrete block or into brick and block wall cavities as shown in Figure 12-10. Vermiculite pours freely into wall heights up to 20 feet without the need for tamping or rodding and subsequent settlement is less than 1 percent. Another advantage of vermiculite is that it won't burn. The coverage capacity of vermiculite masonry insulation is shown in Figure 12-11.

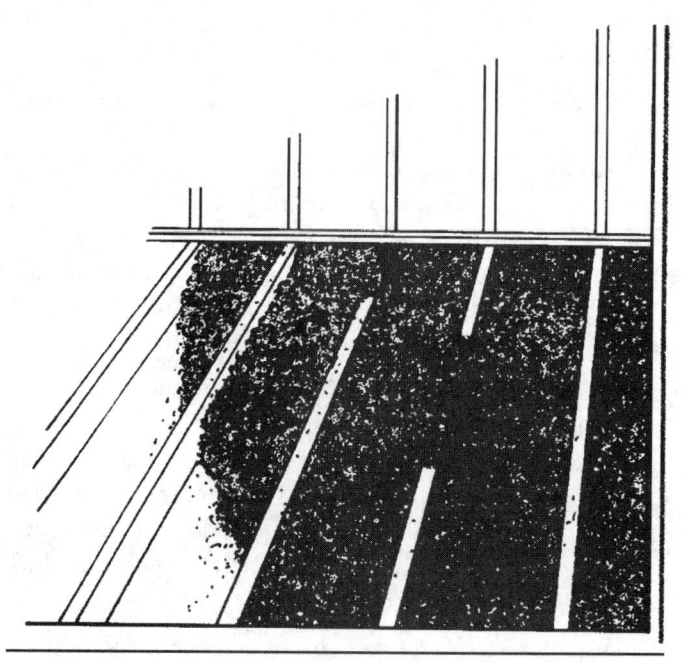

Figure 12-7 Loose fill insulation

Fill thickness (inches)	Number of SF covered by CF @density rating					Manhours per 100 SF of ceiling
	6 lb	7 lb	8 lb	9 lb	10 lb	
1	21.1	18.0	15.9	14.1	13.0	0.6
2	10.6	9.1	8.0	7.1	6.4	0.6
3	7.1	6.1	5.3	4.7	4.2	0.7
4	5.3	4.6	4.0	3.5	3.2	0.8
5	4.2	3.6	3.2	2.8	2.6	0.9
6	3.6	3.0	2.7	2.4	2.2	1.0
7	3.1	2.6	2.3	2.0	1.9	1.1
8	2.6	2.3	2.0	1.8	1.6	1.2

Figure 12-8 Labor and materials for poured ceiling insulation

R-value	Bags per 1000 SF of net area	Maximum SF per bag	Minimum wt (pounds pe SF)	Minimum thickness (inches)
Fiberglass insulation				
R-49	53	19	1.30	17
R-44	45	22	1.15	15
R-38	40	25	1.00	13
R-30	30	33	0.75	10½
R-22	22	45	0.55	7½
R-19	20	50	0.50	6½
R-13	14	73	0.34	4½
R-11	11	90	0.28	4
Rock wool insulation				
R-38	40	25	1.000	17½
R-30	30	33	0.758	13¾
R-22	22	45	0.556	10
R-19	20	51	0.490	8¼
R-13	13	75	0.333	6
R-11	11	90	0.278	5

The above thermal performances are achieved at weights and coverages specified when insulation is installed with pneumatic equipment with a horizontal open blow.

Figure 12-9 R-values and material requirements for blown insulation

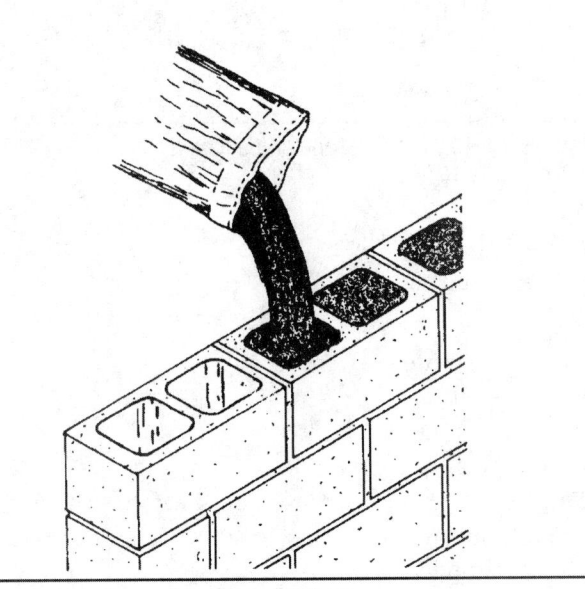

Figure 12-10 Poured cavity insulation

Wall type	Bags per 1000 SF of wall area
8" block	69
12" block	125
1" cavity	21
2" cavity	42
2½" cavity	50
4" cavity	95

Figure 12-11 Coverage capacity of vermiculite insulation (number of 4 CF bags required)

Sprayed Foam Insulation in Walls

Foam insulation is practical for insulating existing masonry cavity walls, curtain walls and stud walls. The best way to put foam insulation into existing walls is to blow it with compressed air into minimum 1-inch-diameter holes drilled between the studs.

You can also use foam insulation in pipe chases to eliminate condensation problems and to reduce noise generated by rushing water. Spray it in the cores of cement block during new construction. You can install foam insulation over a vapor retarder in ceilings, but don't let it cover a recessed light or electrical box. Excessive heat buildup could cause a fire.

Most foam insulation contains urea formaldehyde, although you can get some products without it. You can get a cementitious foam-in-place insulation that's fire-resistant and doesn't give off hazardous gases when burned.

Insulation type	R-value per inch of thickness
Urethane	7.14
Styrofoam	5.41
Glass fiber	4.30
Polystyrene	4.17
Foam glass	2.86
Fiberboard	2.77
Expanded perlite	2.56

Figure 12-12 R-values of various types of rigid insulation

Rigid Insulation

The most common types of rigid board insulation include mineral fiber, glass fiber, polystyrene, urethane, foamed glass, wood fiber boards, compressed particleboard, cork board and Styrofoam. The R-values per inch thickness of various rigid insulation materials are shown in Figure 12-12.

Fiberboard insulation comes in thicknesses through $2^{1}/_{4}$ inches, but the most common sizes are $^{1}/_{2}$ or 1 inch thick. You can get greater thicknesses by building up two or more layers of insulation board. When more than a $1^{1}/_{2}$-inch-thick layer of insulation is required, rather than installing one extra-thick piece, install multiple layers that total the required thickness. Rigid insulation boards made from organic fiber are often used for outside wall sheathing. Some boards come with aluminum foil faces for vapor protection. Standard sizes are 4' x 8' with thicknesses ranging from $^{3}/_{8}$ through $1^{7}/_{8}$ inches.

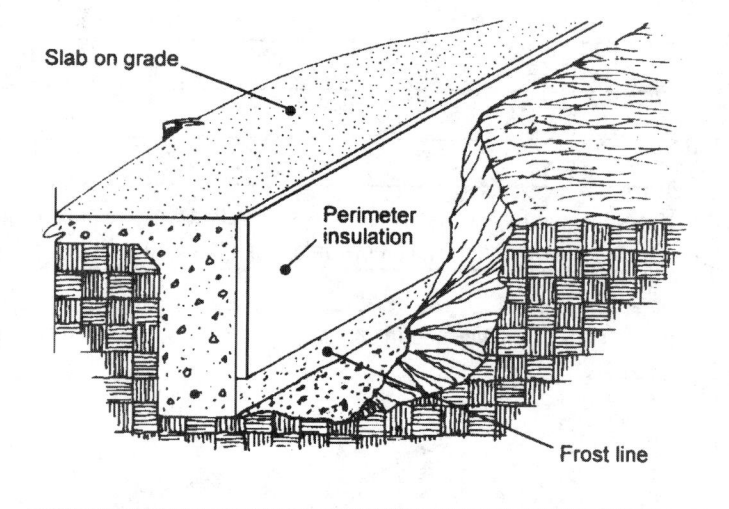

Figure 12-13 Rigid perimeter insulation

Use rigid insulation made of foil-faced polystyrene, polyisocyanurate or urethane for perimeter insulation, as in Figure 12-13. Don't install unprotected polyisocyanurate or urethane below grade, as it will degrade. Asphalt-impregnated cork boards are also popular. Be sure to seal all joints and exposed batts with aluminum tape, and be careful not to tear the foil when backfilling. Apply perimeter insulation with spot-applied adhesives. Rigid perimeter insulation boards are normally 1 to 2 feet wide and 1 to 4 inches thick.

Insulation, Vapor Retarders and Waterproofing

Rigid Styrofoam boards come 16, 24 and 48 inches wide, in lengths of 48 and 96 inches. They're $3/8$ through 4 inches thick. You can also use rigid Styrofoam to insulate slabs, foundation walls, cavity walls and as an exterior sheathing component. Rigid insulation under a slab is installed directly on top of the gravel, with a 6-10 mil poly vapor retarder placed on top, and then the poured concrete, as shown in Figure 12-14. You might think the insulation would crush under the weight of the concrete, resulting in the slab cracking, but since the weight of the slab is distributed over such a large area, this isn't likely. ASTM standards require insulation under a slab be a minimum of 10 PSI, though I recommend you use 15 PSI.

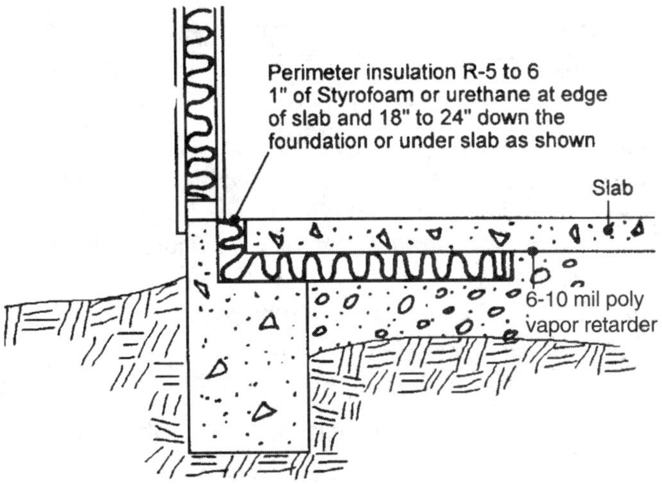

■ **Tapered Rigid Insulation** Tapered rigid insulation usually comes in 2-foot widths with a taper across the 2-foot dimension. The minimum taper required is $1/8$ inch per foot. I recommend a $1/4$-inch taper to provide good drainage. Standard manufactured tapers are listed in Figure 12-15. Again, to build up the insulation to the required thickness, install insulation boards in 1-inch-thick increments beneath the tapered insulation boards.

Figure 12-14 Perimeter insulation under slabs

Incline (inches per foot)	Thickness (high end)	Thickness (low end)
1/8	3/4"	1/2"
	1"	3/4"
	1 1/4"	1"
	1 1/2"	1 1/4"
1/4	1	1/2"
	1 1/2"	1"

Figure 12-15 Tapered insulation thicknesses

Reflective Insulation

Batt and rigid board insulation are available with reflective foil laminated to one or both sides. The reflective material increases the R-value of the insulation two ways. First, it reduces radiant heat loss from inside the building during winter. Second, it reduces radiant heat flow into the building during summer. Reflective faces also serve as a vapor retarder and increase fire resistance.

You can get a reflective insulation with a white-tinted or embossed foil face on one side and a reflective foil face on the other. It's installed in buildings such as warehouses where one face of the insulation is left exposed within the building. A heavy-duty foil-faced insulation is also available for exposed insulation that may be subject to abrasion or impact abuse.

Sprayed-on Roof Insulation

The most widely used sprayed-in-place insulation includes plastic foams such as expanded urethane or polystyrene, isocyanurate foam or fibered masses such as gypsum or perlite. Sprayed-on foams often require a primer coat appropriate for the substrate. Be sure the deck is clean and free of aggregate, sand, curing compounds, oil, grease or any other contaminant that might keep the foam from adhering. Don't spray foam during rain or fog, or in winds of more than 15 miles per hour. Be sure the deck is 50° F or warmer, and follow the manufacturer's instructions.

Sprayed-on plastic foams have several advantages. They're lightweight (less than 1 pound per square foot) and will conform to any roof surface contour. You can spray polyurethane foam into surface irregularities and in varying thicknesses to provide pitch to drain. The insulation provides a monolithic, self-flashing system. There are no joints to seal, and it's easy to patch broken or damaged areas. Sprayed insulation helps keep the deck from changing temperature, so there's less movement, and resists severe wind uplift. On vertical surfaces, such as chimneys, spray to a height of at least 6 inches.

One significant disadvantage of sprayed-on plastic foam is that it burns and gives off dense, black smoke. Other drawbacks are that cracks often occur if the material shrinks, and urethane turns brown after sun exposure, so you have to cover it with a liquid roof covering.

Polystyrene insulation is incompatible with asphalt and pitch. Use a minimum 6-mil-thick polyethylene separation sheet between the insulation and a BUR system.

Lightweight Concrete Fill Insulation

The most widely used poured-in-place insulation is lightweight concrete fill composed of lightweight aggregates and Portland cement. The lightweight aggregate used is either perlite (volcanic glass) or vermiculite (an expanded micaceous silicate mineral). The more aggregate, the greater the insulating value of the fill. But too high a ratio of aggregate to cement will weaken the fill to a point where it isn't considered nailable. Other types of cementitious insulation include cellular Portland cement, poured or sprayed-on gypsum concrete, or gypsum concrete mixed with a wood fiber binder or glass bead binder.

One advantage of lightweight concrete is that it retains heat, and it stores and releases heat slowly.

If a lightweight concrete deck is to receive a built-up roof, mix the cement-to-aggregate ratio 1:4 or 1:6 with a minimum dry density of 22 pounds per cubic foot. The deck must accept and retain nails.

Reducing Heat Loss

A properly insulated and weatherproofed building reduces energy costs and keeps the occupants more comfortable. Insulation also permits the use of smaller heating and cooling units and smaller ductwork. And since those heating and cooling units won't be used as much, they'll last longer. Figure 12-16 shows where to install insulation throughout a house for efficient energy conservation. Required insulation R-values differ not only by the location of the insulation in the building, but on the local climate. The more extreme the local temperatures, the higher

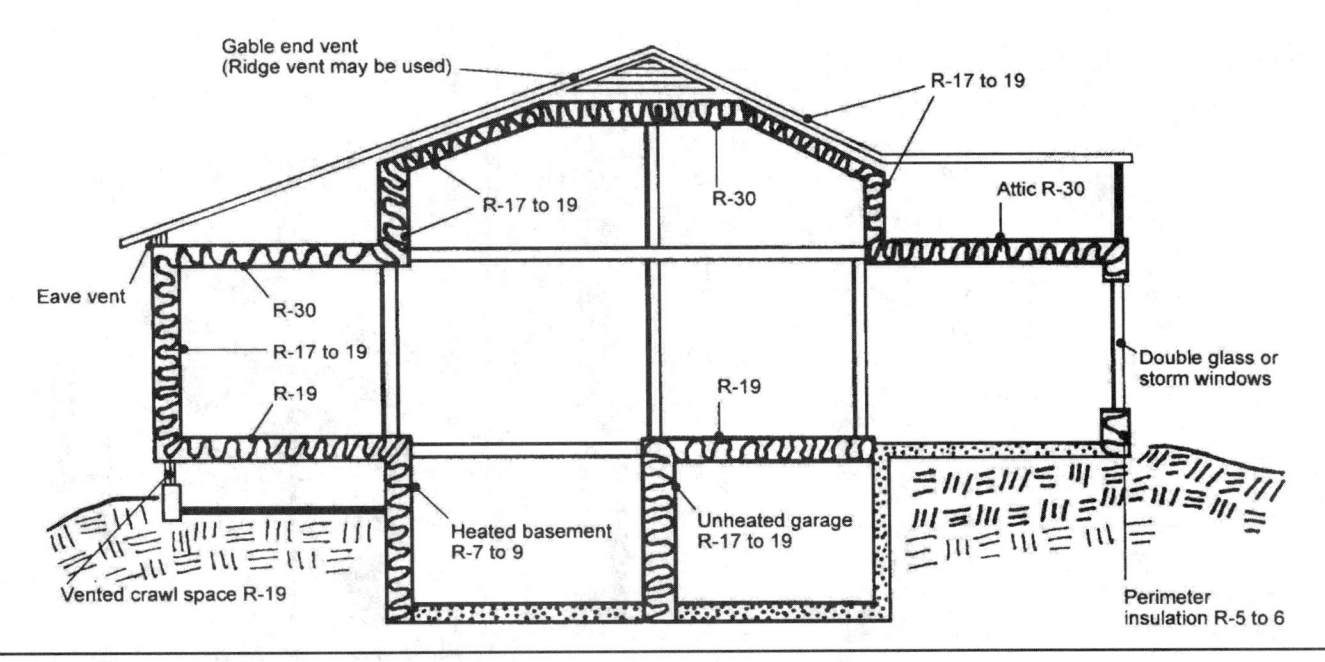

Figure 12-16 Recommended insulation locations

the required R-values. Figure 12-17 is a map showing the various insulation requirements for the continental U.S. But due to both climate change and increasing demands for energy efficiency, the requirements can change; in all cases, upward. Be sure you have, and follow, up-to-date insulation requirements for the area where you're working.

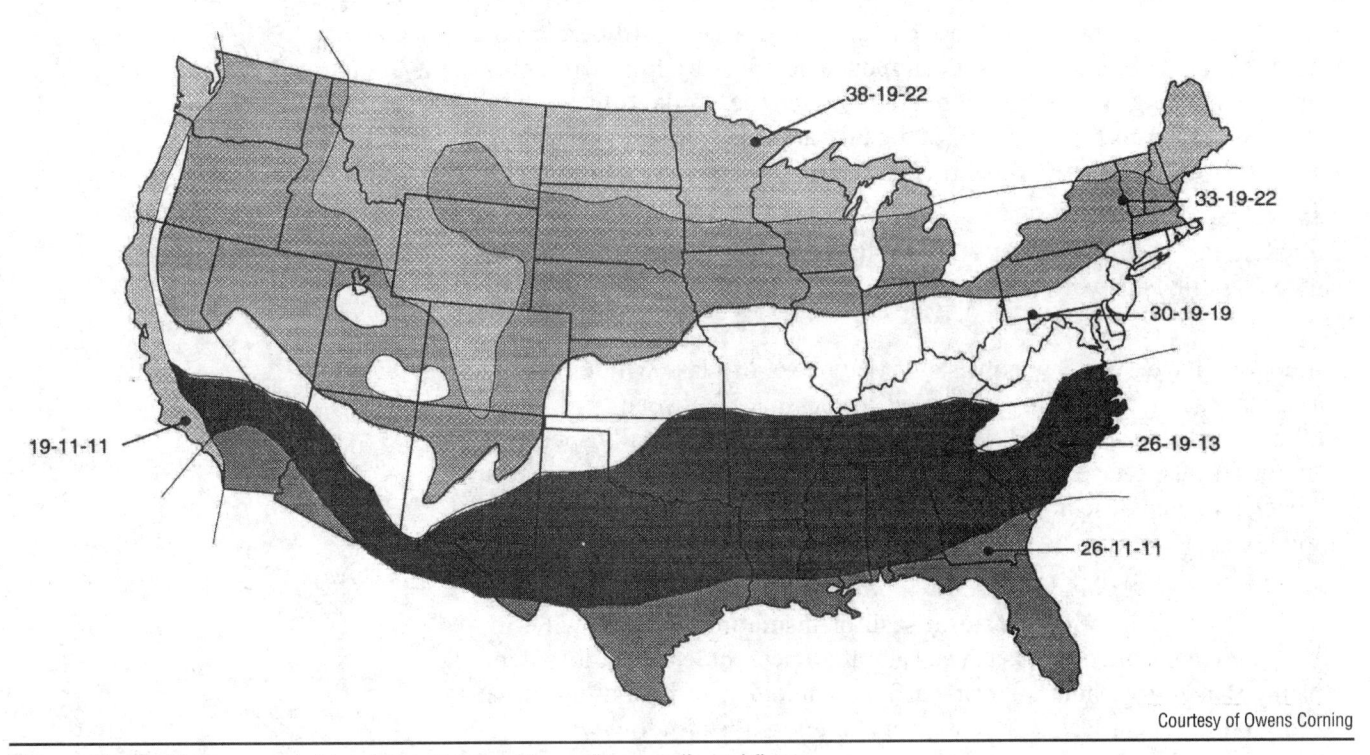

Courtesy of Owens Corning

Figure 12-17 Recommended insulation R-values for ceilings, walls and floors

345

Floors

At slabs on grade, install 1-inch-thick rigid Styrofoam or urethane perimeter insulation (R-5 or R-6) at the edge of the slab and 18 to 24 inches down the inside of the foundation wall, or under the slab as in Figure 12-14.

Floors above heated rooms or heated basements require no insulation, but insulate floors over unheated basements. The R-value of the insulation depends on the climate. Refer to Figure 12-17.

Walls

Outside walls (excluding windows and doors) account for about 13 to 14 percent of the total heat loss in a house. A brick veneer wall with R-11 insulation installed between studs has 69 percent less heat loss than the same type of wall without insulation. You can reduce the heat loss through a masonry wall by as much as 83 percent by installing moisture-resistant insulation (R-7 to R-9) between 2 x 2 furring strips along the inside face of the wall. (Look again at Figure 12-5.)

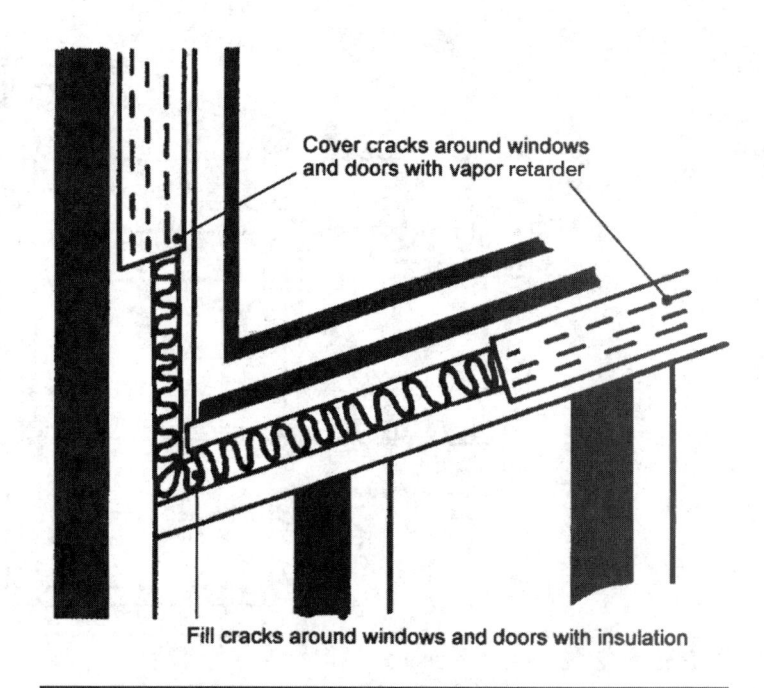

Figure 12-18 Insulating around doors and windows

Windows

Windows and sliding glass doors can account for about 40 percent of the total heat loss in a house. You can reduce heat loss by filling all voids around the opening with insulation. Also, cover the voids with a vapor retarder as shown in Figure 12-18 and caulk all cracks around windows. Later in this chapter there's a chart (Figure 12-23) that shows how much caulk you'll need.

Most homes built in the last 20 years have double-glazed windows; in many jurisdictions they're required on new construction for energy compliance.

A double-glazed window consists of two panes of glass with either air or an inert gas, generally argon, sealed in the space between. The glass and the gas act as insulators. The exterior pane of glass is usually specially coated on its inside face with a reflective, or low-E coating, which reflects heat back into the house in cold months, as well as reducing some heat gain during hot weather.

Where installing double-glazed windows isn't an option, you can reduce heat loss by installing a storm sash or insulating glass. A window with a single pane and storm sash has about 50 percent less heat loss than the same window without a storm sash. A window with insulating glass has about 46 percent less heat loss than a single-pane window without a storm sash.

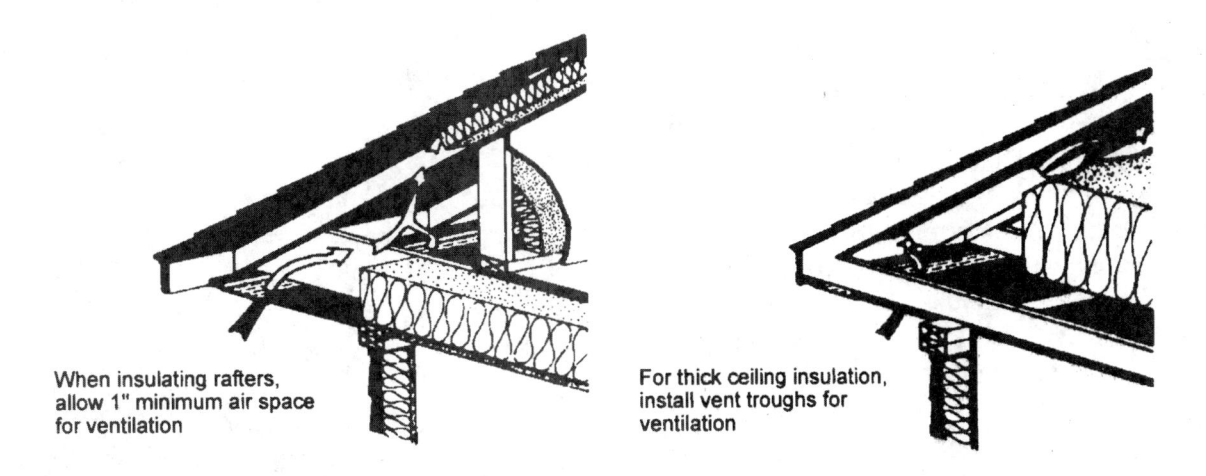

Figure 12-19 Provide for ventilation at eaves

Doors

Outside doors (excluding sliding glass doors) account for about 2 to 3 percent of the total heat loss in a house. As with windows, you can reduce heat loss by adding insulating and vapor retarders, and by caulking.

Also consider installing storm doors or insulated doors. A 1³/₄-inch-thick solid core door and storm door have about 35 percent less heat loss than the same solid core door with no storm door. A urethane-insulated steel door (R-13.8) with no storm door has about 85 percent less heat loss than a 1³/₄-inch-thick solid core door with no storm door. The same urethane-insulated steel door with no storm door has about 76 percent less heat loss than a 1³/₄-inch-thick solid core door with a storm door.

Ceilings

You can insulate a ceiling with batts, blankets or loose fill, or with a combination of these materials. The required R-value of the insulation depends on the local climate. For the sake of comparison, 8-inch-thick insulation (R-25.3) has about 21 percent less heat loss than 6-inch-thick insulation (R-19); 10-inch insulation (R-31.7) has about 35 percent less heat loss than 6-inch insulation; 12-inch insulation (R-38) has about 46 percent less heat loss than 6-inch insulation.

When you install ceiling insulation, don't put insulation over the soffit, especially where soffit vents have been installed. Insulation there can lead to the formation of ice dams in cold climates (see Figure 3-38 back in Chapter 3). Also, if you install batt insulation between rafters, allow at least 1 inch of air space between the insulation and roof deck to provide adequate attic ventilation. That's shown in Figure 12-19.

Roof Insulation

Roof insulation reduces heat loss through the roof in cold weather and heat penetration when it's hot. Roof insulation also prevents condensation from forming on the inside surface of the roof deck. Roof insulation provides a good substrate for roofing felts and helps isolate the roof membrane from stresses caused by deck or structural movement. Insulation reduces deck expansion and contraction. It also keeps heat from passing through built-up roof membranes so the felts don't dry out so fast.

The biggest disadvantage to installing roof insulation beneath a built-up roof is that it causes the BUR system to age faster than it would if installed directly over the deck. That's because, during the heat of the day, the base ply becomes extremely hot since the insulation doesn't let the heat pass through to the inside of the building. At night, the base ply cools quickly, again because the insulation prevents heat from passing through to the BUR system. That extreme temperature range increases expansion and contraction of the system, making it age faster.

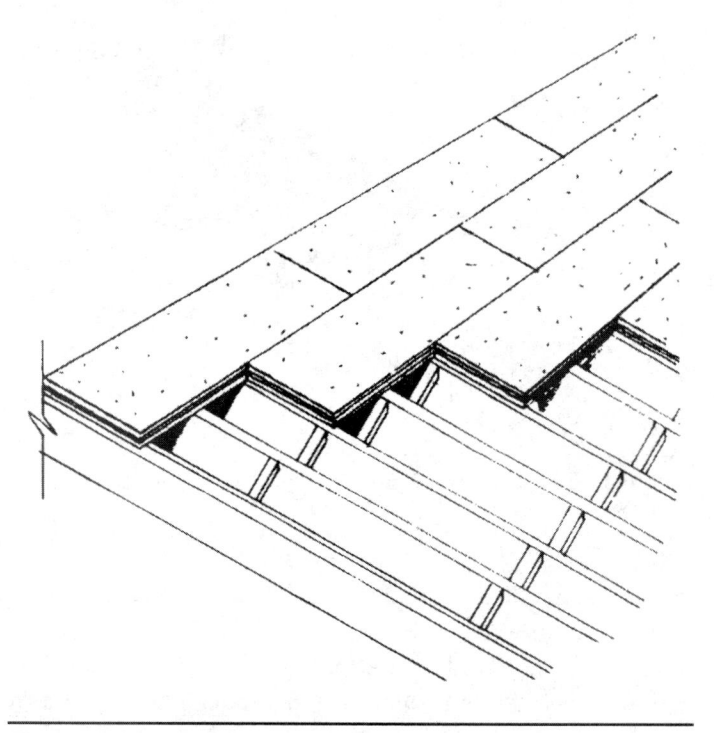

Figure 12-20 Application of rigid insulation

Most roof insulation is rigid or sprayed-on insulation. Common roof insulation products include polyurethane, glass fiber, expanded polystyrene bead board, foam glass, fiberboard and expanded perlite. Of all the insulation products available, polyurethane is the most efficient. But ultraviolet rays rapidly degrade exposed polyurethane, so it's necessary to cover it within 48 hours after application.

■ **Installing Rigid Roof Insulation** Insulation boards are either applied directly to the deck or over a vapor retarder sheet, depending on the amount of humidity expected inside the building. Insulation boards must be strong enough to withstand foot traffic, snow loads and hail impact. The boards should be attached to the deck well enough to resist wind uplift.

Some manufacturers recommend that you install insulation with the long dimension of the board parallel with the long roof dimension and stagger the end joints as in Figure 12-20. Others recommend installing the long dimension of the insulation board perpendicular to the direction of the membrane.

On metal decks, run the long dimension of the board parallel to the ribs and rest the edges firmly on top of a flute. If the insulation needs to be more than $1\frac{1}{2}$ inches thick, install multiple layers. When two or more layers of insulation are used, install the upper layer with end and side joints staggered at least 6 inches from the bottom layer to prevent bitumen from dripping into the building.

Insulation, Vapor Retarders and Waterproofing

To allow for expansion, install all insulation board with a 1-inch gap between the board and any vertical surface such as a wall. Fill all gaps wider than $1/8$ inch between insulation boards with roofing cement and strike the joints smooth. Install 4' x 8' insulation boards wherever possible to reduce the number of joints.

You'll usually anchor insulation board with a solid mopping of hot bitumen. As a rule, don't install rigid insulation with cold-applied adhesives. Note that some materials such as Styrofoam can't withstand hot bitumen so you have to use cold-applied adhesives or mechanical fasteners. Urethane boards can endure hot asphalt (425° F) only very briefly. Use mechanical fasteners over metal decks.

According to some membrane manufacturers, polystyrene insulation will damage bitumen roofing membranes. But you can install fiberboard, perlite board or glass fiber roof insulation with mechanical fasteners over the insulation followed by hot moppings of bitumen and roof membranes. Install fiberboard with mechanical fasteners over Styrofoam, urethane or isocyanurate insulation board followed by hot moppings of bitumen and roof membranes. Some decks aren't suitable for hot bitumen because the bitumen will leak through joints. Then you'll have to apply rosin paper to the deck before you mop to keep bitumen from leaking through.

Don't use foam, sprayed-on insulation or polystyrene boards under fiberglass roofing membranes because they react with asphalt. Use fiberglass, perlite board, wood fiber board or cellular glass instead.

When you use mechanical fasteners to install insulation over a nailable deck, use at least five fasteners per 2' x 4' panel, six fasteners per 3' x 4' panel and 14 fasteners per 4' x 8' panel. Place one in each corner and a random pattern in the field. Use fasteners long enough to penetrate through all the layers of insulation and $3/4$ inch into the deck, but not so long that they poke through the underside of the deck. Drive all nails through tin caps.

When you install insulation over a non-nailable deck, embed the insulation into a solid mopping of hot steep asphalt applied at the rate of 23 pounds per square. Before you mop the asphalt, prime the deck at the rate of one gallon per square.

To prevent felts from slipping on decks steeper than 1 in 12 (asphalt bitumen BUR), or $1/2$ in 12 (coal-tar pitch BUR), nail the rigid board insulation to the deck (if nailable) or install the insulation between treated wood nailing strips (if non-nailable). Refer back to Figure 10-3 in Chapter 10. Over wood decks, install the insulation with large-headed nails long enough to penetrate $1/2$ inch into the deck. Use serrated nails on plywood decks.

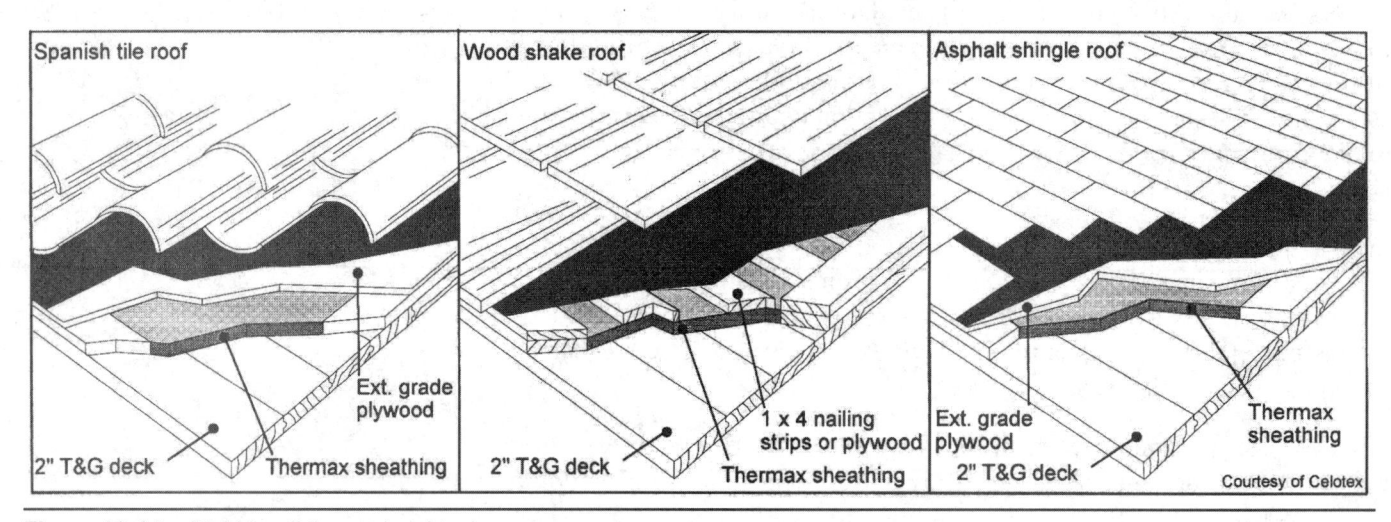

Figure 12-21 Rigid insulation under shingles

You can install rigid insulation under all kinds of shingle roofs, but it's most common over vaulted or exposed-frame ceilings where you can't install insulation in the attic. You must install the insulation over a solid deck. Then install nailing strips or a solid plywood deck before you install the shingles. That's shown in Figure 12-21.

Insulation Values

The *insulation value* of a material describes how well the material conducts or resists heat transfer. Insulation value is expressed in British thermal units (Btu). A Btu is the energy required to raise the temperature of 1 pound of water 1 degree Fahrenheit.

Thermal conductivity (k) measures a material's ability to conduct heat and is the amount of heat (in Btus per hour) transmitted per square foot through one inch of material, per degree difference in temperature (F) on opposite sides of the material. The lower the value of k, the less the heat transmission and therefore the better the insulation.

Most insulation products are rated and stamped with an R-value. The R-value is the total resistance of a material to conduct heat. Provided that you have a homogeneous material, that is, it's the same material all the way through, the total thermal resistance can be defined as:

$$R = \frac{1}{k} \times \text{Material Thickness (inches)}$$

Equation 12-1

A higher value for R means less heat transfer, or better insulation. You can't necessarily relate the thickness of a material to its insulating quality. For example, the R-value of an 18-inch-thick concrete wall is almost identical to that of 1 inch of rock wool insulation.

The coefficient of transmission (U) of a material is defined as:

$$U = \frac{1}{R}$$

Equation 12-2

The lower the U-value, the better the insulation. Both the thermal conductivity (k) and the coefficient of transmission (U) of a material indicate the material's ability to conduct heat. The difference between them is that thermal conductivity is a measure of the heat flow per inch of material thickness, whereas the coefficient of transmission is a measure of heat flow through the entire thickness of the material.

▼ **Example 12-1:** The thermal conductivity (k) of common brick is 5 BTU/hour/SF/inch/° F. Find the total resistance (R) and coefficient of transmission (U) if the brick is 4 inches thick.

From Equation 12-1, the total resistance is:

$$R = \frac{1}{5} \times 4$$
$$= 0.80$$

From Equation 12-2, the coefficient of transmission is:

$$U = \frac{1}{0.80}$$
$$= 1.25$$

▼ **Example 12-2:** Find the coefficient of transmission (U) for a material whose total resistance (R) is 30 Btu/hour/SF/inch/° F.

From Equation 12-2, the coefficient of transmission is:

$$U = \frac{1}{30}$$
$$= 0.033$$

The amount of insulation you need depends on the local climate. Use the map back in Figure 12-17 as a guide to a locality's temperature zone and the R-values for ceiling, wall and floor insulation of buildings within that zone. Check with the local building department in case requirements have changed.

When two or more insulating materials are combined, the total R-value is the sum of the R-value of each individual material. For example, combining a foam board having an insulating value of R-12 and a 2-inch-thick batt with an insulating value of R-6 yields a total insulating value of R-18.

Further, if you install a double layer of the R-12 foam board, you'd add the two R-values of the board together to get an insulating value of R-24. Add to that the two-inch-thick R-6 batt and you have a total R-value of 30. (This isn't true for k-values, which aren't cumulative.)

Total R-value = $R_1 + R_2 + R_3$

| Equation 12-3 |

▼ **Example 12-3**: If a masonry wall having an R-value 2.0 is insulated with a layer of fiberglass insulation with an R-value of 13.0, and finished with drywall having an R-value of 0.5, what is the R-value of the wall?

From Equation 12-3, the R-value of the wall is:

R = 2.0 + 13.0 + 0.5

= 15.5

Vapor Retarders

The first and most important reason to install a vapor retarder is to prevent warm humid air from contacting a cold surface, which results in condensation. The second reason is to reduce the flow of air between the inside of a building and the outdoors.

Use a vapor retarder in locations where the average January temperature is below 40°F. The main disadvantage of vapor retarders is that they can trap moist air within the building. To avoid that, install or recommend ceiling fans, kitchen and bath exhaust fans, and dryer vents to provide adequate air circulation and ventilation. Excess moisture within a building causes such things as lingering odors, dampness, mold, mildew, finish discoloration, peeling paint and decay of framing lumber.

Vapor retarders come in a variety of materials, including asphalt-coated felts, asphalt-laminated papers, aluminum foil and plastic films. Polyethylene film is the most commonly used vapor retarder, although it's hard to work with and tears easily. Polyethylene film comes in 100-foot rolls and in widths of 2, 3, 4, 6, 8, 10, 12, 14, 16 and 20 feet. Film used for vapor retarders is 0.004 or 0.006 inch thick. Tape for sealing joints in the film comes 2 inches wide, 100 feet long and 0.004 inch thick.

There's little room for error when you install vapor retarders. A hole in a vapor retarder is an escape route for moisture, which makes the vapor retarder virtually worthless. Putting a vapor retarder at the wrong location is even worse. Always install a vapor retarder in an exterior wall to the inside (warm side) of the wall to prevent moisture from getting into the wall from inside the building. Apply a porous material which breathes, such as asphalt-saturated felt, to the outside (cold side) of the wall so that moisture can escape. Install roofing felt to the outside that's at least five times more permeable than the vapor retarder. This configuration prevents moist air from condensing and becoming trapped within the wall. I also recommend that 1:5 ratio for sheathing materials used as the outer "skin" of a house.

Never install a vapor retarder on both sides of an exterior wall. No vapor retarder is required within interior partitions.

Measuring Vapor Retarder Effectiveness

The ability of a vapor retarder to resist vapor penetration is measured in *perms* (permeance). One perm equals the transmission of 1 grain (approximately 0.07 grams) of water vapor per square foot per hour per each inch of mercury vapor pressure difference between opposite sides of the vapor retarder. The lower the perm rating, the less the vapor transmission and the better the vapor retarder. To be an effective vapor retarder, the material should have a perm rating below 0.5 perms. Vapor retarders come with a rating of 0.1 perm or less. Polyethylene film has the lowest perm ratings, ranging from 0.02 to 0.08.

Vapor retarders aren't 100 percent efficient in stopping the transmission of water vapor; they only reduce the rate of transmission. This is why they're now called vapor *retarders*, not vapor *barriers*. Since it's impossible to totally prevent the passage of vapor, some moisture will finally penetrate an exterior wall. It's important that this moisture have an escape route. That's why the outer surface (cold side) of an exterior wall must be allowed to breathe.

Vapor Retarders and BUR

Install vapor retarders (only on insulated roofs) to prevent condensation from forming within a BUR system. Condensation causes insulation to swell and then shrink as it dries. Most types of cellular insulation are quickly destroyed by moisture. If this type of insulation gets wet, the roof system is soon completely undermined. That causes the roofing felts to split and then a leaky roof.

Install vapor retarders only on roofs with insulation sandwiched between the deck and the BUR membranes. Install vapor retarders between the deck and the insulation.

Always put vapor retarders over the deck of an insulated roof for any building with a high interior humidity and little or no ventilation. Without a vapor retarder, the insulation will shift the dew point from under the roofing system to within the roofing system. (The dew point is the temperature at which water vapor will condense.) This is one disadvantage of roof insulation. If insulation absorbs any moisture, the vapor retarder and BUR membranes prevent the vapor from escaping. In order to lessen this effect, install insulation board with beveled edges to provide a horizontal passage for trapped water vapor. Also leave a $1/4$-inch gap between insulation boards. Moisture in the insulation is less of a problem when a vented base sheet is installed.

Weatherproofing Existing Homes

Many older homes have had insulation blown into exterior wall cavities, but no vapor retarders were installed. We used to think that this blown-in insulation allowed condensation within the walls, creating the potential for rot and decay. But studies have shown that this condensation isn't actually a problem, because it tends to correct itself. When insulation gets wet, it loses some effectiveness and increases heat loss through the wall. This increases the temperature within the wall, resulting in reduced condensation and increased evaporation.

Because it's not practical to install standard vapor retarder materials in an existing wall, you can use vapor-retarder paint over interior drywall to help block the transmission of interior moisture into the wall.

In homes framed on a pier-and-beam foundation, you can install insulation and vapor retarders in crawl spaces and under the floors. Place puncture-resistant (6-mil) polyethylene sheets over the soil. Hold the sheets in place with bricks and cover the sheets with sand for protection in traffic areas. This reduces the amount of under-floor ventilation required. Figure 12-22 shows how to install under-floor vapor retarders.

When a vapor retarder is installed under the floor in cold climates, install it on the warm side of the floor above the insulation. In climates that require air conditioning in summer, install the vapor retarder beneath the insulation.

I also recommend a vapor retarder over poorly-vented ceilings in homes with flat roofs or cathedral ceilings.

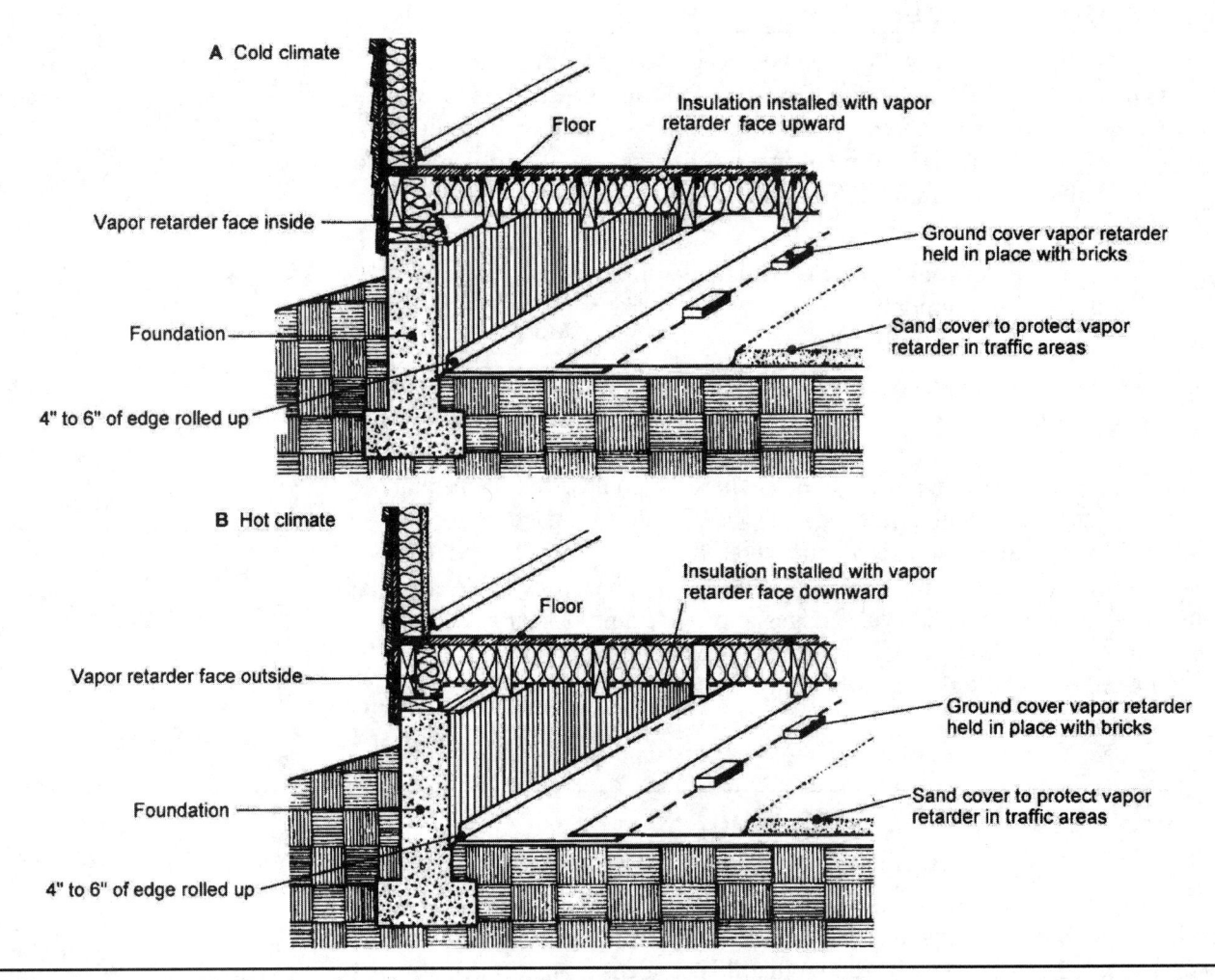

Figure 12-22 Vapor retarder locations for crawl spaces and floors

Caulking and Sealants

In addition to insulation and vapor retarders, you can make any home more energy-efficient by keeping air leakage to a minimum. That's where caulking comes in.

The main difference between a caulk and a sealant is the grade of material. A sealant is a lower grade and I recommend it only for interior work. Most sealants are oil-based and will bleed through latex paint.

There's a wide variety of caulking and sealing compounds on the market, including butyl, latex, polysulfide, polyurethane, solvent acrylic, vinyl acrylic, epoxy and silicone. We'll look at them one at a time.

Butyl caulk and sealant doesn't have good expansion-contraction qualities (10 to 12 percent) and eventually becomes brittle. It does have good resistance to ultraviolet light and weathering. Manufacturers claim is has a life expectancy of 10 to 20 years, depending on quality. It comes in a wide variety of colors and no priming is required.

Acrylic latex caulk is a water-based acrylic compound. Latex caulk shows good resistance to ultraviolet light and weathering. Since it takes latex or oil paint very well, it's often used as a cap bead over other caulking compounds to make them paintable, but wood surfaces require priming to increase adherence. This caulk gives off very little odor and cleanup only requires water. Latex caulk comes in several colors.

Polysulfide caulk and sealant comes either as a one-part product requiring no mixing, or a two-part product. The one-part type requires 6 to 12 months cure time and the two-part variety requires 2 to 4 months. For all practical purposes, it's weatherproof in a few hours. But until it cures, its expansion capabilities aren't at the maximum. Both types allow up to 25 percent expansion and have a life expectancy of 20 years. They require a surface primer on most surfaces, but both have good adhesion. Both have poor-to-fair resistance to ultraviolet light, and only fair-to-good resistance to weather, so they'll crack eventually. They come in a variety of colors, and polysulfide caulks take paint well. A two-component immersion polysulfide sealant is also available to seal swimming pools, reservoirs, dams or other water-submerged environments.

Polyurethane caulk and sealant also comes as a one- or two-part product. Both varieties allow up to 50 percent expansion and have a life expectancy of 15 years. The one-part type requires 2 to 6 months cure time and the two-part variety requires 3 to 8 months. Both require a surface primer on some surfaces. The one-part product (urethane) has good adhesion and the two-part (polyurethane) variety has fair-to-good adhesion. Both types show good resistance to ultraviolet light and to weathering. Both products come in a wide range of colors. A two-component urethane-asphalt sealant is available for sealing horizontal joints in salt water or high-abrasion industrial environments.

Solvent acrylic caulk allows up to 10 percent expansion and has a life expectancy of 10 years. No surface primer is required and adhesion is good. This product exhibits good resistance to ultraviolet light and fair resistance to weathering. Many colors are available and solvent acrylic caulk doesn't require curing.

Silicone caulk and sealant comes with a wide variety of physical properties. Silicone caulk allows expansion ranging from 25 to 100 percent and has a life expectancy of 30 years or more. Most products don't require a primer on most surfaces (primer is recommended over concrete, and some plastics) and all exhibit good adhesion. All products show good resistance to ultraviolet light and to weathering. Most silicone caulk comes in a wide range of colors, but the mildew-resistant kind comes only in white. Due to its expansion-contraction properties, paint does not adhere well to silicone sealants. Most silicone caulks cure in only 7 to 14 days. Silicone caulks are used for a variety of purposes including interior environments of extreme humidity and temperatures, to seal construction joints where movement is extreme, to seal glass, metal and plastic surfaces, and for sealing insulating glass.

Vinyl acrylic caulk can also be used as an adhesive to bond virtually any two materials and no surface primer is required. This product exhibits good resistance to ultraviolet light and to weathering. Vinyl caulk is available in many colors and cures in 48 hours. Don't use this product under water.

Epoxy sealant comes as a two-part product. It sticks to metal, glass, marble, concrete, wood and many other surfaces. Epoxy sealant is normally used to seal joints in high-moisture areas and facilities subject to

	\multicolumn{9}{c}{Joint dimensions (inches)}
	1/8	3/16	1/4	3/8	1/2	5/8	3/4	7/8	1
1/8	1232	821	616	411	307	246	205	176	154
3/16	821	547	411	275	205	164	137	117	103
1/4	616	411	307	205	154	123	103	88	77
3/8	411	275	205	137	103	82	68	58	51
1/2	307	205	154	103	77	62	51	44	39
5/8	246	164	123	82	62	49	41	35	30
3/4	205	137	103	68	51	41	34	29	25
7/8	176	117	88	58	44	35	29	25	22
1	154	103	77	51	39	30	25	22	19

Figure 12-23 Caulking material requirements (linear feet per gallon [231 cubic inches] of caulk)

Insulation, Vapor Retarders and Waterproofing

submersion such as reservoirs, sewer treatment facilities and swimming pools. It's not affected by jet fuel, oil, gasoline, caustics, salts and most acids. The life expectancy varies from 10 to 20 years, depending on service conditions. This product comes in white or gray, with custom colors available.

Caulking compounds come in 11-ounce cartridges, 1-gallon containers, 5-gallon pails and 55-gallon drums. The volume of one tube is 315 cubic centimeters (about 19.22 cubic inches). This means that 1 gallon is about equal to 12 tubes (231 cubic inches). Use Figure 12-23 to find how much caulking material you need. Volumes vary depending on joint design, tooling, backer rod placement and waste.

▼ **Example 12-3:** Assuming a $1/4$-inch-wide by $3/8$-inch-deep bead of caulk applied to both sides of all exterior openings, how much caulk is required for the building diagrammed in Figure 12-24 (on the following page)?

From Figure 12-24, the total linear feet of caulk required is:

3070 Door:	2 Ea x 17 LF/Side x 2 Sides =	36 LF
2630 Window:	1 Ea x 11 LF/Side x 2 Sides =	22 LF
3040 Window:	4 Ea x 14 LF/Side x 2 Sides =	112 LF
4040 Window:	4 Ea x 16 LF/Side x 2 Sides =	128 LF
Total		298 LF

To find how much caulk to buy:

Caulk required = LF ÷ LF/Gal (linear feet per gallon of caulk)

298 LF ÷ 205 LF/Gal = 1.45 gallons

Since there are 12 tubes per gallon:

Caulk required = 1.45 gallons x 12 tubes/gallon
 = 18 tubes

Backer Rods

Install a backer rod of oakum or polyethylene rope before you caulk joints wider than $1/4$ inch. Then moisten your finger to smooth and feather the bead of caulk. Or you can use a small paint brush dipped in soapy water.

Concrete expansion joint sealers consist of a polyurethane-based material applied over a primed backer rod. Apply a sealant tape over the backer rod material before applying the caulking compound to create a barrier between the two materials. Some materials have additional components such as bitumen or rubber for use in joints exposed to fuel spillage.

Roofing Construction & Estimating

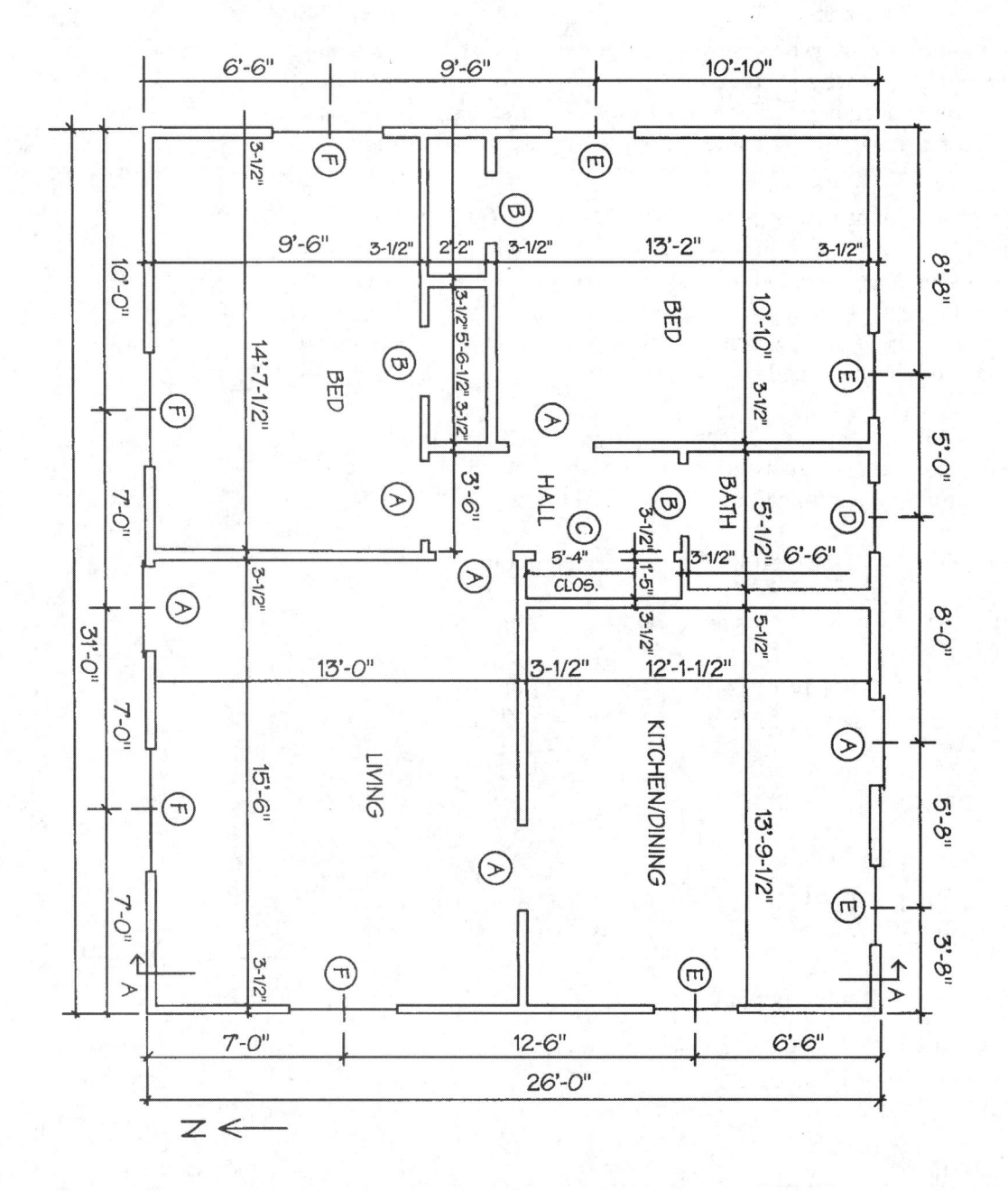

DOOR & WINDOW SCHEDULE	
OPENING	DESCRIPTION
A	3070 DOOR
B	2670 DOOR
C	(PAIR) 2470 DOORS
D	2630 WINDOW
E	3040 WINDOW
F	4040 WINDOW

Illustrated by Tami Atcheson

Figure 12-24 Caulking example

Insulation, Vapor Retarders and Waterproofing

Pipe size (Schedule 40 pipe)	Volume (cu. in.) required per 1" depth	Tubes of caulk required	Gallons of putty required
1"	1.04	0.06	0.005
2"	1.49	0.08	0.007
3"	2.94	0.16	0.010
4"	3.73	0.21	0.020
5"	3.92	0.22	0.020
6"	6.80	0.38	0.030
7"	11.00	0.61	0.050
8"	12.40	0.69	0.060

Figure 12-25 Fireproof caulk or putty required for various pipe penetrations

Fireproof Caulk and Putty

You can get fireproof caulk and putty to seal pipe penetrations and stop the spread of fire, heat, smoke and gas. This synthetic elastomeric one-part material expands by 10,000 percent when heated, and bonds to concrete, metals, wood and plastic. It also seals against water in an unexpanded state. You can get caulk in tubes and putty in quart and 1- or 5-gallon cans. For a three-hour fire rating, a 1-inch depth of caulking is required. You can use Figure 12-25 to find how much fireproof caulking or putty you need.

Foam Sealants

Foam sealant tape is a self-adhesive rolled tape used to seal joints in virtually any type of construction system. It offers protection against weather, vapor, sound and dust. Foam sealant tape is an open-cell polyester polyurethane foam impregnated with neoprene rubber. You can also get butyl-coated PVC foam sealants. The tape comes in widths ranging from 3/8 to 2 inches and in 15-, 20- and 25-foot lengths. Sealant tape will not shrink or dry out and expands to fit the joint in which it is installed.

Polyurethane foam sealants are applied from a container with a hose. This sealant bonds to wood, metal, concrete, brick, glass and most plastics. It withstands expansion and contraction and cures within 24 hours. Foam sealants seal cracks as well as electrical and plumbing penetrations.

Wall Flashing

The roofing subcontractor normally provides wall flashing, but it's usually installed by the general contractor (in wood frame construction) or the masonry subcontractor (in masonry construction).

Wall flashing helps prevent moisture from penetrating outside walls. Moisture penetrates at parapets, sills, projections, recesses and through mortar joints. Without wall flashing, the trapped moisture causes problems, including alkali action from the mortar and dilute sulfuric acid from atmospheric sulfur dioxide. Sulfuric acid is especially harmful to the metal components of a building.

Install wall flashing at each foundation, spandrel, sill, head, parapet, corner and door, plus through-wall flashing at various locations. Those flashings appear in Figures 12-26 and 12-27.

Wall flashing is made from a variety of materials including flexible plastic, asphalt-saturated cotton fabric, saturated or non-saturated woven fiberglass, asphalt felt and cotton fabric, copper, or copper plus lead bonded to asphalt-saturated cotton fabric, fiberglass, flexible rubbery bituminous compound, Kraft paper or polyethylene film. It can also be made of aluminum bonded to and between two layers of asphalt-saturated cotton fabric, polyvinyl chloride resin alloyed with other elastomeric substances, stainless steel, and plain or lead-coated copper. Wall flashing comes in strips, interlocking pieces or in rolls.

Estimating Wall Flashing Material

Take off flashing material by the linear foot for widths up to 12 inches, and by the square foot for widths greater than 12 inches. Some items, like flashing over, under and around doors and windows, are estimated by the piece. When you prepare an estimate, note the flashing gauge and location.

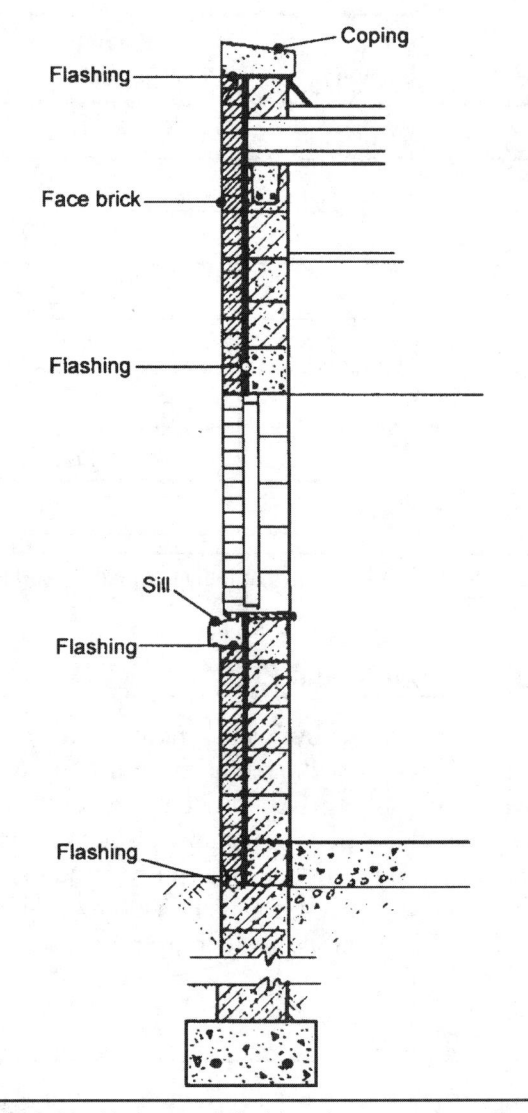

Figure 12-26 Various wall flashings in a masonry wall

Waterproofing

Roofing contractors routinely install waterproofing or dampproofing materials to seal vertical surfaces on concrete or masonry basement walls beneath finish grade. And because the materials and installation methods are so similar to many roofing procedures, I'll discuss other applications as well. Components which come under the heading of waterproofing and dampproofing include shower pans, waterstops, linings for planter boxes, and lining concealed under concrete, terrazzo, ceramic tile, clay pavers and walking surfaces of roofs.

Insulation, Vapor Retarders and Waterproofing

Waterproofing seals out water that's under pressure, called *hydrostatic head:* the pressure produced by the weight of liquid water against the foundation walls and slabs of a structure.

The most common waterproofing is the application of multiple layers of asphalt-saturated roofing felts sandwiched between moppings of hot asphalt or pitch. Other methods include troweled-on mastic, brush- or spray-applied liquid water repellents, single or multiple layers of plastic or rubberized sheet material, or asphalt-coated protective board embedded in cold-applied adhesive. You can buy special waterproof foundation-coating cements in 5-gallon pails or 55-gallon drums.

Cover waterproofing material as soon as possible after installation. If the backfill material contains stones or other sharp objects, install a layer of paper-pulp board over the waterproofing for protection during backfilling operations. If you're careful during compaction to prevent damage to the waterproofing, you can omit the protective board when backfill material is sand or clean loam.

It's often necessary to install a drainage system in addition to waterproofing to prevent standing water from penetrating a basement or foundation wall. A drainage system removes most of the water before it reaches the wall, or lowers the water table near the building to a level below the slab.

That's shown in Figure 12-28. The most commonly used drainage pipe is 4-inch-diameter farm tile, bituminous fiber pipe or perforated plastic pipe. The first comes in 12-inch lengths, the other two in pieces 10 feet long.

Lay at least 2 inches of crushed stone under the pipe and place at least 6 inches over the top of the pipe. Severe groundwater conditions could require the use of well-points and continuous pumping.

Integral Waterproofing

You can use a variety of liquid, paste and powder chemical admixtures to make concrete, mortar and stucco denser and therefore less permeable to water through increased hydration of the cement. This is called the integral method of waterproofing. Estimate quantities for this type of waterproofing according to the cubic yard of concrete poured.

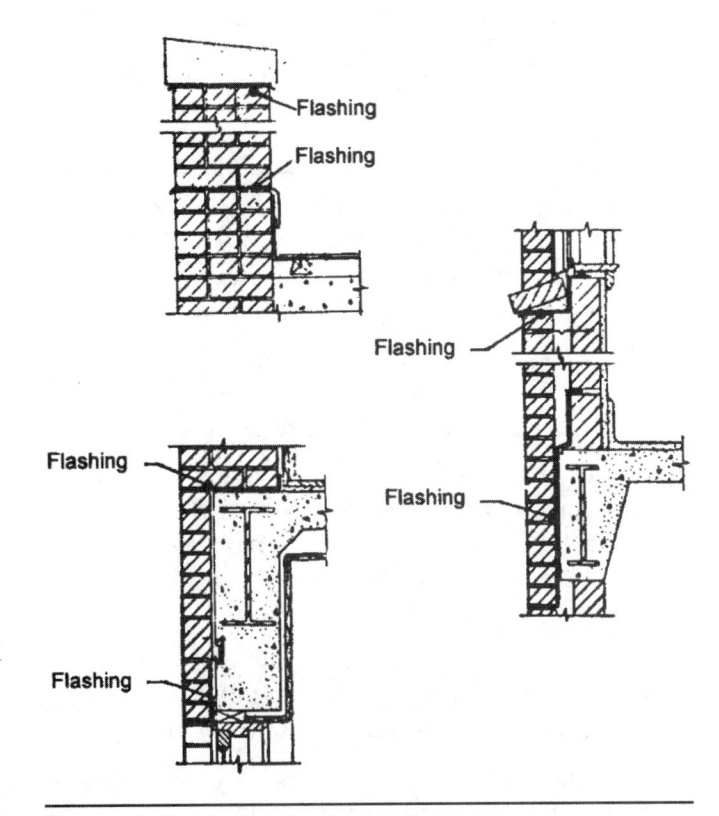

Figure 12-27 Typical flashing details

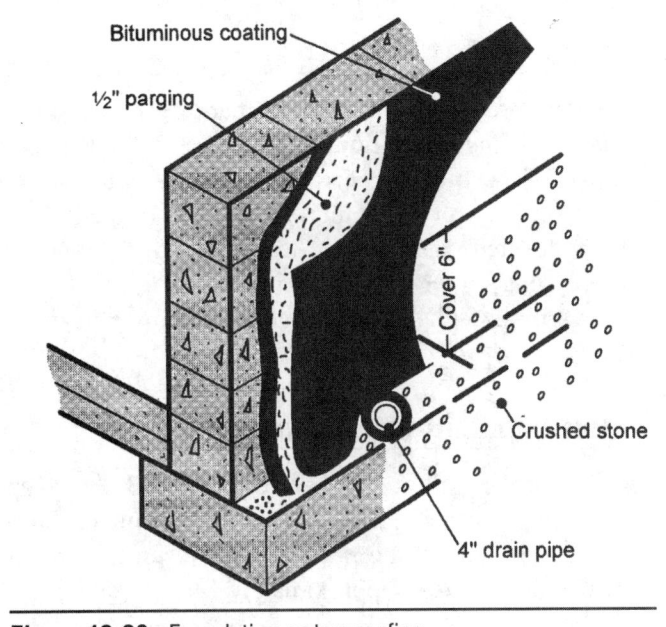

Figure 12-28 Foundation waterproofing

Built-up Waterproofing

Fabrics

Fabrics (cotton, glass fiber or jute) are more elastic than felts and have greater tensile strength. It also takes fewer coats of waterproofing bitumens to reach the desired thickness because fabrics hold the bitumens in place.

These fabrics come in rolls that are 180 feet long (nominal length 150 feet) and 3, 4, 6, 9, 12, 18, 24 and 36 inches wide. Some glass fabric rolls are 48 inches wide.

You can get bitumen-saturated cotton fabric in standard or specification grade. Specification grade contains more cotton. Cotton fabric is the most commonly used material for covering corrugated metal roofs.

Saturated jute fabric (treated burlap) is very coarse so it holds a lot of field-applied bitumen coating material. That results in a heavy build-up of bitumen coating with fewer applications, lowering your labor cost.

Saturated woven glass fabric is widely used with cold-applied bitumens. It has a higher tensile strength than cotton or jute fabric. Standard glass fabric consists of randomly arranged glass fibers bonded with a resinous binder. Heavy-duty glass fabric consists of heavy glass yarn bonded and coated with bitumen.

Membranes

Waterproof membranes are made from a variety of materials, including liquid-applied or sheet elastomeric plastic films, built-up perforated asphalt- or tar-saturated felts (organic or inorganic) or fabrics. Asphalt- or tar-saturated felts are suitable for all subgrade conditions. Apply glass fiber felts to horizontal surfaces only. Other materials include spray-applied glass fiber reinforced bituminous material and composites.

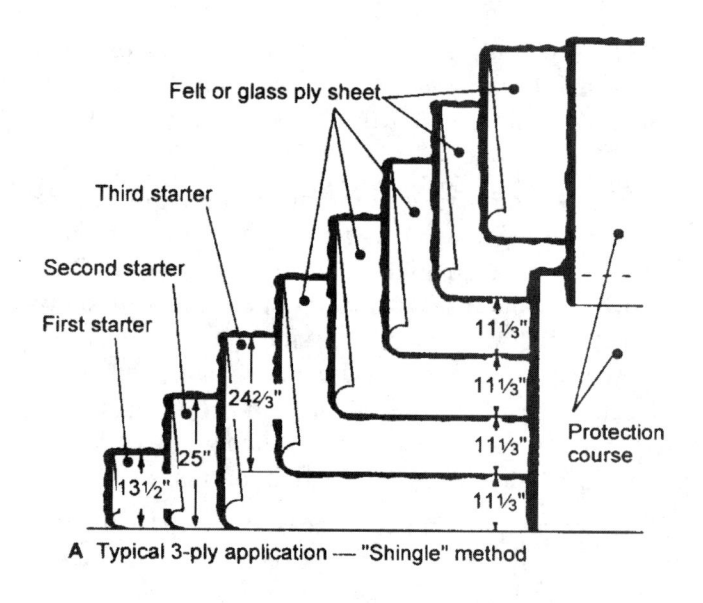

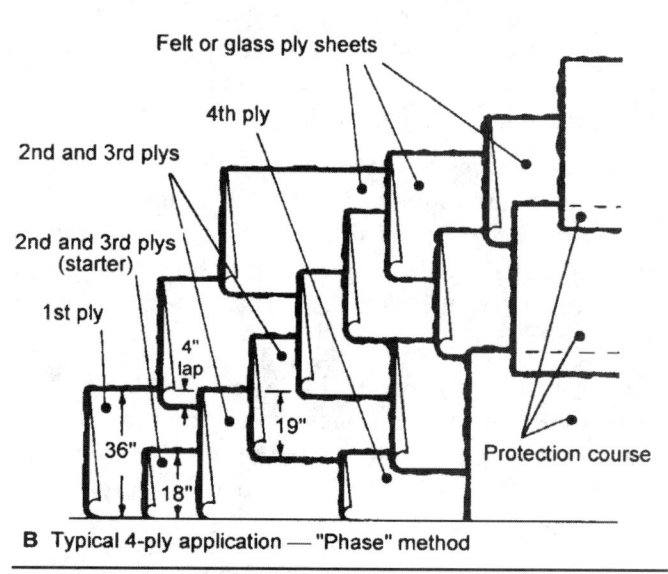

Figure 12-29 Installing built-up membranes

Installing Built-up Membranes

Prime the wall or slab before you install a built-up felt membrane waterproofing system. Use creosote to prime for pitch, and cutback asphalt to prime for asphalt. Apply the primer coat by spray, roller or brush. Apply asphalt primer at the rate of approximately 1 to $1^{1}/_{3}$ gallons per 100 square feet.

Following the primer coat, install the first and subsequent layers of felt into hot interply moppings of bitumen and coat the last layer of felt with hot bitumen. Then cover the entire membrane system with at least $1/_{8}$-inch

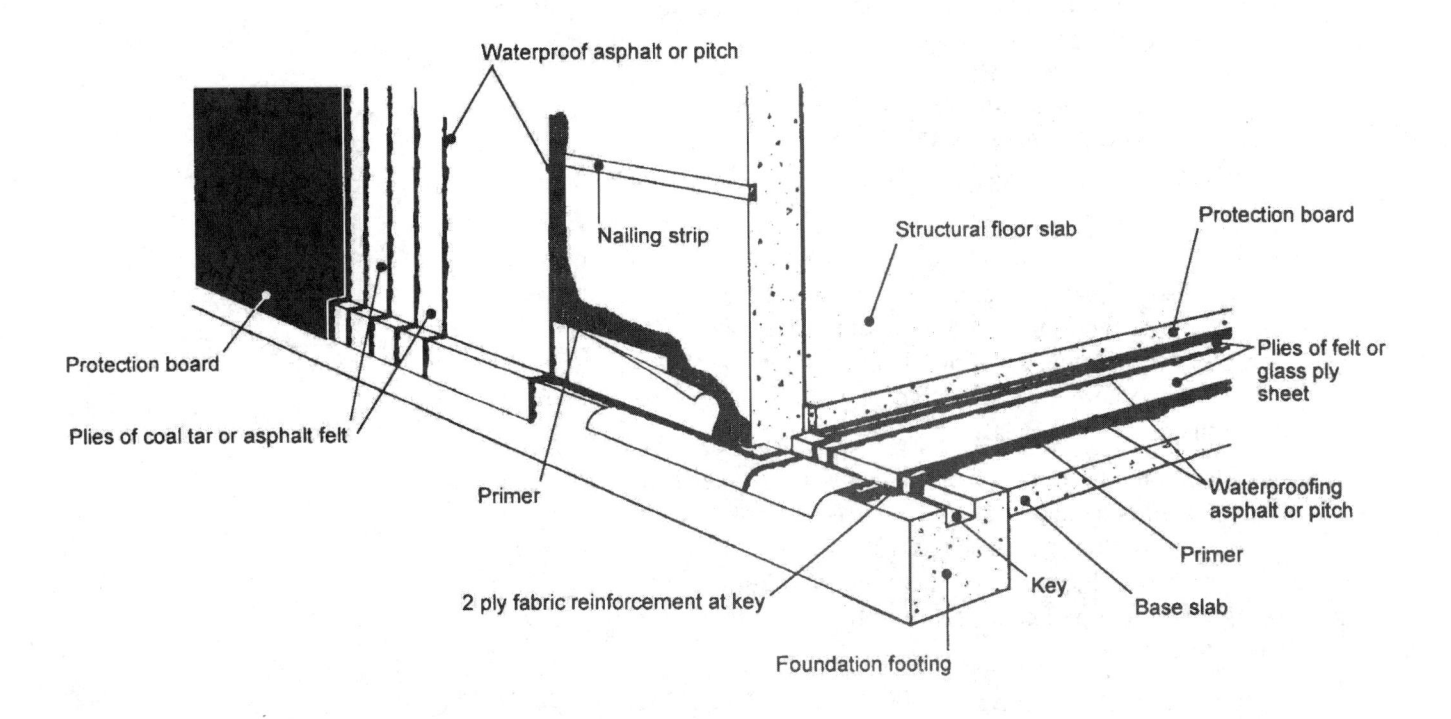

Insulation, Vapor Retarders and Waterproofing

Figure 12-30 Typical slab and foundation waterproofing

protection board embedded in the outer mopping of bitumen. That's shown in Figures 12-29 and 12-30. You can also install the protection board into a cold applied adhesive.

Use dead-level asphalt (Type 1) or coal tar pitch on horizontal or vertical surfaces below grade. Use flat asphalt (Type II) below grade or above grade when temperatures don't exceed 125° F. Use steep asphalt (Type III) above grade on vertical surfaces exposed to direct sunlight. Due to the low softening point of dead-level and flat asphalt or pitch, shelter the finished membrane from the sun during hot weather.

Fibered asphalt mastic contains cutback or emulsified asphalts and mineral fibers that are mixed either at the factory or within a spray nozzle during application. Use a trowel to apply factory-mixed fibered asphalt over a thinned asphaltic primer coat. To reinforce a fibered asphalt membrane coat, embed glass fabric in an asphalt undercoat before you install the membrane. Job-mixed fibered asphalt is especially practical when used over base felts on free-form roofs.

When you install organic felts, apply interply asphalt or pitch at the rate of 23 and 25 pounds per 100 square feet, respectively. With glass fiber felts, use interply asphalt or pitch at the rate of 25 pounds per 100 square feet. You can install cold-applied cutback or emulsified asphalt instead of hot bitumen in temperatures of 40° F and higher, and when dry weather is expected for 24 hours after application. Apply cutback or emulsified bitumens over a thinned primer coat.

In order to allow the waterproofing system to freely expand and contract and thus adapt to building settlement, install the first membrane over spots of cold-applied adhesive. I recommend you use waterproof fabric that's embedded in bitumen instead of felts, even though it's more expensive. However, on relatively flat surfaces (horizontal slope less than 1 in 12), saturated felts are more water-resistant than fabrics.

You can strike a compromise between expense and utility by installing a system composed of felts overlaid by an outer ply of fabric. Whether you use felts or fabrics, reinforce all inside and outside corners with at least two plies of membrane or fabric installed into hot moppings of bitumen. Install the reinforcement either as cushion strips extending 4 inches in each direction from the corner, or by lapping each membrane course (from both sides) around the corner 4 inches.

Built-up membrane waterproofing for slabs on grade is most often a 3-ply "shingle" method (Figure 12-29 A) or a 4-ply "phase" method (Figure 12-29 B). You can use these methods to waterproof foundation walls, also, but only with asphalt felts, not glass. Nail the membranes to the wall with a nailing strip near the top of each course to help hold them in place. That's shown in Figure 12-30.

When you install waterproofing to slabs above grade, prime the sub-slab and install a vapor retarder. Then install two plies of asphalt or glass felts overlaid by another vapor retarder and a protection board. Embed all membranes and protection board in moppings of hot bitumen. This is shown in Figure 12-31. Make sure the topping slab is poured within seven days after you finish the waterproofing system.

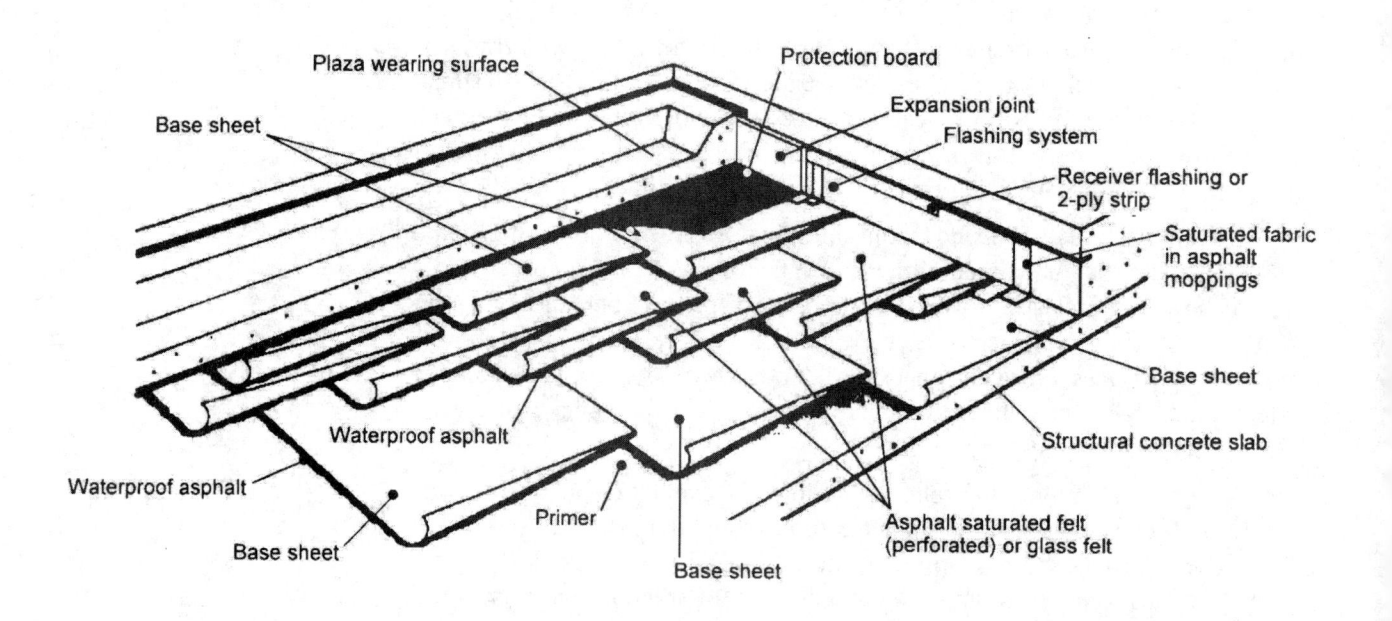

Figure 12-31 Waterproofing a slab above grade

	Head of water (feet)			
	1-3	4-10	11-25	26-50
No. plies felt or glass ply sheets	2	3	4	5
No. bitumen moppings*	3	4	5	6

*Asphalt approximately 23 pounds per mopping (25 pounds with glass ply sheets). Pitch approximately 25 pounds per mopping.

Figure 12-32 Plies of membrane required

Estimating Built-up Waterproofing and Dampproofing

For a 1-ply waterproofing system, allow 12 percent waste for 4-inch side and end laps, and 18 percent for 6-inch laps. You don't have to lap the felts when you install two or more plies. With no laps, you don't have to allow for waste in your estimate because felts come in factory squares (108-square-foot rolls).

On basement walls, start the waterproofing at about 6 inches below the top of the lowest slab. At the top of the system, extend the bitumen moppings 2 inches above finish grade, and extend felts to 2 inches below the finish grade elevation. Use these upper and lower system boundaries when you estimate waterproofing or dampproofing materials.

Apply bitumen with a hand mop at the rate of about 63 pounds per 100 square feet. Use cold-applied asphalt only in areas where it would be hazardous to use hot material or where the coated area is too small to justify using a kettle. Apply cold asphalt by brush or spray at the rate of about $1^{1}/_{2}$ gallons per 100 square feet.

The number of plies of membranes and bitumen moppings required for adequate waterproofing depends on the water pressure against the waterproofing system. Use Figure 12-32 as a guide to determine waterproofing requirements.

Other Types of Waterproofing

Elastomeric Waterproofing Membranes

The most common liquid-applied elastomerics include synthetic rubbers such as chlorophene (neoprene), polyurethane, and polysulfide (Thiokol). Liquid elastomerics must often be applied over a primer coat.

Cured preformed sheet elastomerics normally include butyl (isobutyleneisoprene), chlorophene, neoprene, ethylene propylene (EPDM) and urethane bitumen. Rubberized membrane sheets come in 40-, 30- and 22-gauge thicknesses. Sheets come in standard sizes up to 45' x 100' and larger. Plastic sheets come 0.004, 0.006 and 0.010 inch thick.

Butyl and EPDM elastomeric sheets are resistant to bacteria, fungi, soil acids, ozone and ultraviolet light. Neoprene elastomeric sheets have an additional resistance to oil and certain other chemicals.

You usually apply elastomeric sheets as a single-ply material, so installation goes faster than built-up applications.

Composites

Composites are metal foils or plastic films laminated to fabrics, paper, felts or rubberized asphalt of various types. Plated or coated metals are not considered composites although they produce similar results.

Liquid Plastic Waterproofing

Liquid plastic waterproofing consists of a powdered iron oxide compound which you trowel onto the surface of a concrete wall. Application thickness varies from $5/8$ to 1 inch. Some surfaces, including spalling brick or concrete, require roughening with a bush hammer before you can apply the waterproofing.

Waterproofing Panels

Asphalt-coated protective board comes in 4' x 4' and 4' x 8' sheets in thicknesses of $1/4$, $1/2$, $3/4$ and 1 inch.

Bentonite panels are very popular and cost-efficient waterproofing systems. Bentonite is a clay material that swells when wet, so panels are self-healing even when building settlement causes cracks. You can place the panels within the concrete forms. A biodegradable Kraft paper on the panels prevents them from sticking to the forms. This eliminates the need for the waterproof application after the forms are stripped. Install a protection board when backfilling with coarse material.

You can fasten panels over existing concrete walls with masonry nails. Lap panel edges at least $1 1/2$ inches and staple laps to prevent displacement during concrete placement. You can install this type of panel over vertical walls, under floor slabs and over below-grade roofs. Bentonite panels come in 4' x 4' sheets, $3/16$, $3/8$, $1/2$ and $5/8$ inch thick.

Dampproofing

Use dampproofing materials where there's no hydrostatic head, but where you want to prevent moisture from entering due to capillary action resulting from occasional exposure. Such conditions occur where there's good drainage or where the moisture content of the soil is low. The most common types of dampproofing include layers of bitumen, heavy asphaltic paint or other liquid water repellents. These coatings are sometimes called *hydrolithic coatings*.

In below-grade dampproofing, apply water-resistant coatings to exterior surfaces. The coating penetrates and seals pores to form a continuous protective film. Above grade, apply protective coatings over the inside surfaces of walls before plastering or tiling, or to the outside of back-up walls. Colorless sealers are available for coating exposed masonry surfaces.

Bituminous Dampproofing

The most common form of dampproofing consists of two or more coats of cold- or hot-applied asphalt or pitch installed over a primer coat. You can also use troweled-on hot fibrous asphalt. Install a layer of asphalt-saturated felt over the outer coat to protect the dampproofing from damage during backfilling. Some manufacturers recommend that you install a protection board over dampproofing materials, but don't count on dampproofing materials to act as an adhesive for the protection course. Install glass fabric to reinforce asphalt dampproofing applications when you expect ground movement or vibration.

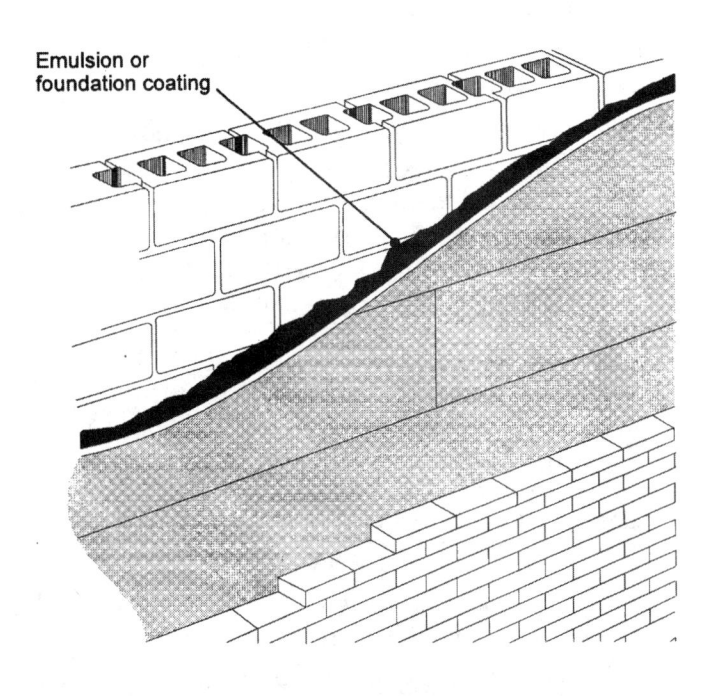

Figure 12-33 Dampproofing a cavity wall

You can use emulsified asphalts for dampproofing, but don't apply these products in cold weather (below 40° F) or when you expect rain within 24 hours after application. You can apply fibrated emulsion coatings over "green" or damp concrete, but apply cutback asphalts over dry surfaces only. Brush or spray emulsion coatings in two coats at an application rate of 1 to 3 gallons per 100 square feet, per coat, depending on the amount of dampproofing desired.

You can also use emulsion coatings to dampproof masonry cavity walls as shown in Figure 12-33. You can use troweled bituminous mastic in cavity-wall construction or at foundation walls. Apply the mastic at the rate of 8 to 9 gallons per 100 square feet to provide an $1/8$-inch coat (wet). When installed over foundation walls, cover the mastic with a protection board before backfilling.

Dampproofing and Waterproofing Existing Walls

Apply plaster bond paint to the inside or outside of an exterior basement wall, followed by plaster. You don't need lath when you use plaster bond paint. The number of coats and covering capacity of the paint will vary, depending on the porosity of the surface of the wall. Labor cost varies depending on the consistency of the paint, the condition of the wall and the application method.

You can apply waterproof wax into cut and repointed mortar joints in an existing masonry wall. The labor varies depending on the type of mortar originally used.

When you install certain types of waterproofing or dampproofing materials over a brick or block masonry wall, the wall is sometimes given a cement *parge coat*, $1/2$-inch-thick cement plaster coat applied by plasterers. Follow the parge coat with a bituminous coating applied at the rate of 1 gallon per 100 square feet. Refer back to Figure 12-28. The parge coat takes about 1.9 bags of masonry cement and 0.23 tons of sand per 100 square feet of wall.

There are many specially formulated cement (hydrolithic) coatings now available that provide a waterproof envelope in interior applications. Hydrolithic coatings are formulated using various additives to improve different properties. These additives include:

- purified iron particles and various chemical agents that improve workability and oxidation.

- compounds that promote crystalline growth within concrete capillaries, where only the interior side of an existing building can be treated.

Some hydrolithic cement coatings are installed over layers of reinforced glass fiber fabric. Your budget, the architect, and the degree of protection you need determine the number of fabric layers required. Apply hydrolithic coatings to surfaces using a brush, roller or trowel.

Penetrating acrylic or quartz carbide sealers are available for sealing and protecting existing architectural concrete, masonry, block or stone. You can also use transparent silicone-based sealers over masonry surfaces to protect the surface and minimize efflorescence. They also help minimize damage due to freeze-thaw cycles.

There are sealers made of a mixture of organosilane and ethyl alcohol. This type of sealer is resistant to salt which contributes to concrete spalling and the corrosion of reinforcing steel.

Epoxy Waterproof Coatings

Epoxy coatings consist of a mixture of epoxy and emery. Use epoxy coatings on concrete floors to provide traction and to improve chemical and wear resistance. Likely candidates for this application include chemical, food, beverage and meat processing and industrial plants.

Coal tar epoxy coatings are available for application to floors and walls in sewer treatment facilities. You can also use epoxy materials which are made for patching, sealing and waterproofing cracks in existing concrete structures.

Now that we've covered all the information you need for roofing and waterproofing in new construction, let's look at the types of liquid-applied roof coatings available today and where and how to apply and estimate each. That's the subject of the next chapter.

13 Roof Coatings

▶ Conventional roofing and roof coatings are two distinct subjects. The materials are different and the application techniques are different. True, some roof coatings are designed for application over new roofing. But most roof coatings are applied to extend the life of an existing roof. If a roof surface and the membrane are still sound, applying a modern roof coating can make tear-off unnecessary. That's a major advantage for any owner.

Both roofing and roof coatings serve the same purpose – to protect the building and building contents below. Roofing and roof coatings require many of the same tools. And if you're licensed and insured to do roofing work, you're qualified to accept roof-coating work also. As a professional roofer, you should be prepared to offer clients roof coatings as an option. In practice, most roofing companies are also roof-coating companies.

Conventional roofing such as wood shingles, roll or hot asphalt, tile and slate have been used for centuries. Modern roof coatings were developed in the last half of the 20th century. Chemists created liquid coatings that (1) bond securely with nearly any firm substrate, (2) resist the elements, especially ultraviolet (UV) radiation from the sun, (3) absorb and transmit less heat from the sun, (4) expand and contract without breaking (elastomeric), and (5) have a life expectancy of 20 years or more.

This chapter will explore four common liquid roof coatings:

- Sprayed polyurethane foam (SPF) roofing
- Liquid EPDM rubber roofing
- Silicone roofing
- Fiberglass GRP roofing

We're not going to discuss polyurethane coatings. Poly coatings have been used as roofing, but the cost is considerably higher than other elastomerics. Poly coatings are a good choice for coating the interior of planters and as a top coat for areas under steady foot or vehicle traffic. But poly coatings are considerably more difficult to apply correctly. Even the slightest moisture trapped under a polyurethane top coat can puncture the surface when roof temperatures climb. As a roof coating, polyurethane isn't competitive with silicone or EPDM.

The term *roof coating* can be applied to any material intended to coat a roof. Many acrylic paints are sold as roof coating. These products can improve the appearance or thermal characteristics of a wood, metal or asphalt roof, but acrylics are water-based, have low tensile strength, and don't stand up when water ponds on the roof. That's true even if you see terms like *aluminized* or *siliconized* or *rubber based* in the product description. Because acrylics are water-based, some portion erodes in every rainstorm. Acrylics aren't intended to waterproof a low-slope industrial roof for 20 years or more. When we use the term "roof coating" in this chapter, we're not referring to acrylics.

Dozens of manufacturers offer elastomeric low-slope roof coatings with 95-percent-or-more solids content and that come with a 20-year warranty. Many are sold under a trade name rather than the product description, such as EPDM or silicone. Some are intended as a finish top coat for a new roof assembly. Others can add years to the useful life of an existing roof surface. All low-slope roof coatings require specific application techniques. Some manufacturers offer certification classes and grant recognition to roofers trained in applying their products. Factory certification has clear advantages. Manufacturers may extend a warranty only if work is done by a contractor certified in that specific brand of roof coating.

These warranties can be important to owners. For example, many roof-coating manufacturers offer a 20-year warranty. The typical cost of an entirely new roof system with a similar 20-year warranty would be at least three times as much. The advantage of roof coating is obvious.

If you're serious about bidding on roof-coating work, the best choice is to find a local manufacturer or distributor and develop skill working with products offered by that vendor. Many coating manufacturers are regional. They offer coatings developed for the local climate. That makes good sense. Roof coatings that work well in Hawaii are unlikely to be the best choice for a roof in Oregon, Arizona or Vermont. No roof coating is ideal for every situation and every climate.

The *Cool Roof Rating Council* maintains a sortable list of roof coating manufacturers and products at https://coolroofs.org/directory. Most listed products show both reflectance (the fraction of solar energy reflected by the coating) and the emittance (ability of the roof coating to radiate absorbed heat). Ratings show both the initial reflectance and aged reflectance (after three years). Manufacturer websites are listed so you can get more information on any product.

This chapter isn't a substitute for a class offered by a manufacturer of the coating you're planning to apply. Neither will this chapter cover every type of coating offered by every manufacturer for every type of job. There are too many good choices. Consider this chapter a starting point for developing experience with roof coatings. We'll suggest good professional practice for any low-slope roof-coating job.

Proper Curing is Basic

Elastomeric roof coatings have to cure completely to perform as intended. Curing slows in cold weather and may stop entirely when frost begins to form. Many roof coatings need warm weather (above 55° F) to cure properly. But there are exceptions. For example, the manufacturer of GE Enduris claims their product can be applied at temperatures down to 0° F. EPDM coatings stop curing when the temperature falls below 55° F but will resume curing when the air temperature warms again. Labels on the coatings you apply will have more details.

Moisture also delays curing. Don't schedule a roof-coating job any time rain is expected. Many roof coatings are waterproof four to six hours after application. But don't plan to apply any roof coating in the rain. Even condensation can affect performance of some roof coatings. On a cold morning, it's common to see dew accumulate on a roof surface. A roof coating applied over condensed moisture isn't going to set properly and will likely lose adhesion.

Don't be concerned about rain or freezing temperatures once the coating has cured. All common roof coatings are formulated to repel water and survive freezing temperatures once the coating is fully cured.

Exposure to full sun increases the cure rate. Coating applied in shady spots won't cure as quickly as coating applied in the sun. Even at 55° F, coating shaded by A/C equipment, parapet walls, a chimney or trees will take longer to fully cure. If you've got a job with many shady spots, try to finish application early in the afternoon before sunlight is lost.

Advantages of Roof Coatings

Modern roof coatings are designed to reflect up to 80 percent of solar radiation back into the atmosphere. That offers two advantages. First, the less heat retained, the less heat damage to the substrate – both the applied roofing and sheathing under the roof surface. Second, when less heat is absorbed by the roof, less heat will be radiated into the building interior. That's an important consideration if interiors are cooled in the summer. Savings vary with climate and type of roof under the coating, but experience suggests that a reflective white roof coating can reduce the cooling load of a building by as much as 5 percent. That's a significant saving for an owner accustomed to spending hundreds of dollars a month on electricity for summer cooling.

In full sun on a summer day, asphalt, built-up, wood, metal and slate roofing can reach 150° F. Nearly any reflective white roof coating can reduce the surface temperature by 50° F or more.

Ultraviolet light from the sun erodes most organic materials. Leave a cloth lounge chair in the sun for a few summers and you'll see the fabric fade and eventually shred or rip. Most organic materials do the same – lose color and strength as UV rays penetrate the material. So too with most roofing materials. Exposure to sunlight weakens, discolors and shrinks organic roofing materials such as wood, composition shingles and built-up roofing. Stone, concrete, metal and slate roofing are exceptions. But these surfaces come with a disadvantage. They're excellent heat reservoirs, absorbing heat during the day and releasing that heat into the building interior after sundown. That's OK during the winter heating season but very undesirable during the summer cooling season.

The best roof coatings have two advantages. They're immune to UV radiation and they're *elastomeric*, which means they can expand and contract with temperature changes without cracking or chipping. That helps keep water above the roof surface where it can't do serious damage. Moisture that penetrates the roof surface will encourage mold, mildew and rot in wood framing.

Not every low-slope roofing material is a candidate for roof coating. Built-up roofing with gravel embedded in the top coat usually has to be torn off rather than re-coated. But metal roof decks, single-ply roofing and modified bitumen roofing are all good candidates for re-coating with rubber roofing.

How to Apply Elastomeric Roof Coating

Every type of roof coating has advantages and disadvantages. Application methods vary. The best guide for applying any roof coating will be instructions provided by the manufacturer. But the suggestions that follow will be a good guide on most jobs.

Step 1: Every coating job begins with an evaluation of the existing roof. Most low-slope roofs have drainage problems – puddles that form after a rain. Even if the roof was built to drain perfectly, subsidence or shrinkage of the framing below can create "bird baths" that retain water after a rain. Any puddle that doesn't evaporate within two days after a rain is a roofing problem. If you see stains from puddled water, consider filling these bird baths before applying any coating. Your roofing dealer can recommend a product to level the surface. Many roof coatings can be applied extra thick to level roof depressions.

Look for physical damage to the roof.

- *Alligatoring* is caused by UV rays and is the most common type of damage on roof surfaces. Deep cracks in the roof surface should be filled and overlaid with poly mesh and patching material before roof coating begins.

- Some birds collect nesting material from roof surfaces, leaving voids that have to be either filled or bridged. In some climates, insects bore into the roof surface to deposit their larvae, which then hatch into grubs. Birds searching for these grubs peck holes in the roof surface.

- Roof moss is a problem in many areas. Moss retains moisture, adds to the roof load and creates an uneven surface that's hard to coat evenly. Some moss can be removed by pressure washing. If that doesn't work, you'll have to scrape it off the roof.

- If an antenna or seasonal decorations have been placed on the roof, expect to find nail or screw holes through the roof surface. Holes larger than ½-inch in diameter should be patched before roofing begins.

- Wipe your finger over any roof sections that appear coated with a white powdery substance. If the powder wipes off, it's evidence of chalking. Pressure washing will usually remove loose chalk. Consider sealing the surface with primer or emulsion if the surface remains heavily chalked.

- Check sheet metal flashing for rust and deterioration. Plan to replace any flashing that's loose. The most problem-prone area in any roof is where dissimilar materials join, such as where roofing meets stucco or masonry. Joints like these should be flashed with metal or composite material to allow for expansion and contraction as surface temperatures change. Moisture on the surface above should drain over the metal flashing and fall harmlessly on the coated surface below.

- Check roof drains and the gutter system. Recommend repairs if the drainage system is obstructed or defective.

- Check the area around any vent for grease residue. Most kitchen exhaust fans have a filter which catches grease before air is discharged. A dirty filter can let grease escape up the vent and onto the roof. Grease that accumulates on the roof surface can prevent proper bonding of the new roof coat. Remove grease with a solvent. Then suggest to the owner that the vent filter be cleaned or replaced occasionally to keep acidic grease from eroding the new roof surface.

- Look for evidence of recent electrical, plumbing or HVAC work on the roof. Repairs made recently may have left holes or gaps in the roof surface. If the owner has any plans for construction that require penetration of the roof, it's best to have these repairs done before roof coating begins.

- If you see blisters in the roof surface, make repairs. Small blisters can be sliced open and pressed firmly against the sheathing below. Large blisters should be cut away and filled with the recommended patching material.

- Check for "spongy" sections of roof deck. Weak sheathing panels should be replaced before coating begins. Most roofing and roof-coating jobs on older buildings require at least some work on the roof deck – typically where moisture has penetrated into the sheathing. Roof deck that gets wet and stays wet long after a rain will support mold. If you see signs of mold (discoloration), apply a primer such as *Kilz* before roof coating starts. Mold remediation may require evaluation by a licensed professional and is beyond the scope of this manual.

Step 2: When inspection of the roof surface is complete, begin figuring the surface area to be coated. Include vertical areas such as curbs or parapet walls around the roof perimeter. Calculating the area for roof coating work is about the same as figuring the area for any roofing job. The primary difference is that roof coatings are more expensive. Doublecheck your calculations.

Step 3: Decide on the coating product that's best for the job. Consider the performance rating of the coating, the cost, the condition of the existing roof, the slope, the exposure, and the climate at the job site. One note of caution: Some owners' associations restrict the choice of roof colors. For example, some associations don't allow brilliant white roof coatings in spite of their clear advantages. Most roof coatings can be tinted to match whatever color or LRV (light reflectance value) an owners' association demands. Get the association's approval of roof color and LRV before ordering materials.

Step 4: Clean the roof. Loose material, debris, dirt and grease will prevent a good bond. Remove any loose or flaking waterproof coating. Scrape or vacuum off roofing granules or rock dress. Scrape rusted metal with a wire brush. Then clean the roof thoroughly with a pressure washer, washing brush or scrubber. Don't cut into or disturb sound roofing material. Too much pressure applied in vulnerable areas can open seams or erode the surface. Some contractors add soap or detergent to the wash water to break up dirt and grease. Be sure any soap residue is removed completely before you start coating.

When done, sweep away puddled water so the roof will dry evenly. Plan to wait at least 24 hours after pressure washing before doing any patching or coating. If the coating manufacturer recommends a primer or sealer for the surface you're coating, this is the time to apply that primer.

Step 5: Repair damaged areas – anything that interrupts the smooth, even roof surface. Level depressions and voids with filler or patching compound compatible with the coating you've selected. Use woven polyester and patching compound to bridge holes and imperfections and to reinforce seams in the existing roof. Poly membrane comes in rolls. When coated with patching compound, the membrane forms a stiff bridge that can span holes and cracks up to ½-inch wide.

Cut the poly membrane to lap over joints, seams, gaps, blisters and holes in the roof surface. Use a brush or trowel to butter the patch area with a first coat of elastomeric cement. The thickness of the cement should be ⅛ inch to ¼ inch. Embed polyester fabric in the freshly applied cement, then lightly brush the polyester fabric to fully saturate the material. Smooth out any wrinkles or bubbles. Wait a few minutes, then apply a second coat of elastomeric cement. The polyester fabric should be completely covered with cement.

If any of the surface is permeable or absorbent, apply a waterproofing primer or elastomeric emulsion sealer compatible with the coating you plan to use. Most sealers can be applied by brush, roller or spray. Make sure all patches and filled and sealed areas are completely dry before you start coating.

Step 6: Don't schedule application of the coating until there's a promise of good weather for the next two days. Sunny days are best – no rain, no frost and a daytime high of at least 55° F.

Mix each bucket of coating thoroughly. When mixed, the roof-coating material should have a uniform consistency and color. Remove and discard any "skin" that forms at the top of the bucket. Application can be by brush, roller, squeegee or spray. No matter how you apply roof coating, a smooth, even application is essential. If you're using a brush or roller, make long, even strokes. Don't go over the same area too many times. For a roller, select a ⅜-inch or ½-inch nap. On most jobs, a 9-inch-long roller is ideal. If using a sprayer, an airless spray rig that yields 3,300 P.S.I. at the nozzle is usually a good choice.

Many manufacturers recommend two light coats rather than a single thicker coat. Follow the manufacturer's recommendation on coverage – gallons per 100 square feet. A coating applied too thick doesn't make a better job; a coating too thin risks premature failure. Allow plenty of curing time between the first and second coats; most manufacturers recommend 24 hours. If you see areas where the coating needs touch-up, wait 24 hours before re-coating.

On parapet walls or any vertical surface, the same applies. Several thin coats are better than a single thick coat, which is likely to run down a vertical wall and leave ripples in the cured surface. Even on a flat roof, you'll probably find vertical curbs that have to be coated. Two coats, each 5 to 7 mils thick, will minimize running. Applying multiple coats takes more time, but the result will be a smooth, watertight, monolithic surface with no wasted material.

Sprayed Polyurethane Foam (SPF) Roofing

Polyurethane foam has been used as an insulation material for many years. It's dependable, affordable, and if properly maintained, can last almost indefinitely. SPF is a combination of isocyanate and polyurethane. When used as roof coating, the two components are heated and then pumped through a spray gun, which combines the ingredients and then atomizes the mixture.

SPF provides a seamless membrane covering for the entire roof. The foam can be sprayed onto virtually any surface, even irregularly shaped roofs with protrusions. Like any elastomeric, SPF must be applied over a clean, dry surface. The foam weighs about 50 lbs. per square and gives an R-value of 7.14 per inch of thickness. Foam also acts as an air barrier, preventing infiltration into or out of the building. Another advantage with foam roofing: SPF can be used to build up low areas that would otherwise collect water after a rain.

SPF foam changes color and deteriorates if exposed to UV rays or moisture. Use an elastomeric top coat to protect SPF from damage. The top coat has to be elastomeric to stretch and shrink as the SPF foam below expands and contracts during the course of a day. The top coat should be tough enough to stand up under any likely weather conditions and occasional light foot traffic. Unlike silicone roofing, EPDM can, if necessary, be painted with latex paint.

Liquid EPDM Rubber Roofing

EPDM (ethylene propylene diene terpolymer monomer) is an elastomeric sold as both a liquid and in solid sheets. In liquid form, it can be used either for making repairs on an existing roof or to cover the entire surface. EPDM coating on a roof can be expected to last 20 years or more; some have lasted over 50 years. EPDM resists heat and cold, has a reflective surface that can lower the cost of summer cooling, doesn't crack or flake or become brittle, and can be applied over almost any substrate except hypalon or TPO (thermoplastic olefin).

Even at low temperatures, EPDM remains flexible. It resists ponding water, ozone, UV radiation, acids, alkalis and hail. The only maintenance needed on an EPDM roof is occasional debris removal. If necessary, it can be painted with latex paint. Silicone roofing can't be painted.

Although it doesn't cure in cold weather; needing a temperature of at least 55° F to begin curing, EPDM roofing can still be applied at low temperatures. Even if the temperature falls below 55° F at night, curing will begin again when the temperature rises above 55°. An EPDM coat will cure in four to six hours, depending on temperature and humidity. As a general rule, only one coat of EPDM roofing is needed. Weight added to the roof is minor. If a second or touchup coat is required, wait five to seven days before application.

EPDM is a good choice for re-coating nearly any residential low-slope roof, and can be applied to concrete, foam, coated or galvanized metal, PVC, vinyl, wood, fabric, weathered aluminum or fiberglass. Application over copper flashing is permissible only if the copper has weathered at least five years.

Application procedure is similar to what's described earlier in this chapter. Start by repairing any leaks, small cracks or holes in the roof. Once repairs are complete, remove all debris and clean the roof. Some roofing substrates require a coat of primer before EPDM is applied. Follow the manufacturer's recommendations.

If the roof is less than 5,000 square feet, you'll probably want to apply EPDM with hand tools: squeegee, paint roller with a long handle and a paint brush for coating roof penetrations and flashings. If your job is larger than 5,000 square feet, or is standing seam, corrugated or has an irregular surface, use an airless sprayer. Never use an air-powered sprayer, as air pressure can cause bubbles in the mix. EPDM will cover about 40 square feet per gallon with a 20-mil thickness. Use mineral spirits (paint thinner) for cleanup.

EPDM roofing is also available in sheets, in widths up to 50 feet and lengths up to 200 feet, with the same properties as the liquid. Details and application methods, as well as quantity estimating and a sample labor and material cost estimate, appear in Chapter 11, *Elastomeric Roofing*.

EPDM and silicone roofing are about equal in resisting harsh weather, ozone and UV rays, with EPDM roofing reputedly slightly better at resisting abrasion and rips. However, EPDM may have a shorter life expectancy than silicone.

Silicone Roofing

Silicone roofing (Figure 13-1) was developed as a top coat for sprayed foam roofing. A silicone top coat protects the foam from UV and moisture, an advantages that soon became clear to roofing contractors. Silicone coating that could protect a foam roof could protect nearly any type of roof surface.

Applied and cured properly, silicone roof coating creates a seamless and waterproof surface that won't degrade, chalk or crack due to ozone or UV rays, and that resists damage from water ponding. The smooth surface is resistant to mold, mildew and staining. It's very flexible when cured and has great adhesion to the substrate. Application can be with a sprayer, roller or brush. Silicone roof coating can be used either to repair roof leaks or to coat an entire roof.

The ideal application temperature is 55° F. At lower temperatures the cure time is longer. Silicone can be applied at temperatures as high as 120° F. Delay application if rain is expected within an hour or two or if the air temperature is within 5° F of the dew point.

Silicone roofing will usually be the best choice on low-slope roofs in harsh climates where the roof is exposed to hot, wet weather and intense UV radiation. Even as it ages, silicone remains reflective and immune to UV rays.

Courtesy of Gaco Western

Figure 13-1 Gaco Silicone

Figure 13-2 Silicone roofing job, before and after

It's also a good choice for coating over SPF roofs, aged asphalt, aged EPDM, concrete, fiberglass, metal, modified bitumen and composite cap sheets. See the before and after pictures of a silicone roofing job in Figure 13-2. In addition to flat or sloped roofs, silicone can be used on RVs, mobile homes and campers, and has an estimated material life of over 50 years. Most manufacturers recommend that silicone roofing be applied by a roofing professional.

As with any rubberized roofing, the surface to be coated should be clean, dry, and free of any oil, grease or dirt. Power-wash the roof with plain water. Then saturate the surface with a power-wash solution made for silicone roof jobs. Let the solution soak for 15 minutes, then sweep the surface with a push broom or brush. Follow with another quick power-wash using plain water, then use a leaf blower to remove excess moisture and reduce the drying time.

Figure 13-3 A Gaco Adhesion Test Kit

Figure 13-4 Liquid Roof Tape

Roof Coatings

Apply an epoxy primer to increase adhesion. Two primer coats may be needed. Each coat will dry in six to eight hours, depending on the weather.

To be sure the surface is ready for a silicone coating, test by applying silicone roofing to a small area and laying a cloth strip in the silicone. To test to be sure the surface is ready for a silicone coating, apply silicone roofing to a small area and lay a cloth strip in the silicone. Then brush on another coat of silicone, leaving a 1-inch strip uncoated. Let that cure for 24 hours. When fully cured, remove the cloth strip. If the coating under the test strip adheres to the substrate, you can expect good adhesion. Figure 13-3 shows a Gaco adhesion test kit.

Prepare the roof for coating by sealing all joints, fasteners, rooftop penetrations and rough spots. Liquid roof tape (Figure 13-4) is made for this purpose and can be applied with a brush or putty knife. To add extra strength, apply 4-inch polyester deck tape (Figure 13-5) over seams, joints, flashings, vents, drains, fasteners and dissimilar surfaces. Overcoat the tape with liquid roof tape. Allow four to five hours for the liquid to dry. Use a seam sealer over roofs with seams, gaps and overlaps.

On smooth surfaces, two coats of silicone are usually recommended. For small areas, simply brush or roll silicone onto the surface. For larger areas, apply silicone with an airless sprayer, shown in Figure 13-6. To get a more-even coat thickness, apply the second coat at right-angles to the first. To be sure the second coat covers evenly, add some color tint to the base coat. That makes it easy to spot any thin spots in the top coat.

Depending on the temperature and humidity, each coat will be tack-free within an hour. Before applying any second coat, be sure the first coat is completely dry and cured. Typical curing time is three to 12 hours, depending on temperature and humidity.

Coverage per gallon will vary with the type of material, thickness and temperature of the roof surface and roof coating, application method (brush, spray, roller, broom or squeegee), coarseness of the roof surface and experience of the roofer. Even roller knap thickness will affect coverage.

On a smooth-surface roof, two 11-mil coats will provide a 22-mil-thick layer. On the first coat, coverage will be less, as some coating will be absorbed by small cracks or voids in the roof surface. The second coat should go on more easily and coverage per gallon will be more, as little of the coating will be absorbed into the roof surface. At 22-mil thickness, you'll need about one gallon of silicone roof coat for each 50 square feet of roof coated. On a textured surface the coverage rate will be considerably less. The manufacturer of the coating you select is sure to offer more precise figures.

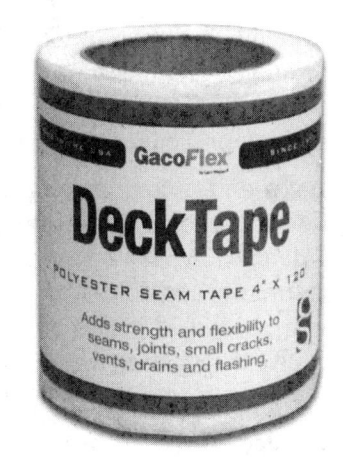

Courtesy of Gaco Western

Figure 13-5 Polyester Deck Tape

Courtesy of Gaco Western

Figure 13-6 Airless Sprayer

If there are walkways on the roof, you can create a visible path by applying a silicone walkway tinted to a different color. See Figure 13-7. Then sprinkle walkpad granules onto the wet coating at the rate of about ½ pound per 100 square feet. Allow this to dry for at least four hours. Use naphtha or mineral spirits for cleanup. Figure 13-8 shows a walkpad applied to a roof.

Silicone roofing requires very little maintenance. But removing debris and mildew occasionally will preserve the reflective property of silicone. Use a garden hose with a spray nozzle and a soft bristle deck brush. If you use a pressure washer, don't exceed 2,000 P.S.I. and use a 40-degree nozzle. Keep the tip at least 18 inches off the deck. If the roof gets stained by leaves or soil, consider recoating the stained area.

Blisters in the roof surface are usually caused by moisture trapped in the substrate. Slice off the blister and remove any loose or poorly fitting coating. Let the surface dry, then re-apply the same roof coating.

Any rubberized roof coating can be damaged when rooftop equipment has to be installed or serviced. Recommend that service personnel lay plywood panels on the roof surface before moving any heavy equipment.

If snow has to be removed from the roof, use plastic shovels. It's best to keep snow blowers and metal snow shovels off the roof.

Damage to the surface can be repaired by spot treatment with more silicone roof cover. To preserve the warranty coverage, the roof may have to be inspected annually by a roof-coating professional.

Figure 13-7 Silicone Walk Pad

Figure 13-8 Walkpad Application

Roof Coatings

Fiberglass (GRP) Roofing

If a portion of the roof will serve as a walkway or sitting area, consider applying a glass-reinforced plastic roof coat. GRP is available in many colors and textures, and creates a seamless, leakproof surface that resists impact, chemicals and fire, and that won't show wear from foot traffic. It also serves as a thermal and electrical insulator. GRP costs more than other roof coatings, but it can be expected to last for 35 years or more.

All roofing has to be removed before GRP is installed. OSB and T&G plywood make the best substrate for GRP decks. Allow expansion space between the deck and any abutting walls. If you're laying a new plywood deck, allow a ⅛-inch gap between panels.

Sweep the deck clean. Then wipe the OSB or plywood with acetone. Install masking tape or gaffer tape "bandages" over deck joints. Lay chopped strands of fiberglass mat over all joints, then brush fiberglass resin over the strips of mat. When the resin has hardened, sand lightly to remove any rough spots. Then lay chopped-strand fiberglass matting over the entire deck, lapping edges by 2 inches and apply more resin over these mats. When fully set, sand any rough spots with 40- to 60-grit sandpaper. The system should cure overnight. Once hardened, the surface can be painted.

Estimating Elastomeric Roof-Coating Work

Estimating Material Costs

The cost of materials will be the area you have to cover divided by the coverage per gallon times the cost per gallon. One coat is standard for residential jobs; commercial work may require two coats to reach the mil thickness specified by the manufacturer. Where color from the existing roof surface is likely to bleed through the new coating, many manufacturers recommend laying down a coat of primer under the new roof coat.

Estimating Labor for Roof Coatings

These estimates will apply on many low-slope elastomeric roofing jobs. Figures include the time required for typical job setup at the start of the job and cleanup when work is done. The steps are those described in detail earlier in the chapter.

Step 1: Allow one hour to complete a survey of job requirements.

Step 2: Allow one hour to calculate the roof surface to be coated.

Step 3: Allow two hours to select, order and take delivery of materials needed for the job.

Step 4: Cleaning the roof.

If roof granules, rock ballast or loose roof debris have to be scraped off and hauled away, figure one manhour for each 250 square feet.

Figure one manhour to sweep and pressure wash each thousand square feet of roof surface.

Figure two manhours per 100 linear feet if rusted metal flashing has to be wire-brushed.

Step 5: Repair damaged areas.

Remove and replace roof sheathing: two manhours per panel 4' x 8' or smaller.

Leveling depressions: Allow 20 minutes to fill and finish each depression. A heavily ponded 6-foot-diameter depression can require up to an hour to brush and wash clean and then build up with poly fabric and roof coating.

Bridging voids and holes: Figure five minutes for each area that requires patching.

Reinforcing seams: Figure 75 linear feet per manhour.

Waterproofing with primer or elastomeric emulsion: Brush at 150 S.F. per manhour. Roller at 500 S.F. per manhour.

Step 6: Apply roof coating.

Brush: 120 linear feet per manhour.
Roller: 200 square feet per manhour (two coats)
Spray: 400 square feet per manhour
Spray SPF 1 inch thick: 350 square feet per manhour
Spray SPF 2 inches thick: 200 square feet per manhour

Example Estimate

For this example, let's assume you have to cover an existing 3,000-square-foot (30 square) low-slope roof that has a smooth, sound surface. Let's also assume this is one of those rare jobs with no complications, no deteriorated sheathing, no parapet walls, no access problems or the like. The job requires sweeping and pressure-washing the surface, a coat of primer and then rolling on two coats of silicone roof coating.

Assume, for the purpose of this example, we'll use primer that costs $50 per gallon and silicone coating costing $60 per gallon when bought in a 5-gallon bucket. We'll also assume one gallon of primer will cover 500 square feet and one gallon of silicone coating applied in two coats will cover 50 square feet at 22-mil thickness. We expect coverage of 90 square feet per gallon for the first coat and 112 square feet per gallon for the second. Numbers are rounded to make calculation easier.

Primer:
3,000 divided by 500 (6 gallons) x $50 = $300

Silicone coating, first coat:
3,000 divided by 90 (33.3 gallons) x $60 = $1,998

Silicone coating, second coat:
3,000 divided by 112 (26.7 gallons) x $60 = $1,602

Total material cost: $3,900

Labor Costs for Roof Coating

We'll use manhour estimates from the *National Construction Estimator*:

Labor to sweep and pressure-wash 3,000 sf at 1,000 sf per manhour will require 3 manhours.

Then the primer will be rolled on at 500 sf per manhour:

3,000 sf divided by 500 = 6 manhours

We'll roll on two coats of liquid silicone at an average of 200 sf per manhour:

3,000 sf divided by 200 = 15 manhours

Total manhours: 3 + 6 + 15 = 24

Using a roofing crew costing $35.99 per manhour (including taxes and insurance):

24 x $35.99 = $863.76

Our estimate of labor and material cost for this job will be:

$3,900 + $863.76 = $4,763.76.

Then add to your estimate time needed to: (1) walk the job, (2) figure the area, (3) calculate the materials required, (4) order and take delivery of materials, (5) mobilize materials, equipment and tradespeople to the roof, (5) clean up and (6) demobilize when work is complete.

These are your "hard costs". Your selling price has to include overhead, profit, contingency and supervision expense. On a typical job, hard costs may be only 60% of your selling price.

You won't find many projects as easy to estimate as this example. Most jobs will include at least some of the following:

- Deteriorated sheathing to be replaced
- Loose seams to be reinforced with poly mesh and seam sealer
- Depressions to be filled
- Corroded flashing that has to be brushed clean or replaced
- Voids and holes to be patched with poly mesh and an extra layer of coating
- Parapet walls, roof vents and installed equipment that requires hand brushing of primer and roof coat
- Poor roof access. Consider carefully how you will move tools, supplies and debris on and off the roof.

Each of these tasks will add to the cost of the job and must be included in your estimate. If the roof has a granule or rock ballast topping, that has to be shoveled and brushed off the roof and hauled to a legal landfill. Dump fees are usually assessed per ton. If the roof has a rock ballast surface, expect to shovel and brush off about 400 pounds of rock per square. On a 30-square job, that's 6 tons of granules. Disposing of this much weight can be expensive. Check fees with your local landfill operator.

Tools of the Trade

All roof coating "pros" need specialized tools. Most roofing jobs will require at least a few of the tools on this list.

Deck moisture scanners can detect moisture under the roof surface. Not all leaks are obvious. Use a moisture scanner to find places where water may have damaged the membrane, insulation, sheathing or framing. This can be important where building codes restrict or prohibit roof coating if more than 25 percent of the roof area has detectable moisture under the surface. The Florida building code prohibits applying any roof coating if the surface below has more than 5 percent moisture content in the membrane or more than 8 percent moisture in the insulation package.

Thermal cameras can be useful on commercial jobs. Thermal images can detect leaks of hot or cold air. Any river of heat or cold passing through a roof raises questions: Why is that happening? Is that what we want? If not, what can we do to stop wasting resources or delay premature failure?

Mil thickness gauge will be essential if job specs require a minimum coating thickness in mils. Hand-held ultrasonic thickness gauges are available from many roofing suppliers.

Core cutting tools can take a sample of the existing roof deck. Take a core sample to identify the condition and materials used in the existing roof deck.

Industrial grade scissors have a stainless steel blade that cleans easily, stays sharp and cuts poly fabric easily.

Leaf blower/vac is useful when removing loose debris (vacuum mode) and dispersing puddles (blower mode) or when tree branches overhang the roof. Use a leaf blower to clear away debris that falls ahead of your coating crew.

Seam rollers are useful when you need extra weight to ensure good adhesion on seams, in corners, around flashing and in tight areas.

Tank spreaders can save time on some jobs. Smaller tank spreaders have a capacity of about 15 gallons. Larger spreaders hold up to 40 gallons. Working from a tank spreader tends to even out coverage.

Sprayers can apply either single-component or two-component coatings. The airless sprayer you select should be matched to the coverage and pressure recommended by the coating manufacturer. Some sprayer manufacturers offer applicator training for contractors who buy their equipment.

Squeegee spreaders are used to distribute roof coating on a smooth surface. Applicators pour roof coating material in lines across the roof, spread the lines with a squeegee and then roll the spread material into a smooth top coat.

Roller or Spray?

Both have advantages and disadvantages. If either roller or spray is recommended for the roof coating you select, that's your best choice. Absent any recommendation, here are points to consider:

- Some roof coatings (such as SPF) must be applied with a sprayer.
- Spray application is faster, but setup and cleanup take longer. And you have the cost of the spray rig.
- Good spray operators need training, practice, and experience. This is especially true when applying silicone coatings or when system pressure, fluid viscosity, and spray distance are critical.
- You may need more than one sprayer. Neoprene roof coatings usually require specialized spray rigs.
- Airless spray rigs are dependable if well-maintained. But a breakdown can spoil your day.
- On smaller jobs, a roller may be a better choice because you don't have the lengthy setup and cleanup work a spray rig requires.
- With a roller, there's no heavy equipment to lug around.
- Application with a roller requires little extra training.
- Roller application is by far the best choice on windy days (over 10 MPH).

Granted, most roof coating contractors use spray equipment. But some very professional roof coating contractors prefer a roller on most jobs. They feel a rolled-on job gives their customers a better-quality roof. Back-rolling the sprayed surface may yield a smoother, more even mil thickness.

14 Roofing Repair and Maintenance

▶ Leaks can occur almost anywhere on a roof, and for any number of reasons. Severe weather which brings rain, snow, wind, and hail is hard on roofs. Falling objects, foot traffic, and accumulated debris all take a toll. And some roofs leak because they're just plain worn out. Proper installation is the best prevention for many kinds of damage, and helps keep a roof sound over many years. Maintenance is the second best. Offer your customers periodic follow-up inspection and maintenance services, and you'll detect most roof damage before it becomes a major disaster.

People usually only notice a roof leak after water begins to drip through the ceiling. But the source of that leak is rarely directly above where you see the evidence. Often, water comes in under flashing or shingles, then runs over or below the sheathing and down a rafter. Sometimes it even runs the length of a rafter and into an outside wall. That kind of leak can go unnoticed until it's done a lot of damage.

Finding the Source of Leaks

When you look for a leak on an asphalt shingle roof, start by looking for worn shingles with dark patches on them. This indicates the loss of surface granules. Also watch for curled or damaged shingles.

On wood shingle or shake roofs, look for missing or rotted shingles. Watch for cupped or curled shingles which could allow the entry of wind-driven rain under the shingle.

On built-up roofs, search for deterioration or delamination of felts, especially in low spots where ponds occur. Look for blisters, and inspect for loose or deteriorated flashing and gravel stops.

On tile roofs, watch for broken or missing tiles. Remove and replace broken tiles and replace missing ones.

On metal roofs, inspect for loose or damaged panels. Remove and replace panels which aren't sound.

On any roof with a valley, clean out all debris trapped in the valley. Search for worn or broken flashing material, or flashing material installed off-center.

Examine every roof for loose or missing ridge shingles. Those are more susceptible to loosening than field shingles. Around flashing and vent pipes, look for cracks and gaps in roofing cement. At chimneys, watch for cracked and loose mortar at the cap flashing.

Attic Inspection

When you're trying to locate a roof leak from the attic, start above the ceiling leak and follow water stains on the deck to the potential source. Darken the attic and watch for a ray of sunlight which could indicate a nail hole. Also, have a helper hose down the roof while you watch for a leak from inside the attic. If you find a nail hole, mark the location by pushing a nail or straightened coat hanger through the hole, where you'll be able to see it from atop the roof.

At Chimneys

Leaks often occur around chimneys when water runs down the flue or through the mortar on top of the chimney (the seal cap), detouring through an open joint and onto the sheathing. You can prevent this by installing a flue cap or by sealing the flue cap with roofing cement, followed by an application of felt painted with aluminum paint. Sometimes, a new layer of mortar is required at the top of the chimney. You can prevent water from entering mortar joints by tuck-pointing the joints and applying a brush coating of clear silicone sealer.

You can keep water from entering around flashing with a heavy application of plastic cement, followed by a strip of roofing felt painted with aluminum paint. Use roofing tape layered with asphalt cement to patch flashing and asphalt roofing materials. The tape is made from asphalt-saturated cotton, fiberglass, or some other porous fabric. Roofing tape comes in rolls up to 50 yards long and 4 to 36 inches wide.

Leaks occur around counterflashing during blowing rain because the counterflashing doesn't lap far enough over the base flashing. Loose mortar above or in flashing joints can also allow leaks at counterflashing. Scrape the joints clean to a depth of $1/2$ inch and fill them with butyl rubber caulk or with fresh mortar.

At Eaves

Roof leaks occurring at the eaves of a roof are usually the result of a roofing contractor installing shingles with inadequate overhang. The end result is the rotting of fascia trim, roof deck and soffit. If the roof is framed with a closed cornice, water can also seep in and damage the outside walls, floor covering and wood floors.

At Valleys

Figure 14-1 Roof penetration in a valley

Valleys are especially vulnerable to leakage. You can prevent most of these problems during installation. For instance, clip (*dub off*) sharp shingle corners in the valley. (Refer back to Chapters 4 and 6 for details about how to shingle a valley.) If you don't cut the corners, any water flowing along the edge of the shingle can hit a sharp corner and run along the top of the shingle until it finds a place to leak. The location of the leak is often far from the valley, so it's hard to find and repair. That's why it's sometimes a waste of time to try to find a leak by searching the attic for evidence of water entry.

Other valley problems that result from poor installation include:

- roof penetrations in a valley, as shown in Figure 14-1
- valley flashing installed off-center
- full-lace valley with incorrect installation of overlapping shingles

Leaves and other debris collecting in a valley or gutter can also cause a valley to leak. Water dams up and backs up under shingles adjacent to the valley. Also, debris retains moisture, which promotes the growth of fungus and mildew. Either will degrade shingles and metal. You can prevent this by periodically cleaning the valley and gutter, especially during autumn when leaves fall.

To repair a break in an asphalt shingle valley flashed with roll roofing, apply a coating of roofing cement. Repair an opening in a metal valley by embedding aluminum patching tape into roofing cement. To prevent potential leaks in valleys, embed the corners of all shingles within a valley in roofing cement, then caulk the edges of the shingles.

A valley must sometimes be replaced because of hail damage or sloppy installation. If the valley is 90-pound roll roofing, leave the old valley intact and just add to it. Apply a ¼-inch-thick layer of roofing cement, then slide a new valley strip over the old valley and under loosened shingles adjacent to the valley, starting at the bottom of the valley. Renail the loose shingles and replace any missing ones.

At Vent Flashings

Leaks around vent flashings are very common and are usually the result of improper shingling around the flashing. Install shingles *under* the lower end of the flashing instead of on top. Otherwise, water can be diverted under the shingles. Also, dirt and granules can collect under the tabs, leading to rusting and holes in the metal flashing.

Leaks at vent pipes can also result from roofing cement that has dried and cracked. Remove all loose material and reapply roofing cement.

On tile roofs, leaks at vent pipes are usually caused by mortar that's cracked around the base of the vent. As a temporary remedy, apply roofing cement over the problem area. However, a long-term solution is to seal the base of the vent with new mortar.

Miscellaneous Leaks

When a leak occurs at mid-roof, the cause is often an exposed nail hole, a joint occurring in a low spot or a joint in the shingles where the exposure is greater than 5 inches. One type of leak that's easy to find from inside the attic is one caused by nails driven through the shingles and deck where the homeowner installed Christmas decorations or antenna guy wires on the roof.

One big source of leaks and roof damage is low tree branches that sweep across the roof during high winds. The damage is even greater when branches sag under the weight of snow or rain.

Repairing Leaks

Asphalt Shingle Repair

You can spot repair minor cracks or worn areas on asphalt shingles by applying roofing cement and sprinkling on loose granules and rubbing them into the cement. Either buy tubes of granules from a roofing supplier, or make your own by rubbing two shingles together. If appearance isn't important, you can leave off the mineral granules.

If an asphalt shingle is curled, apply a coating of roofing cement beneath the shingle, and press the shingle into the cement. If you have to use nails to keep the shingle in place, cover the nail holes with roofing cement.

If only a single tab is damaged, raise the tab with a flat pry bar and remove the nails. You might also have to loosen the shingle above the damaged tab. Cut out the damaged tab and nail down a replacement. You can buy special asphalt plastic roofing cement for sealing and patching cracks and breaks in roofing. It comes in 1-quart or 1-gallon cans, or in 5-gallon pails. Apply this cement over nail holes and the cut joint.

You can use this same procedure to replace an entire 3-tab shingle. I recommend that you replace the entire shingle, never just a single tab. Either use a hacksaw to cut nails which hold shingles immediately above the damaged shingle, or notch the top of the new shingle to fit around the nails. Cement the new shingle in place. If you notch the shingle, cement the overlying shingle on top of it.

You can also repair a damaged shingle by installing galvanized sheet metal beneath the shingle and nailing the metal into a bed of roofing cement. Cover the nail heads with plastic cement.

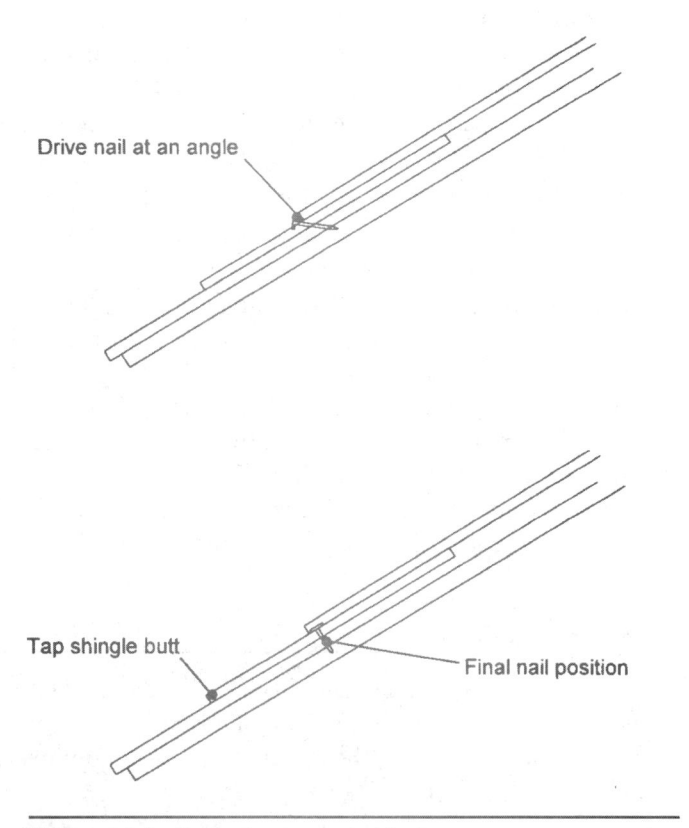

Figure 14-2 Replacing a wood shingle

Warm weather is the best time to repair asphalt shingles. When they're warm, asphalt shingles become pliable, so you're less likely to damage them. Never apply plastic cement to the bottom edge of a shingle to seal it, since this will allow water to be trapped and back up to cause a leak.

Repairing Wood Shingles and Shakes

You can patch small cracks in wood shingles with roofing cement. Small cracks in shakes are rarely a source for roof leaks since the underlayment (not the shake) provides the primary weather protection.

To replace a wood shingle or shake, use a hacksaw to cut the nails that hold it in place. Then you can usually pull it out by hand. Trim a new shingle to size, push it into place and nail it at an angle, just below the overlying shingle. Tap the butt of the new shingle to align it with the existing shingles and hide the nail. See Figure 14-2. Cover all exposed nail heads with roofing cement.

The most common leak on a wood shingle roof occurs where joints or splits line up in three adjacent courses. Splits in an individual wood shingle can also be a source of a leak. The easiest way to repair leaks caused by splits or joints is to install a flat piece of galvanized metal under the joints and splits. To keep the metal sheet in place, make a 90-degree bend in the bottom edge, just enough to give it friction to stay where you put it — about $1/2$ inch. Don't nail the metal because the nails will gradually work loose and leave another opening for water to get in.

Wind-blown rain can enter under bowed or cupped wood shingles or shakes. To eliminate this problem, split the shingle with a chisel and remove a $1/4$-inch-wide sliver, then nail down the two resulting shingles with a nail on each side of the split. Cover the nail heads with roofing cement and drive the nails as close to the top edge of the shingle as possible. Don't nail the shingle near the butt because that causes the shingle to raise at the tapered end and dislodge the shingle above it.

Slate Repairs

To replace a broken slate, insert the new slate underneath the two overlying courses and nail it into place. See Figure 14-3. Use these nail location guidelines to place nails correctly:

- 5 inches down from the head of the *new* slate
- 2 inches down from the butt of the *second* overlying course
- through the vertical joint between slates in the *first* overlying course

In Figure 14-3, the nail is covered with a 3-inch by 8-inch copper strip. This metal strip is bent so that it acts like a flat spring to hold the new slate tightly in place beneath the overlying courses.

Be sure to use a slate that's compatible with the original roof. Match the color as closely as possible. Use a slate with the same weathering properties (weathering vs. unfading) and surface texture as the existing slates. If you're able to identify the quarry that made the original slates and they're still operating, order the new and replacement slates from the same place.

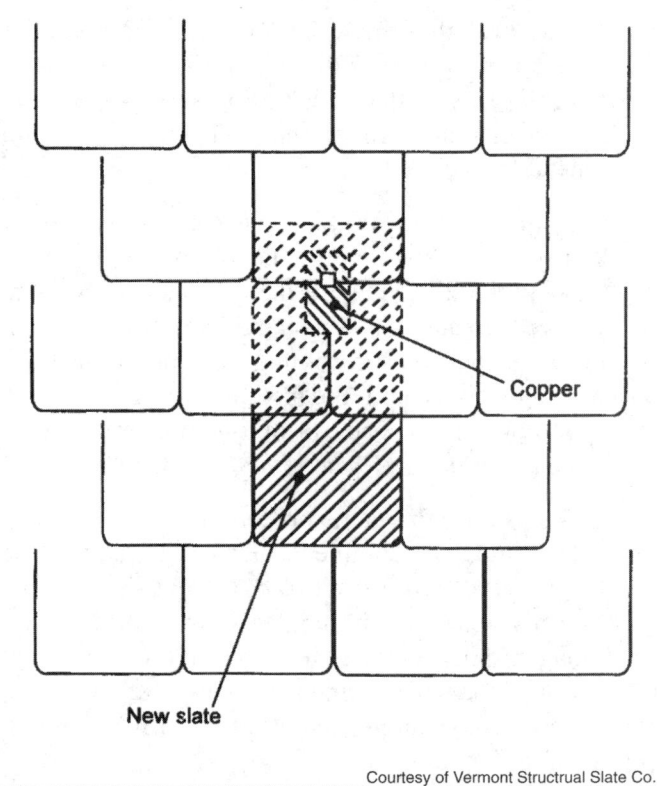

Courtesy of Vermont Structrual Slate Co.

Figure 14-3 Proper method of inserting a new slate

Roof Maintenance

Encourage your customers to regularly inspect and maintain their roofs (or hire you to do it). Here are some tips to help make a roof look good, perform well and last for years:

- Never paint or coat a roof to change the color.
- Keep roof surfaces and gutters clean so water will drain quickly and freely.

- Never let water from a downspout pour directly onto a roof below.

- Keep trees trimmed so they don't rub against the roof covering. Also, keep climbing plants trimmed back from the roof.

- Remove snow and ice from the roof carefully so you don't damage the roof. Use a broom with an extension pole. Never climb onto a wet or snow-covered roof.

- Being walked on is never good for a roof. Keep it to a minimum. Remind service people to be careful and watch them to make sure they do so.

- Make sure that accessories such as antenna wire or anchors are made of noncorrosive material to prevent metal discoloration or "iron staining" on the roof.

- Never pressure-clean an asphalt shingle roof. The pressure will remove granules and each treatment will take three years off the life of the roof.

Algae Discoloration

The discoloration of asphalt roofing materials due to algae is a common problem in humid parts of the country. You can see evidence of this condition (often called *fungus growth*) in Figure 14-4. If left unchecked, algae can turn a light-colored asphalt shingle roof to dark brown or black. Algae also discolors other roof coverings, including wood shingles, shakes, tile, and built-up roofs. In humid areas, consider using one of the wide variety of algae-resistant roofing products now available.

Figure 14-4 Shingle discoloration caused by algae

A discolored roof can be lightened, at least temporarily, by bleaching. First, cover any plants that may be affected by the run-off. Sponge or fog spray a dilute mixture of chlorine bleach (about 50/50) onto the roof surface, then immediately rinse it down with clear water.

Efflorescence on Concrete Tile Roofs

Efflorescence (or *lime bloom*) sometimes occurs on concrete products. The free lime in concrete goes through various chemical reactions during manufacture, resulting in a white chalky deposit on the surface of the tile. This deposit is only temporary. Within a year, rainwater dissolves and carries away the deposit, restoring the tiles to an even appearance. Efflorescence is purely superficial and in no way affects the strength or other functional properties of the roof tile.

Assessing Hail Damage

A common cause of homeowners' insurance claims is hail damage to a roof. If the roof is actually damaged, it should be repaired or replaced. But some homeowners assume that they're entitled to a new roof after any hailstorm, whether the roof has been damaged or not. If the homeowner pushes the issue to the point of filing a lawsuit, the insurance company will often replace the roof to avoid a trial, since juries are seldom sympathetic to an insurance company.

The sad result, however, is that we all help pay to replace the undamaged roof. Insurance companies base their premiums on the cost of doing business in a local area. If there is an inordinate number of claims they have to pay out on, premiums will go up.

Here is a guide to assessing hail damage that's fair to both the homeowner and the insurance company.

Hail Damage to Asphalt Shingles

Hail damage to asphalt shingles leaves such obvious signs that a simple visual inspection of the roof lets you determine whether a roof is damaged or not. Here's what to look for:

- Dark spots on the shingles' surface. These spots mark hail impact points and they look dark because the impact pops the mineral granules loose from the shingles' surface. Spotting is easy to see on light colored shingles, but it's not always obvious on darker color shingles.

- Pitting that's visible on the shingles' surface. The pit locations correspond to the dark spots because they are also a result of the hail impact. Pitting is visible on shingles no matter what color they are.

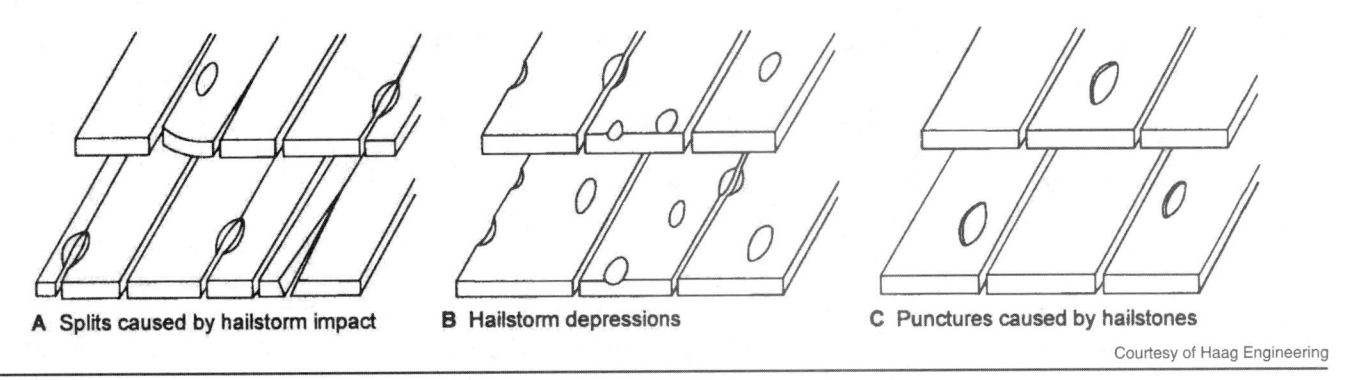

A Splits caused by hailstorm impact **B** Hailstorm depressions **C** Punctures caused by hailstones

Courtesy of Haag Engineering

Figure 14-5 Shingle damage caused by hail

- When touched, the pits and dark spots feel soft, like a bruise or soft spot on an apple.

Hail damage is basically cosmetic, although it can shorten the life of the roof if it's severe. Some insurance companies pay a percentage of the total value of the roof for repair.

Hail Damage to Wood Roofs

Hailstones vary in size, shape and hardness, so impact marks vary in size, shape and depth. Wind-driven hailstones do the most damage, so the windward side of a roof sustains more damage than the leeward side. Examine roof penetrations such as roof vents, soil stacks and heater flues to see an indication of the nature of the hailstones. If thin-sheet components aren't dented, the hail probably didn't damage the shingles or shakes.

Hail damage (versus normal weathering) on a wood shingle or shake roof isn't always easy to identify. To arrive at some uniform way to assess hail damage on wood roofs, Haag Engineering Company of Dallas, Texas has cataloged damage data by observing shingled test panels exposed to various simulated hailstorms. Testing has been conducted since 1963.

One important observation in the Haag study is that flatgrain and slashgrain shingles tend to split and warp more than edgegrain shingles. Curled or warped shingles are more apt to be split by hail impact than flat ones. Hailstones don't significantly affect sound shingles. And edgegrain shingles aren't as likely to split, because they don't warp as much. But when shingles do split, the characteristics of the split are the same for flatgrain, slashgrain, or edgegrain shingles.

When hail causes splits, the splits nearly always match impact marks, and the exposed surface of the split is the color of fresh wood throughout. See Figure 14-5 A. The exception is hailstone impact against a warped shingle, which sometimes causes a split slightly offset from the impact mark. In this case, thumb pressure in the hail-caused dent opens the split to reveal fresh-colored wood.

Assess roof damage as soon as possible after a storm because hail-caused splits are most obvious then. Impact marks fade as surface wood fibers recover. Inspect the roof when it's dry because it's easier to see new splits then. That's because wet wood swells, so the splits tend to close up. Note that splits caused by hail occur at the moment of impact, and hail-caused depressions which don't cause splitting at the time will not cause future splitting (Figure 14-5 B).

Hailstones can sometimes actually puncture a shingle. Punctures are usually rounded and the exposed underlying wood has a fresh wood color (Figure 14-5 C).

■ **Natural Weathering** After 10 years' weather exposure, about a third of all edgegrain shingles, and about two-thirds of slashgrain and flatgrain shingles, split naturally. These splits usually begin, and are widest, at the butt of a shingle, though some splits originate at nail holes.

Natural splits normally have rounded edges and eroded interiors. These splits usually have a V-shaped cross-section and don't extend the full depth of the shingle. Wood near the bottom of the split often has a fresher wood color because it hasn't been exposed to the weather for as long.

The extent of weathering depends on how steep the roof slope is as well as the roof slope direction. South-facing shingles installed on a low roof slope will weather fastest and the most.

■ **Repairing Wood Roofs** Shingles with hail-caused splits which aren't aligned with joints or weathered splits normally need not be repaired. You can repair a wood shingle or shake by replacing it, or by under-shimming the damaged shingle or shake with a galvanized steel sheet, aluminum sheet or a roofing felt shim. A roof is considered beyond economical repair when the repair cost exceeds 80 percent of the replacement cost. This equates roughly to 30 hail-caused splits per square for cedar shingles, and 25 hail-caused splits per square for cedar shakes.

It isn't practical to repair old, badly-weathered roofs. In general, it's better to replace shingle roofs older than 20 years, shake roofs older than 25 years, and roofs damaged by severe weather conditions.

Roofing Demolition

If a tear-off is necessary, strip shingles down to the deck. When you remove any roof, start the tear-off at the ridge and work down the slope, as shown in Figure 14-6. This method is especially necessary when you remove wood shingles or shakes installed over spaced sheathing. Otherwise, broken material will fall through the open sheathing into the attic.

Figure 14-6 Remove shingles starting at the ridge

Dislodge wood shingles, shakes, slate and metal roofing panels with a crowbar, removing two to three courses at a time. Remove tile shingles by hand. When you remove asphalt shingles, roll roofing or a built-up roof, use a flat spade shovel to pry up material. File a notch into the blade of the shovel so you can remove shingle nails along with the shingles. If you don't remove the nails, parts of the old shingles will remain under the nails and make the deck surface uneven and bumpy. And the old nails will pop up through new underlayment unless you use roll roofing for underlayment.

If the shingles don't fall apart when you remove them, stack them into bundles for removal. If the shingles are old and brittle, collect and remove them with a scoop shovel and carry them to the edge of the roof with a wheelbarrow.

Some roofers prefer to take off the paper and asphalt shingles all at the same time. Then they can roll the shingles into the paper and roll the whole lot down the roof as a unit. This method reduces the amount of mess left to clean up at the end.

If you have to tear off more than one layer of shingles, remove only one layer at a time. Use a spud or a spud hammer (Figure 14-7) to remove a gravel roof. Spread tarps along the ground under the edges of the roof to help collect debris falling off the roof.

Figure 14-7 Spud hammer

If you must remove siding during a tear-off, for instance at a dormer wall, mark and store the pieces in an ordered way, so you can reinstall each piece in its original location. Use existing nail holes as a guide when replacing the material.

After you've removed the old shingles and underlayment, level and repair the deck, if necessary. Replace rotted, damaged and warped sheathing, or delaminated plywood. Cover all large cracks, knotholes, loose knots and resinous areas with sheet metal patches nailed to the deck. Remove all loose or protruding nails or hammer them into the deck. Some roofing contractors recommend that gaps between spaced sheathing be filled in with boards, or that the entire roof area be covered with plywood. You don't necessarily need a solid deck under new shakes, and in humid climates it's a good idea to keep the spaced sheathing.

If the fascia board or trim is broken or rotted, or if the drip edge is broken or rusted, now is the time to replace them. To install a new drip edge when you're re-roofing over existing shingles, trim the old shingles back from the edge of the roof. If the old roof is wood shingle, cut the shingles back from the eaves and rakes $5^1/_2$ inches and install a 1 x 6 in their place over the new drip edge. That's shown in Figure 14-8. Also, remove the old hip and ridge units. If there's a concealed metal cover underneath, don't attempt to remove it because it would probably take a lot of shingles with it. Instead, look ahead to Figure 14-13 to see how to install beveled siding over each side of the ridge.

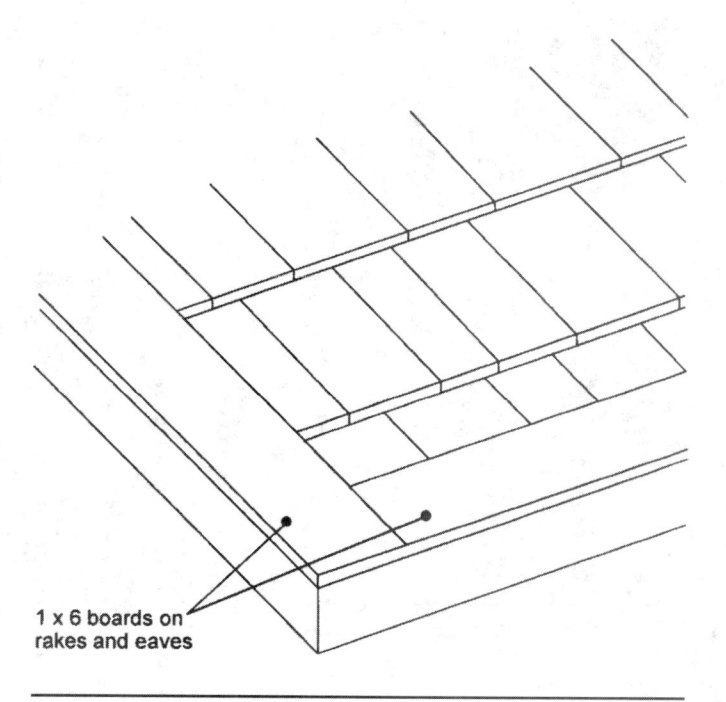

Figure 14-8 New edging strips

Here's a list of pointers for various re-roofing situations:

- When you re-roof a roof that has ribbon courses, remove the entire top shingle (not just the exposed part) of each ribbon course by hand to avoid humps on the new roof.

- On asphalt shingle re-roofs, old flashings can often be raised. When this isn't possible, cut new shingles to fit around the old flashing, and seal the flashing with plastic cement. Paint the cement and vent with aluminum paint.

- When you tear off a gravel roof, you'll usually have to replace the vent flashings. New metal flashings and valleys are a must when you re-roof with wood shingles or shakes. That's because the bond between wood and plastic cement is only temporary, since wood will soak up water and release the plastic cement.

- Carefully remove any metal flashing demolished during a roof tear-off and use it as a pattern for new flashing.

- If metal counterflashing at the chimney and other vertical surfaces hasn't deteriorated, try to temporarily bend it up and out of your way and reuse it.

- Use a flat pry bar to remove a gutter and reassemble it on the ground the same way it was installed at the eaves. When you reinstall the gutter, drive the spikes into fresh wood, instead of into the old hole.

- If the roof deck is warped, but otherwise in good condition, straighten the deck by driving nails through the old shingles into the rafters.

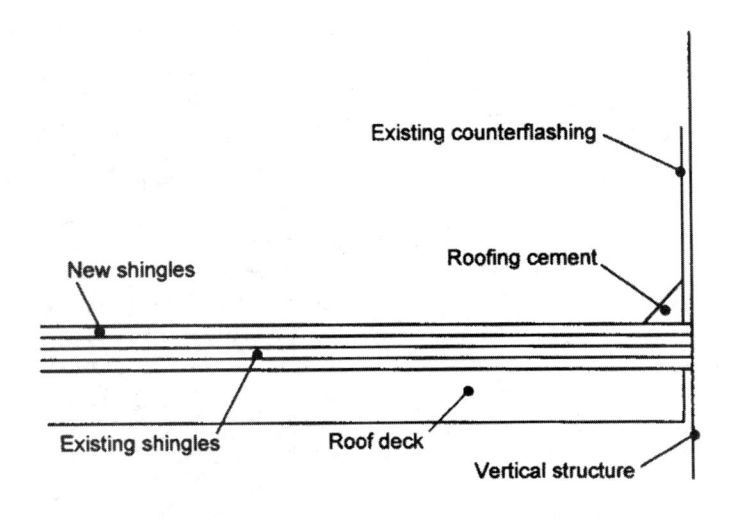

Figure 14-9 Sealing a roof-wall juncture with plastic cement

New Flashing

On a re-roof, try to install new flashing under existing metal counterflashing if you can. But sometimes that's impossible. Then you'll have to replace both the flashing and the counterflashing. That adds considerably to the cost. Here's a relatively inexpensive alternative. Seal the top and outside surfaces of the new metal flashing with roofing cement (also referred to as *bull*), then lay in a 4-inch-wide strip of felt and smooth it into the roofing cement. Then coat the felt with a $1/8$-inch-thick layer of roofing cement and spray the coating with a good grade of aluminum paint.

If you don't install new metal flashing, seal the juncture between the horizontal and vertical surfaces with a $1^{1}/_{2}$-inch-wide bead of roofing cement applied at a 45-degree angle with respect to both surfaces. See Figure 14-9. Cover the joint with aluminum paint for longer-lasting protection.

When it's not economically feasible to install new counterflashing at a chimney on a wood shingle or shake roof, you must install step flashing. On a shake roof, install metal flashing 13 inches long and extend it 4 inches up the wall and 4 inches under the shingle. On a wood shingle roof, install metal flashing 7 inches long and extend it 2 inches up the wall and 3 inches under the shingle. Reinforce the juncture with a felt strip, then cement and paint the joint as described above.

When you install flashing against a vertical side wall using asphalt shingles over old asphalt shingles, terminate the new shingles within $1/4$ inch of the existing flashing and embed the shingle ends in a 3-inch-wide bed of roofing cement. Then seal the joint with a bead of cement.

When you install flashing against a vertical side wall using asphalt shingles over old wood shingles, nail an 8-inch-wide strip of 50-pound smooth roll roofing over the old roof at the juncture of the shingles and the vertical wall. Drive nails in two rows on 4-inch centers along both edges

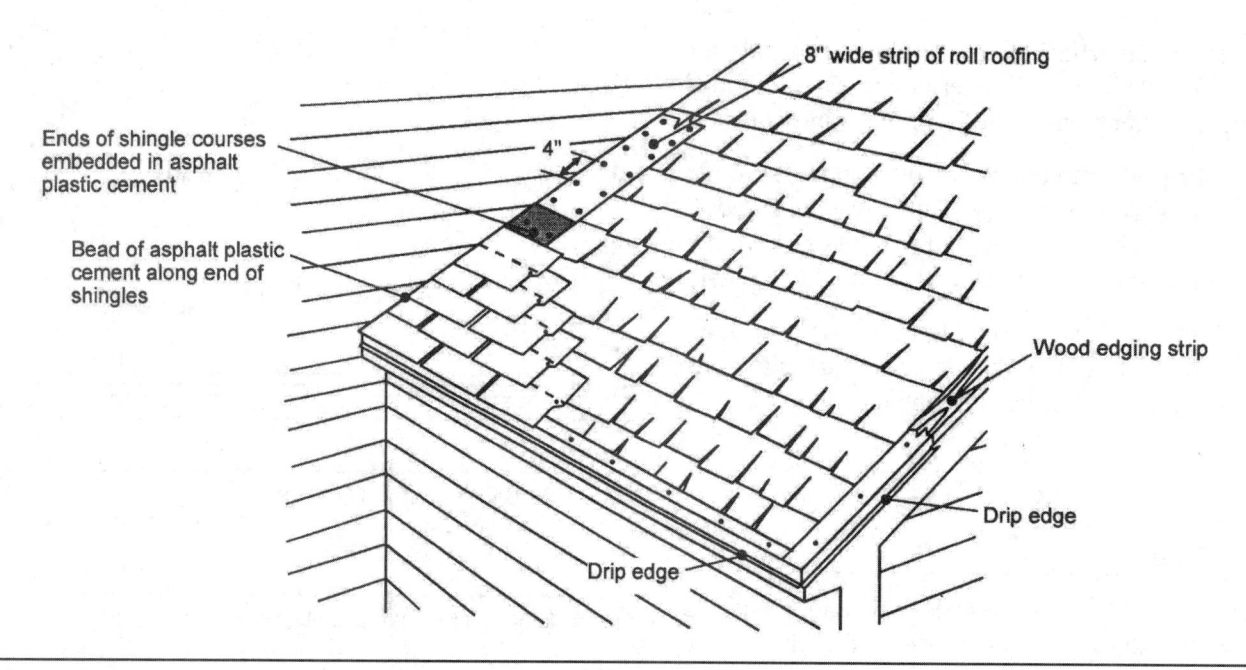

Figure 14-10 Flashing against a vertical wall when re-roofing over wood shingles

of the strip. Always use roofing nails when you re-roof over an existing roof. Embed the ends of new shingles in roofing cement applied to the strip, then seal the end joint with a bead of roofing cement, as in Figure 14-10.

When you install a metal valley on a shingle re-roof, don't use water guards at the metal edges because it won't be flat enough to produce a smooth roof.

If an existing asphalt shingle roof has an open valley, build up the exposed part of the existing valley to the level of the existing shingles by installing a 90-pound mineral-surfaced roll roofing filler strip. Then you can install a new open valley, woven valley or a closed-cut valley.

Re-Roofing

Whether or not you must tear off the existing roof covering depends on the type of roof or the number of layers or weight of the existing roof covering. Re-roofing with no tear-off obviously takes less time and no underlayment is required between roofs. Also, the existing roof provides additional protection and insulation.

Three layers of shingles is usually the most a roof can support. Most building inspectors as well as many insurance companies follow this rule. However, be cautious about relying too much on that three-layer guideline. There are many jurisdictions whose building code doesn't permit more than one roof covering. There, you *have* to tear off the old roof. And, there are plenty of roofs that aren't structurally sound enough to carry the load of two layers of shingles, let alone three. The load-bearing capacity of a roof depends on many factors, including the rafter size, strength and spacing and the strength of the roof sheathing material. Also, remember that the per square weight of shingles varies not only with shingle type, but also with the shingle exposure; the less the exposure, the greater the weight of the roofing. Some high-end shingles are not only extra heavy, but are laid with a short exposure, resulting in an extremely high weight per square. You don't want to put a second layer of this on top of another unless you're really sure of the strength of the structure.

If you're dealing with a wood roof, keep in mind that wood shingles and shakes absorb water. A wood roof that's wet weighs far more than it does when it's dry. If you have any doubt about the ability of a particular roof to support another layer, consult a structural engineer. Of course, it might be cheaper to just tear off the existing roof, especially since the engineer might say to tear it off anyway.

Incompatible Substrates

The existing roof covering determines not only whether or not tear-off is necessary, but also what roof covering options are available when no tear-off is required.

You must completely remove certain types of shingle roofs before you install a new roof covering. Because of the irregular surface, never attempt to install new shingles over a shake roof. Never install new shingles over tiles, slates or metal panels because it's too hard to nail down the new roof. A built-up roof with an aggregate surface isn't often resurfaced with shingles due to its rough surface. Also, the roof slope is usually too low for most shingles and the combined weight of aggregate, and new roofing is often too much for the roof frame. *Always* ask a structural engineer before you take on a job like this.

Asphalt Strip Shingles over Asphalt Shingles

You can install new asphalt strip shingles over old ones if you first smooth the substrate by nailing down warped shingles and replacing missing ones. Split a warped shingle and nail down the resulting two shingles. A new asphalt shingle roof will sag over missing shingles.

When you re-roof with asphalt shingles over asphalt shingles, the easiest way is to match the existing shingle pattern. So if you're replacing a roof that has old-style shingles like T-lock, giant individual or hexagonal shingles, you're almost certainly going to have to do a tear-off. I doubt if you can get those shingles any more.

When you re-roof over asphalt shingles, no underlayment is required because the original roof serves that purpose. When you re-roof over existing wood shingles, some building codes require that you install 30-pound felt or synthetic underlayment between the existing and new roofs.

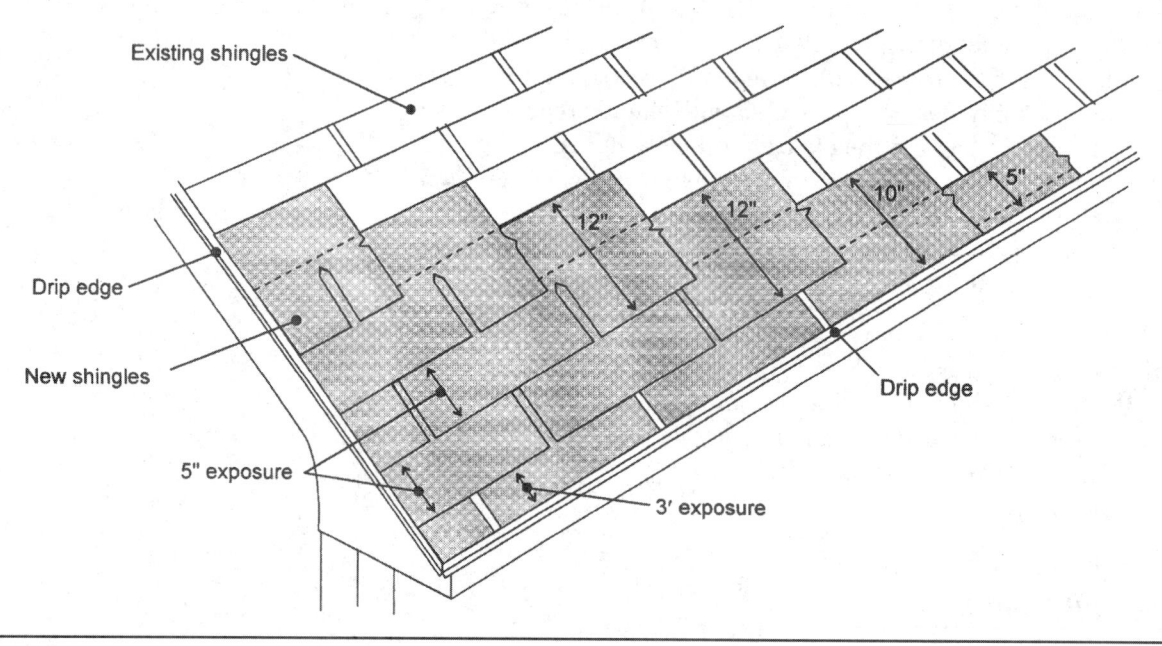

Figure 14-11 Application of new asphalt shingles over existing asphalt shingles

The "Butting-Up" Method

When re-roofing over wood, asphalt shingles or roll roofing, install the new roof using the "butt-up" (or "butt and run") method. This means you install the tops of the new shingles flush against the butts of the old shingles, as shown in Figure 14-11. I recommend that you use this method because it's the easiest way to avoid all the problems associated with excessive shingle buildup. A re-roof with excess shingle buildup is unattractive. Furthermore, such a roof is more easily damaged by hail and foot traffic.

For these same reasons, it's equally important to offset overlaps when you re-roof with roll roofing over roll roofing.

When re-roofing using three-tab or strip shingles using the butt-up method, you can install the starter course one of two ways. The preferred method is to cut the shingle tabs and tops to make a strip whose width is equal to the exposure of the old shingles (normally 5 inches). Install the resulting strip over the exposed part of the first row of shingles of the old roof. If you're using self-sealing shingles, locate the factory-applied adhesive strips adjacent to the eaves. Remove about 3 inches from the end of the first starter-course shingle to prevent the cutouts and joints of the first course of shingles from being aligned over the joints of the existing starter-course shingles.

Overlay the new starter course by a first course of 10-inch-wide shingles made by cutting 2 inches from the shingle tops. This course will cover the new 5-inch starter course plus the 5-inch exposed part of the second row of shingles of the old roof. That's also shown in Figure 14-11. An easier, but less desirable, method is to cut and install two rows of 10-inch-wide shingles for both the starter course and first shingle course. You'll get a bulge along the bottom of the original second course of shingles. It's also less wind-resistant along the eaves. In either case, apply a spot of roofing cement under each tab of the first course of shingles for added wind resistance.

Install succeeding shingle courses using full-width shingles with their heads butted up against the butts of the old shingles. The full-width shingles will be 2 inches lower than those installed on the old roof. The exposure of the first shingle course is 3 inches and that of the succeeding courses is 5 inches. The difference in exposure isn't apparent, especially if gutters are installed at the eaves.

If the exposure of the old roof is greater than 5 inches or if the old roof is crooked horizontally, remove the old shingles. If the exposure is less than 5 inches, the quantity of new shingles required will be greater than that required for a standard 5-inch exposure roof. Here's Equation 4-2 from Chapter 4:

$$\text{Percentage-of-Increase Factor} = \frac{\text{Recommended Exposure}}{\text{Actual Exposure}}$$

Use this formula to find out how many shingles you'll need. For example, a 4-inch-exposure roof will require:

$$\text{Percentage-of-Increase Factor} = \frac{5 \text{ in.}}{4 \text{ in.}}$$
$$= 1.25$$

That's 25 percent more shingles. Using the same formula, a 4½-inch-exposure roof will require 11 percent more shingles.

▼ **Example 14-1:** A 30-square re-roof is required over strip shingles with a 4-inch exposure. Assuming that the recommended exposure is 5 inches, find the number of squares of shingles required to cover the roof with strip shingles with a 4-inch exposure.

From the formula above, the Percentage-of-Increase Factor is 25 percent. So you'll need 30 squares x 1.25, or 37.5 squares of shingles.

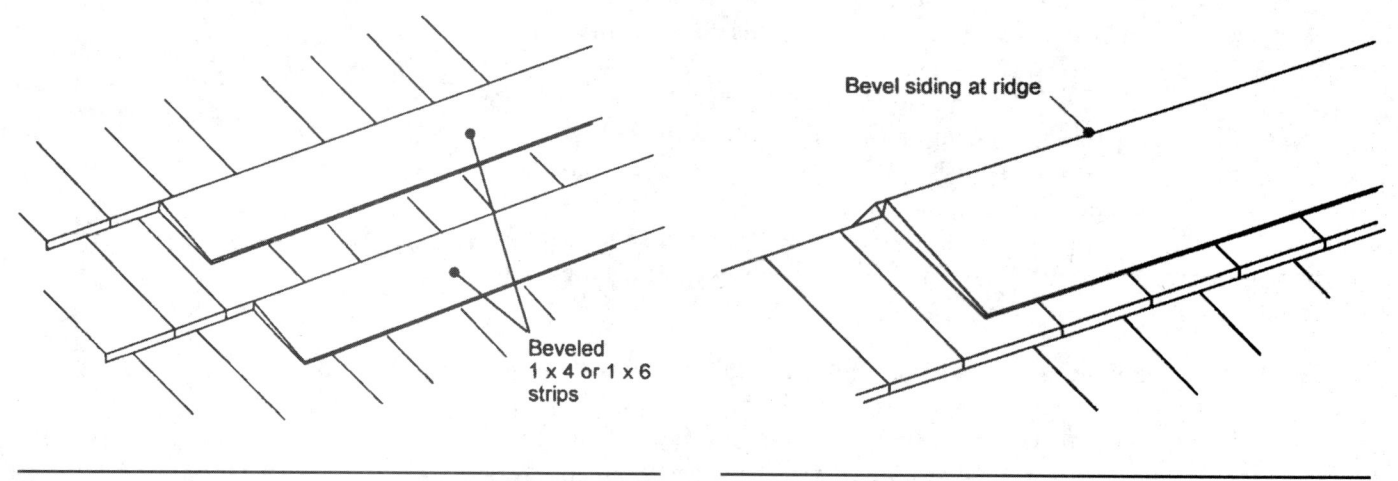

Figure 14-12 "Horse feathers" of feathering strips

Figure 14-13 Beveled siding board at ridge

Asphalt Shingles over Wood Shingles

If wood shingles don't provide a good nailing surface for asphalt strip shingles, but are otherwise in good condition, smooth and improve the surface by installing beveled 1 x 4s or 1 x 6s ("horse feathers" or feathering strips) against the old shingle butts. That's shown in Figure 14-12. Also, put a beveled siding board on each side of the ridge (Figure 14-13).

If shingles and trim at the eaves and rake are badly weathered in areas subject to high winds, you'll also need to trim the wood shingles back from the eaves and rake so you can install 1 x 4 or 1 x 6 edging strips (see Figure 14-8). Install a new drip edge at the same time. This lumber provides a smooth surface as well as a nailing surface for the new roof.

Asphalt Shingles over Built-up or Roll Roofing

You can install asphalt shingles over a built-up roof after you scrape off the aggregate surface, and provided the roof slope is at least 2 in 12. There's no need to reseal the BUR if the felts are in good shape. If there's rigid insulation between the sheathing and the felts, install a plywood nailing substrate over the insulation before you apply new shingles.

You can install asphalt strip shingles over roll roofing, provided the old surface is smooth and the roof slope is at least 2 in 12. Split any buckles or blisters and nail the roll roofing flat against the sheathing. Nail down any lapped joints that have separated.

Re-roofing with Wood Shingles and Shakes

You can apply wood shingles over any type of asphalt shingle, roll roofing or smooth-surface built-up roofing, provided the roof has a slope of at

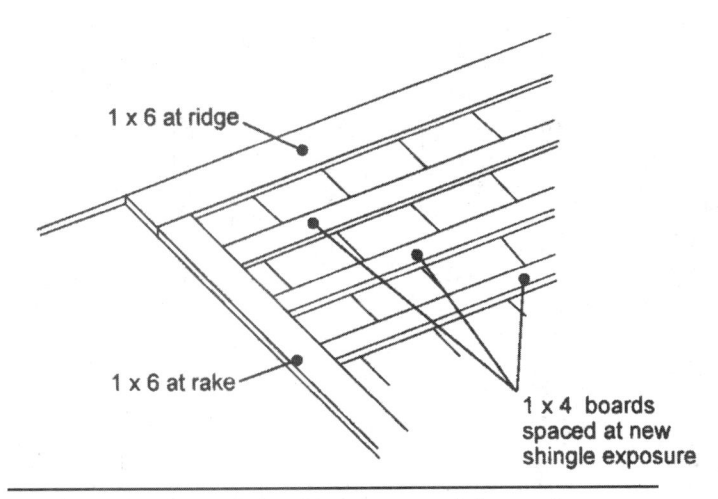

Figure 14-14 Spaced sheathing over an existing roof

least 3 in 12, and the substrate is capable of holding nails. You can install wood shingles over an aggregate-surfaced roof as long as the shingles are installed over spaced sheathing applied to the top of the old roof, as in Figure 14-14. In addition to spaced sheathing, you'll need to install a 1 x 6 along the eaves, rakes and on each side of the ridge. Also install 1 x 4s along the edges of the valleys in order to receive new metal valley flashing. Spaced sheathing has the added advantage of providing good air circulation.

You can apply new wood shingles over old wood shingles. If the surface is rough, install feathering strips against the old shingle butts (Figure 14-12).

You can install shakes over any type of asphalt shingles, wood shingles, roll roofing or smooth-surface built-up roofing that's not leaking, provided the roof slope is at least 4 in 12.

Re-Roofing with Roll Roofing

You can install roll roofing over roll roofing, provided the old surface is smooth. Never install roll roofing over any type of asphalt shingle or over an aggregate-surfaced built-up roof.

Re-Roofing with Metal Panels

You can install metal panels over all types of asphalt shingles, wood shingles, roll roofing and built-up roofs, provided that the roof has an adequate slope. Check the manufacturer's specifications.

Re-Roofing with Tiles

Be careful here. Some homeowners whose roof is asphalt shingles or roll roofing want a tile or slate roof. This can be done; you can install roofing tile over all types of asphalt shingles, roll roofing and built-up roofs if the old roof is fairly even or can economically be made so, and if the roof has adequate slope. But here's the kicker: a tile or slate roof is much heavier than asphalt shingles and requires heavy-duty trusses and thick decking material. If the house was designed to hold asphalt shingles or other relatively lightweight roofing, it's unlikely the builder would have gone to the unnecessary expense of heavy-duty trusses or extra-strong decking.

A look into the attic will tell you if the trusses or framing are heavy-duty or not. If you're in doubt or if the homeowner insists, bring in an engineer.

Re-Roofing with Slate

You can install slate over all types of asphalt shingles, roll roofing, wood shingles and built-up roofs. The roof slope must be at least 4 in 12 (unless flat-roof construction is desired) and the roof frame and sheathing must be heavy-duty, strong enough to carry the added load.

Slates used to re-roof over wood shingles must be long enough to span two courses of wood shingles in each course. Spanning two courses of wood shingles provides the slates with two points of support. That is, each slate will rest on two wood shingle butts. A shorter slate, spanning a single course of wood shingles, has a single point of support and wouldn't lie well. The slates you use for this re-roof job measure at least 18 inches long. You install them with four punched nail holes instead of the usual two. Although their length makes these slates heavier and more difficult to handle, their only other disadvantage is a tendency to break under foot traffic.

Estimating Re-Roofing

The material quantities required for a re-roof are about the same as those required for a new roof. When you re-roof (as compared to roofing over a new wood deck), there's more labor involved due to the demolition and shingle trimming required. But this may be offset by the fact that it's usually not necessary to install felt under a re-roof.

When you estimate a re-roof or tear-off and re-roof project, work out your estimate while you're on the roof. That way you can assess the old roof, including the condition and number of layers of shingles. Remember that it's easier to remove nails than staples, so notice how the old roof was attached. You can also check the condition of framing members, sheathing, fascia, drip edge, vents, vent caps and flashing. If the deck has deteriorated to the extent that it will no longer hold a nail, you'll have to replace that as well.

When you're going to tear off a roof, it's sometimes very hard to judge ahead of time how much sheathing you'll have to replace. This is especially true at the eaves, since you can't see that area from inside the attic. Estimate those repairs on a "cost-plus" basis.

You can often spot rotten deck areas by walking over the roof. A rotted deck will sag under your weight and will feel bouncy as you walk over it. That will be most apparent near the ridge, or on the north side of a house where the sun doesn't shine as directly.

Attic Ventilation

Insulation, weatherstripping and caulking make a home more airtight and confine water vapor in the house. Eventually, most of the water vapor passes through the ceiling and accumulates in the attic. In cold weather, in a poorly-ventilated attic, this warm moist air condenses when it reaches the cold underside of the roof sheathing. In hot weather, hot moist air tries to escape through gaps in the roof covering. The end result can be a buckled and rotten deck, deteriorated underlayment and blistered shingles. Many re-roofing jobs result from inadequate attic ventilation.

With proper ventilation, air circulates freely throughout the attic, carrying away the water vapor before it can condense. Never cover vent openings during cold weather, and make sure that soffit vents are not blocked by insulation.

Putting a new roof on a building which lacks proper ventilation is like painting over a rotten board. Call the owner's attention to any ventilation problem. That will save you money in the long run because you'll avoid expensive call-backs. It will also establish your reputation for know-how and quality workmanship. And besides that, a ventilation problem may void the shingle manufacturer's warranty.

Good attic ventilation must be not only adequate in area relative to the attic floor area, but balanced. The air intake (at or near the soffits) and the air exhaust (at or near the ridge) should be approximately equal.

In figuring the area of vent openings required, the standard practice, and minimum recommended by the FHA, is 1 square foot of clear opening per 300 square feet of attic surface *provided* certain conditions are met, such as a vapor retarder is installed and at least 40 percent and no more than 50 percent of the required ventilating area is provided by ventilators located in the upper portion of the attic or rafter space. Otherwise, the rule is 1 square foot per 150 square feet of attic surface.

I recommend you go with the 1-to-150 ratio regardless, as you can't be certain an attic meets all the conditions required for the lesser vent area, and because more ventilation is almost always better, as long as it's properly located. An inadequately ventilated attic is going to create all kinds of problems for the homeowner, and ultimately, for you.

Keep in mind that this is *net free ventilating area* (NFVA) and excludes the reduction caused by any louvers or screen required to keep rain and pests out. Screens vastly reduce the free area of the vent; for example, a 4-square-foot opening protected with louvers and mesh would actually only have 2 square feet of clear opening. Figure 14-15 shows by how much to increase the area of openings to account for louvers and screens.

▼ **Example 14-2:** Assume a ventilation system allowing 1 square foot of net free ventilating area for each 150 square feet of attic surface and two outlet ventilators with No. 16 mesh and plain metal louvers. Find the gross ventilation area required for each outlet ventilator for the building diagrammed in Figure 14-16 *Ventilation example*.

Obstructions in ventilators, louvers and screens[1]	Multiply required net area in square feet by:
1/4 inch mesh hardware cloth	1
1/8 inch mesh screen	1 1/4
No. 16 mesh insect screen (with or without plain metal louvers)	2
Wood louvers and 1/4 inch mesh hardware cloth[2]	2
Wood louvers and 1/8 inch mesh screen[2]	2 1/4
Wood louvers and No. 16 mesh insect screen[2]	3

[1] In crawl-space ventilators, screen openings should be no larger than 1/4 inch; in attic spaces no larger than 1/8 inch.
[2] If metal louvers have drip edges that reduce the opening, use same ratio as shown for wood louvers.

Figure 14-15 Ventilating area increase required for louvers and screens in crawl spaces and attics

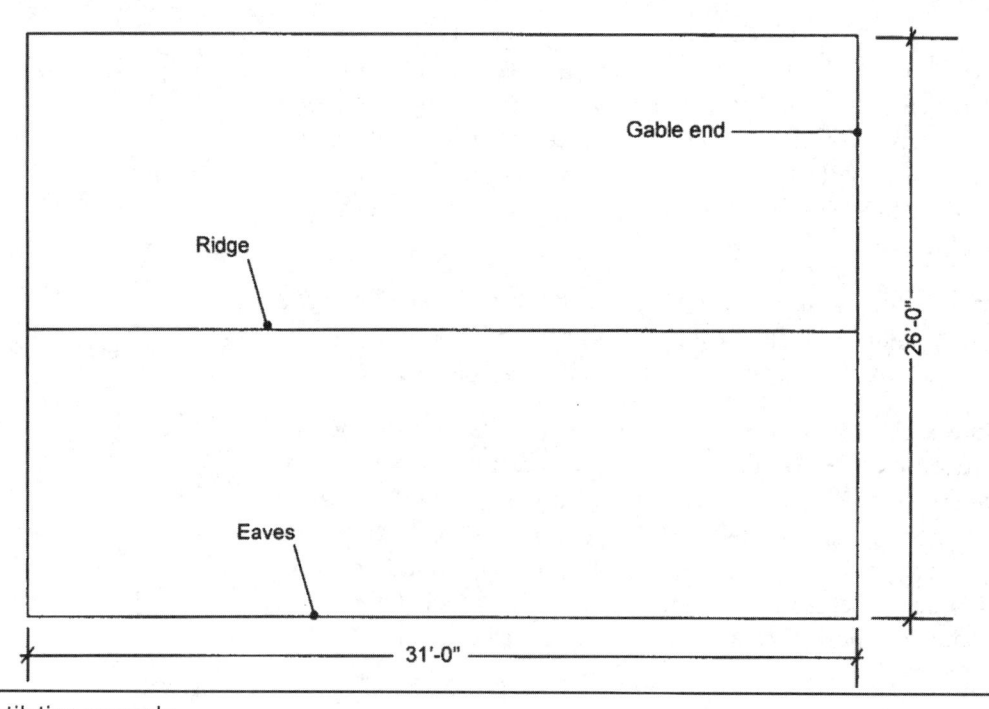

Figure 14-16 Ventilation example

Roofing Repair and Maintenance

$$\text{Minimum net ventilation area} = \frac{31' \times 26'}{150}$$

$$= 5.37 \text{ square feet}$$

From Figure 14-15 the minimum gross area required is:

Minimum gross ventilation area = 5.37 x 2
= 10.74 square feet

For two outlet ventilators, each ventilator requires:

$$\text{Gross ventilation area per outlet vent} = \frac{10.74}{2}$$

= 5.37 square feet, or 773 square inches
(5.37 x 144 square inches per SF)

I also recommend you use a double-louvered soffit vent next to the fascia board in combination with a continuous ridge vent, as shown in Figure 14-17.

Gutters and Downspouts

Install gutters at the eaves to collect runoff water and divert it away from the building. Without a gutter system, eaves trim is subject to rotting. Also, runoff erodes planting beds, and heavy runoff might beat the plants themselves to death. Even worse, runoff water can saturate soil next to the foundation, leading to moisture penetration through the foundation or basement wall.

The most common materials for gutters and downspouts are aluminum, galvanized steel, and vinyl. Materials less commonly used include copper or zinc. You can buy painted metal gutters, but they also come unfinished so you can paint them to match the building. Use at least 28-gauge metal or 16-ounce copper for drainage systems.

Gutters generally come 4, 5 or 6 inches wide. As a rule of thumb, you can use 4-inch-wide gutters on roofs with drainage areas up to 750 square feet. Roof drainage areas between 750 and 1,500 square feet require 5-inch gutters and roof drainage areas

Figure 14-17 Continuous ridge vent

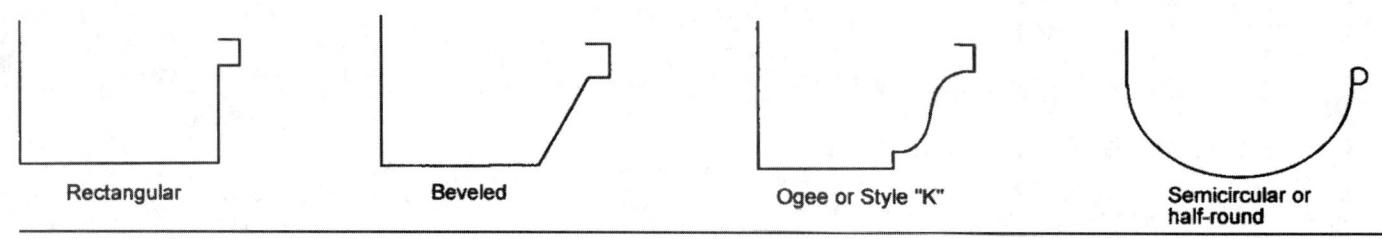

| Rectangular | Beveled | Ogee or Style "K" | Semicircular or half-round |

Figure 14-18 Gutter cross sections

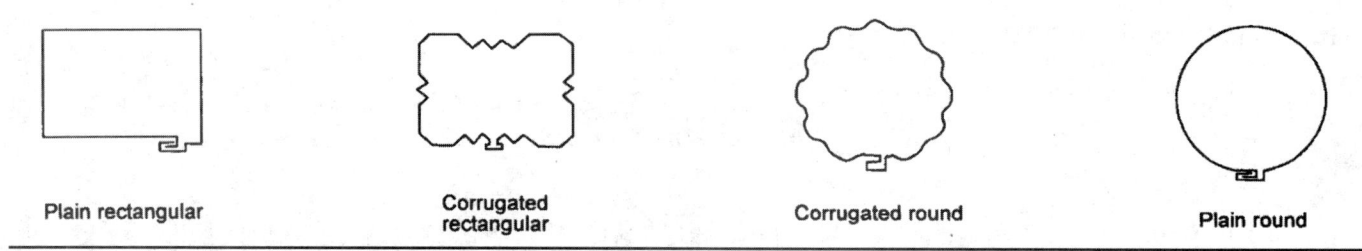

| Plain rectangular | Corrugated rectangular | Corrugated round | Plain round |

Figure 14-19 Downspout cross sections

greater than 1,500 square feet require 6-inch gutters. In very wet climates, use gutters one size larger than these rules of thumb. Figure 14-18 shows several gutter shapes.

Downspouts (also called *leaders* or *conductors*) are usually rectangular or circular, many with corrugations for added strength. They're shown in Figure 14-19. Corrugations also provide room for expansion if ice forms in the downspout. For the same reason, rectangular downspouts are better than round ones in places where the temperature drops below freezing.

There are several ways to fasten gutters and downspouts to a building. You can use hangers, straps, or spikes, as shown in Figure 14-20. You generally space gutter hangers at 3-foot centers, but in heavy snow areas reduce this spacing to $1^1/_2$ feet. Don't use strap-type gutter hangers in heavy snow areas as they can tear loose under the weight of snow and ice and take the shingles with them.

The spike and spacer tube hanging system has the advantage of allowing the metal gutter to expand and contract freely.

Install downspout straps at 6-foot centers. Use a minimum of two straps, one at the gooseneck and one at the bottom elbow (or shoe) to secure the downspout to the wall. Solder joints in gutters and downspouts, although some specifications only require joints to be caulked with silicone or butyl rubber caulking compound.

Install downspouts large enough to handle all rainwater runoff. A general rule for residential construction is to allow 1 square inch of downspout end area for each 100 square feet of roof area. A 7-square-inch downspout end area is the minimum recommended.

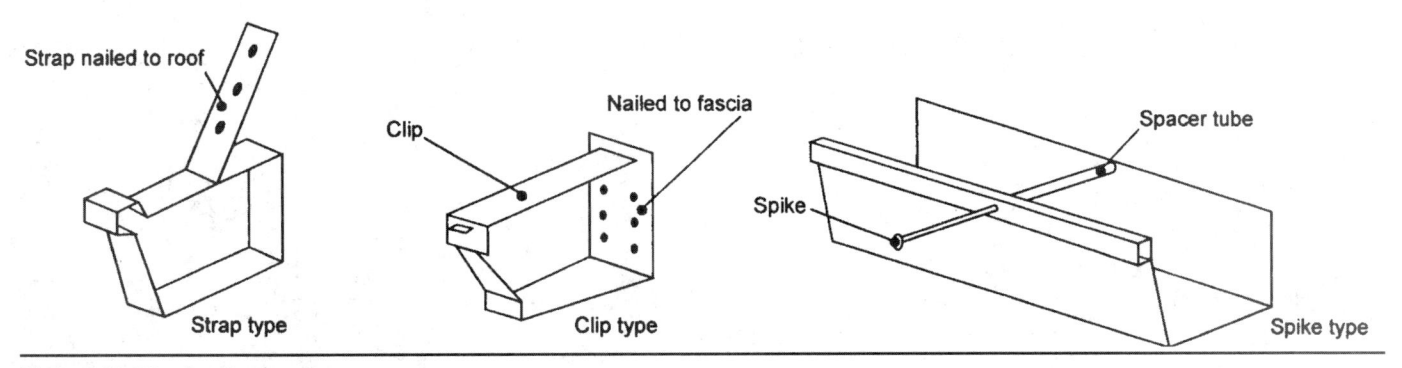

Figure 14-20 Guttter hangers

Slope metal gutters to drain toward downspouts with a slope of at least $1/16$ inch of fall per linear foot of gutter run. Allow a maximum run of 10 feet from the high point of the gutter to the downspout if possible, and try to limit runs to 30 feet. If this isn't possible, increase the leader area by 1 square inch for each 20 feet of run between downspouts. On long gutter runs, you can center the high point of the gutter between the downspouts, and slope the gutter toward them in both directions. On long straight gutter runs, install expansion joints on 60-foot centers (minimum).

You can also get a molded gutter with a sloping inner lining. You can set a molded gutter true with the eaves line so it doesn't look crooked.

If more than one gutter run empties into a downspout, provide a leader head as shown in Figure 14-21. Don't install downspouts carrying water from a larger and higher roof so they discharge onto a lower and smaller roof. Instead, let each gutter drain to the ground. A concrete splash block at the base of the downspout helps divert water away from the building. If a splash pan is used, install one with a back width at least 4 inches wider than the downspout.

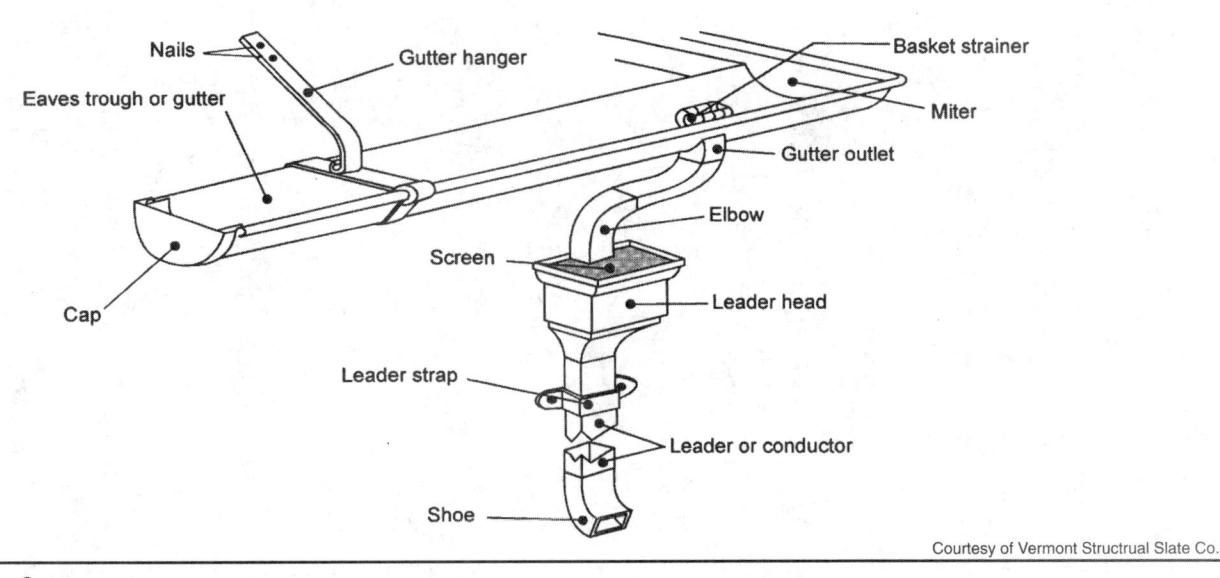

Courtesy of Vermont Structrual Slate Co.

Figure 14-21 Gutter parts

Estimating Gutter and Downspout Systems

Take off gutter and downspout material by the linear foot. These come in 10-foot lengths. You buy gutter and downspout accessories by the piece. These accessories include inside and outside miters, slip joint connectors, end caps, end pieces, elbows, pipe straps, gutter hangers, basket strainers, and leader heads. Also order splash diverters as in Figure 14-22. The concrete splash blocks placed beneath each downspout are usually installed by the general contractor. If you've agreed to provide them, however, be sure you remember to price and include them in your estimate.

Gutter System Maintenance

Clean debris from gutters and downspouts frequently, especially in the spring and fall. This is very important, because a clogged gutter can fill with water and freeze, causing ice dams. You can keep downspouts from clogging by installing gutter baskets or strainers, as shown in Figure 14-21.

Replace or repair sagging or broken straps with solder, bolts or screws. Fill small holes with epoxy resin and seal larger holes with adhesive-backed aluminum tape. Caulk leaking joints with silicone or butyl rubber caulking compound.

Figure 14-22 Splash diverter

Now you know how to install and repair all types of roofing. That just leaves one very important part of your business to cover in the final chapter — how to estimate (and maximize) production rates.

15 Estimating (and Maximizing) Production Rates

▶ Production rates and unit costs are very hard to predict because of the many variables that affect production. If you don't already have them, start right now keeping good records of your crews' jobs and their production rates. They're the most reliable indicator of your future production rates. Until you have your own rates to go by, you can use published costs. For the examples in this chapter, we'll use figures from the *National Construction Estimator*, published by Craftsman Book Company. There's an order form for this estimating manual and database, including an estimating program, and other construction manuals and software, bound into the back of this book.

Labor Unit Prices

Unit prices based on observation of your own crews are your best predictor of labor costs on your next job. But whether you use your own records or estimating guides, you'll have to consider job complexity and other factors when setting your labor unit prices. You can determine any daily or hourly labor unit price (LUP) by:

$$\text{Labor Unit Price} = \frac{\text{Total Daily Crew Cost}}{\text{Daily Crew Production}} \quad \boxed{\text{Equation 15-1}}$$

or

$$\text{Labor Unit Price} = \frac{\text{Total Hourly Crew Cost}}{\text{Hourly Crew Production}} \quad \boxed{\text{Equation 15-2}}$$

▼ **Example 15-1:** Find the labor unit price in cost per square for installing fiberglass shingles. Assume daily crew production of 6.36 squares per day with a crew of one roofer costing $40.00 per hour and one laborer costing $30.00 per hour. The hourly crew cost is $70.00 per hour ($40.00 + $30.00).

Total Daily Crew Cost = 8 hours each x $70.00 per hour

= $560.00

From Equation 15-1, the labor unit price is:

$$\text{LUP} = \frac{\$560.00}{6.36 \text{ SQ}} = \$88.05 \text{ per square}$$

Production Based on Labor Unit Prices

If you know the labor unit price, then you can project the production rate as follows:

$$\text{Daily Production} = \frac{\text{Daily Crew Cost}}{\text{LUP}}$$

Equation 15-3

or

$$\text{Hourly Production} = \frac{\text{Hourly Crew Cost}}{\text{LUP}}$$

Equation 15-4

▼ **Example 15-2:** Assume a labor unit price of $88.05 per square with a crew costing $70.00 per hour. Find the approximate daily production rate you can expect for installing fiberglass shingles:

Total Daily Crew Cost = 8 hours x $70.00 per hour

= $560.00

From Equation 15-3:

$$\text{Daily Production} = \frac{\$560.00/\text{Day}}{\$88.05/\text{SQ}} = 6.36 \text{ squares per day}$$

Project Duration

You can predict the total project duration for a given item of work with the following formula:

$$\text{Project Duration (Days)} = \frac{\text{Total Material}}{\text{Daily Production}}$$

Equation 15-5

or

Estimating (and Maximizing) Production Rates

$$\text{Project Duration (Hours)} = \frac{\text{Total Material}}{\text{Hourly Production}}$$

Equation 15-6

▼ **Example 15-3:** At a production rate of 6.36 squares per day, find how much time it will take to install 27 squares of fiberglass shingles.

From Equation 15-5:

$$\text{Project Duration} = \frac{27 \text{ SQ}}{6.36 \text{ SQ/Day}} = 4.25 \text{ days}$$

You can find production rates for time required per unit installed as follows:

$$\text{Production (Crew-days/Unit)} = \frac{\text{LUP}}{\text{Daily Crew Cost}}$$

Equation 15-7

or

$$\text{Production (Crew-hours/Unit)} = \frac{\text{LUP}}{\text{Hourly Crew Cost}}$$

Equation 15-8

▼ **Example 15-4:** Find the production time required per square of shingles installed by a crew working at a total cost of $70.00 per hour, and a LUP of $88.05 per square.

From Equation 15-8, you find total hours per square by:

$$\text{Production (Crew-hours/Unit)} = \frac{\$88.05/\text{SQ}}{\$70.00/\text{Hr.}} = 1.26 \text{ hours/square}$$

Use this formula to find production rates in crew-hours per unit:

$$\text{Production (Crew-hours/Unit)} = \frac{\text{Crew-hours/Day}}{\text{Daily Production}}$$

Equation 15-9

▼ **Example 15-5:** Find how many crew-hours per square are required at $88.05 per square for a crew costing $70.00 per hour.

Use Equation 15-9:

$$\text{Production (Crew-hours/Unit)} = \frac{16 \text{ Hr./Day}}{6.36 \text{ SQ/Day}}$$

$$= 2.52 \text{ crew-hours per square}$$

Required Production Rates

Use the following formula to find the production rate required to finish a given work item within its labor budget:

$$\text{Required Daily Production} = \frac{\text{Daily Crew Cost} \times \text{Total Material}}{\text{Total Labor Budget}}$$

Equation 15-10

or

$$\text{Required Hourly Production} = \frac{\text{Hourly Crew Cost x Total Material}}{\text{Total Labor Budget}}$$

Equation 15-11

▼ **Example 15-6:** Assuming a crew cost of $560.00 per day, use Equation 15-10 to find the production rate required to install 27 squares of shingles within a labor budget of $2,376.00.

$$\text{Required Production} = \frac{\$560.00/\text{Day} \times 27 \text{ SQ}}{\$2,376.00} = 6.36 \text{ squares per day}$$

Prefabrication for Multiple Uses

Some work items such as scaffolding can be prefabricated and reused many times. In those cases, you can calculate the total labor cost as follows:

$$\text{Total Labor Cost} = \frac{\text{Labor Cost to Prefabricate}}{\text{Number of Uses}} + \text{Labor Cost to Install}$$

Equation 15-12

Adjusting Labor Costs

Some labor manuals give only an hourly cost for a particular trade and a labor unit price for a specific task. There are two ways that you can tailor unit prices (the actual wage you pay plus burden) to fit your own situation:

Option 1: (Using Equation 15-2)

$$\text{Actual LUP} = \frac{\text{Actual Hourly Costs}}{\text{Production (Units/Hour)}}$$

where (from Equation 15-4):

$$\text{Production (Units/Hour)} = \frac{\text{Published Hourly Costs}}{\text{Published LUP}}$$

Option 2:

$$\text{Actual LUP} = \frac{\text{Actual Hourly Costs}}{\text{Published Hourly Costs}} \times \text{Published LUP}$$

Equation 15-13

▼ **Example 15-7:** Use both options above to find the actual labor unit price to install fiberglass shingles where the published costs are $65.90 per square with a crew costing $70.00 per hour when your crew cost is only $65.00 per hour.

Solution:

Option 1 (using Equation 15-4):

$$\text{Production (Units/Hour)} = \frac{\$70.00/\text{Hr.}}{\$65.90/\text{SQ}} = 1.0622 \text{ SQ/Hr.}$$

From Equation 15-2:

$$\text{Actual LUP} = \frac{\$65.00/\text{Hr.}}{1.0622 \text{ SQ/Hr.}}$$

$$= \$61.19/\text{SQ}$$

Option 2 (using Equation 15-13):

$$\text{Actual LUP} = \frac{\$65.00/\text{Hr.}}{\$70.00/\text{Hr.}} \times \$65.90/\text{SQ}$$

$$= \$61.19/\text{SQ}$$

Obviously, Option 2 is the simpler method.

Estimating with Published Prices

The *National Construction Estimator* (*NCE*) is an encyclopedia of building costs that's updated each year. The program produces an estimate report, using the contents of the book as a database.

The *NCE* is divided into two parts. The Residential Division contains costs for homes and apartments with a wood or masonry frame. The Industrial and Commercial Division contains other construction costs not covered in the Residential Division. The Residential Division is arranged in alphabetical order by construction trade and type of material. The Industrial and Commercial Division follows the 16-section Construction Specification Institute (CSI) format.

Here's how to understand published costs like those in the *NCE* and use them to price your jobs.

Published labor costs are an average based on nationwide surveys for the various trades. Hourly wage rates are higher for industrial and commercial work than for residential construction, reflecting the fact that tradespeople on commercial and industrial jobs are often paid more per hour than tradespeople on residential jobs. The labor costs shown include taxes and insurance (the "contractor's burden") but don't include markup (overhead and profit) or supervision costs. Material costs include normal waste and coverage loss. Sections identified as "subcontract" include both the subcontractor's markup and supervision expense and are intended to reflect prices commonly quoted by subcontractor specialists.

Craft Code	Cost Per Manhour	Crew Composition	Craft Code	Cost Per Manhour	Crew Composition
B1	$33.85	1 laborer and 1 carpenter	BR	$35.97	1 lather
B2	$35.02	1 laborer, 2 carpenters	BS	$33.27	1 marble setter
B3	$32.67	2 laborers, 1 carpenter	CF	$35.55	1 cement mason
B4	$37.45	1 laborer 1 operating engineer 1 reinforcing iron worker	CT	$35.35	1 mosaic & terrazzo worker
			D1	$36.54	1 drywall installer 1 drywall taper
B5	$37.06	1 laborer, 1 carpenter 1 cement mason 1 operating engineer 1 reinforcing iron worker	DI	$36.47	1 drywall installer
			DT	$36.60	1 drywall taper
			HC	$29.39	1 plasterer helper
			OE	$42.71	1 operating engineer
B6	$32.93	1 laborer, 1 cement mason	P1	$36.89	1 laborer, 1 plumber
B7	$30.72	1 laborer, 1 truck driver	PM	$43.47	1 plumber
B8	$36.51	1 laborer 1 operating engineer	PP	$34.27	1 painter, 1 laborer
B9	$33.06	1 bricklayer 1 bricklayer's helper	PR	$37.46	1 plasterer
			PT	$38.22	1 painter
BB	$37.97	1 bricklayer	R1	$35.99	1 roofer, 1 laborer
BC	$37.38	1 carpenter	RI	$39.33	1 reinforcing iron worker
BE	$40.63	1 electrician	RR	$41.66	1 roofer
BF	$34.28	1 floor layer	SW	$42.16	1 sheet metal worker
BG	$36.13	1 glazier	T1	$32.99	1 tile layer, 1 laborer
BH	$28.15	1 bricklayer's helper	TL	$35.66	1 tile layer
BL	$30.31	1 laborer	TR	$31.13	1 truck driver

Figure 15-1   Craft codes definition

In the *NCE*, manhours per unit installed and the craft performing the work (craft code) are listed in the "Craft @ Hrs" column. Figure 15-1 is part of the section in the *NCE* which defines craft codes. Figure 15-2 is part of the residential roofing section from the *NCE*.

The average manhour cost is calculated by dividing the total hourly crew cost by the number of crew personnel. For example, an R1 crew consists of one roofer and one laborer. These workers cost $41.66 and $30.31 per hour, respectively. So the average cost per manhour is:

$$\text{Average Cost per Manhour} = \frac{\$41.66 + \$30.31}{2}$$
$$= \$35.99 \text{ per manhour}$$

The labor unit costs in the *NCE* are the result of multiplying the installation time (in manhours) by the average cost per manhour. For example, the labor cost listed in Figure 15-2 for installing asphalt shingles is $65.90 per square. This figure is calculated:

Labor Cost Per Unit = 1.83 Manhours/Square x $35.99/Manhour

= $65.86 per square

Fiberglass base sheet ASTM D-4601. 300 square foot roll. Nailed down
  Per 100 square feet   R1@.300   Sq   17.70   10.80   28.50
Modified bitumen adhesive. Rubberized. Use to adhere SBS modified bitumen membranes and rolled roofing. Blind nailing cement. Brush grade.
  5 gallons   —   Ea   78.90   —   78.90
Cold process and lap cement. Bonds plies of built-up roofing and gravel to bare spots on existing gravel roof. Restores severely worn roof. Gallon covers 50 square feet.
  0.9 gallon   —   Ea   14.10   —   14.10
  4.75 gallons   —   Ea   44.10   —   44.10
Cold-Ap® built-up roofing cement, Henry. Bonds layers of conventional built-up roofing or SBS base and cap sheets. Replaces hot asphalt in built-up roofing. For embedding polyester fabric, gravel and granules. Not recommended for saturated felt. ASTM D3019 Type III. Two gallons cover 100 square feet.
  0.9 gallon   —   Ea   18.20   —   18.20
  4.75 gallons   —   Ea   70.90   —   70.90
Cold-process and lap cement. ASTM D3019 Type III. Two gallons cover 100 square feet.
  5 gallons   —   Ea   56.50   —   56.50

Composition Roofing Shingles Material costs include 5% waste. Labor costs include roof loading and typical cutting and fitting for roofs of average complexity. Class A shingles have the highest fire rating.

3-tab asphalt fiberglass® shingles, Three bundles per square (100 square feet). 5" exposure, Class A fire rating. 60 MPH wind resistance.
  20 year, no algae-resistance   R1@1.83   Sq   57.30   65.90   123.20
  25-30 year, algae-resistant   R1@1.83   Sq   85.30   65.90   151.20
Architectural grade laminated shingles, Owens Corning, GAF and similar. Three bundles per square (100 square feet). Class A fire rating. 110 MPH to 130 MPH wind resistance. Limited lifetime warranty.
  No algae resistance   R1@1.83   Sq   107.00   65.90   172.90
  Algae resistant, 110mph   R1@1.83   Sq   100.00   65.90   165.90
  Algae resistant, 130mph   R1@1.83   Sq   107.00   65.90   172.90
Impact resistant, architectural grade laminated shingles, Class 4, limited lifetime warranty, incl algae-resistance, 130MPH wind resistance, Class A fire. Three bundles per square (100 square feet). Class A fire rating. 110 MPH to 130 MPH wind resistance.
  Class 4 impact rated, lifetime warranty, 130MPH   R1@1.83   Sq   119.00   65.90   184.90
Premium grade laminated shingles, Owens Corning, GAF and similar. Limited lifetime warranty, algae-resistant, Class A fire rating. 110 MPH to 130 MPH wind resistance. Three bundles per square (100 square feet).
  5-inch exposure, 110MPH   R1@1.83   Sq   181.00   65.90   246.90
  5-inch exposure, cool roof Title 24, 130MPH   R1@1.83   Sq   216.00   65.90   281.90
  Premium, to 8-inch exposure, 130mph   R1@1.83   Sq   304.00   65.90   369.90
Hip and ridge shingles with sealant, Per linear foot of hip or ridge.
  ProEdge hip and ridge   R1@.028   LF   1.67   1.01   2.68
  StormMaster hip and ridge   R1@.028   LF   1.79   1.01   2.80
  Decorative ridge   R1@.028   LF   2.54   1.01   3.55
Shingle starter strip, Peel 'n' stick SBS membrane. Rubberized to seal around nails and resist wind lift.
  9" x 33' roll   R1@.460   Ea   20.60   16.60   37.20
  Per linear foot   R1@.014   LF   .62   .50   1.12
Allowance for felt, flashing, fasteners, and vents.
  Typical cost   —   Sq   22.00   —   22.00

Roofing Papers See also Building Paper. Labor laying paper on 3-in-12 pitch roof.
Asphalt roofing felt (115.5 SF covers 100 SF), labor to roll and mop
  15 lb (432 SF roll)   R1@.490   Sq   5.75   17.60   23.35
  30 lb (216 SF roll)   R1@.500   Sq   11.50   18.00   29.50

Figure 15-2 Residential roofing cost data

For simplicity, this cost is rounded to $65.90 per square. The loss of accuracy is insignificant.

To find out how many units one worker can finish in an 8-hour day, divide the manhours per unit into 8 hours. To find how many units a crew can complete, multiply the units for one worker by the number of crew members.

For example, the production rate for asphalt shingle installation is 1.83 manhours per square for an R1 crew (one roofer and one laborer). The daily production of each crew member is:

Daily Production (per man) = 8 Hours/Day ÷ 1.83 Manhours/Square

= 4.37 squares per day

The daily crew production is:

Daily Crew Production = 4.37 Squares/Man-Day x 2 workers

= 8.74 squares per day

Adjusting Labor Costs

Figure 15-3 is a section of the *NCE* chart which defines labor costs. Here's how to customize the labor costs given in a published cost book if your wage rates are different from those in the book.

Start with the taxable benefits you offer. Assume workers on your payroll get one week of vacation and one week of sick leave each year. Convert these benefits into hours. Your figures should look like this:

40 vacation hours + 40 sick leave hours = 80 taxable leave hours

Then add the regular work hours for the year:

80 taxable leave hours + 2,000 regular hours = 2,080 total hours

Multiply these hours by the base wage per hour. If you pay roofers $25.00 per hour, the calculation would be:

2,080 hours x $25.00 per hour = $52,000 per year

Next calculate the payroll tax and insurance rate for each trade. If you know the rates that apply to your employees, use those rates. If not, use the rates from a published price guide. For this example, we'll use 43% (the rate for roofers in the *NCE*, Figure 15-3, in the column *Insurance and employer taxes*). To increase the annual taxable wage by 43%, multiply by 1.4300:

$52,000 per year x 1.43 tax & insurance rate = $74,360.00 annual cost

Then add the cost of non-taxable benefits. Suppose your company has no pension or profit sharing plan but does contribute towards medical insurance for employees. Assume that your contribution toward your roofer's insurance is $250 per month, or $3,000 per year.

Estimating (and Maximizing) Production Rates

Residential Division

Craft	1 Base wage per hour	2 Taxable fringe benefits (@5.56% of base wage)	3 Insurance and employer taxes (%)	4 Insurance and employer taxes ($)	5 Non-taxable fringe benefits (@4.91% of base wage)	6 Total hourly cost used in this book
Bricklayer	$27.76	$1.54	24.94%	$7.31	$1.36	$37.97
Bricklayer's Helper	20.58	1.14	24.94	5.42	1.01	28.15
Building Laborer	21.00	1.17	32.07	7.11	1.03	30.31
Carpenter	26.11	1.45	30.97	8.54	1.28	37.38
Cement Mason	26.40	1.47	22.89	6.38	1.30	35.55
Drywall installer	27.02	1.50	23.22	6.62	1.33	36.47
Drywall Taper	27.11	1.51	23.22	6.65	1.33	36.60
Electrician	30.97	1.72	19.65	6.42	1.52	40.63
Floor Layer	25.34	1.41	23.52	6.29	1.24	34.28
Glazier	26.34	1.46	25.34	7.04	1.29	36.13
Lather	27.11	1.51	21.05	6.02	1.33	35.97
Marble Setter	25.06	1.39	21.13	5.59	1.23	33.27
Millwright	26.56	1.48	21.02	5.89	1.30	35.23
Mosiac & Terrazzo Worker	26.62	1.48	21.13	5.94	1.31	35.35
Operating Engineer	31.25	1.74	24.83	8.19	1.53	42.71
Painter	28.02	1.56	24.55	7.26	1.38	38.22
Plasterer	26.74	1.49	28.05	7.92	1.31	37.46
Plasterer Helper	20.98	1.17	28.05	6.21	1.03	29.39
Plumber	32.01	1.78	24.00	8.11	1.57	43.47
Reinforcing Ironworker	28.06	1.56	28.12	8.33	1.38	39.33
Roofer	26.73	1.49	43.00	12.13	1.31	41.66
Sheet Metal Worker	30.66	1.70	25.62	8.29	1.51	42.16
Sprinkler Fitter	31.46	1.75	24.76	8.22	1.55	42.98
Tile Layer	26.86	1.49	21.13	5.99	1.32	35.66
Truck Driver	22.60	1.26	25.83	6.16	1.11	31.13

Hourly Labor Cost

The labor costs shown in Column 6 were used to compute the manhour costs for crews on page 7 and the figures in the "Labor" column of the Residential Division of this manual. Figures in the "Labor" column of the Industrial and Commercial Division of this book were computed using the hourly costs shown on page 309. All labor costs are in U.S. dollars per manhour.

It's important that you understand what's included in the figures in each of the six columns above. Here's an explanation:

Column 1, the base wage per hour, is the craftsman's hourly wage. These figures are representative of what many contractors will be paying craftsmen working on residential construction in 2021.

Column 2, taxable fringe benefits, includes vacation pay, sick leave and other taxable benefits. These fringe benefits average 5.56% of the base wage for many construction contractors. This benefit is in addition to the base wage.

Column 3, insurance and employer taxes in percent, shows the insurance and tax rate for construction trades. The cost of insurance in this column includes workers' compensation and contractor's casualty and liability coverage. Insurance rates vary widely from state to state and depend on a contractor's loss experience. Typical rates are shown in the Insurance section

Figure 15-3 Labor costs

$74,360 annual cost + $3,000 medical plan = $77,360 total annual cost

Divide this total annual cost by the actual hours worked in a year. This is your total hourly labor cost including all benefits, taxes and insurance. Assume your roofer will work 2,000 hours a year:

$77,360 ÷ 2,000 = $38.68 per hour

Finally, find your modification factor for the labor costs. Divide your total hourly labor cost by the total hourly cost shown in the published costs. For the roofer in our example, the figure in the *NCE* is $41.66:

$38.68 ÷ $41.66 = 0.928

Your modification factor is 93 percent. Multiply labor costs for roofers in the cost book by 0.93 to find your estimated cost.

If You Don't Know the Labor Rate

On some estimates you may not know what actual labor rates will apply. In that case, use both labor and material figures in the cost book without making any adjustment. But to make them more accurate, apply the area modification factor for your region. The *NCE* lists percentage guidelines you use to adjust the prices to fit the part of the country where you live. When you've compiled all labor, equipment and material costs, multiply the totals by the appropriate factor in the area modification table. Figure 15-4 shows part of the *NCE*'s area modification table.

Roofing Labor Tips

Study each job carefully before you assign costs. The number and sizes of roof penetrations, the number of hips and especially the number of valleys, all affect production rates. Structures such as dormers which interrupt the shingle pattern require you to project the shingle pattern and tie in remote shingle courses. That increases your time. When tie-ins are required, the shingle pattern affects production. For example, with 3-tab shingles the 6-inch pattern is the easiest to use when making a tie-in.

You're the only one who can judge how efficient and dependable your workers are, and how well they work together. To a certain point, a larger crew will produce more than a smaller crew. But if the crew is too large management problems and workspace limitations will actually hinder production.

The strength and stamina of individual crew members can also affect production rates. But strength isn't as important as basic common sense and a positive attitude toward the work. Workers' attitudes can change daily, or even hourly. A disgruntled or frustrated worker slows everyone down, and bad attitudes can be contagious.

Maximizing Production Rates

Job complexity and the skill and experience of your workers probably influence productivity more than anything else. But there are many other factors that can affect your production costs.

■ **Work Habits** Place as many shingles as possible without changing position — but don't try to shingle the whole roof from one position. Try to find a happy medium. Also, lay out partial bundles of shingles at strategic

Estimating (and Maximizing) Production Rates

Area Modification Factors

Construction costs are higher in some areas than in other areas. Add or deduct the percentages shown on the following pages to adapt the costs in this book to your job site. Adjust your cost estimate by the appropriate percentages in this table to find the estimated cost for the site selected. Where 0% is shown, it means no modification is required.

Modification factors are listed alphabetically by state and province. Areas within each state are listed alphabetically. For convenience, one representative city is identified in each three-digit zip or range of zips. Percentages are based on the average of all data points in the table. Factors listed for each state and province are the average of all data points in that state or province. Figures for three-digit zips are the average of all five-digit zips in that area. Figures in the Total column are the weighted average of factors for Labor, Material and Equipment.

The National Estimator program will apply an area modification factor for any five-digit zip you select. Click Utilities. Click Options. Then select the Area Modification Factors tab.

These percentages are composites of many costs and will not necessarily be accurate when estimating the cost of any particular part of a building. But when used to modify costs for an entire structure, they should improve the accuracy of your estimates.

Location	Zip	Mat.	Lab.	Equip.	Total Wtd. Avg.
Alabama Average		-1	-7	0	-4%
Anniston	362	0	-13	-1	-6%
Auburn	368	-1	-8	0	-4%
Bellamy	369	-2	13	-1	5%
Birmingham	350-352	-3	8	-1	2%
Dothan	363	-1	-13	0	-7%
Evergreen	364	-2	-20	-1	-10%
Gadsden	359	-4	-15	-1	-9%
Huntsville	358	1	-3	0	-1%
Jasper	355	-2	-16	-1	-8%
Mobile	365-366	-1	-3	0	-2%
Montgomery	360-361	-1	-3	0	-2%
Scottsboro	357	0	-8	0	-4%
Selma	367	-1	-10	0	-5%
Sheffield	356	-1	1	0	0%
Tuscaloosa	354	1	-9	0	-4%
Alaska Average		14	33	4	23%
Anchorage	995	16	38	5	26%
Fairbanks	997	16	40	5	27%
Juneau	998	18	20	6	19%
Ketchikan	999	3	36	1	18%
King Salmon	996	16	32	5	23%
Arizona Average		1	-9	0	-4%
Chambers	865	1	-19	0	-8%
Douglas	855	0	-18	0	-8%
Flagstaff	860	2	-17	1	-7%
Kingman	864	1	-11	0	-5%
Mesa	852	1	5	0	3%
Phoenix	850	1	6	0	3%
Prescott	863	3	-16	1	-6%
Show Low	859	2	-18	1	-7%
Tucson	856-857	0	-10	0	-5%
Yuma	853	0	5	0	2%
Arkansas Average		-2	-13	0	-7%
Batesville	725	0	-20	0	-9%
Camden	717	-4	1	-1	-2%
Fayetteville	727	0	-8	0	-4%
Fort Smith	729	-1	-14	0	-7%
Harrison	726	-1	-25	0	-12%
Hope	718	-3	-15	-1	-8%
Hot Springs	719	-2	-25	-1	-13%
Jonesboro	724	-1	-18	0	-9%
Little Rock	720-722	-1	-6	0	-3%
Pine Bluff	716	-4	-19	-1	-11%
Russellville	728	0	-9	0	-4%
West Memphis	723	-3	-1	-1	-2%
California Average		2	16	1	9%
Alhambra	917-918	3	15	1	8%
Bakersfield	932-933	1	5	0	2%
El Centro	922	1	-1	0	0%
Eureka	955	1	13	0	7%
Fresno	936-938	0	-5	0	-2%
Herlong	961	2	17	1	9%
Inglewood	902-905	3	16	1	9%
Irvine	926-927	3	24	1	13%
Lompoc	934	3	2	1	3%

Location	Zip	Mat.	Lab.	Equip.	Total Wtd. Avg.
Long Beach	907-908	3	17	1	9%
Los Angeles	900-901	3	13	1	8%
Marysville	959	1	18	0	9%
Modesto	953	1	2	0	1%
Mojave	935	0	11	0	5%
Novato	949	3	36	1	18%
Oakland	945-947	3	48	1	24%
Orange	928	3	22	1	12%
Oxnard	930	3	1	1	2%
Pasadena	910-912	4	16	1	9%
Rancho Cordova	956-957	2	6	1	4%
Redding	960	1	-8	0	-3%
Richmond	948	2	35	1	17%
Riverside	925	1	7	0	4%
Sacramento	958	1	6	0	3%
Salinas	939	3	-1	1	1%
San Bernardino	923-924	0	4	0	2%
San Diego	919-921	3	13	1	8%
San Francisco	941	3	55	1	27%
San Jose	950-951	3	33	1	17%
San Mateo	943-944	4	40	1	21%
Santa Barbara	931	3	11	1	7%
Santa Rosa	954	3	32	1	16%
Stockton	952	2	7	1	4%
Sunnyvale	940	3	39	1	20%
Van Nuys	913-916	3	14	1	8%
Whittier	906	3	14	1	8%
Colorado Average		2	1	1	1%
Aurora	800-801	3	11	1	7%
Boulder	803-804	3	5	1	4%
Colorado Springs	808-809	2	-3	1	0%
Denver	802	3	13	1	8%
Durango	813	1	-3	0	-1%
Fort Morgan	807	2	-6	1	-2%
Glenwood Springs	816	2	6	1	4%
Grand Junction	814-815	1	-1	0	0%
Greeley	806	3	8	1	5%
Longmont	805	3	1	1	2%
Pagosa Springs	811	0	-9	0	-4%
Pueblo	810	-1	2	0	0%
Salida	812	2	-15	1	-6%
Connecticut Average		1	16	0	8%
Bridgeport	066	0	13	0	6%
Bristol	060	1	24	0	12%
Fairfield	064	2	17	1	9%
Hartford	061	0	23	0	11%
New Haven	065	1	15	0	7%
Norwich	063	0	7	0	3%
Stamford	068-069	4	21	1	12%
Waterbury	067	1	12	0	6%
West Hartford	062	1	9	0	5%
Delaware Average		1	3	0	2%
Dover	199	1	-9	0	-4%
Newark	197	2	10	1	6%
Wilmington	198	0	9	0	4%

Location	Zip	Mat.	Lab.	Equip.	Total Wtd. Avg.
District of Columbia Average					
Washington	200-205	2	23	1	12%
Florida Average		0	-10	0	-5%
Altamonte Springs	327	-1	-6	0	-3%
Bradenton	342	0	-12	0	-6%
Brooksville	346	0	-16	0	-7%
Daytona Beach	321	-2	-18	-1	-9%
Fort Lauderdale	333	3	1	1	2%
Fort Myers	339	0	-12	0	-6%
Fort Pierce	349	0	-20	-1	-10%
Gainesville	326	-1	-18	0	-9%
Jacksonville	322	-1	-3	0	-2%
Lakeland	338	-3	-13	-1	-8%
Melbourne	329	-2	-15	-1	-8%
Miami	330-332	2	-1	1	1%
Naples	341	3	-8	1	-2%
Ocala	344	-3	-23	-1	-12%
Orlando	328	0	2	0	1%
Panama City	324	-2	-21	-1	-11%
Pensacola	325	-1	-17	0	-8%
Saint Augustine	320	-1	-4	0	-2%
Saint Cloud	347	0	-5	0	-2%
St Petersburg	337	0	-12	0	-6%
Tallahassee	323	0	-13	0	-6%
Tampa	335-336	-1	-1	0	-1%
West Palm Beach	334	1	-5	0	-2%
Georgia Average		-1	-7	0	-4%
Albany	317	-2	-10	-1	-6%
Athens	306	0	-11	0	-5%
Atlanta	303	3	23	1	12%
Augusta	308-309	-2	-2	-1	-2%
Buford	305	0	-5	0	-2%
Calhoun	307	-1	-19	0	-9%
Columbus	318-319	-1	-6	0	-3%
Dublin/Fort Valley	310	-3	-13	-1	-8%
Hinesville	313	-2	-11	-1	-6%
Kings Bay	315	-2	-19	-1	-10%
Macon	312	-2	-7	-1	-4%
Marietta	300-302	1	8	0	4%
Savannah	314	-1	-7	0	-4%
Statesboro	304	-2	-21	-1	-11%
Valdosta	316	-1	-1	0	-1%
Hawaii Average		17	25	6	20%
Aliamanu	968	17	29	6	22%
Ewa	967	17	23	6	20%
Halawa Heights	967	17	23	6	20%
Hilo	967	17	23	6	20%
Honolulu	968	17	29	6	22%
Kailua	968	17	29	6	22%
Lualualei	967	17	23	6	20%
Mililani Town	967	17	23	6	20%
Pearl City	967	17	23	6	20%
Wahiawa	967	17	23	6	20%
Waianae	967	17	23	6	20%
Wailuku (Maui)	967	17	23	6	20%

Figure 15-4 Area modification table

423

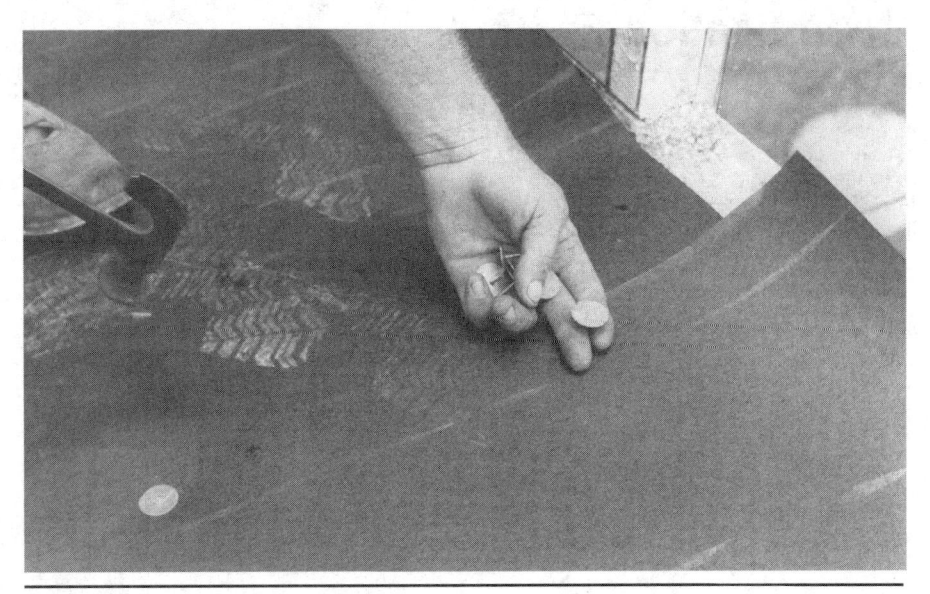

Figure 15-5 How to hold a roofing nail

locations along the roof so that they can be easily reached when needed. When you install any type of shingle, apply four or five courses at a time without changing body position. And most roofers I work with think it's more comfortable to sit on your hip than to kneel.

Hold roofing nails between your index finger and middle finger as shown in Figure 15-5. If you're right-handed, start shingling in the lower left corner of the roof. Work up and out toward the right with your left leg curled under you. Also, nail each shingle from left to right. Use the exposure gauge on your roofer's hatchet to set shingle exposure. Unless the roof has been lined out with horizontal and vertical chalk lines, check both ends of a long shingle for proper alignment. Shingle high-traffic areas (like the area immediately above a ladder) last.

In hard-to-nail places, dab the head of the hatchet in a bit of roofing cement, then embed the nail head in the cement. The cement will hold the nail in place as though it were magnetized to the hatchet.

Place a stack of 3-tab shingles to your left (if you are right-handed) with the tabs facing you. This way you can pick up and place each shingle in one motion and keep your right hand on the hatchet at all times. Notice the worker in Figure 15-6.

■ **Tools** Use proper roofing tools and keep all cutting tools sharp. Don't use a straightedge for a guide when you cut shingles. This takes too much time. Learn to make straight, accurate cuts freehand. Cut asphalt shingles by scoring from the back side and then bend the shingle until it breaks. Cutting through the mineral surface will quickly dull a cutting tool.

It may seem obvious that power tools like pneumatic nailers and electric saws and drills will increase production. But that's only true if the job is

Estimating (and Maximizing) Production Rates

Figure 15-6 Positioning the body and shingles

big enough to justify moving and setting up the power equipment. Also, equipment such as mobile placers aren't cost-effective unless they're used enough to justify their purchase and maintenance costs. It's often wiser to rent than to own.

■ **Materials** All things being equal, a large, heavy shingle is harder to install than a light one. But, the greater the exposure, the faster you'll cover the roof. Lower grades of wood shingles are narrower than the higher grades, reducing productivity and raising labor costs.

■ **Roof Slope** The steeper the roof or the higher the roof above the ground, the harder and more expensive it will be to shingle. On roofs steeper than 6 in 12, you'll need toe boards (Figure 15-7). Also, scaffold rental adds significantly to your job cost.

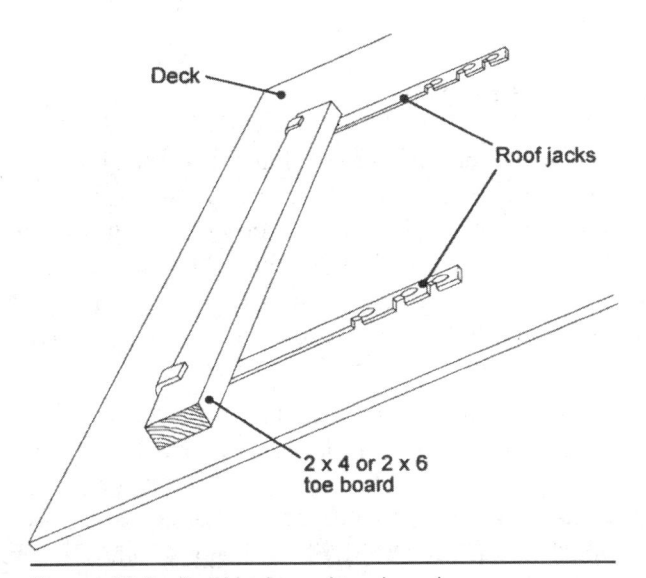

Figure 15-7 Roof jacks and toe board

■ **Job Organization** When a job is poorly planned, the specs are unclear, or the plans themselves are sloppy, it's hard for workers to follow a smooth routine. There's often a lot of debate, and the workers design the job as they go along. Change orders, especially those that require work to be demolished and rebuilt or relocated, delay the job momentum. And if the building owner can't make timely decisions, the job will suffer.

Production is always lower on remodeling projects where tenants remain in the property. Accident prevention, noise and dust control, and job scheduling are problems that must be anticipated under those conditions.

■ **Other Contractors** The work of subcontractors who precede you on the job is beyond your control, but can still make your life miserable. An uneven or out-of-square slab adds time and headaches to the framer's job. If the framer installs rafters off-center, that makes it difficult to attach sheathing properly. As a result the eaves and rake won't be square and that makes your job difficult.

■ **Weather** High winds hamper installation of sheet material, and wet weather prohibits the use of power tools. During hot weather, try to shingle areas exposed to the sun early in the day and work in shaded areas during the hotter time of day. Sit on freshly installed shingles which are still cool, not on the part of the roof where the sun's been beating down for hours. During cold weather, try to follow the sun throughout the day.

■ **Protect the Surroundings** Protect existing siding by laying 4 x 8 sheets of plywood or waferboard against the building. You can also lay this material over shrubs and the lawn for protection. If bitumen drips onto siding, clean it with kerosene or tar remover sold by auto supply shops.

Accident Prevention

Accidents and their prevention are major labor cost factors for roofers. Remember that roofing contractors pay the highest workers' compensation insurance rates of all the construction trades in the nation because of their high accident rate. Help keep the lid on insurance premiums by making sure your people know — and follow — the safety rules. Accidents will happen, but don't contribute to them by your own negligence, carelessness, or poor supervision.

Here's a mini-manual of safety tips you should follow yourself, and pass on to your crews. These procedures will make the work go faster and safer.

■ **Apparel and Protective Gear** Wear loose and non-binding clothing, but be watchful of loose shirttails or baggy clothes around power equipment and ladders. Some roofers buy their work clothes from a second-hand store and discard them after they get too dirty to tolerate. Clothes worn while roofing will eventually ruin a washing machine.

Wear protective eye goggles when you use power equipment for cutting, or around hot or caustic materials. Wear protective gloves when you're working with hot bitumen. It's also a good idea to wear gloves when you work with sheet metal.

Wear soft-soled shoes to prevent slipping. That also helps prevent damage to shingles. Never step on loose granules or on a shingle that isn't nailed down. To prevent slipping and to protect your pants, cut a piece of inner tube long enough to reach from the hip to the knee and slip it around your leg. Use a couple of old belts to hold it on. Or sit on a thick foam rubber pad to keep you from sliding off the roof.

Estimating (and Maximizing) Production Rates

■ **Ladders and Scaffolding** Regularly inspect ladders to see that they are solidly put together, including rungs and locking devices. When you use an extension ladder, be sure at least one rung is above the edge of the roof, and that the top of the ladder is securely tied to a support. Whenever you can, rest the ladder against a horizontal part of the roof like the eaves. Avoid leaning it against sloping surfaces like the gable ends.

Be sure to use ladder safety devices; in many instances an OSHA requirement. Roofers generally don't fall off ladders; the ladder falls with them on it. There are now easy-to-use tubular metal devices that firmly clamp the upper part of the ladder to the fascia, not only keeping it from slipping, but holding the ladder slightly away from the edge of the shingles, preventing damage to the drip edge. We used to put a bundle of shingles on the edge of the roof and lean the ladder against that to protect the drip edge, but these devices are easier and much safer.

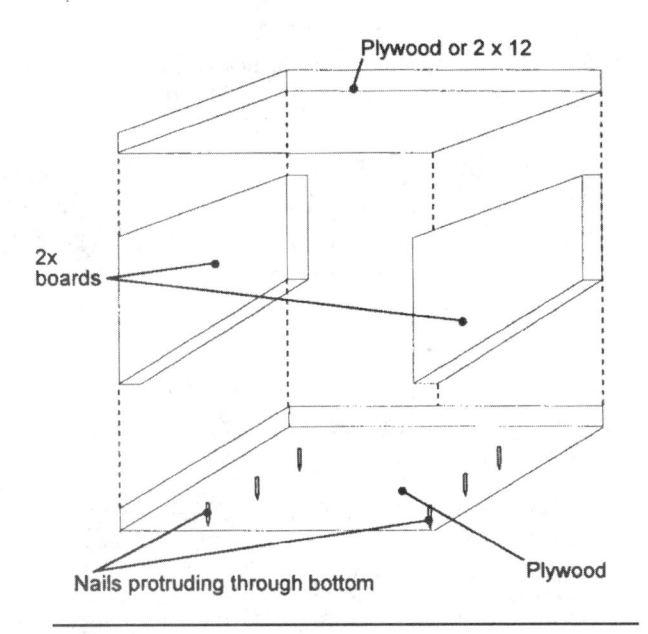

Figure 15-8 Wooden roofer's seat

Set the base of the ladder away from the roof's edge a distance equal to ¼ the working height of the ladder. Set it on level ground, and if shimming is necessary to level a leg, use plywood sheets, not blocks. Some roofers glue rubber or carpet to the ladder feet to prevent the ladder from slipping, or set the ladder feet on a sheet of plywood or waferboard.

We used to feel quite safe with the feet of the ladder pushing into the soft soil of a lawn. But now, many lawns are artificial turf. The ground under it is packed hard, and the feet of a ladder won't sink in. Add to that the fact that artificial turf is much more slippery than grass and you have a disaster waiting to happen. There are devices that firmly anchor the feet of the ladder to the ground to keep it from slipping. You may choose not to spend the 30 seconds it takes to attach these safety devices, but you may pay for it with months in the hospital.

Don't lean out on the ladder, and don't overload yourself or the ladder. When you use wood board planking over ladders or tube scaffolding, make sure the boards extend beyond the supports in the event that the boards bend under weight. You don't want the boards falling through the supports. Put rails on the outside of scaffolds to prevent falls — also an OSHA requirement. Keep metal scaffolding away from power lines.

Especially on steep roofs, apply shingles along the eaves while standing on some type of scaffolding until the roof covering has progressed to the point where you can install roof jacks (Figure 15-7) or use a roofer's seat (Figure 15-8). If it isn't possible to set up scaffolding while you work along the eaves, tie yourself to a rope that is securely fastened to a fixed object such as a tree located on the opposite side of the building. Don't make the mistake of one poor do-it-yourselfer I heard of. There wasn't a tree in his front yard, so he tied the rope to the rear bumper of his car in the driveway. His second mistake was to not tell his wife, who needed something from the store!

■ **Keep Your Eyes Open** Always watch for loose shingles and debris. Be wary of release paper that's applied to certain roll roofing products such as ice shield material; the release paper is very slippery. And fiberglass felts have very little tear strength. If you walk over them, they tear away from the nails easily. Look out for slippery areas, moss and wet leaves. Don't walk on a wet roof.

When you remove an old roof, work from the top down and periodically clear the debris. Rope off the dump zone, and avoid stepping on exposed nails.

Keep off a roof during a lightning storm. Don't touch wires crossing over the roof. If you use cranes or forklifts to raise roofing materials, watch out for overhead power lines.

■ **Watch That Pot** Keep an eye on kettle temperature since bitumens can explode if they're overheated.

■ **"High Anxiety"** Be patient with apprentices, especially when it comes to fear of heights. Don't encourage anyone to do anything they feel is dangerous. People can usually overcome this fear, but for some it takes longer than for others. A friend of mine had a rough time qualifying for a position with the local volunteer fire department because heights panicked him. This same man now teaches firefighting to new recruits in one of the largest cities in Florida.

That's a real success story. And I hope this book will contribute to your success, by opening up new horizons and helping you make your roofing business more efficient and profitable.

Appendix

A Roof-slope factors for determing rafter lengths (common, jack, hips and valleys)

B Valley length factors

C Roofing equations used in this book

Appendix A Roof-slope factors for determining rafter lengths (common, jack, hips and valleys)

(1) Roof Slope	(2) Common or jack rafters (factor x run = actual length)	(3) Hips or valleys (factor x run = actual length)	(4) Hips or valleys (factor x plan length = actual length)
1 in 12	1.004	1.417	1.002
2 in 12	1.014	1.424	1.007
3 in 12	1.031	1.436	1.015
4 in 12	1.054	1.453	1.027
5 in 12	1.083	1.474	1.042
6 in 12	1.118	1.500	1.061
7 in 12	1.158	1.530	1.082
8 in 12	1.202	1.564	1.106
9 in 12	1.250	1.601	1.132
10 in 12	1.302	1.642	1.161
11 in 12	1.357	1.685	1.191
12 in 12	1.414	1.732	1.225
13 in 12	1.474	1.782	1.260
14 in 12	1.537	1.833	1.296
15 in 12	1.601	1.888	1.335
16 in 12	1.667	1.944	1.375
17 in 12	1.734	2.002	1.416
18 in 12	1.803	2.062	1.458
19 in 12	1.873	2.123	1.501
20 in 12	1.944	2.186	1.546
21 in 12	2.016	2.250	1.591
22 in 12	2.088	2.315	1.637
23 in 12	2.162	2.382	1.684
24 in 12	2.236	2.450	1.732

Appendix B Valley length factors

Factor x Run of Low-Sloped Roof = Actual Length of Valley

Slope of Steep Roof

Low Roof Slope	1/12	2/12	3/12	4/12	5/12	6/12	7/12	8/12	9/12	10/12	11/12	12/12	13/12	14/12	15/12	16/12	17/12	18/12	19/12	20/12	21/12	22/12	23/12	24/12
1 in 12	1.417	1.123	1.058	1.035	1.023	1.017	1.014	1.011	1.010	1.009	1.008	1.007	1.006	1.006	1.006	1.005	1.005	1.005	1.005	1.005	1.004	1.004	1.004	1.004
2 in 12		1.424	1.213	1.134	1.090	1.067	1.053	1.044	1.038	1.033	1.030	1.027	1.025	1.024	1.023	1.021	1.021	1.020	1.019	1.019	1.018	1.018	1.018	1.017
3 "			1.436	1.275	1.193	1.146	1.116	1.097	1.083	1.074	1.066	1.061	1.056	1.053	1.050	1.048	1.046	1.044	1.043	1.042	1.041	1.040	1.039	1.038
4 "				1.453	1.323	1.247	1.199	1.167	1.144	1.127	1.115	1.106	1.098	1.092	1.087	1.083	1.080	1.077	1.075	1.073	1.071	1.070	1.068	1.067
5 "					1.474	1.367	1.298	1.251	1.217	1.193	1.175	1.161	1.150	1.141	1.133	1.128	1.123	1.118	1.115	1.112	1.109	1.107	1.105	1.103
6 "						1.500	1.409	1.346	1.302	1.269	1.244	1.225	1.210	1.197	1.187	1.179	1.172	1.167	1.162	1.158	1.154	1.151	1.148	1.146
7 "							1.530	1.451	1.395	1.353	1.321	1.296	1.277	1.261	1.248	1.238	1.229	1.221	1.215	1.209	1.205	1.201	1.197	1.194
8 "								1.564	1.495	1.444	1.405	1.374	1.350	1.331	1.315	1.302	1.291	1.281	1.273	1.267	1.261	1.256	1.251	1.247
9 "									1.601	1.540	1.494	1.458	1.429	1.406	1.387	1.371	1.357	1.346	1.337	1.328	1.321	1.315	1.310	1.305
10 "										1.642	1.588	1.546	1.512	1.485	1.462	1.444	1.428	1.415	1.404	1.394	1.386	1.379	1.372	1.367
11 "											1.685	1.637	1.599	1.568	1.542	1.521	1.503	1.488	1.475	1.464	1.455	1.446	1.438	1.432
12 "												1.732	1.689	1.654	1.625	1.601	1.581	1.563	1.549	1.536	1.525	1.516	1.507	1.500
13 "													1.782	1.742	1.710	1.683	1.661	1.642	1.625	1.611	1.599	1.588	1.579	1.571
14 "														1.833	1.798	1.768	1.743	1.722	1.704	1.688	1.675	1.663	1.653	1.644
15 "															1.888	1.855	1.828	1.805	1.785	1.768	1.753	1.740	1.729	1.718
16 "																1.944	1.914	1.889	1.867	1.849	1.833	1.818	1.806	1.795
17 "																	2.002	1.975	1.951	1.931	1.914	1.898	1.885	1.873
18 "																		2.062	2.037	2.015	1.996	1.980	1.965	1.953
19 "																			2.123	2.100	2.080	2.062	2.047	2.033
20 "																				2.186	2.164	2.146	2.129	2.115
21 "																					2.250	2.230	2.213	2.197
22 "																						2.315	2.297	2.281
23 "																							2.382	2.365
24 "																								2.450

Appendix C   Roofing equations used in this book

Key to abbreviations

E = exposure (Chapter 5)
E = coefficient of linear expansion (Chapter 10)
FS = factory squares
HL = head lap
L = length
LF = linear feet
LSW = last strips width
LUP = labor unit price

N = number of ribbon courses
R = recess (Chapter 1)
R = ridge length (Chapter 5)
SF = square feet
T = change in temperature
TL = top lap
W = width

Chapter 1   Measuring and Calculating Roofs

Equation 1-1: Area of a level rectangular roof = L x W

Equation 1-2: Perimeter of an irregular roof = L + W + L + W + R + R
or 2L + 2W + 2R,
or 2(L + W + R)

Equation 1-3: Perimeter of a rectangular roof = 2(L + W)

Equation 1-4: Slope = $\dfrac{\text{Total Rise}}{\text{Total Run}}$ = $\dfrac{\text{Rise (in inches)}}{12 \text{ (inches per foot of run)}}$

Equation 1-5: Pitch = $\dfrac{\text{Total Rise}}{\text{Total Span}}$

Equation 1-6: Slope = 2 x Pitch

Equation 1-7: Perimeter of an irregular sloped roof = P = 2(L + W + R)

Equation 1-8: Perimeter of a rectangular sloped roof = P = 2(L + W)

Equation 1-9: Perimeter of a gable roof = 2(Length + Actual Width)

Equation 1-10: Actual Width = 2(Run x Roof-Slope Factor)

Equation 1-11: Perimeter of a gable roof = 2[L + (W x Roof-Slope Factor)]

Equation 1-12: Allowance factor = $\dfrac{\text{Area Covered (including allowances)}}{\text{Net Roof Area}}$

Equation 1-13: Actual (Net) Roof Area = Roof Plan Area x Roof-Slope Factor

Equation 1-14: Plan Length (Hip or Valley) = 1.414 x Run

Equation 1-15: $\text{Ridge} = L - [2 \times (\frac{W}{2})] = L - W$

Chapter 3   Underlayment on Sloping Roofs

Equation 3-1: Lap Area (ridge, hip, and valley) = Total Length x 1 SF/LF

Equation 3-2: Overcut Area (gables) = LF of Rake x 0.34 SF/LF

Equation 3-3: $\text{Coverage (SF/FS)} = \frac{(\text{Roll Length} - \text{End Lap}) \times \text{Exposure}}{\text{Number of FS per Roll}}$

Equation 3-4: $\text{Exposure} = \frac{\text{Roll Width} - \text{Top Lap}}{\text{Number of Plies}}$

Equation 3-5: $\text{FS/Sq.} = \frac{100 \text{ SF/Square}}{\text{Coverage/FS}}$

Chapter 4   Asphalt Shingles

Equation 4-1: $\text{Shingles/Square} = \frac{100 \text{ SF}}{\text{Shingle Length (in.)} \times \text{Exposure (in.)}} \times 144 \text{ sq. in./SF}$

Equation 4-2: $\text{Percentage-of-Increase Factor} = \frac{\text{Recommended Exposure}}{\text{Actual Exposure}}$

Equation 4-3: $\text{Courses} = \frac{\text{Dimension of Structure}}{\text{Exposure}}$

Equation 4-4: $\text{Exposure} = \frac{\text{Dimension of Structure}}{\text{Number of Courses}}$

Equation 4-5: Top Lap = W - E

Equation 4-6: Top Lap = E + HL

Equation 4-7: Head Lap = TL - E

Equation 4-8: Head Lap = W - 2E

Equation 4-9: Exposure = $\dfrac{W - HL}{2}$

Equation 4-10: Starter Course (SF) = Eaves (LF) x Exposed Area (SF/LF)

Equation 4-11: Area (SF/LF) = $\dfrac{\text{Exposure (in.)}}{12}$

Equation 4-12: Number of Eaves Shingles = $\dfrac{\text{Eaves Length (ft.)}}{\text{Shingle Length (ft.)}}$

Equation 4-13: Net Allowance (ridge and hip units) = SF Required - SF Salvaged

Chapter 5 Mineral-Surfaced Roll Roofing

Equation 5-1: Starter-Strip Waste at Ridge (single coverage) = 4 x E x (L - R - E)

Equation 5-2: Starter-Strip Waste at Ridge (double coverage) = 5.67' x (L - R - 1.417')

Equation 5-3: Starter-Strip Waste at Hips = 4 x E x (Run - E) x Roof-Slope Factor

Equation 5-4: Starter-Strip Waste at Hips (17" exposure)
= 5.67' x (Run - 1.417') x Roof-Slope Factor

Equation 5-5: Waste at Starter Strip & Hips
= 4 x E x [(L - R - E) + (Run - E)] x Roof-Slope Factor

Equation 5-6: Waste at Starter Strip & Hips (17" exposure)
= 5.67' x [(L - R - 1.417') + (Run - 1,417')] x Roof-Slope Factor

Equation 5-7: Waste at Starter Strip & Hips (double coverage, perpendicular to eaves)
= 4 x E x (Run - E) x Roof-slope factor

Equation 5-8: Last strip width (LSW) = E x the decimal part
(numbers to the right of the decimal point)
of the expression: $\dfrac{\text{W/2 x Roof–slope factor}}{E}$

Equation 5-9: LSW = E x the decimal part of the expression: $\dfrac{L}{E}$

Equation 5-10: Waste (SF/LF) = E - (2 X LSW)
when LSW is equal to or less than half the exposure

Equation 5-11: Waste (SF/LF) = 2 x (E - LSW)
when LSW is greater than half the exposure

Chapter 6 Wood Shingles and Shakes

Equation 6-1: Hip and Ridge Units $= \dfrac{\text{Total LF Ridge and Hips}}{\text{Exposure (ft.)}}$

Equation 6-2: Total Units $= \dfrac{\text{Total LF Ridge and Hips}}{\text{Exposure (ft.)}} + \text{Number of Hips}$

Chapter 8 Slate Roofing

Equation 8-1: Squares Ordered per Square Covered equal:

$$\dfrac{\text{Slates per 100 SF at your head lap}}{\text{Slates per 100 SF at 3-inch head lap}} \times \text{Roof area (in squares)}$$

Equation 8-2: Hip and ridge slates $= 2 \times \dfrac{\text{LF of ridge and hip}}{\text{Exposure}}$

Chapter 9 Metal Roofing and Siding

Equation 9-1: Waste Factor $= \dfrac{\text{Gross Panel Area}}{\text{Net Panel Area}}$

Equation 9-2: Panels per Square $= \dfrac{100 \text{ SF}}{\text{Net Panel Area (SF)}}$

Equation 9-3: Panels per Square $= \dfrac{14{,}400 \text{ sq. in.}}{\text{Net Panel Area (sq. in.)}}$

Chapter 10 Built-up Roofing

Equation 10-1: Exposure $= \dfrac{34 \text{ inches}}{\text{No. of Plies}}$

Equation 10-2: Change of Length $= E \times T \times L$

Chapter 12 Insulation, Vapor Retarders and Waterproofing

Equation 12-1: $R = \dfrac{1}{k} \times \text{Material Thickness (inches)}$

Equation 12-2: U (coeffecient of transmission) $= \dfrac{1}{R}$

Equation 12-3: Total R-value $= R_1 + R_2 + R_3$

Chapter 15 Estimating (and Maximizing) Production Rates

Equation 15-1: $\text{Labor Unit Price} = \dfrac{\text{Total Daily Crew Cost}}{\text{Daily Crew Production}}$

Equation 15-2: $\text{Labor Unit Price} = \dfrac{\text{Total Hourly Crew Cost}}{\text{Hourly Crew Production}}$

Equation 15-3: $\text{Daily Production} = \dfrac{\text{Daily Crew Cost}}{\text{LUP}}$

Equation 15-4: $\text{Hourly Production} = \dfrac{\text{Hourly Crew Cost}}{\text{LUP}}$

Equation 15-5: $\text{Project Duration (Days)} = \dfrac{\text{Total Material}}{\text{Daily Production}}$

Equation 15-6: $\text{Project Duration (Hours)} = \dfrac{\text{Total Material}}{\text{Hourly Production}}$

Equation 15-7: $\text{Production (Crew-days/Unit)} = \dfrac{\text{LUP}}{\text{Daily Crew Cost}}$

Equation 15-8: $\text{Production (Crew-hours/Unit)} = \dfrac{\text{LUP}}{\text{Hourly Crew Cost}}$

Equation 15-9: $\text{Production (Crew-hours/Unit)} = \dfrac{\text{Crew-hours/Day}}{\text{Daily Production}}$

Equation 15-10: $\text{Required Daily Production} = \dfrac{\text{Daily Crew Cost} \times \text{Total Material}}{\text{Total Labor Budget}}$

Equation 15-11: $\text{Required Hourly Production} = \dfrac{\text{Hourly Crew Cost} \times \text{Total Material}}{\text{Total Labor Budget}}$

Equation 15-12: $\text{Total Labor Cost} = \dfrac{\text{Labor Cost to Prefabricate}}{\text{Number of Uses}} + \text{Labor Cost to Install}$

Equation 15-13: $\text{Actual LUP} = \dfrac{\text{Actual Hourly Costs}}{\text{Published Hourly Costs}} \times \text{Published LUP}$

Index

A

Accident prevention 426-428
Acid-flux solder 239
Acrylic
   carbide sealers 368
   caulk 355-356
   emulsion 296
   polymer binder 216
Acrylic roof coatings....................370
Acrylic-based adhesive 292
Actual length, common rafters......15
Actual width, roofs......................17
Adhesion, caulk.................. 355-356
Adhesive
   acrylic-based 292
   bonding 326-327
   cold-process 324
   heat activated 73
   neoprene 292
   strips.........................78, 103, 402
Aggregate
   built-up roof 277, 301-302, 304
   crushed stone............ 293-294, 325
   elastomeric roofing ... 320, 322-325
   grading 291, 293
   moisture content 293
   underlayment............................36
Aggregate-surfaced roofs........42, 86
Air circulation
   ......30, 38, 162, 323, 352, 405, 407
Algae-resistant roofing.................393
Alligatoring................. 315-316, 317
Allowance factor.....................18, 43
Alloy 245, 259-260, 268-269, 272
Aluminum
   coatings 277, 294, 303, 305
   composites..............................330
   flange.....................................309
   flashing....................................64
   foil vapor retarder 352
   paint..........................388, 398-399
   pigment flakes 296
   roof coating 296-297
   shim.......................................396
   shingles 273
   soldering................................259
   tape..............................389, 412
Aluminum-core shingles.............273
Aluminum-faced insulation ...339, 342
American method..........................76
Anchor
   bar..............................326, 329
   clips......................................249
Anti-ponding metal 192-193
Antimony 269
   solder....................................258
Apex tile.........................202, 216
   trim.......................................203
Apprentices...............................428
Apron flashing 158-159, 236, 252
Area
   level roof 7-8
   sloped roof 17-21
Asbestos 292
Asphalt
   bitumen 283, 319, 321, 344, 349
   bituminous....................... 328-329
   cement 291
   dead-level......... 297, 306, 324-325
   emulsion....................291, 294, 367
   flux 36
   mastic..............................291, 367
   plastic cement.........................95
   primer.....................94, 282, 286
Asphalt shingles 71-120
   buckling..................................102
   built-up roof 285
   bundles............................108, 109
   cap 89-90
   colors............................... 74-76
   components 71-72
   coursing 79-81, 105-106
   dimensional 75
   estimating costs 119-120
   estimating quantities... 105, 109-118
   fiberglass 72
   flashing............................. 93-101
   free-tab 78-79
   giant.......................................76
   high-slope applications 84-85
   hip units......................... 110-112
   individual 76
   installation......................... 76-83
   interlocking.............................76
   laminated 75
   life expectancy 71
   low-slope applications 84-85
   organic...................................72
   patterns 79-82
   per square requirements ... 104-105
   plies 109
   random-tab 75
   re-roof, over 401
   re-roofing with 401-404
   removing 397
   repairing 390
   ridge units..................... 109-111
   self-sealing 73
   slate roofing 235, 236
   slope requirements 74
   square-butt............................75
   square-tab...............................75
   stapling 103-104
   starter course 78, 109-119
   storage 73-74
   stretch 24
   strip 75-76
   tabs 75
   thick-butt 75
   three-tab 75-76
   triple-tab 75
   two-tab 76, 103
   underlayment........... 43, 61, 66-68
Asphalt-faced insulation 324
Asphalt-fibrated emulsion............291
Asphalt-laminated paper 352
Asphalt-saturated cotton 360, 388
Asphalt-treated lumber................327
Asphaltic
   paint.....................................367
   primer......................296, 362-363
Attic
   inspection......................388, 389
   insulation..............................347
   ventilation 29, 347, 407-409
Average manhour costs................418

B

Back-nailing..........................48, 283
Backer rod 369
Bacteria 314, 366
Baffles 56, 188
Ballast 293, 322-325
Ballasted elastomeric system ....
.... 323, 328-329
Bar-anchored system.................329
Barge board............................201
Barge tiles 198
Base flashing
   asphalt shingles 93-96, 96
   built-up roof 284, 302-303
   elastomeric roofing 333
   slate roofing 236
Base mat................................ 71-72
Base sheet
   built-up roof ....
   ....283-284, 288-289, 294-295, 300
   coated 284
   fiberglass 36-37
   roll roofing 121-122
   underlayment.................... 36-37, 47
   vapor retarder 279, 284-285
   vented 284, 317-318, 353
   waterproofing.........................364
Basement walls 360-361
Basket strainer............... 411-412
Batt insulation 337-339, 347
Battens
   counter................... 190, 192-193
   expansion 254-256
   integral 249
   metal roofing.................... 249-250
   seams................256-257, 269-271
   slotted 191
   spaced 161-162
   strip 327
   tile roofing.................................
   ...190-192, 194, 196-197, 212, 214
   wood 254, 256
Bauxite 293
Bell eaves 161-162
Bellows 309
Bentonite clay 291
Beveled
   gutter....................................410
   siding............................398, 404
Binders, plastic...........................36
Bird holes 194-195
Birdstop 193
Bismuth, solder 258
Bitumen
   asphalt 289, 362-365
   built-up roof ....
   ....278-279, 281-310, 313, 315-316
   built-up roof 318
   coal-tar 288-304, 301-302, 363
   cold-applied..........................291
   cutback in 291, 362-363
   storage 291
   types.....................................290
Bitumen trap............................304
Bitumen-saturated cotton 289, 362
Bituminous
   fiber pipe 361
   paint..............................66, 182, 235
Black steel 266
Blanket insulation 338-339, 347
Bleaching 394
Blind nail method......................127
Blisters
   .... 284, 290, 314-317, 319, 388, 404, 407
Blisters, roof surface 373, 380
Blocking...........................28, 29
Blown insulation 339-341
Body-colored tile 188
Bond lines 75
Bonding adhesive...............326-327
Boston hip 230-231, 242
Brass, solder 258
Brazing....................................235
Bright plate............................260
British thermal units (Btu) 350
Broken slates 392
Bubble level 12
Bubbles 316-317
Buckled deck...........................29
Buckling underlayment.........61, 123
Building code 23, 30, 36-38, 46,
   61, 64-65, 69, 157, 184, 190, 194, 250
Building insulation....................337
Building paper.................. 178-181
Building protection 426
Built-up roofs 277-331
   aggregate-surfaced ... 277, 293-294
   base sheets..................... 284-286
   cold-process 292
   estimating costs 312-313
   estimating quantities 310-312
   fasteners 280
   flashing..........297-298, 302-305
   four-ply 289, 295-296
   insulation....................279, 344, 347
   life expectancy 277
   membranes 286-289
   metal roofing over....................273
   mineral-surfaced 277, 286
   repairs........................... 315-318
   roll-roof 295
   slope 278
   smooth-surfaced........ 277, 294-295
   substrates......... 24, 27, 36, 278-283
   testing...................................313
   three-ply 288, 295
   vapor barriers 353
   warranties.......................... 314-315
Bull..399
Butt thickness.................. 151-152
Butted corner.........................177
Butting-up method 402-403
Butyl
   caulk........................355, 410, 412
   rubber 322, 366

C

Caliche 293
Canoe valley..................... 233-234
Cant strip
   built-up roof 285, 299, 306, 307
   slate roofing 226, 230
   tile roofing..................... 192-193
   wood shingles & shakes............182
Cap bead...................................355
Cap sheet
   built-up roof ....
   ....285, 302, 305, 307, 315
   modified asphalt............296, 302

roll roofing 121
tile roofing 190
underlayment 37
Cap shingles 89-90
Cap strip 326
Capillary action 366
Capping in 35
Caps, tin 44, 349
Cathedral ceiling 192, 354
Caulk 355-359
acrylic 355-356
adhesion 355-356
butyl 355, 410, 412
doors 346
epoxy 355
estimating quantities 356-357
fireproof 357
latex 355
life expectancy 355
masonry 96
polysulfide 355
polyurethane 355
roof repair 389, 407, 412
silicone 355-356
solvent acrylic 356
vinyl acrylic 356
windows 346
Caulking tubes 357, 359
Ceiling
insulation 60-62, 338-340, 350
joists 25
ventilation 354
Cellular glass insulation 349
Cement
hydrolithic 368
lightweight insulating 283, 319, 325, 344
roofing 39, 46-48, 61, 66-68, 78-79, 85-86
smudges 134
Cement-fiber boards 281
Cementing, roofing 41
Centipoise 291
Ceramic granules 301, 321
Certi-guard 149
shakes 149
shingles 149
Certi-last 150
damaged 391-392, 394
edge grain 151, 395-396
fire-resistant 150
flatgrain 151, 395-396
pre-cut 156-157
rebutted 175-176
shingles 150
slashgrain 151, 395-396
Certi-sawn 153
covering capacity 164-165
exposures 156
fire-retardant 149
grading 153
hand-split 165
machine-grooved 153
repairing 392
resawn 165
sheathing 24, 29-32
sidewall 175
staggered pattern 175
starter-finish 156
straight-split 153, 164-165
tapersplit 165
underlayment 35
valley flashing 69
Change orders 425
Chimney flashing
asphalt shingles 93-97
metal roofing 251-252
roofing repair 388, 399
slate roofing 236-237
tile roofing 210, 212
wood shingles & shakes 159
Chipper 201
Chlorinated polyethylene 322
Chlorine bleach 394
Chlorosulfonated polyethylene 321, 322, 328
Chromium 270
Cleats ... 235, 237, 239, 258-259, 271-272
Clip hangers 411
Clips 239, 268
anchor 249
edge 29
panel 249
storm 196-197, 200, 203
Closed valley 232, 233
Closed-cut valley 87-89
Closure caps 248
Closure strip
metal roofing 248, 250, 252
tile roofing 193
Coal-tar bitumen 289-306
elastomeric membranes 322
epoxy coatings 368
Coated
base sheets 284
metals 366
roll 61
sheet 121, 286, 300-301
Coatings, roof 369-385
Coefficient
of expansion 254, 256, 270, 309
of transmission 351
Coke 289
Cold asphalt emulsion 291
Cold-applied bitumen 122, 123, 286, 362, 363, 367
Cold-process
adhesive 324
roofing 46, 292
Cold-rolled
copper 271
metal roofing 246
Collar beam 25
Color patterning 34
Colorless sealers 367
Comb ridge 229-230
Combing slate 229-230
Commercial standard slate 222
Common rafter 13
Compressed particleboard 342
Compression insulation 298
Concealed clip panels 268
Concealed-nail method 134, 138
Concentrated roof loads 32
Concrete cant 299
Concrete form lumber 26
Condensation 29, 68-69, 352-354
Conductor 411
Continuous
cleat 285
clip 251
flashing 97-98, 133, 207-210, 251-252
vent 409
Controlled-flow drainage system 93, 301
Cool Roof Rating Council 370
Copper
composites 329
cornice temper 271
flashing 64-67, 69, 232, 235-237, 271
roofing ... 40-41, 253-254, 256, 271-274
seams 235
sheet metal 232
shingles 273
solder 258
waste factors 271
Copper-bearing steel 272
Cork board 342
Corner
laced 178, 184
woven 177
Corner flashing 159, 176, 208-209, 360
Cornice temper copper 271
Cornice, open 30
Corrugated
roofing 275
siding 275
Corrugated metal
downspouts 410
panels 30, 247-248, 267
roofs 24, 30, 251, 264, 275
Cotton fabric 362
Counter battens 190, 192-193
Counterflashing
asphalt shingles 98
built-up roof 284, 302-304, 308
metal roofing 261
roofing repair 389, 399
slate roofing 236
tile roofing 208-212
Coursing roll roofing 126-132
Coxcomb ridge 230
CPE 322, 328
CPE elastomeric 328
Craft codes 418
Cranes 428
Creosote 362
Creosote-treated lumber 327
Crickets 93-95, 158, 210, 212, 236, 252
flashing 93-95, 158, 210, 212
valley 95
Cross seams 258-259, 271-272
Cross-bond method 198
Crossover 308, 309
Crushed tile 293
Crushed stone 293, 325
CSPE 322, 328
elasticity 319
Cupped shingles 392
Curb
flange 308, 309
flashing 305, 327
roof 259
Curing, roof coatings 371
Curled shingles 391, 395
Cushion strips 364
Cutback bitumen 289, 291, 294, 362-363, 367
Cutouts 73, 75, 86, 103, 402
Cutting allowance 18, 42

D

Daily production 414
Damaged sheathing 398
Damaged tabs 391
Dampproofing 360, 367-368
bituminous 366-367
cavity wall 367
existing wall 367-368
hydrolithic 367-368
Dead load 27-28, 301, 323
Dead soft copper 271
Dead-level
asphalt 297, 306, 324-325
roof 289
Decay-resistant wood shingles 152
Decking
concrete 282
defects 278
false 162-163
fasteners 280
fluted metal 247
grades 32
gypsum 281, 284, 286, 319, 344
gypsum concrete 280
heavy-load bearing 301
lightweight insulating 279-280
metal 163-164, 279
metal roofing ... 296, 348, 388, 397, 401
nailable 283, 286, 294-296
non-nailable 282-283, 286, 294-296, 300-301, 318, 325
perforated steel 281
precast concrete 301
roof 23-24, 32, 74
staggered joints 26, 29
structural cement fiber 286
structural wood fiber 282
tongue and groove 26
ventilation 281
wet-fill 279, 303
wood 282, 286
Delays 425
Demolition, roofs 396-400
asphalt shingles 396
flashings 398
gravel 398
metal 397
roll 397
shake 396-397
slate 397
tile 397
wood shingles & shakes 396, 397
Demolition, siding 398
Deterioration
deck 29, 35, 407
flashing 388
metal roofing 245
underlayment 29
Diagonal
method 29, 83
pattern 82-84
trails 84
Dimensional shingles 84
Direct-nail method 46
Discoloration 393
asphalt shingles 73, 393-394
concrete tiles 394
Diverters
splash 63-65, 66, 412
water 203
Dolomite 293
Door flashing 176, 360
Doors
insulating 346
storm 346
Dormer ridge 87
Dormers 92-93
asphalt shingles 81-83, 92-93
estimating labor 422
flashing 209-211, 236-237
siding, removing 398
slate roofing 236-237
underlayment 63
wood shingles & shakes 159
Double-coursing 175, 179-180

Double-coverage
roll roofing.......122, 132, 134, 139
underlayment.................. 39-40, 50
Downspouts................ 393, 410-412
Drain slots...................................191
Drainage............. 278, 301, 409-410
Drip edge
asphalt shingles...........................76
estimating quantities 42-43
metal roofing...................248, 259
roll roofing.......................127, 131
roofing repair 398, 400, 406
tile roofing............... 190, 192-193
underlayment........... 35, 41-42, 43
wood shingles & shakes...........154
Dry hip...............................202, 203
Dryer vents.................................352
Drying in.......................................35
Dubbing...86
Dump zones 428
Duration of project......................414
Dutch lap method..........................76
Dutch weave................................175
estimating quantities 175
fasteners 157
flashing............. 158-159, 182-184
gauged......................................175
installation................................175
insulation, rigid 161-162
low-slope applications 160-161
mansard roof 161, 162
nails...157
panels 184-185
patterns.....................................175
roof junctures 182-184
sheathing under............. 24, 29-30
sidewall 175
staggered patterns....................175
steep-slope applications 161
underlayment.............................39
valleys............63, 68-69, 156-157
white cedar 175

E

Eaves
bell....................................161, 162
closures 347
estimating....................................18
flashing.................. 35, 60-61, 304
length...58
protection ...30, 31, 56, 68-69, 154-155
roll roofing 123, 127, 136
roofing repair 389
swept 161-162
trough 411
underlayment........................ 43-44
units..110
ventilation 347
Edge clips.....................................29
Edge grain wood shingles ....
..........................150, 152, 395, 396
Edge strips... 128, 284-285, 299, 404
Edging, roof..................................41
Efflorescence, on tile 368
Elastomeric roof coatings ... 372-375
application........................ 372-375
Elastomeric roofing............ 319-335
ballasted 322-329
composite 329-330
waterproofing...........................362
Elbow 410-412
Electric thermal wire.....................62
Electrolysis.................................235
Emery..368
Emulsified asphalt.......................363
Emulsifier....................................291
Enamel-coated shingles 273
End caps......................................412
End covers..................................248
End laps..... 44-45, 46, 265, 270, 295
End-wall flashing........................273
Envelope strip.....................285, 299
EPDM 193, 320, 322-335
equipment stand 306-307
estimating............................. 69-70
estimating costs............... 334-335
estimating quantities 334
failure.......................................317
fiberglass.........................302, 360
flashing.....................................333
flat roof penetration.......... 305-307
foundation 360
front-wall....................................98
fully-adhered 322-329
head 360
hypalon............................322, 328
insulation......................... 325-326
life expectancy 324
liquid 322
mechanically-fastened...... 322-330
metal...............................193, 196
neoprene...................................322
rubber..............................193, 196
tile roofing............... 192-194, 196
EPDM elastomeric............. 324-327
epoxy sealant............................356
fiberglass shingles............... 38-39
hypalon elastomeric 328
polysulfide caulk......................355
polyurethane caulk...................355
roll roofing 121
silicone caulk 355-356
slate roofing 39-40, 221
solvent acrylic caulk 355
tile roofing................................187
EPDM elastomeric systems 324-328
flashing.....................................320
waterproofing...........................366
EPDM rubber roofing, liquid...........
.... 376-377
Epoxy
caulk...355
resin..412
sealant, life expectancy 357
Equations, roofing.............. 432-436
Equiviscous temperature............291
Estimating costs, labor........ 413-428
asphalt shingles........................120
built-up roofs................... 312-313
elastomeric roofing 335
insulation..................................340
metal roofing............................275
roll roofing...............................147
slate roofing 244
wood shingles & shakes...........186
Estimating costs, materials
asphalt shingles 119-120
built-up roofs................... 312-313
elastomeric roofing 334
metal roofing............................275
roll roofing 147
slate roofing 244
wood shingles & shakes...........186
Estimating material quantities
asphalt shingles 108-109
built-up roofs................... 310-312
caulk................................ 356-357
drip edge.......................... 42-43
elastomeric roofing 334
flashing................................. 69-70
gutters.......................................412
hip units........... 110-112, 216, 218
interlayment 57-60
metal roofing & siding..................
..........................259-262, 264-267
ridge units........................ 109-111
roll roofing 134-146
roofing repair 406-407
shakes 166-175
sheathing 34
slate roofing 239-244
tile roofing....................... 215-216
underlayment...................... 49-56
valley flashing...................... 69-70
waterproofing...........................365
wood shingles & shakes... 166-175
Ethyl alcohol 368
Exceeding dead load 27
Exhaust fans 352
Expanded insulation
polystyrene...............................348
urethane........................... 342-344
Expansion..
... 296, 323, 348, 355-357, 410, 411
batten.......................................254
cleat...................................249, 256
coefficient 254, 256, 271, 309
joint sealers 357
joints................ 302, 307-309, 364
Exposed insulation............. 342-343
Exposed-frame ceiling 350
Exposed-nail method ....
.................125, 129, 131, 134, 137
Exposure gauge..........................424
Exposure, weather
asphalt shingles 73-74, 103
built-up roof 287-289
decking 30-31
re-roofing 402- 403
roll roofing 122
sheathing 30-32
underlayment.................45, 56, 57
wood shingles & shakes...........164

F

Fabrics................................286, 289
composites........................329, 366
polyester........... 289, 321-322, 328
waterproofing...................362, 364
Face grain, plywood......................27
Factors
allowance 18, 42-43
hip-slope.............................. 14-15
percentage-of-increase 215, 403
roof-slope 430
valley length............................431
valley-slope.......................... 14-15
Factory square................... 36-37, 50
Fancy-butt wood shingles 176
Fantail hip 230-231
Farm tile.....................................361
Fascia 44-45
cover...304
raised 193, 196
rake..24
roofing repair 398, 406
sloping..14
tile roofing........ 192-193, 196, 197
wood shingles & shakes... 154-155
Fastener spacing
nails 27-29
staples..29
Fasteners
asphalt shingles 102-103
built-up roof 279-283
decking.....................................280
elastomeric roofing ....
..................322, 325-327, 328-329
mechanical... 279-283, 328-329, 349
metal roofs and siding..............259
slate roofing 234-235
stagger nailing............................78
staggered nailing.............103, 128
tile roofing....................... 194-196
wood shingles & shakes...........157
Feathering strips.................404, 405
Felt 35-41
asphalt-saturated 36, 121, 277
buckling.............................. 37-38
coated 36, 277, 284, 286
exposure 288
organic......................... 36-37, 287
roofing...................................26, 31
shim..396
slate-faced 295
strips................................297, 305
tar-saturated...............36, 277, 301
tearing 37
warped..26
waterproofing.................. 362-365
Fiberboard
high-density.............................325
insulation.................283, 284, 348
rigid 299
Fibered asphalt mastic.................363
Fiberglass
asphalt-saturated 36-37, 287
corrugated sheets.....................267
fabric 289, 362, 368
felts, built-up roof ....
........................286, 287, 296, 312
felts, waterproofing...349, 363, 364
flashing....................................302
insulation.......... 325, 338-341, 352
insulation, built-up roof ....
...............................281, 284, 301
shingles 24, 38-39
underlayment....................... 36-37
Fiberglass (GRP) roofing............381
preparation for.........................381
Fibrated emulsion........................367
Field slates 227, 242-244
Fire-resistant
asphalt shingles 72
elastomeric roofs 328
felts.................................287, 293
Fire-retardant, wood shingles &
shakes.......................................149
Fishmouths................. 290, 316-317
Flame-resistant insulation 339
Flange
curb 309
low-profile................................309
rubber 98
straight.............................308, 309
vent................... 101-102, 158, 210
Flash point..................................291
Flashing
aluminum 64-65
apron 158-159, 236, 252
asphalt shingles 93-101
built-up roof......297-298, 302-305
cement 47, 333
continuous ....
.......97-98, 133, 207-210, 251-252
copper...64-67, 69, 232, 235, 237, 371
corners...... 159, 176, 208-209, 360
cricket.......... 93-95, 158, 210, 212

curb 305, 327
demolition 398
deteriorated 406
door 176, 360
dormer 209-210, 236-237
eaves 35, 60-61, 304
edge 132
elastomeric roofing 333
end-wall...................................273
step 158-159
window......................................176
Flashing, base
asphalt shingles 93-96
built-up roof 284, 302-303
elastomeric roofing 333
slate roofing 236
Flashing, cap
asphalt shingles 93-97
built-up roof 285, 303
slate roofing 235, 236
Flashing, chimney
asphalt shingles 93-97
metal roofing 251-252
roofing repair 388, 399
slate roofing 236-237
tile roofing......... 158-159, 210, 212
Flashing, counter
asphalt shingles 98
built-up roof 284, 302-304, 308
metal roofing 261
roofing repair 389, 399
slate roofing 236
tile roofing....................... 208-212
Flashing, roll roofing ... 124, 132-133
built-up roof 291
repair 399
underlayment......................64, 67
Flashing, valley
asphalt shingles 76
metal roofing 250-251
roll roofing 124
tile roofing 190, 203-206
underlayment........ 35, 49-50, 63-68
wood shingles & shakes 154
Flashing, vent
asphalt shingles 98, 101-105
elastomeric roofing 333
roll roofing 132-133
slate roofing 238
tile roofing.................. 210, 213-214
Flat asphalt..................................363
Flat seam, metal roofing ....
...............................254-255, 270-272
Flat
slate 222, 225
spade......................................397
Flatgrain wood shingles ....
........................150, 152, 395, 396
Flexible asphalt 72
Flood coat................... 291, 293-294
Floor insulation ... 345-346, 351, 354
Flue 388
Fluted metal deck......................247
Fly rafter......................................24
Foam insulation..............................
...278, 320, 332, 342-344, 351-352
glass............................329, 342, 348
Foam, sprayed polyurethane
(SPF)............................. 375-376
Foil-faced insulation 338-339
Foot traffic
built-up roof 278, 293
elastomeric roofing 321
metal roofing 254
roofing repair 387, 402, 406
slate roofing 225

underlayment......................35, 61
Forklifts....................................428
Foundation flashing 360
Foundation, pier and beam........354
Four-ply
built-up roof 288, 295-296
waterproofing 364-365
Framing, ladder...........................23
Free lime 394
Free-tab shingles 78-79
Freedom Gray 269
Freezebacks........................ 60-61
Full-lace valley..........................389
Fully-adhered system...326, 329, 334
Fungus
built-up roof 314
roofing repair 389, 393
waterproofing.........................366
wood shingles & shakes... 149-150
Furring strips...........................182
Fusion......................................290

G

Gable
molding..................................154
overhang..................................30
roof landing..............................33
Gaco roof coatings............... 377-380
Galvalume.........................245, 272
Galvanic corrosion 259
Galvanized metal
drip edges 41
flashing.....................................64
metal roofing 268
roofing 272-273
roofing repair 391-392, 396
Galvanized steel shingles............273
Gauged shingle patterns 175
GE Enduris 371
Glass
bead binder............................355
fiber mats 36
insulating...............................346
Glaze coat, built-up roof ....
.... 290, 293-294, 297
Glazed tile.................................188
Gooseneck................................410
Graded mineral aggregate ...291, 293
Graduated roof, slate ....
.... 222-224, 227, 232
Gravel.............. 293, 324-326, 398
Gravel stop 42
built-up roof 285, 304
elastomeric roofing 326
Green
concrete.................................367
lumber.....................................26
Gross
roof area 17-18
vent area 408-409
GRP (fiberglass) roofing............381
Gutters
basket 412
built-up roof 278
debris..............................389, 412
estimating quantities 412
half-round..............................410
hangers 410-411
molded..................................411
ogee.....................................410
outlet....................................411
rectangular...........................410
roofing repair ... 392, 399, 410-412
round...................................410

run 411
spikes........................... 410-411
straps 410-412
wood shingles & shakes...........155
Gypsum
decks...............281, 284, 286, 319
insulation................. 294, 343-344

H

Hail damage
built-up roof 294-295
elastomeric roofing 321
metal roofing 273
repairing 390, 394-396
Half pattern 79
Half-round gutter 410
Hand lugs.................................196
Handsplit wood shakes 152, 165
Hard
copper....................................271
lead................................235, 246
Head flashing 360
Head lap
asphalt shingles 108-109
slate roofing 224-229, 239, 240
Head lugs.................................189
Headers 60-61
Heat
aging.....................................329
loss, interior..............................60
transmission 350-351
Heat-activated adhesive 73
Heavy-load bearing decks...........301
Hexagonal shingles 401
High-density fiberboard 325
High-tin solder 259
High-wind precautions
asphalt shingles 73, 90-92
built-up roof 294
elastomeric roofing ....
.... 325-326, 331-332
roofing repair 390
underlayment..........................46
Hip
caps.....................................273
covers... 126, 127, 129, 130, 132, 159
flashing...................160, 230, 398
underlayment..................45, 50
Hip (jack) rafter................ 13-14, 21
Hip roof landing..........................34
Hip units
asphalt shingles 89-91, 110-112
gross roof area......................18, 50
roll roofing 135
roofing repair 398
slate roofing 231, 241
tile roofing........ 201-203, 216, 218
wood shingles & shakes...160, 168
Hip-ridge junctures 90-91
Hip-slope factors 14-15
Homeowners' associations,
restrictions.............................374
Hooded pitch pan 305-306
Horizontal supports......................25
Horse feathers 182
Hot-dip process 245-246
Hot-rolled, metal roofing 246
Hourly production......................414
Humidity, relative.......................29
Hurricane precautions 46, 196
Hydration 361
Hydrolithic cement....................368
Hydrostatic head 361, 366
Hypalon roofing 324, 328

I

Ice damage precautions
asphalt shingles 84, 93
gutters...............................410, 412
ice dams 60-62
ice shields 60-61
insulation..............................347
sheathing................................24
underlayment................. 60-62
Impact marks............................395
Indium, solder 258
Inert fillers................................301
Infrared radiation 328, 339
Injuries, roofing........................320
Installation, TPO roofing 331-332
Insulated doors.........................346
Insulating glass.........................346
Insulation...........................337-352
aluminum-faced 339, 342
asphalt-faced 324
batt....................... 337-339, 347
blanket.................. 338-339, 347
blown...................339-341, 353
building 337
ceiling.....................................
........60-62, 338-340, 345-347, 407
cellular glass.........................349
compressed............................298
fiberboard 283-284, 348
flame resistant 339
floor................ 345-346, 351, 354
foam..........................
.... 320, 329, 332, 342-344, 349
foil-faced...........338-339, 342-343
installation................. 338-340
lightweight 279-280
lightweight concrete fill 344
loose fill 339-341
masonry wall 339-340
mineral fiber 342
mobile home................... 338-339
paper-faced.................. 338-339
perimeter 342-343
perlite 339-341
phenol formaldehyde 325
polystyrene...... 320, 329, 342-344, 349
polyurethane........... 321, 342-344
poured 339-341
rigid 161, 163, 282, 320,
324, 332, 342-343, 348-350, 404
rigid-fit 338
rock wool 338-341, 351
roof.......................................278,
284-287, 296-297, 300-301, 307,
324-326, 328-329, 347-349
slab 343
sound-control batts................339
sprayed 359-360
sprayed foam 342-344
stops 283, 304
taper......................................306
tapered.................................343
unfaced 338-339
vermiculite 339-341
wall............322, 338-343, 345-346
Insulation, R-values
blown fiberglass 340
blown rock wool 339-341
calculating 351-352
fiberglass batt 338-339
loose fill 339-340
recommendations 345
rigid 342
rock wool 338

440

Insurance
  homeowner's... 24, 187, 221, 394-395
  workers' compensation 426
Integral
  battens 249
  standing seam 259
Interior
  heat loss................................60
  partitions 352
Interlayment ....
  ....32, 39-40, 56-57, 152, 154
  coverage 57
  estimating quantities 57-60
Interply bitumen.......... 313, 362-363
Iron
  oxide..................................366
  shingles 273
  staining 393
Isobutylene-isoprene elastomers ...366
Isocyanurate foam
  insulation................. 342-343, 349

J

J-bead 209
  lead 69
  leaks 390
  non-metal 67-68
  pan 200, 207-209, 212
  parapet wall 304
  pipe 305
Jack rafter.......................... 13-14
  run 21
Job
  complexity.....................413, 422
  delays 425
  organization..........................425
Joint spacing, plywood.................29
Joint tape 327
Joints
  expansion 302, 307-309, 364
  malleted 255
  mitered 159
  staggered 26, 29
Joists......................................25
Junctures 182
  apex 184
  concave 183
  convex 183
Jute 289, 362

K

K-values 352
Kaolin.....................................294
Kerosene 426
Kettle 291, 320, 365
Keys 75
Kiln-dried lumber.........................26
Knots 155

L

L-brackets 163
Labor costs 413-428
  adjusting 416, 420-423
  asphalt shingles 120
  built-up roof 312-313
  elastomeric roofing 335
  estimating 413-428
  metal roofing........................275
  per unit 413-414
  production rates........... 413-414
  published 417-420

reducing................................422
  roll roofing 147
  roof coatings 383-384
  slate roofing 244
  wood shingles & shakes...........186
Laced corner.....................178, 184
Lacing 56-57
Ladder framing............................23
Ladders...........................42, 427
Laminated, roof decking 32
Lap
  areas 50
  caulk 322, 325-327, 329
  cement 322, 328
  head....108-109, 224-229, 239-240
Last strip width 140
Latex caulk..............................355
Lath strips..... 179-180, 190-191, 230
Lead
  apron 207
  flange........................210, 213
  flashing 64, 68-69
  roofing...............................40
  saddle 68
  shield 235
  skirt 66, 203-206, 213
  sleeve..................................98
  soaker 200-203
  strips 235
Lead-tin alloy 260
Leaders..................278, 410, 411
  head 411, 412
  strap 411
Leaks
  asphalt shingles 101
  eaves 389
  chimneys 388
  flashings 60-61, 63, 68, 389-390
  mid-roof 390
  repairing 390-392
  roll roofing 133
  tile roofing 198
  underlayment........................35
  valleys 389
  vent....................................390
  wood shingles & shakes...387, 391
Life expectancy asphalt shingles....71
  built-up roofing 277
  butyl caulk.........................355
Lifespan..................................275
Light reflectivity, TPO roofing ...331
Lightning................................428
Lightweight insulating concrete......
  ....283, 319, 325, 344
Limestone........................72, 293
Limited service warranty 314-315
Liquid
  elastomers 365-366
  emulsions 291
  water repellents 367
Liquid EPDM rubber roofing...........
  376-377
Live load 27-28
Loads
  dead............. 27-28, 301, 324-325
  live..............................27-28
  uneven 32
Lock seam 237, 259
Long-life roofs................. 38, 121
Lookout rafters................... 23-24
Loose insulation 347
  knots 27
  mortar 389
  shingles 390
Loosely-laid systems..................
  ....320, 324, 329, 330
Louvers 408

Low-profile flange 309
Low-slope roofs ....
  ....28, 35, 150, 152, 289
Lugs.........................188-189, 196
Lumber
  asphalt treated 327
  concrete forms.......................26
  creosote-treated 327
  green...................................26
  kiln-dried 26
  pressure-treated 192
  scrap 26

M

Machine-grooved shakes 153, 176
Malleted joints 255
Manhours, estimating labor
  asphalt shingles 120
  built-up roofs................. 312-313
  costs.......................... 413-428
  elastomeric roofing 335
  insulation...........................340
  metal roofing.......................275
  roll roofing 147
  slate roofing 244
  tile roofing 219
  wood shingles & shakes..........186
Mansard roofs ...161-162, 185, 246-247
Marble chips.............................293
Masonry
  asphalt shingles 96
  built-up roofing 299, 303
  insulation..................... 339-340
  waterproofing 360, 367-368
  wood shingles & shakes..........182
Mastic.....................................294
  asphalt 291, 363, 367
Material costs, roof coatings 381
Material waste..............................5
Maximum span for plywood.........28
Mean temperature 24
Measuring roofs 5-6
Mechanical fasteners
  built-up roof 279, 281-283
  elastomeric roofing 322-330
  insulation...........................349
Membrane
  built-up roof 286-289
  elastomeric 322
  vented 284
Membrane roofing 277
Metal
  clips............................. 65-67
  closures 191, 196
  coatings 245-246
  decks 279
  eaves drip 197, 201
  foil composites 366
  patches........................26, 398
  valley flashing 63-64, 389
Metal roof decks 296, 348
  roofing repair.........388, 397, 401
Metal roof overhangs 248
Metal roofing 245-275
  aluminum 267-268
  batten seams 249
  cold-rolled 246
  copper..................... 271-274
  corrugated.............. 24, 30, 275
  decking requirements.............247
  estimating labor costs............275
  fasteners 259
  flashings 250-252
  galvanized 260-268, 272

hot-rolled 246
  lead...................................269
  life expectancy 245
  Monel 269
  ribbed 24, 29, 261-263
  ridges 248
  stainless steel......................270
  standing seams 248
  valleys 250
Metal roofing, estimating quantities
  copper..................... 269-271
  corrugated sheets.................275
  per square 259-262
Metal roofing waste factors... 259-260
  copper...............................271
Metal shingle fasteners 273
Metal shingles 24, 272-274
  underlayment............... 274-275
Metal siding 259, 268
  aluminum 272-274
  estimating quantities 259-262
  siding panels.................. 261-263
Micrometer.............................321
Mid-span rafters 33
Mildew 149, 352, 389
Mineral
  colloids 292
  fiber insulation 342
  fillers 296
  granules ... 71-72, 121, 294, 296, 390
  spirits 296
  stabilizers 72
  wool..................................338
Mineral-surfaced roll roofing.........
  121-137
Minimum
  live load............................27
  roof slope 31, 38-40
  span, plywood 28
  underlayment............... 38-40
Miter............................. 411-412
Mitered
  corner 177-178
  hip slates 230-231
  joints................................159
  tiles 200-201
Mobile home insulation 338-339
Mobile placers.........................425
Modification factor 421-423
Modified bitumen asphalt (MBA) ....
  .... 122-123, 285
Moisture content of aggregate 293
Mold.....................................352
Molded gutter..........................411
Monel 245, 269
Mopping, spot ....
  ....282, 284, 296, 318, 323
Mortar
  built-up roofing 303
  dampproofing.....................368
  flashing.................. 95-97, 360
  roofing repairs...............388, 390
  slate roofing 232
  tile roofing ... 188, 193-202, 207, 214
  waterproofing 361
Mortar-set method.....................48
Moss, roof 149-150, 373
Muriatic acid 246

N

Nail, sizes...............................27
Nail hole.....................388, 390, 391
Nailable decks, built-up roofs ....
  .... 295-296, 299, 305, 318

441

Nailing flange.................304-305
Nailing strips
  built-up roof.............................299
  sheathing..............................31, 35
  slate roofing........................229, 234
  wood shingles & shakes...........161
Nailing, plywood................................29
National Construction Estimator
  asphalt shingles........................120
  built-up roofs.............................312
  elastomeric roofing..................335
  metal roofing.............................275
  production rates...............413-414
  published costs................417-423
  roll roofing................................147
  sheathing....................................34
  slate roofing..............................244
  tile roofing................................219
  unit prices..................................413
  wood shingles & shakes...........186
Negative method............................7-8
Neoprene.........................334, 365-366
  adhesive....................................292
  sealants.....................................359
Neoprene-hypalon roofing.............322
Net roof area...17-19, 21, 51, 58, 146
Net vent area.................................409
Nickel...270
No-tab shingles.......................90, 103
Non-metal valley flashing.........65-67
Non-nailable decks
  built-up roof
    ....282, 294-296, 300-301, 318
  elastomeric roofing..................325
Non-skid surface............................321
Nose lugs...............................188-189
Nosing..41
Number of eaves units..................110
Number of slates required............240

O

Ogee gutter....................................410
Open
  cornice..30
  valley...........................203, 233, 400
Organic
  felts.................................36-37, 297
  hydrocarbon.............................289
  shingles...............................71-72
  solvents....................................291
Organosilane sealers.....................368
Outside design temperature..............
  ....24-25, 285
Over-walling..................................180
Overcut allowance............................51
Overhang
  asphalt shingles...76, 79, 83, 90, 110
  metal roofs................................248
  roll roofing.................................123
  roof..................21, 23, 30, 60-61
  shakes.................................154, 167
  slates..226
  tiles..190
  underlayment..............................44
  wood shingles....................154, 167
Overhead..417
Overhead power lines....................428
Overlaps...51
Oxidation.........................296, 315, 368
Ozone..............................328, 329, 366

P

Paint
  aluminum.................388, 398, 399
  bituminous....................66, 182, 235
  plaster bond.............................367
  vapor-retarder..........................354
Painting galvanized metal............246
Pan flashing, tile roofing..................
  ....196, 198, 200, 207-209, 212
Panel clips......................................249
Panels
  concealed clip..........................268
  corrugated metal..........................
    ....29, 248, 265-266, 267, 281
  job-fabricated seams.........253-258
  pre-cut metal.....................258-260
  uncoated metal........................266
  vented metal............................325
Paper, building....................178-181
  asphalt-laminated....................352
  rosin..............40, 283, 286, 300, 349
  rosin-sized.......................286, 295
Paper-faced insulation..........338-339
Paper pulp board..........................361
Parapet wall flashing
  built-up roofs..............296, 302-304
  elastomeric roofing..................327
  metal roofing............................252
  underlayment..............................35
Parge coat......................................368
Particleboard, compressed...........342
Paver blocks..................................325
Pavers.....................................323, 325
Penetrating oils.............................292
Percentage-of-increase factor..........
  ....215, 403
Perforated
  pipe...361
  steel deck.................................281
Perimeter insulation.....................346
Perimeter of roof..............8-9, 17, 43
Periphery..8
Perlite board....................284, 299, 349
Phasing, built-up roof...................297
Phenol formaldehyde insulation...325
PIB..322
Pier-and-beam foundation............354
Pine shingles.................................150
Pipe
  bituminous fiber......................361
  perforated.................................361
Pipe
  flashing....................................305
  penetrations....................286, 359
Pitch...................11, 74, 224-225
  pan...................................306-307
Pitch bitumen....................................
  ....10, 278, 283, 319, 328, 344
Pits in shingles..............................394
Plan area...................................7, 19
Plan length......................................21
Planks, precast..............................283
Plaster
  bond paint.................................367
  lath....................................228, 231
Plastic
  binders.......................................36
  corrugated sheets....................267
  films.........................329, 352, 366
  foam insulation...............343-344
Plastic-bitumen composites.........330
Plasticizers....................................301
Plate, bright..................................260
Plenum..285
Ply sheets..37

Plywood
  built-up roof..............278, 282, 286
  elastomeric roofing..........325-327
  grades...................................27-28
  grain direction.............................29
  rigid insulation.........................349
  sheathing............................24, 27-28
  underlayment..............................43
Plywood panels
  clips......................................28-29
  spans..27
  thickness....................................28
  tongue and groove....................29
Polyester
  fabric..289
  polyurethane foam sealant.......359
Polyester-reinforced fabrics.............
  ....322, 328
Polyethylene
  film.............................352-353, 360
  rope..357
Polyisobutylene.............................322
Polyisocyanurate insulation..............
  ....325, 342
Polysiloxane...................................321
Polystyrene beads.........................340
Polystyrene
  foam insulation........................320
  insulation.............329, 342-344, 349
Polysulfide caulk...........................355
Polyurethane
  caulk...355
  coatings....................................370
  insulation.................321, 342-344
Polyvinyl chloride..................322, 329
Ponded water
  built-up roof.....278, 296, 301, 314
  elastomeric roofing..................320
Porous substrates.........................287
Positive method............................7-8
Poured insulation.................339-341
Power lines, overhead..................428
Power tools..........................424, 426
Pre-cut metal panels............258-260
Precast
  concrete decks........................301
  planks.......................................283
Precoating.....................................235
Precut shingle.....................156-157
Prefabricated units
  apex tile...................................202
  hip....................................112-119
  ridge................................112-113
Prefabrication...............................416
Preformed
  drip edge..........................248-249
  pans..................................255-256
Pressure-cleaning.........................393
Pressure-treated
  lumber......................................192
  wood shingles.........................150
Prices, labor unit..........................416
Production rates..................413-414
  maximizing......................422-426
  required..................................415
Profit..417
Protection board
  built-up roofs...........................282
  waterproofing.......362-364, 366-367
Protective gear.............................426
Protractor..12
Published
  hourly wages...........................416
  labor unit prices.....................416
Purified iron particles..................368
Purlins............25, 164, 247-249, 265

PVC..319, 322
  flashing....................................304
  roof edge.............304-305, 331-332
  saddle..68
  sealers......................................292
  shake roof............................64, 69
  sidewall shingles......................273
  silicone.....................................321
  skylights...................................301
  slate roofing....................235-239
  soil stack............................98, 333
  spandrel...................................360
  step...399
  through-wall............................360
  tin.......................................64, 235
  traffic pads.......................300-301
  upper roof edge flashing..........132
  urethane..................................321
  vinyl..304
  wall..184
  waste factors............................334
  wood shingles & shakes...............
    ....64, 69, 182-184

Q

Quartz carbide sealer....................368

R

R-values, insulation................338-347
  blown fiberglass......................340
  blown rock wool..............339-341
  calculating........................351-352
  fiberglass..........................338-341
  loose fill............................339-340
  recommendations....................345
  rigid...342
  rock wool...............338, 339-341
Racking......................................83-84
Radiant heat..................................343
Rafter spacing..........................27, 29
Rafters...13
  actual lengths....................14-15
  common.....................................13
  fly..24
  heel-cut......................................14
  hip......................................13-14, 21
  hip jack................................13-14
  lookout................................23-24
  mid-span....................................33
  plan length.................................14
  ridge...13
  run.......................................14-21
  tail-cut.......................................14
  types....................................13-14
  valley jack...........................13-14
Raised fascia........................193, 196
Rake
  fascia...24
  tiles.................198-200, 203, 216
Rakes..23, 24
  roll roofing.................124-129, 136
  shingles................78-80, 154, 167
  underlayment....................41-43, 51
Random
  pattern...........................79-80, 84
  slates.................................222, 227
Re-roofing............................400-407
  asphalt shingles...............400-404
  built-up roof............................294
  estimating........................406-407
  metal roofing...........................405
  roll roofing...............................405

slate roofing 405-406
tile roofing 405-406
wood shingles & shakes ... 404-405
Re-walling 180
Rebutted shingles 175
Recess 9
Recommended live loads 28
Recovery board 318
Rectangular gutter 410
Redwood 261
Redwood shingles 150
Reglet 327
Reinforced
  mineral fillers 296
  polyester sheets 295, 321-322
Rejointed shingles 176
Rejuvenation 315
Relative humidity 29
Relief vents 281, 284, 307
Repairing
  asphalt shingles 390-391
  built-up roof 315-318
  shakes 391
  slates 392
  splits 396
  valleys 389
  wood shingles 391
Resaturant 292, 315
Resawn shakes 152, 165
Residual moisture 319
Resin, epoxy 412
Ribbon courses 180-181, 398
Ridge 13, 45, 80-82
  board 13, 25
  caps, metal roofing 250
  covers
    flashing 160, 200
    length, equation 22
    metal roofing 259
    rafter 13
    roll 90
    roll roofing ... 126-127, 129-130, 132
    saddle 203
    sagging 24
    seam 250
    shingles 388
    slates 223
    tiles 192-193, 202, 207, 216
    vents 259, 409
Ridge units
  asphalt shingles ....
    .... 90-93, 109-111, 112-113
  estimating quantities 109-111
  roll roofing 135
  roof area 18
  underlayment 50
  wood shingles & shakes ... 160, 168
Rigid fiberboard 299
Rigid insulation 342-343
  built-up roofs 282
  elastomeric roofing ...320, 324, 332
  roofing repair 404
  wood shingles & shakes ... 163-164
Rise, roof 11-12
Rock wool insulation 338, 351
Roll roofing 121-137
  built-up 286, 295
  concealed nail (blind nail) ....
    .... 124, 127-129
  double-coverage 130-133
  estimating costs 147
  estimating quantities 134-146
  exposed nail 124-127
  flashing 124, 132-133
  hip and ridge units ....
    .... 126, 129-130, 132

installation 123-134
live expectancy 121
modified bitumen asphalt (MBA) ...
  122-123
pattern-edge 122
re-roofing 405-406
selvage 122, 130
sheathing under 24
shed-roof 132
single-coverage 124
split-sheet 122, 130
starter course 76
storage 122
waste factors 134-146
Roll roofing, as flashing
  roofing repair 399-400
  tile roofing 204, 210
  underlayment 64, 67
Roll valley metal 64
Roll waterproofing 35
Roof area 17-19
Roof coatings 369-385
  acrylic 370
  advantages 369, 371-372
  application tools 385
  benefits 369
  certification 370
  common usage 369
  curing 371
  differences from 369
  elastomeric, application ... 372-375
  emittance 370
  labor estimates 381-382
  life expectancy 369
  material costs 381
  ratings 370
  reflectance 370
  selling price 383
  tools 384-385
  types 369
  warranty 370
Roof
  curb 259
  deck 23-24, 32
  drainage areas 409-410
Roof drains
  built-up roofs ....
    .... 278, 302, 306-307, 311
  elastomeric roofing 333
Roof
  built-up ....
    ...... 278, 284-287, 296-297, 301, 307
  edge flashing ....
    .... 132, 304-305, 331-332
  edging 41
  elastomeric 324-326, 328-329
  frame 23
  inspection 387, 392
  insulation 347-348
  juncture 182
  load 27, 32
  maintenance 387, 392-394
  moss 373
  overhang 21, 23, 30, 60-61
  perimeter 8-9, 17, 326-327, 331
  periphery 8
  pitch 11-12
  relief vent 281
  rise 11-12
  run 11-12
  sectioning 20
  surface blisters 373
  supports 25
  types 10
  walkway 300-301, 319
  wood shingles & shakes ... 162-164

Roof penetrations
  asphalt shingles 86
  built-up roofs 296, 302, 304-306
  elastomeric 327
  roofing repair 389-390, 395
  underlayment 49-50
  span 11
  structure 32
Roof slope
  area 17-21
  asphalt shingles 74
  measuring 6-10
  minimum 31, 38-40
  roll roofing 130
  underlayment 36-40, 46, 51-52
  varying 63
Roof-slope factors 15, 17, 430
Roofer's hatchet 81, 424
Roofing
  equations 432-436
  felts 26, 31
  injuries 320
  silicone 377, 380
  tape 388
  TPO 330
Roofinox 269
Roofs
  level 7
  loading 32-34
  mansard 162, 184-185
  shed 97
Rosin
  flux 258
  paper 40, 283, 286, 300, 349
Rosin-sized paper 286, 295
Rotten deck 29
Round
  gutter 410
  valley 232-233
Rounding off 49
Rubber
  closure 196
  vent flange 98, 251
Rubber-vinyl composites 329
Rubberized
  asphalt 122, 329, 366
  composites 329
  membrane 366
Run
  rafter 21
  roof 11-12

S

Saddle flashing 68, 93, 158-159
Saddle hip 230-231, 242
  strip 230
Saddle ridge 228, 230-231
  strip 230
Sagging roof 24
Salvaged shingles 110-112
Saturant 72
Scaffolding 427
Scheduling 425
Schmid
  top 187, 190, 198
  underlayment 50
Scoria 293
Scrap lumber 26
Screen 411
Screened vent 408
Seal cap 388
Sealant 355-357
  epoxy 356
  foam 359

neoprene 359
tape 357, 359
Sealed underlayment system ....
  .... 46-48, 192, 198, 207-210
Sealers
  organosilene 368
  quartz carbide 368
Seam solvent 329
Seam-welding, TPO roofing ....
  .... 330, 332
Seams, metal roofing 249-259
  batten 256-257, 272
  cleats 255-256, 258-259
  cross 258-259, 261-262, 271-272
  flat 254-255, 261-266, 271
  integral standing 259
  job-fabricated 253-258
  lock 237, 259
  riveted 253-254
  soldered 253-255, 258-259
  standing 237, 255-256
  welded 259
Sectioning a roof 20
Seismic
  movement 307
  zones 24
Selvage
  edge 122
  starter strip 130-131
  strip 130
  waste 135
Separation sheet 319, 323, 328
Settlement 96
Shading 84
Shake
  bundles 152, 164-166
  spacing 155
  waste 166-180
Shakes 149-186
  installation 154-164
  machine-grooved 176
Shale 294
Sheathing 23-32
  deflection 278-279
  eaves 30-31, 154-155
  estimating quantities 34
  plywood 27-29
  skip 29
  solid ....
    ... 24-32, 76, 124, 160-161, 190, 265
  spaced ....
    ...29-31, 40, 56, 154-155, 162, 405
  support 29
  waferboard 29
Sheet lead 269
Sheet metal 232
  copper 271
  galvanized 261-262
  gauge 245
Sheets 245
  plastic corrugated 267
  reinforced polyester ... 295, 321-322
Shingle
  nails 102-103
  patterning 79-82
  patterns, asphalt 79-81
  siding 175-182, 184
  tabs 75
  undercoursing 175-180
Shingles
  fancy-butt 176
  installation 154-164
  rejointed 176
Shiplap boards 26
Shoe 410-411

Side
   lap 265, 267, 269, 279, 287
   seam 258
Sidewall shingles
   bundles 184
   coursing 178-180
   double-coursing 179-180
   exposure 179-182
   installation 176-178
   ribbon coursing 180
   single-coursing 178, 180
   spacing 178
   staggered coursing 180-181
Siding
   flashing with 96-97
   metal 259-261, 268
   shingle 175-182, 184
Silicone
   caulk 355-356
   rubber 321
   sealant 216, 356, 368, 388
Silicone roofing 377, 380
   application temperatures 377
   coverage 379
   surface damage 380
Silver, solder 258
Single-coverage
   roll roofing 122, 132
   underlayment 39-40, 43, 50, 192
Skylight flashing 305
Slag 293-294, 302
Slashgrain wood shingles ....
   .... 150, 152, 395-396
Slate 221-234
   colors 221
   combing 229-230
   commercial standard 222
   coursing 223-224
   damaged 392
   estimating costs 244
   estimating quantities 239-244
   exposure 223-224
   fasteners 234-235
   flashing 235-239
   grades 222
   graduated 222-224, 227, 232
   head lap 224-229
   hips 230-231
   life expectancy 37-38, 221
   mitered-hip 230, 231
   nailing 230
   random 224, 227
   re-roofing 406
   repair 392
   ridges 223, 228-230
   sizes 222
   standard 222-226
   textural 222-224, 232
   thickness 222
   tools 239
   trade names 222
   trapezoidal 230
   triangular 230
   under-eaves 223, 241
   underlayment 224
   unfading 222
   unpunched 223
   valleys 232-234
   weathering 222
   weight 225-226
Slate jointing 224, 227
Slate roofs 35-40, 64, 269
Slate-surfaced felt 295
Slater's punch 239
Slating nails 235

Slip joint connectors 412
Slip sheet 286, 319, 323-325, 329
Slippery surfaces 427-428
Slope 10, 12-13
   degrees 12
   variable 20
Sloping fascia 14
Slotted corrugated panels 281
Smooth-surfaced roof ....
   .... 304, 325-326, 404-405
Snow loading 247, 254, 278, 348
Snow precautions
   elastomeric roofing 320
   flashing 60, 93
   interlayment 56
   maintenance 393
   tile roofing 192
   wood shingles & shakes ....
   .... 154-156, 163
Soffit 185, 246-247
   ventilation 60, 162, 407, 409
Soft copper 235, 271
Solder 235, 237-238, 410, 412
   acid-flux 239
   high-tin 259
   rosin-flux 235
   tin-lead 258
Solid sheathing ....
   .... 24-32, 76, 124, 160, 190, 404
Solvent acrylic caulk 355-356
Solvent
   organic 291
   seam 329
Sound
   control batts 339
   transmission 339
Spaced battens 161-162
Spaced sheathing ....
   .... 29-31, 56, 154-155, 162
   over solid sheathing 32, 405
Spacer tube 410-411
Spacers 31, 160-161
Spacing, sheathing boards 30
Span 11, 268
Span rating 28
Spandrel flashing 360
Splash
   block 411-412
   diverter 63-65, 412
Splice covers 308, 309
Sponge 202, 207
Spongy roof deck 374
Spot mopping
   built-up roof 282, 284, 296, 318
   elastomeric roofing 323
Sprayed insulation 348, 349
Sprayed Polyurethane Foam (SPF) ...
   375-376
Spud 397
Squangle 12
Square 18
   factory 50
Square-edged boards 26, 30
Stacking 107
Stagger nailing 78, 103, 128
Staggered
   butts 75
   joints, decks 26, 29
Stainless steel
   composites 329
   roofing 40
Standing seams, metal roofing ....
   .... 255-256, 270-272
Staples 103-104, 157, 406
   sizes 27-29

Starter course
   asphalt shingles 76-78, 109-119
   estimating 18
   roofing repair 402-403
   underlayment 56-58
   wood shingles & shakes ....
   .... 154-155, 161, 167
Starter roll 76, 78
Starter strip 193
   allowance 135
Starter-finish shake 156
Statement of compliance 314
Steel
   deck 164
   door 347
Steel-core shingles 273
Steep asphalt
   built-up roof 279, 286, 299, 316
   insulation 349
   underlayment 48
Step flashing
   asphalt shingles 94, 96-98
   roofing repair 399
   stop, gravel 285, 304, 326
   wood shingles & shakes ... 158-159
Storm
   clips 196-197, 200, 203
   door 346-347
   sash 346
Story pole 176-177
Straight flange 308, 309
Straight-bond method, tile 198
Straight-split shakes 152, 164, 165
Straight-up method, asphalt, shingle ...
   83-84
Strainers 412
Strapping 162-163, 283-284
Stretch, asphalt shingles 24
Strip
   copper 271
   saddle ridge 228
Strip-shingle roofs 63
Structural
   cement fiber deck 286
   corrugated glass 267
   expansion joints 307
   wood fiber deck 283
Struts 24-25
Stucco 97-98, 182, 208-209, 361
Subcontractors 426
Sulfur dioxide 360
Sulfuric acid 360
Supervision 417
Support spacing 27-29
Supports, horizontal 25
Surface primer 355-356
Surface-coated tile 188
Sweat sheet 44
Swept eaves 161-162
Synthetic
   benefits 37
   disadvantages 38
   rubber 328, 365
   tear-resistance 37
   underlayment 37-38, 274-275

T

T-bevel 12-13
T-lock
   asphalt shingles 76
   roofing repair 401
Tab notches 75
Tabs, shingle 75
Tape measure 5

Tapered
   edge strips 299
   insulation 306
Tapersplit shakes 152-153, 165
Tar
   saturant 36-37
   strip 73
Tarred felt 36
Tear-resistance, synthetic
   underlayment 37
Tedlar sheets 316
Tees 308
Temporary roof 300
Terne metal roofing & siding ....
   269-270
   drip edge 40-41
Terne plate 269-270
Terne-coated stainless steel 269
Textural slate 222-224, 232
Thermal
   conductivity 350-351
   expansion 254
Thermoplastic polyolefin (TPO) ....
   330-333
Three-ply waterproofing 364-365
Three-tab shingles
   .... 75-76, 110, 422, 424
Through-wall flashing 360
Tie-in 422
Tile gaps 198
Tile roofing 187-229
   accessories 216
   barge 198
   barrel 197-198
   body-colored 188
   clay 187
   concrete 187
   direct-nail 194
   estimating quantities 215-216
   farm 361
   fasteners 194-196
   glazed 188
   installation 190-207
   life expectancy 38, 187
   loading 33-34
   mitered 200-201
   mortar finishing 200, 207
   mortar mix 216
   mortar-set 194, 196-197
   prefabricated apex 202
   rake 198-200, 203, 216
   re-roofing 405-406
   repairing 388
   replacement 214
   ridge 192-193, 202, 207, 216
   shapes 187
   straight-bond method 198
   surface coated 188
   trimming 201
   under-eaves 193-194
   underlayment 35-38, 64
   unglazed 188
Tin
   caps 44, 349
   flashing 235
   roofing 260
   solder 258
   tags 44, 47, 48, 211
Tinning 235
Titanium dioxide 328
Tongue-and-groove boards 26, 32
   deck 26
   plywood 29
Top lap
   asphalt shingles 108
   tile roofing 187, 190, 198

underlayment.................................45
Topping slab............................364
Torching...................................122
TPO roofing 330-333
  benefits............................. 330-331
  composition.............................330
  cons 332-333
  installation...................... 331-332
  life expectancy 333
  light-reflectivity 331
  seam-welding 330, 332
Traffic pads 300-301
Trap rock....................................72
Trapezoidal slates.......................230
Tree sap.....................................26
Triangular slates........................230
Trowel 196, 197, 207
Two-tab shingles........................103

U

U-value....................................351
Unpunched slate.......................225
UL ratings 72
  weights 72
  wind-resistant 73
Ultraviolet light................348, 355
  built-up roof 296, 315
  caulking 355-356
  elastomeric roofing ....
  .... 320-321, 328-329
  insulation...............................348
  waterproofing..........................366
Umbrella, metal........................306
Uncoated
  felts.......................................287
  metal panels 266
Under-eaves slates 226-227, 241
Under-shimming 396
Undercoursing shingles....... 175-180
Underfloor ventilation................354
Underlayment........................ 35-70
  buckling.............................61, 123
  coursing 43
  double-coverage....... 39-40, 43, 50
  end laps 45-46
  estimating quantities 49-56
  installation......................... 43-49
  laps 45-46, 50, 61, 67-68
  metal shingles 274-275
  minimum 37
  non-sealed ....
  .... 46-47, 189, 203, 205, 207-210
  overhang..................................44
  recommended................... 36-40
  requirements.................... 38-40
  saturated felt..................... 36-37
  saturated fiberglass....................36
  sealed............46-48, 204, 206-211
  single coverage...39-40, 43, 50, 192
  synthetic 37-38, 274-275
  unsealed system 65, 205
  valley..........................44-45, 67-68
  waste factors............ 50, 51-52, 57
  weights 36-39
Uneven roof loads 32
Unfaced insulation 338-339
Unfading slate 222
Unglazed tile 188
Uniform live loads 28
Unsaturated felt 323

Unsealed underlayment system.........
....65, 205
Upper roof edge flashing 132
Uprights............................... 24-25
Urethane..............321, 342, 349, 355

V

V-beam.....................................268
V-beam sheets..........................268
V-ridge 200-201
  weights 187-188
V-ridge tile 200-202
  canoe 233-234
  clip...205
  closed 63, 232, 234
  closed-cut 63-66
  half-lace............................. 63, 66
  open.................................63, 232
  radius....................................234
  roll roofing 124, 127
  round 232-233
  underlayment..................... 44-45
  valley blocks 232, 234
  woven 63-66
V-ridge tiles 202
Valley debris 63-65, 86, 388
Valley flashing 63-68
  asphalt shingles 76
  estimating............................ 69-70
  installing, metal................... 67-70
  installing, non-metal 65-67
  metal................ 63-64, 204, 232
  metal roofing 251
  roll roofing 124
  tile roofing............. 190, 203-206
  underlayment.......35, 49-50, 63-68
  wood shingles & shakes...........154
Valley jack rafter................. 13-14
Valley length factors 22, 431
Valley rafter 13-14
  plan length.............................21
Valley-ridge juncture 203, 205
Valley-slope factors 14-15
Vapor retarder 352-354
  aluminum foil..........................352
  batt insulation 338-339
  built-up roof ....
  ...... 281, 284-286, 300, 301, 353
  crawl space, location................354
  disadvantage..........................352
  effectiveness...........................353
  elastomeric roofing 324
  floor location..........................354
  foil...352
  insulating doors & windows 347
  materials................................352
  placement 352-354
  rigid insulation 343
  tile roofing................. 208-209, 211
  wood shingles & shakes... 163-164
Vapor retarder paint 354
Varying roof slopes....................63
Vaulted ceiling 350
Vent
  collar.......................................98
  flange................ 101-102, 158, 210
  flange, rubber..................98, 251
  screen...................................408
  sleeve.....................................101
  space.......................................61
  stack 101, 158, 333

Vent flashing
  asphalt shingles 98, 101-105
  boot.......................................333
  elastomeric roofing 333
  jack..............................210, 213
  louvers...................................408
  roll roofing 132-133
  slate roofing 238
  tile roofing............. 210, 213-214
Vent pipe
  asphalt shingles 98-101
  built-up roof 305
  roll roofing 133
  roofing repair 388
  slate roofing 238
  tile roofing............................210
Vented
  base sheet 298, 353
  membrane..............................284
  metal panels 325
Ventilation........ 29, 60-61, 192, 281
  attic 407-409
  underfloor..............................354
Ventilator 238
Vents
  dryer.....................................362
  gross area 408
  relief 281, 284, 307
  ridge 259, 409
  screened................................408
Vermiculite insulation.......... 339-341
Vinyl acrylic
  caulk.....................................356
  roofing 322
Viscosity.................................291
Volcanic glass 344
Vulcanized rubber roofing 324

W

Waferboard 29
  sheathing 24
Wage rates.............. 417, 420-421
Walking in......................... 48, 123
Walkpad..................................380
Walkways......................319, 380
Wall
  flashing.................. 184, 360-361
  insulation..............................342
Warped
  felts..26
  shingles.................................401
Warranty 5, 38, 327, 328, 330
Waste...5
Waste factors..............114, 119, 260
Water
  course 189
  cut-offs 297
  dam.................................249, 389
  diverter.................................203
  jackets....................................75
  lines.......................................75
  repellents..............................361
  shield...............................60, 93
Water guards
  metal roofing.........................250
  slate roofing 232
  tile roofing........ 200, 203-204, 207
  valley flashing 64, 65-67
Water-retaining roofs... 293, 301-302
Waterproof wax 368
Waterproofing 360-361
  bentonite...............................366

  built-up............................ 362-365
  cold-applied................... 362-365
  composite..............................366
  elastomeric 320, 365-366
  epoxy....................................368
  estimating quantities 365
  installation..................... 360-365
  integral.................................361
  liquid plastic..........................366
  phase method 364
  roll..35
  shingle method 364
Waterstops..............................360
Weather..................................426
Weather checks 188-189
Weathering slate......................222
Weatherstripping......................407
Weep holes 197, 201-202
Weight
  built-up roof 26
  clay tile 27
  roof deck 27
  roofing felt 27
  shakes...................................152
Welding...................................270
White
  cedar shingles........................175
  metal.....................................269
Wicking...................................37
Wind
  block.............................. 203-204
  damage..................................390
Wind-resistant roofing
  asphalt shingles 90-92
  built-up roof 294
  elastomeric roofing ....
  ....323, 326, 331-332
  roofing repair 390
  shingles..................................73
  underlayment..........................46
Windows 60-61, 237
  casing...................................176
  flashing.................................176
Wood
  batten strips..........................256
  ceiling....................................32
  closure strips.........................253
  nailers............283-284, 299, 327
  resins...............................35, 330
Wood shingles.................. 149-186
  bell eaves........................161, 162
  bundles............................152, 167
  covering capacity 164, 166
  decking, metal.......................163
Wood starter strip....................192
Workers' compensation insurance....
....426
Workspace limitations 422
Woven
  corner...................................177
  fiberglass flashing 360
  valley............................... 86-89
Wye tile............................202, 203

XYZ

Z-bar................................ 208-209
Z-bar flashing 208-209
Z-closure 250
Zinc
  flashing..................................64
  napthenate 150

Practical References for Builders

National Estimator Cloud

Generate professional construction estimates for all residential and commercial construction from your internet browser. Includes 10 Craftsman construction cost databases, over 40,000 labor and material costs for construction, in an easy-to-use format. Cost estimates are well-organized and thoroughly indexed to speed and simplify writing estimates for nearly any residential or light commercial construction project – new construction, improvement or repair. Convert the bid to an invoice – in either QuickBooks Desktop or QuickBooks Online. Access your estimates from anywhere and on any device with a Web browser. Monthly and one-time billing options available.
Visit https://craftsman-book.com/national-estimator-cloud for more details.

Home Building Mistakes & Fixes

This is an encyclopedia of practical fixes for real-world home building and repair problems. There's never an end to "surprises" when you're in the business of building and fixing homes, yet there's little published on how to deal with construction that went wrong — where out-of-square or non-standard or jerry-rigged turns what should be a simple job into a nightmare. This manual describes jaw-dropping building mistakes that actually occurred, from disastrous misunderstandings over property lines, through basement floors leveled with an out-of-level instrument, to a house collapse when a siding crew removed the old siding. You'll learn the pitfalls the painless way, and real-world working solutions for the problems every contractor finds in a home building or repair jobsite. Includes dozens of those "surprises" and the author's step-by-step, clearly illustrated tips, tricks and workarounds for dealing with them. **384 pages, 8½ x 11, $52.50**
eBook (PDF) also available; $26.25 at www.craftsman-book.com

Estimating Home Building Costs Revised

Accurate estimates are the foundation of a successful construction business. Leave an item out of your original estimate and it can take the profit out of your entire job. This practical guide to estimating home construction costs has been updated with *Excel* estimating forms and worksheets with active cells that ensure accurate and complete estimates for your residential projects. Load the enclosed CD-ROM into your computer and create your own estimate as you follow along with the step-by-step techniques in this book. Clear, simple instructions show how to estimate labor and material costs for each stage of construction, from site clearing to figuring your markup and profit. Every chapter includes a sample cost estimate worksheet that lists all the materials to be estimated. Even shows how to figure your markup and profit to arrive at a price.
336 pages, 8½ x 11, $38.00
Also available as an eBook (PDF), $19.00 at www.craftsman-book.com

Craftsman's Construction Installation Encyclopedia

Step-by-step installation instructions for just about any residential construction, remodeling or repair task, arranged alphabetically, from Acoustic tile to Wood flooring. Includes hundreds of illustrations that show how to build, install, or remodel each part of the job, as well as manhour tables for each work item so you can estimate and bid with confidence. Also includes a CD-ROM with all the material in the book, handy look-up features, and the ability to capture and print out for your crew the instructions and diagrams for any job. **792 pages, 8½ x 11, $65.00**
Also available as an eBook (PDF), $32.50 at www.craftsman-book.com

Construction Forms for Contractors

This practical guide contains 78 practical forms, letters and checklists, guaranteed to help you streamline your office, organize your jobsites, gather and organize records and documents, keep a handle on your subs, reduce estimating errors, administer change orders and lien issues, monitor crew productivity, track your equipment use, and more. Includes accounting forms, change order forms, forms for customers, estimating forms, field work forms, HR forms, lien forms, office forms, bids and proposals, subcontracts, and more. All are also on the CD-ROM included, in Excel spreadsheets, as formatted Rich Text that you can fill out on your computer, and as PDFs. **360 pages, 8½ x 11, $48.50**
Also available as an eBook (PDF), $24.25 at www.craftsman-book.com

Handbook of Construction Contracting, Volume 1

Everything you need to know to start and run your construction business; the pros and cons of each type of contracting, the records you'll need to keep, and how to read and understand house plans and specs so you find any problems before the actual work begins. All aspects of construction are covered in detail, including all-weather wood foundations, practical math for the job site, and elementary surveying.
416 pages, 8½ x 11, $32.75

Handbook of Construction Contracting, Volume 2

Everything you need to know to keep your construction business profitable; different methods of estimating, keeping and controlling costs, estimating excavation, concrete, masonry, rough carpentry, roof covering, insulation, doors and windows, exterior finishes, specialty finishes, scheduling work flow, managing workers, advertising and sales, spec building and land development, and selecting the best legal structure for your business.
320 pages, 8½ x 11, $33.75

Contractor's Guide to QuickBooks by Online Accounting

This book is designed to help a contractor, bookkeeper and their accountant set up and use QuickBooks Desktop specifically for the construction industry. No use re-inventing the wheel, we have used this system with contractors for over 30 years. It works and is now the national standard. By following the steps we outlined in the book you, too, can set up a good system for job costing as well as financial reporting.
156 pages, 8½ x 11, $68.50

National Home Improvement Estimator

Current labor and material prices for home improvement projects. Provides manhours for each job, recommended crew size, and the labor cost for removal and installation work. Material prices are current. Gives step-by-step instructions for the work, with helpful diagrams, and home improvement shortcuts and tips from experts.
548 pages, 8½ x 11, $98.75. Revised annually
Also available as an eBook (PDF), $49.38 at
www.craftsman-book.com

Construction Contract Writer

Relying on a "one-size-fits-all" boilerplate construction contract to fit your jobs can be dangerous - almost as dangerous as a handshake agreement. Construction Contract Writer lets you draft a contract in minutes that precisely fits your needs and the particular job, and meets both state and federal requirements. You just answer a series of questions - like an interview - to construct a legal contract for each project you take on. Anticipate where disputes could arise and settle them in the contract before they happen. Include the warranty protection you intend, the payment schedule, and create subcontracts from the prime contract by just clicking a box. Includes a feedback button to an attorney on the Craftsman staff to help should you get stumped - *No extra charge.*
$149.95. Download the Construction Contract Writer at:
www.constructioncontractwriter.com

Profits in Buying & Renovating Homes

Step-by-step instructions for selecting, repairing, improving, and selling highly profitable "fixer-uppers." Shows which price ranges offer the highest profit-to-investment ratios, which neighborhoods offer the best return, practical directions for repairs, and tips on dealing with buyers, sellers, and real estate agents. Shows you how to determine your profit before you buy, what "bargains" to avoid, and how to make simple, profitable, inexpensive upgrades. **304 pages, 8½ x 11, $24.75**

Construction Estimating Reference Data eBook

Provides the 300 most useful manhour tables for practically every item of construction. Labor requirements are listed for sitework, concrete work, masonry, steel, carpentry, thermal and moisture protection, doors and windows, finishes, mechanical and electrical. Each section details the work being estimated and gives appropriate crew size and equipment needed.
Available only as an eBook (PDF), $29.50 at www.craftsman-book.com

National Appraisal Estimator

An Online Appraisal Estimating Service. Produce credible single-family residence appraisals – in as little as five minutes. A smart resource for appraisers using the cost approach. Reports consider all significant cost variables and both physical and functional depreciation.

Visit www.craftsman-book.com/national-appraisal-estimator-online-software for more information.

National Construction Estimator

Current building costs for residential, commercial, and industrial construction. Estimated prices for every common building material. Provides manhours, recommended crew, and gives the labor cost for installation.
672 pages, 8½ x 11, $97.50. Revised annually
Also available as an eBook (PDF), $48.75 at www.craftsman-book.com

Markup & Profit: A Contractor's Guide Revisited

In order to succeed in a construction business, you have to be able to price your jobs to cover all labor, material and overhead expenses, and make a decent profit. But calculating markup is only part of the picture. If you're going to beat the odds and stay in business - profitably, you also need to know how to write good contracts, manage your crews, work with subcontractors and collect on your work. This book covers the business basics of running a construction company, whether you're a general or specialty contractor working in remodeling, new construction or commercial work. The principles outlined here apply to all construction-related businesses. You'll find tried and tested formulas to guarantee profits, with step-by-step instructions and easy-to-follow examples to help you learn how to operate your business successfully. Includes a link to free downloads of blank forms and checklists used in this book.
336 pages, 8½ x 11, $59.50
Also available as an eBook (EPUB, MOBI for Kindle), $39.95 at www.craftsman-book.com

Rough Framing Carpentry

If you'd like to make good money working outdoors as a framer, this is the book for you. Here you'll find shortcuts to laying out studs; speed cutting blocks, trimmers and plates by eye; quickly building and blocking rake walls; installing ceiling backing, ceiling joists, and truss joists; cutting and assembling hip trusses and California fills; arches and drop ceilings — all with production line procedures that save you time and help you make more money. Over 100 on-the-job photos of how to do it right and what can go wrong. **304 pages, 8½ x 11, $26.50**

Residential Wood Framing Construction Quick-Card

This 6-page laminated card covers the construction essentials for wood framing based on the new 2021 International Residential Code (*IRC*).
6 pages, 8½ x 11, $9.95

Contractor's Survival Manual Revised

The "real skinny" on the down-and-dirty survival skills that no one likes to talk about - unique, unconventional ways to get through a debt crisis: what to do when the bills can't be paid, finding money and buying time, conserving income, transferring debt, setting payment priorities, cash float techniques, dealing with judgments and liens, and laying the foundation for recovery. Here you'll find out how to survive a downturn and the key things you can do to pave the road to success. Have this book as your insurance policy; when hard times come to your business it will be your guide. **336 pages, 8½ x 11, $38.00**
Also available as an eBook (PDF), $19.00 at www.craftsman-book.com

Paper Contracting: The How-To of Construction Management Contracting

Risk, and the headaches that go with it, have always been a major part of any construction project — risk of loss, negative cash flow, construction claims, regulations, excessive changes, disputes, slow pay — sometimes you'll make money, and often you won't. But many contractors today are avoiding almost all of that risk by working under a construction management contract, where they are simply a paid consultant to the owner, running the job, but leaving him the risk. This manual is the how-to of construction management contracting. You'll learn how the process works, how to get started as a CM contractor, what the job entails, how to deal with the issues that come up, when to step back, and how to get the job completed on time and on budget. Includes a link to free downloads of CM contracts legal in each state. **272 pages, 8½ x 11, $55.50**
eBook (PDF) also available; $27.75 at www.craftsman-book.com

National Repair & Remodeling Estimator

The complete pricing guide for dwelling reconstruction costs. Reliable, specific data you can apply on every repair and remodeling job. Up-to-date material costs and labor figures based on thousands of jobs across the country. Provides recommended crew sizes; average production rates; exact material, equipment, and labor costs; a total unit cost and a total price including overhead and profit. Separate listings for high- and low-volume builders, so prices shown are specific for any size business. Estimating tips specific to repair and remodeling work to make your bids complete, realistic, and profitable.
528 pages, 8½ x 11, $98.50. Revised annually
Also available as an eBook (PDF), $49.25 at www.craftsman-book.com

The Art of Roof Cutting Series DVD Library - Basic

A master framer demonstrates how to build hip, gable, gambrel and shed roofs, valleys, fake valleys and other details. As a skilled craftsman you will appreciate the value of these comprehensive roof framing videos. Used in technical schools and highly recommended by building contractors.
DVD - 2 Hrs. $79.95

The Art of Roof Cutting Series DVD Library - Advanced

A master framer demonstrates how to build octagons, roofs on variable plate heights, complex hip and gable roofs, hip to valley extensions, irregular angled rafters and more. As a skilled craftsman, you will appreciate the value of these comprehensive roof framing videos. Used in technical schools and highly recommended by building contractors.
DVD - 2.5 Hrs. $79.95

National Renovation & Insurance Repair Estimator

Current prices in dollars and cents for hard-to-find items needed on most insurance, repair, remodeling, and renovation jobs. All price items include labor, material, and equipment breakouts, plus special charts that tell you exactly how these costs are calculated.
488 pages, 8½ x 11, $99.50. Revised annually
Also available as an eBook (PDF), $49.75 at www.craftsman-book.com

Renovating & Restyling Older Homes

Any builder can turn a run-down old house into a showcase of perfection — if the customer has unlimited funds to spend. Unfortunately, most customers are on a tight budget. They usually want more improvements than they can afford — and they expect you to deliver. This book shows how to add economical improvements that can increase the property value by two, five or even ten times the cost of the remodel. Sound impossible? Here you'll find the secrets of a builder who has been putting these techniques to work on Victorian and Craftsman-style houses for twenty years. You'll see what to repair, what to replace and what to leave, so you can remodel or restyle older homes for the least amount of money and the greatest increase in value. **416 pages, 8½ x 11, $33.50**

Contractor's Plain-English Legal Guide

For today's contractors, legal problems are like snakes in the swamp — you might not see them, but you know they're there. This book tells you where the snakes are hiding and directs you to the safe path. With the directions in this easy-to-read handbook you're less likely to need a $200-an-hour lawyer. Includes simple directions for starting your business, writing contracts that cover just about any eventuality, collecting what's owed you, filing liens, protecting yourself from unethical subcontractors, and more. For about the price of 15 minutes in a lawyer's office, you'll have a guide that will make many of those visits unnecessary. Includes a CD-ROM with blank copies of all the forms and contracts in the book. **272 pages, 8½ x 11, $49.50**

Residential Roof Assemblies Quick-Card

This six-page, full-color card quickly covers roofing information used in architectural plans and engineering drawings. Includes several quick lookup tables and pictures. Based on the 2021 International Residential Code (*IRC*). **6 pages, 8½ x 11, $9.95**

How to Succeed With Your Own Construction Business

Everything you need to start your own construction business: setting up the paperwork, finding the work, advertising, using contracts, dealing with lenders, estimating, scheduling, finding and keeping good employees, keeping the books, and coping with success. If you're considering starting your own construction business, all the knowledge, tips, and blank forms you need are here. **336 pages, 8½ x 11, $28.50**
Also available as an eBook (PDF), $14.25 at www.craftsman-book.com

Builder's Guide to Accounting Revised

Step-by-step, easy-to-follow guidelines for setting up and maintaining records for your building business. This practical guide to all accounting methods shows how to meet state and federal accounting requirements, explains the new depreciation rules, and describes how the Tax Reform Act can affect the way you keep records. Full of charts, diagrams, simple directions and examples, to help you keep track of where your money is going. Recommended reading for many state contractor's exams. Each chapter ends with a set of test questions, and a CD-ROM included FREE has all the questions in interactive self-test software. Use the Study Mode to make studying for the exam much easier, and Exam Mode to practice your skills. **360 pages, 8½ x 11, $51.50**
eBook (PDF) also available, $25.75 at www.craftsman-book.com

National Building Cost Manual

Square-foot costs for residential, commercial, industrial, military, schools, greenhouses, churches and farm buildings. Includes important variables that can make any building unique from a cost standpoint. Quickly work up a reliable budget estimate based on actual materials and design features, area, shape, wall height, number of floors, and support requirements. Now includes free download of Craftsman's easy-to-use software that calculates total in-place cost estimates or appraisals. Use the regional cost adjustment factors provided to tailor the estimate to any jobsite in the U.S. Then view, print, email or save the detailed PDF report as needed.
280 pages, 8½ x 11, $98.00. Revised annually
Also available as an eBook (PDF), $49.00 at www.craftsman-book.com

Craftsman eLibrary

Craftsman's eLibrary license gives you immediate access to 60+ PDF eBooks in our bookstore for 12 full months!
You pay only one low price. $129.99.
Visit **www.craftsman-book.com** for more details.

Shear Walls & Sheathing Lateral Loads Quick-Card (Bldrs Book)

This 6-page card quickly covers the construction essentials you need in loads and shear walls based on the new 2018 *International Residential Code* (*IRC*) for loads, shear walls & sheathing. Special sections focus on specific essentials for each area of framing. **6 pages, 8½ x 11, $7.95**

Building Code Compliance for Contractors & Inspectors

An answer book for both contractors and building inspectors, this manual explains what it takes to pass inspections under the 2009 International Residential Code. It includes a code checklist for every trade, covering some of the most common reasons why inspectors reject residential work — footings, foundations, slabs, framing, sheathing, plumbing, electrical, HVAC, energy conservation and final inspection. The requirement for each item on the checklist is explained, and the code section cited so you can look it up or show it to the inspector. Knowing in advance what the inspector wants to see gives you an (almost unfair) advantage. To pass inspection, do your own pre-inspection before the inspector arrives. If your work requires getting permits and passing inspections, put this manual to work on your next job. If you're considering a career in code enforcement, this can be your guidebook. **8½ x 11, 232 pages, $32.50**
eBook (PDF) also available; $16.25 at www.craftsman-book.com

Craftsman Book Company
6058 Corte del Cedro
Carlsbad, CA 92011

☎ **Call us.**
1-800-829-8123
Fax (760) 438-0398

In A Hurry?
We accept phone orders charged to your
○ Visa, ○ MasterCard, ○ Discover or ○ American Express

Name _____
e-mail address (for order tracking and special offers)
Company _____
Address _____
City/State/Zip _____ ○ This is a residence
Total enclosed_____(In California add 7.5% tax)
We pay shipping when your check covers your order in full.

Card#_____
Exp.date_____ CVV#_____ Initials_____

Tax Deductible: Treasury regulations make these references tax deductible when used in your work. Save the canceled check or charge card statement as your receipt.

Order online http://www.craftsman-book.com

10-Day Money Back Guarantee

- ○ 79.95 Art of Roof Cutting Series DVD - Basic
- ○ 79.95 Art of Roof Cutting Series DVD - Advanced
- ○ 51.50 Builder's Guide to Accounting Revised
- ○ 32.50 Building Code Compliance for Contractors & Inspectors
- ○ 48.50 Construction Forms for Contractors with a CD-ROM
- ○ 68.50 Contractor's Guide to Quickbooks by Online Accounting
- ○ 49.50 Contractor's Plain-English Legal Guide
- ○ 38.00 Contractor's Survival Manual Revised
- ○ 65.00 Craftsman's Construction Installation Encyclopedia
- ○ 38.00 Estimating Home Building Costs Revised
- ○ 32.75 Handbook of Construction Contracting Volume 1
- ○ 33.75 Handbook of Construction Contracting Volume 2
- ○ 52.50 Home Building Mistakes & Fixes
- ○ 28.50 How to Succeed w/Your Own Construction Business
- ○ 59.50 Markup & Profit: A Contractor's Guide Revisited

- ○ 98.00 National Building Cost Manual
- ○ 97.50 National Construction Estimator
- ○ 98.75 National Home Improvement Estimator
- ○ 99.50 National Renovation & Insurance Repair Estimator
- ○ 98.50 National Repair & Remodeling Estimator
- ○ 55.50 Paper Contracting: The How-To of Constr. Management Contracting
- ○ 24.75 Profits in Buying & Renovating Homes
- ○ 33.50 Renovating & Restyling Older Homes
- ○ 9.95 Residential Roof Assemblies Quick-Card
- ○ 9.95 Residential Wood Framing Construction Quick-Cards
- ○ 26.50 Rough Framing Carpentry
- ○ 7.95 Shear Walls & Sheathing Lateral Loads Quick-Card
- ○ 62.50 Roofing Construction & Estimating, Revised

Prices subject to change without notice